PARTICLE PHYSICS
at the Year of 250th
Anniversary of
Moscow University

Faculty of Physics of Moscow State University

INTERREGIONAL CENTRE FOR ADVANCED STUDIES

Proceedings of the 12th Lomonosov
Conference on Elementary Particle Physics

PARTICLE PHYSICS at the Year of 250th Anniversary of Moscow University

Moscow, Russia 25 – 31 August 2005

Editor

Alexander I Studenikin

Department of Theoretical Physics
Moscow State University, Russia

World Scientific

NEW JERSEY • LONDON • SINGAPORE • BEIJING • SHANGHAI • HONG KONG • TAIPEI • CHENNAI

Published by

World Scientific Publishing Co. Pte. Ltd.
5 Toh Tuck Link, Singapore 596224
USA office: 27 Warren Street, Suite 401-402, Hackensack, NJ 07601
UK office: 57 Shelton Street, Covent Garden, London WC2H 9HE

British Library Cataloguing-in-Publication Data
A catalogue record for this book is available from the British Library.

PARTICLE PHYSICS AT THE YEAR OF 250TH ANNIVERSARY OF MOSCOW UNIVERSITY
Proceedings of the 12th Lomonosov Conference on Elementary Particle Physics

ISBN 981-270-067-6

Printed in Singapore by World Scientific Printers (S) Pte Ltd

Moscow State University
Faculty of Physics
Interregional Centre for Advanced Studies

Dedicated to the 250th Anniversary of Moscow State University

TWELFTH LOMONOSOV CONFERENCE ON ELEMENTARY PARTICLE PHYSICS

Mikhail Lomonosov
1711-1765

Moscow, August 25-31, 2005

Sponsors

Russian Foundation for Basic Research
Russian Agency for Science and Innovation
Russian Academy of Sciences
Russian Agency for Atomic Energy

Supporting Institutions

Faculty of Physics of Moscow State University
Skobeltsyn Institute of Nuclear Physics, Moscow State University
Interregional Centre for Advanced Studies
Joint Institute for Nuclear Research (Dubna)
Institute of Theoretical and Experimental Physics (Moscow)
Institute for Nuclear Research (Moscow)
Budker Institute of Nuclear Physics (Novosibirsk)

International Advisory Committee

E.Akhmedov (ICTP, Trieste & Kurchatov Inst.,Moscow),
S.Belayev (Kurchatov Inst.,Moscow),
V.Berezinsky (LNGS, Gran Sasso),
S.Bilenky (JINR, Dubna),
J.Bleimaier (Princeton),
M.Danilov (ITEP, Moscow),
G.Diambrini-Palazzi (Univ. of Rome),
A.Dolgov (INFN, Ferrara & ITEP, Moscow),
V.Kadyshevsky (JINR, Dubna),
S.Kapitza (EAPS, Moscow)
A.Logunov (IHEP, Protvino),
V.Matveev (INR, Moscow),
P.Nowosad (Univ. of Sao Paulo),
L.Okun (ITEP, Moscow),
V.Rubakov (INR, Moscow),
D.Shirkov (JINR, Dubna),
J.Silk (Univ. of Oxford),
A.Skrinsky (INP, Novosibirsk),
A.Slavnov (MSU & Steklov Math.Inst, Moscow)
A.Smirnov (ICTP, Trieste & INR, Moscow),
P.Spillantini (INFN, Florence),

Organizing Committee

V.Bagrov (Tomsk State Univ.),
V.Belokurov (MSU), V.Braginsky (MSU),
A.Egorov (ICAS, Moscow), D.Galtsov (MSU),
A.Grigoriev (MSU & ICAS, Moscow),
P.Kashkarov (MSU), A.Kataev (INR, Moscow),
O.Khrustalev (MSU), V.Mikhailin (MSU & ICAS, Moscow)
A.Mourao (IST/CENTRA, Lisbon),
N.Narozhny (MEPHI, Moscow),
A.Nikishov (Lebedev Physical Inst., Moscow),
N.Nikiforova (MSU), V.Ritus (Lebedev Physical Inst., Moscow),
Yu.Popov (MSU) , V.Savrin (MSU),
D.Shirkov (JINR, Dubna), Yu.Simonov (ITEP, Moscow),
A.Sissakian (JINR,Dubna),
A.Studenikin (MSU & ICAS, Moscow),
V.Trukhin (MSU)

Moscow State University
Interregional Centre for Advanced Studies
Centre of International Educational Projects
Ministry of Education and Science of Russia

SIXTH INTERNATIONAL MEETING ON PROBLEMS OF INTELLIGENTSIA

"INTELLIGENTSIA and VIOLENCE: Responses to Repression and Terrorism"

Moscow, August 31, 2005

Presidium of the Meeting

V.A.Sadovnichy (MSU) - Chairman
V.V.Belokurov (MSU)
J.Bleimaier (Princeton)
G.Diambrini-Palazzi (Universiry of Rome)
V.G.Kadyshevsky (JINR)
S.P.Kapitza (Russian Academy of Sciensies)
N.S.Khrustaleva (Ministry of Education and Science, Russia)
A.I.Studenikin (MSU & ICAS) - Vice Chairman
V.I.Trukhin (MSU)

FOREWORD

The 12th Lomonosov Conference on Elementary Particle Physics was held at the Moscow State University (Moscow, Russia) on August 25-31, 2005. The conference was dedicated to the 250th Anniversary of the Moscow State University.

The conference was organized by the Faculty of Physics of the Moscow State University and the Interregional Centre for Advanced Studies and supported by the Joint Institute for Nuclear Research (Dubna), the Institute of Theoretical and Experimental Physics (Moscow), the Institute for Nuclear Research (Moscow), the Budker Institute of Nuclear Physics (Novosibirsk) and the Skobeltsyn Institute of Nuclear Physics (Moscow State University). The Russian Foundation for Basic Research, the Russian Agency for Science and Innovation, the Russian Academy of Sciences and the Russian Agency for Atomic Energy sponsored the conference.

It was more than twenty years ago when the first of the series of conferences (from 1993 called the "Lomonosov Conferences"), was held at the Department of Theoretical Physics of the Moscow State University (June 1983, Moscow). The second conference was held in Kishinev, Republic of Moldavia, USSR (May 1985).

After the four years break this series was resumed on a new conceptual basis for the conference programme focus. During the preparation of the third conference (that was held in Maykop, Russia, 1989) a desire to broaden the programme to include more general issues in particle physics became apparent. During the conference of the year 1992 held in Yaroslavl it was proposed by myself and approved by numerous participants that these irregularly held meetings should be transformed into regular events under the title "Lomonosov Conferences on Elementary Particle Physics". Since then at subsequent meetings of this series a wide variety of interesting things both in theory and experiment of particle physics, field theory, astrophysics, gravitation and cosmology were included into the programmes. It was also decided to enlarge the number of institutions that would take part in preparation of future conferences.

Mikhail Lomonosov (1711-1765), a brilliant Russian encyclopaedias of the era of the Russian Empress Catherine the 2nd, was world renowned for his distinguished contributions in the fields of science and art. He also helped establish the high school educational system in Russia. The Moscow State University was founded in 1755 based on his plan and initiative, and the University now bears the name of Lomonosov.

The 6th Lomonosov Conference on Elementary Particle Physics (1993) and all of the subsequent conferences of this series were held at the Moscow State University on each of the odd years. Publication of the volume "Particle Physics, Gauge Fields and Astrophysics" containing articles written on the basis of presentations at the 5th and 6th Lomonosov Conferences was supported by the Accademia Nazionale dei Lincei (Rome, 1994). Proceedings of the 7th and 8th Lomonosov Conference (entitled "Problems of Fundamental Physics" and "Elementary Particle Physics") were published by the Interregional Centre for Advanced Studies (Moscow, 1997 and 1999). Proceedings of the 9th, 10th and 11th Lomonosov Conferences (entitled "Particle Physics at the Start of the New

Millennium", "Frontiers of Particle Physics" and "Particle Physics in Laboratory, Space and Universe") were published by World Scientific Publishing Co. (Singapore) in 2001, 2003 and 2005, correspondently.

The physics programme of the 12th Lomonosov Conference on Elementary Particle Physics (August, 2005) included review and original talks on wide range of items such as neutrino and astroparticle physics, electroweak theory, fundamental symmetries, tests of standard model and beyond, heavy quark physics, non-perturbative QCD, quantum gravity effects, physics at the future accelerators. Totally there were more than 300 participants with 107 talks including 34 plenary (30 min) talks, 38 session (25-20 min) talks and 35 brief (15 min) reports. One of the goals of the conference was to bring together scientists, both theoreticians and experimentalists, working in different fields, so that no parallel sessions were organized at the conference.

Following the tradition that has started in 1995, each of the Lomonosov Conferences on particle physics has been accompanied by a conference on problems of intellectuals. The 6th International Meeting on Problems of Intelligentsia held during the 12th Lomonosov Conference (August 31, 2003) was dedicated to discussions on the issue "Intelligentsia and Violence: Responses to Repression and Terrorism ". Three papers on this subject are included into this volume.

The Round table discussion on "Neutrino Physics and Astrophysics" was held during the last day of the 12th Lomonosov Conference. The main results of the of the Round Table discussion were summarized in the **Memorandum** on **"Neutrino and Astroparticle Physics"** approved by the participants of the conference:

"The progress in particle physics comes from both the high energy frontier and precision experiments. This applies to accelerator and non-accelerator physics. In the last years, field like neutrino physics, astroparticle physics and cosmology have had an spectacular development.

One may anticipate that these clues to the knowledge of nature will develop more along the XXI century, particular when taking into account the longer time periods involved in the construction of higher energy facilities.

There has been important progress in neutrino and astroparticle physics achieved during the last several years:

- The non-vanishing neutrino mass and flavour violation has been observed in neutrino oscillation experiments.
- The validity of the Standard Solar Model has also been proven.
- The non-zero neutrino mass can have an important impact on cosmology, in particular, for our understanding of the baryon asymmetry of the universe. On the other hand, the upper boundary of the sum of three neutrino masses can now be constrained on the level of the order of 1 eV from cosmology.
- Observations of tritium beta-decay have lowered the neutrino upper mass limit to the level of 2.1 eV.
- Double beta decay experiments have reached a sensitivity $\sim$ (0.5 – 1) eV for effective Majorana mass of the neutrino.

World-wide recognition of the obtained results has been evidenced by two Nobel Prizes which have been recently awarded for research in neutrino and astroparticle physics.

Further progress in the study of the fundamental properties of neutrinos will open the window to a new physics. Application of these studies could also play a very important role in our understanding of the inner structure of stellar cores as well as of the early stages of evolution of the universe. Studies of geo neutrinos also open promising possibilities for the future.

More accurate measurements of neutrino characteristics will make further progress in the field possible. Our conference also focused on the need to train specialized manpower in this field for the future".

The success of the 12th Lomonosov Conference was due in a large part to contributions of the International Advisory Committee and Organizing Committee. On behalf of these Committees I would like to warmly thank the session chairpersons, the speakers and all of the participants of the 12th Lomonosov Conference and the 6th International Meeting on Problems of Intelligentsia.

We are grateful to the Rector of the Moscow State University, Victor Sadovnichy, the Vice Rector of the Moscow State University, Vladimir Belokurov, the Dean of the Faculty of Physics, Vladimir Trukhin, the Director of the Skobeltsyn Institute of Nuclear Physics, Mikhail Panasyuk, the Directors of the Joint Institute for Nuclear Research, Vladimir Kadyshevsky and Alexey Sissakian, the Director of the Institute for Nuclear Research, Victor Matveev, the Director of the Budker Institute of Nuclear Physics, Alexander Skrinsky, the Vice Director of the Institute of Theoretical and Experimental Physics, Mikhael Danilov, and the Vice Deans of the Faculty of Physics of the Moscow State University, Anatoly Kozar and Pavel Kashkarov for the support in organizing these two conferences.

Special thanks are due to Inna Bilenkina and Alexander Suvorinov (the Russian Agency for Science and Innovations), Nelli Khrustaleva (the Russian Agency of Education), Boris F. Myasoedov (the Russian Academy of Sciences) and Oleg Patarakin (the Russian Agency for Atomic Energy) for their valuable help.

I would like to thank Concezio Bozzi, Alexander Dolgov, Andrey Kataev, Catherine Leluc, Lev Okun, Alexander Olshevsky and Alexey Smirnov for their help in planning of the scientific programme of the conference and inviting speakers for the topical sessions of the meeting.

Furthermore, I am very pleased to mention Alexander Grigoriev, the Scientific Secretary of the conference, Andrey Egorov, Artyem Ivanov, Sergey Shinkevich, and Marina Mescheraykova, Maria Moiseeva, Olya Moiseeva, Nastay Sutormina for their very efficient work in preparing and running the meeting.

These Proceedings were prepared for publication at the Interregional Centre for Advanced Studies with support by the Russian Foundation for Basic Research, the Russian Agencies for Science and Innovations and Education, the Russian Academy of Sciences and the Russian Agency for Atomic Energy.

Alexander Studenikin

CONTENTS

New Developments in Quantum Field Theory

THE CONCEPT OF MASS IN THE EINSTEIN YEAR

L.B. Okun[a]
State Research Center, Institute for Theoretical and Experimental Physics, 117218, Moscow, Russia

Abstract. Various facets of the concept of mass are discussed. The masses of elementary particles and the search for higgs. The masses of hadrons. The pedagogical virus of relativistic mass.

1 From "Principia" to Large Hadron Collider (LHC)

The term "mass' was introduced into mechanics by Newton in 1687 in his "*Principia*" [1]. He defined it as the amount of matter. The generally accepted definition of matter does not exist even today. Some authors of physics text-books do not consider photons – particles of light – as particles of matter, because they are massless. For the same reason they do not consider as matter the electromagnetic field. It is not quite clear whether they consider as matter almost massless neutrinos, which usually move with velocity close to that of light. Of course it is impossible to collect a handful of neutrinos similarly to a handful of coins. But in many other respects both photons and neutrinos behave like classical particles, while the electromagnetic field is the basis of our understanding of the structure of atoms. On the other hand, the so-called weak bosons W^+, W^-, Z^0 are often not considered as particles of matter because they are too heavy and too short-lived.

Even more unusual are such particles as gluons and quarks. Unlike atoms, nucleons, and leptons, they do not exist in a free state: they are permanently confined inside nucleons and other hadrons.

There is no doubt that the problem of mass is one of the key problems of modern physics. Though there is no common opinion even among the experts what is the essence of this problem. For most of particle theorists, as well as members of LHC community, the solution of the problem is connected with the quest and discovery of the higgs – scalar boson which in the Standard Model is responsible for the masses of leptons and quarks and their electroweak messengers: W and Z. The discovery of higgs and the study of higgs sector might elucidate the problem of the pattern of hierarchy of masses of leptons and quarks: from milli electron Volts for neutrinos to about 180 GeV for t-quark. For many physicists it is a QCD problem: how light quarks and massless gluons form massive nucleons and atomic nuclei. Still for majority of confused students and science journalists there is no difference between mass of a body m and its energy E divided by c^2: they believe in the "most famous formula $E = mc^2$".

If higgs exists, its discovery will depend on the funding of the particle physics. In 1993 the termination of the SSC project sent the quest for the higgs into a painful knockdown. The decision not to order in 1995 a few dozen of extra superconducting cavities prevented, a few years later, LEP II from crossing the 115 GeV threshold for the mass of the higgs.

If we are lucky and higgs is discovered around year 2010 at LHC, then the next instrument needed to understand what keeps the masses of the higgs below 1 TeV scale, is ILC (International linear collider). This machine would provide a clean environment for the study of higgs production and decays. It could also be used for discovery and study of light supersymmetric particles (SUSY). A prototype of ILC was suggested a few years ago by DESY as the project TESLA. There was no doubt that if funded, TESLA would work, but the funding was not provided by the German government. The new variant of ILC envisions increasing the maximal center of mass

[a] e-mail: okun@itep.ru

energy of colliding electron and positron from) 0.5 TeV to 1 TeV. If everything goes well, ILC can start before 2020.

Further increase of energy, to say, 5 TeV, would call for a machine of the type of CLIC (Compact linear collider) the project of which is under discussion at CERN for more than a decade. In this machine the role of clystrons is supposed to play a low energy but very high current "decelerator" the energy of which would be pumped into the high energy accelerator part of CLIC. Unlike situation with ILC, even the mere feasibility of CLIC is not clear now. Special experimental research to ascertain the feasibility is going on at CERN.

The discussion of higgses, neutrinos and QCD in connection with the fundamental problems of mass is often accompanied and even overshadowed by a "pseudoproblem" of the so-called "relativistic mass" (see section 5).

2 Mass in Newtonian Mechanics

The more basic is a physical notion, the more difficult to define it in words. A good example give the 1960s editions of "Encyclopedia Britannica" where energy is defined in terms of work, while the entry "work" refers to labour and professional unions. Most people have intuitive notions of space and time. Every physicist has intuitive notions of energy, mass, and momentum. But practically everybody has difficulties in casting these notions into words without using mathematics.

Though the definition of mass ("Definition I: The quantity of matter is the measure of the same, arising from its density and bulk conjointly") given by Newton in his "*Principia*" [1] was so unclear that scholars are discussing its logical consistency even today, the equations of Newtonian mechanics are absolutely self-consistent. Mass m enters in the relations of velocity $\mathbf{v} = d\mathbf{r}/dt$ and momentum $\mathbf{p}$:

$$\mathbf{p} = m\mathbf{v} \ , \tag{1}$$

as well as acceleration $\mathbf{a} = d\mathbf{v}/dt$ and force $\mathbf{F}$:

$$\mathbf{F} = d\mathbf{p}/dt = m\mathbf{a} \ . \tag{2}$$

It also enters in the equation defining the force of gravity with which a body with mass m_1 at point $\mathbf{r}_1$ attracts another body with mass m_2 at point $\mathbf{r}_2$:

$$\mathbf{F}_g = -Gm_1m_2\mathbf{r}/r^3 \ . \tag{3}$$

Here $\mathbf{r} = \mathbf{r}_2 - \mathbf{r}_1$, $r = |\mathbf{r}|$, while G is the famous Newton constant:

$$G = 6.67 \cdot 10^{-11}\mathrm{m}^3\mathrm{kg}^{-1}\mathrm{s}^{-2} \ . \tag{4}$$

The kinetic energy of a body is defined as

$$E_k = \mathbf{p}^2/2m = m\mathbf{v}^2/2 \ . \tag{5}$$

The potential gravitational energy:

$$U_g = -Gm_1m_2/r \ , \tag{6}$$

while the total energy in this case is

$$E = E_k + U_g \ . \tag{7}$$

The total energy is conserved. When a stone falls on the earth, its potential energy decreases (becomes more negative), kinetic energy increases: so that the total energy does not change. When the stone hits the ground, its kinetic energy is shared by the ambient molecules raising the local temperature.

One of the greatest achievements of the XIX century was the formulation of the laws of conservation of energy and momentum in all kinds of processes.

At the beginning of the XX century it was realized that conservation of energy is predetermined by uniformity of time, while conservation of momentum – by uniformity of space.

But let us return to the notion of force. People strongly felt the force of gravity throughout the history of mankind, but only in XVII century the equations (3) and (6) were formulated.

An important notion in this formulation is the notion of gravitational potential φ_g. The gravitational potential of a body with mass m_1 is

$$\varphi_g = -\frac{Gm_1}{r} \tag{8}$$

Thus the potential energy of a body with mass m_2 in a potential (8) is

$$U_g = m_2\varphi_g \ , \tag{9}$$

which coincides with eq. (6).

A century later similar equations were formulated for another long-range interaction, the electrical one:

$$\mathbf{F}_e = e_1 e_2 \mathbf{r}/r^3 \ , \tag{10}$$

$$U_e = e_1 e_2/r \ , \tag{11}$$

$$\varphi_e = e_1/r \ , \tag{12}$$

$$U_e = e_2\varphi_e \ . \tag{13}$$

In these equations, which define the Coulomb force, Coulomb potential energy, and Coulomb potential respectively, e_1 and e_2 are electrical charges of two bodies.

An important role in the theory of electricity is played by the strength of electric field $\mathbf{E}$. Charge e_1 creates field with strength

$$\mathbf{E} = e_1\mathbf{r}/r^3 \ . \tag{14}$$

Thus

$$\mathbf{F}_e = e_2\mathbf{E} \tag{15}$$

As most of matter around us is electrically neutral, the electrical interaction was known for centuries only as a kind of trifle. Unlike mass m, which is always positive, the charge e has two varieties: positive and negative. Two charges of opposite sign attract each other, while those of the same sign are repelling. Protons residing in the nucleus of an atom have positive charge, the charge of electrons, which form atomic shells, is negative. As a result the atom is electrically neutral.

Electrical interaction and its ramifications determine the main features of atoms, molecules, condensed matter, and biological cells. Gravitational interaction is too feeble to play any role at that level. To see this consider an electron and a proton. Their masses, respectively, are

$$m_e = 9.1 \cdot 10^{-31}\ \text{kg} \ , \quad m_p = 1.7 \cdot 10^{-27}\ \text{kg} \ . \tag{16}$$

Their electric charges:

$$e_p = -e_e = e = 4.8 \cdot 10^{-10}\ \text{esu} \ , \tag{17}$$

where esu denotes electrostatic unit:

$$(1\text{ esu})^2 = 10^{-9}\text{ kg m}^3\text{s}^{-2} \quad . \tag{18}$$

Hence

$$-e_e e_p = 2.3 \cdot 0^{-28}\text{ kg m}^3\text{s}^{-2} \quad . \tag{19}$$

On the other hand

$$G m_e m_p = 1.03 \cdot 10^{-67}\text{ kg m}^3\text{s}^{-2} \quad , \tag{20}$$

Thus in an atom the gravitational force is $\sim 10^{-40}$ of electric one.

The importance of gravity for our every day life is caused by the huge number of atoms in the earth, and hence by its very large mass:

$$M = 5.98 \cdot 10^{24}\text{ kg} \quad . \tag{21}$$

Taking into account the value of the radius of the earth

$$R = 6.38 \cdot 10^6\text{ m} \quad , \tag{22}$$

we find the gravitational force of the earth acting on a body with mass m close to its surface:

$$\mathbf{F}_g = m\,\mathbf{g} \quad , \tag{23}$$

where $\mathbf{g}$ is acceleration directed towards the center of the earth:

$$g = |\mathbf{g}| = \frac{6.67 \cdot 10^{-11} \cdot 5.98 \cdot 10^{24}}{(6.38 \cdot 10^6)^2}\text{m s}^{-2} = 9.8\text{ m s}^{-2} \quad . \tag{24}$$

Let us note that for gravity $\mathbf{g}$ plays the role of the strength of gravitational field, which is analogous to the role of $\mathbf{E}$ for electricity. The acceleration $\mathbf{g}$ does not depend on the mass or any other properties of the attracted body. In that sense it is universal. This universality was established by Galileo early in the 17th century in his experiments with balls rolling down an inclined plane. One can see this plane, with little bells ringing when a ball passes them, in Florence. (Apocryphal history tells that Galileo had discovered universality of g by dropping balls from the Tower of Pisa.)

Gravitational and electric interactions are the only long-range interactions the existence of which has been established. Two other fundamental interactions, referred to as strong and weak, have very short ranges: 10^{-15} m and 10^{-18} m respectively. Their first manifestations were discovered about a century ago in the form of radioactivity. Their further study has lead to new disciplines: nuclear physics and the physics of elementary particles.

Quite often you might find in the literature, in the discussion of Newtonian mechanics, the terms "inertial mass" m_i and "gravitational mass" m_g. The former is used in equations (1), (2), (5), defining the inertial properties of bodies. The latter is used in equations (3), (6), (23), describing the gravitational interaction. After introducing these terms a special law of nature is formulated:

$$m_i = m_g \quad , \tag{25}$$

which is called upon to explain the universality of g.

However Galileo had discovered this universality before the notion of mass was introduced by Newton, while from Newton equations (1), (2), (3) the universality of g follows without additional assumptions. Thus the notions and notations m_i and m_g are simply redundant. As we will see later, their introduction is not only redundant, but contradicts the General Theory of Relativity, which explains why the same mass m enters equations (1) - (3).

The advocates of m_i and m_g argue by considering the possibility that in the future the more precise experimental tests might discover a small violation of Galilean universality. But that would mean that a new feeble long-range force exists in nature. In literature this hypothetical force is often referred to as a "fifth force" (in addition to the four established ones). When and if the "fifth force" is discovered it should be carefully studied. But at present it should not confuse the exposition of well established physical laws. Especially confusing and harmful are m_i and m_g in the text-books for students.

At that point it is appropriate to summarize the properties of mass in Newtonian mechanics:

1. Mass is a measure of the amount of matter.
2. Mass of a body is a measure of its inertia.
3. Masses of bodies are sources of their gravitational attraction to each other.
4. Mass of a composite body is equal to the sum of masses of the bodies that constitute it; mathematically that means that mass is additive.
5. Mass of an isolated body or isolated system of bodies is conserved: it does not change with time.
6. Mass of a body does not change in the transition from one reference frame to another.

3 Mass in Special Relativity

Of great conceptual importance in modern science is the principle of relativity first stated by Galileo: A rectilinear motion of a physical system with constant velocity relative to any external object is unobservable within the system itself. The essence of this principle was beautifully exposed in the famous book "Dialogue Concerning the Chief World Systems – Ptolemaic and Copernican", published in 1632 [2]:

"Shut yourself up with some friend in the main cabin below decks on some large ship, and have with you there some flies, butterflies, and other small flying animals. Have a large bowl of water with some fish in it; hang up a bottle that empties drop by drop into a wide vessel beneath it. With the ship standing still, observe carefully how the little animals fly with equal speed to all sides of the cabin. The fish swim indifferently in all directions; the drops fall into the vessel beneath; and, in throwing something to your friend, you need throw it no more strongly in one direction than another, the distances being equal: jumping with your feet together, you pass equal spaces in every direction. When you have observed all these things carefully (though there is no doubt that when the ship is standing still everything must happen in this way). Have the ship proceed with any speed you like, so long as the motion is uniform and not fluctuating this way and that, you will discover not the least change in all the effects named, nor could you tell from any of them whether the ship was moving or standing still. In jumping, you will pass on the floor the same spaces as before, nor will you make larger jumps toward the stern than toward the prow even though the ship is moving quite rapidly, despite the fact that during the time you are in the air the floor under you will be going in a direction opposite to your jump. In throwing something to your companion, you will need no more force to get it to him whether he is in the direction of the bow or the stern, with yourself situated opposite. The droplets will fall as before into the vessel beneath without dropping toward the stern, although while the drops are in the air the ship runs many spans. The fish in their water will swim toward the front of their bowl with no more effort than toward the back, and will go with equal ease to bait placed anywhere around the edges of the bowl. Finally the butterflies and flies will continue their flights indifferently toward every side, nor will it ever happen that they are concentrated toward the stern, as if tired out from keeping up with the course of the ship, from which they will have been separated during long intervals by keeping themselves in the air. And if smoke is made by burning some incense, it will be seen going up in the form of a little cloud, remaining still and moving no more toward one side than the other. The cause of all these correspondences of effects is the fact that the ship's motion is common to all the things contained in it, and to the air also. That is why I said you should be below decks; for if this took place above in the open air, which would not follow the course of the ship, more or less noticeable differences would be seen in some of the effects noted."

Sometimes one can hear that the ship of Galileo was discussed two centuries earlier by cardinal Nicolaus Cusanus (1401 – 1464) in his book "De docta ignorata" ("On the scientific ignorance") published in 1440. Indeed, one can read in volume II, at the beginning of chapter XII "The properties of the earth":

"It is clear to us that the earth is actually moving, though we do not see this, as we feel the movement only through comparison with a point at rest. Somebody on a ship in the middle of waters, without knowing that water is flowing and without seeing the shores, how could he ascertain that the ship is moving?" [3] [b]. The relative character of motion is expressed in these lines quite clearly. But the cabin of Galileo's ship is full of various phenomena and experiments, proving that observable effects look the same in any inertial reference frame. At this point we define an inertial reference frame, as that which moves rectilinearly with constant velocity with respect to the stars. We shall give a more accurate definition when considering General Theory of Relativity.

If the velocity of the ship is $\mathbf{u}$ and it moves along the axis x, the coordinates of two inertial frames are connected by equations:

$$t' = t\ , \quad x' = x + ut\ , \quad y' = y\ , \quad z' = z\ , \tag{26}$$

where $u = |\mathbf{u}|$, the primed coordinates refer to the shore, while unprimed to the ship. From the definition of velocity $\mathbf{v} = d\mathbf{r}/dt$ one easily sees that

$$\mathbf{v}' = \mathbf{v} + \mathbf{u} \tag{27}$$

and that $\mathbf{v}'$, $\mathbf{p}'$, $\mathbf{a}'$, $\mathbf{F}'$, $\mathbf{F}'_g$, E'_k, U'_g, $\mathbf{F}'_e$, U'_e satisfy the same equations (1) - (11) as their unprimed analogues.

Galilean principle of relativity is the quintessence of Newtonian mechanics. Nevertheless the latter is called non-relativistic mechanics, as opposed to Einsteinian mechanics which is called relativistic. This is one of many examples of lack of complete consistency in the language of physics which is a natural product of its evolution. The point is that Newtonian mechanics satisfies the Galilean principle of relativity only partially. The cabin of the original Galilean ship did not contain apparatuses that were able to measure the velocity of light. This velocity was first established in 1676 by Danish astronomer O. Roemer (1644 - 1710), who deduced from the observations of the moons of Jupiter, performed by J. Cassini, that it is $2.4 \cdot 10^5$ km s^{-1}. Further measurements during three centuries established its present value: $c = 3 \cdot 10^5$ km s^{-1}.

Of greatest importance was the discovery made two centuries later by American physicists A. Michelson and E. Morley. By using a special two arm rotating interferometer they established that the velocity of light did not depend on the angle between

---

[b] I am grateful to Peter Zerwas for arousing my interest to Nicolaus Cusanus.

the light ray and the vector of velocity of the earth on its journey around the sun. In this experiment the earth itself played the role of Galilean ship. That result signalled that the simple law (27) of addition of velocities is not valid for light.

This, in its turn, meant that the coordinate transformations (26) - (27) should be changed when v and/or u (due to relative character of velocity) are of the order of c.

This change had been performed by H. Lorentz (1904) [4], H. Poincare (1905) [5,6] and A. Einstein (1905) [7, 8]. Lorentz considered deformation of electron moving through the so called ether, filling all the universe, and introduced primed spatial and time coordinates, as purely auxiliary quantities. Poincare and Einstein wrote transformations between primed and unprimed coordinates:

$$t' = (t + ux/c^2)\gamma \ , \ \ x' = (x + ut)\gamma \ , \ \ y' = y \ , \ \ z' = z \ , \tag{28}$$

where

$$\gamma = 1/\sqrt{1 - u^2/c^2} \ . \tag{29}$$

They were called Lorentz transformations by Poincare and later by Einstein.

Poincare believed in ether and considered that the remaining problem is to understand it. Einstein simply dispensed with ether, he considered transformations (28) - (29) as a direct expressions of properties of space and time. Galilean relativity of inertial motion resulted in relativity of simultaneity, of time, and of length.

Proceeding from his article [7] Einstein came [8] to a fundamental conclusion that a body at rest has rest-energy E_0:

$$E_0 = mc^2 \ . \tag{30}$$

Here m is the mass of the body, while index 0 in E_0 indicates that this is the energy in the body's rest frame.

In 1906 M. Planck explicitly wrote the expressions of total energy E and momentum $\mathbf{p}$ of a body with arbitrary value of its velocity $\mathbf{v}$:

$$E = mc^2\gamma \ , \ \ \mathbf{p} = m\mathbf{v}\gamma \ , \tag{31}$$

where

$$\gamma = (1 - v^2/c^2)^{-1/2} \ . \tag{32}$$

These expressions can be easily derived by assuming that E and $\mathbf{p}$ transform in the same way as t and $\mathbf{r}$, each pair $(E, \mathbf{p}c)$ and $(t, \mathbf{r}/c)$ forming a four-dimensional vector. Indeed, by applying Lorentz transformations to a body at rest, taking into account relation (30) and writing v instead of u, we come to (31) - (32). Of course, the isotropy of space should be also accounted for.

The notion of four-dimensional space-time was introduced in 1908 by G. Minkowski [10]. While four-vectors transform under Lorentz transformations (rotations in Minkowskian pseudo-Euclidian space), their squares are Lorentz-invariant:

$$\tau^2 = t^2 - (\mathbf{r}/c)^2 \ , \tag{33}$$

$$m^2c^4 = E^2 - (\mathbf{p}c)^2 \ . \tag{34}$$

Here τ is the so-called proper time, while m, as before, is the mass of a body. But now it acquires a new meaning, which was absent in Newtonian mechanics. (Note that for a body at rest ($\mathbf{p} = 0$) one recovers from eq. (34) the relation (30) between mass and rest-energy.)

It is impossible to discuss the concept of mass without explicitly basing the discussion on the achievements of XX century physics and especially on the notion of elementary particles such as electrons, photons and neutrinos, less elementary, such as protons and neutrons (in which quarks and gluons are confined), or composite, such as atoms and atomic nuclei. It is firmly established that all particles of a given kind (for instance all electrons) are identical and hence have exactly the same value of mass. The same refers to protons and neutrons.

Atoms and atomic nuclei ask for further stipulations because each of these composite systems exists not only in its ground state, but can be brought to one of its numerous excited states (energy levels). For instance, a hydrogen atom is a bound system of a proton and electron, attracted to each other by Coulomb force (10). As proton is approximately two thousand times heavier than electron, one usually speaks about electron moving in the Coulomb potential (11) of proton. According to the laws of quantum mechanics this movement is quantized, forming a system of levels. The lowest level is stable, the excited ones are unstable. Electron jumps from a higher level to a lower one by emitting a quantum of light - photon. Finally it reaches the ground level.

The energy in atomic, nuclear and particle physics is measured in electron Volts: 1 eV is the energy which electron gains by traversing a potential of 1 Volt; 1 keV $= 10^3$ eV, 1 MeV $= 10^6$ eV, 1 GeV $= 10^9$ eV, 1 TeV $= 10^{12}$ eV, 1 PeV $= 10^{15}$ eV, 1 EeV $= 10^{18}$ eV.

The binding energy of electron at the ground level of hydrogen atom is 13.6 eV. Due to relation between rest-energy and mass it is convenient to use as a unit of mass 1 eV/c^2. The mass of electron $m_e = 0.511$ MeV/c^2, the mass of proton $m_p = 0.938$ GeV/c^2. The mass of a hydrogen atom in its ground state is by 13.6 eV/c^2 smaller than the sum $m_e + m_p$. This mass difference is often referred to as defect of mass.

As c is a universal constant, it is appropriate to use it in relativistic physics as a unit of velocity and hence to put $c = 1$ in all above values of masses and defects of mass. In what follows we will use as units of mass eV and its derivatives: keV, MeV, GeV, etc.

One eV is a tiny unit when compared with Joule (J) or kilogram:

$$1\ \mathrm{J} = 6.24 \cdot 10^{18}\ \mathrm{eV} \quad , \quad 1\ \mathrm{eV} = 1.6 \cdot 10^{-19}\ \mathrm{J} \ , \tag{35}$$

$$1\ \mathrm{kg} = 5.61 \cdot 10^{35}\ \mathrm{eV} \quad , \quad 1\ \mathrm{eV} = 1.78 \cdot 10^{-36}\ \mathrm{kg} \ . \tag{36}$$

However it is four orders of magnitude larger than one degree of Kelvin (K).

$$1\ \mathrm{K} = 0.86 \cdot 10^{-4}\ \mathrm{eV} \quad , \quad 1\ \mathrm{eV} = 1.16 \cdot 10^{4}\ \mathrm{K} \tag{37}$$

(In eq. (37) we put dimensional Boltzmann factor k equal to unity, taking into account that kT, where T is temperature, characterizes the mean energy of an ensemble of particles.) Let us estimate the relative change of mass in a few everyday processes.

The light from the sun is absorbed by vegetation on the earth to produce carbohydrates via reaction of photosynthesis:

$$\mathrm{light} + 6\mathrm{CO_2} + 6\mathrm{H_2O} = 6\mathrm{O_2} + \mathrm{C_6H_{12}O_6} \ .$$

The total energy of light required to produce one molecule of $C_6H_{12}O_6$ is about 4.9 eV. This does not mean that the photons are massive. They are massless, but the kinetic energy of photons is transformed into the rest energy of carbohydrates.

A combustion of methane in the gas burner of a kitchen stove:

$$\mathrm{CH_4} + 2\mathrm{O_2} \to \mathrm{CO_2} + 2\mathrm{H_2O} \ . \tag{38}$$

In this reaction 35.6 MJ of heat is released per cubic meter of methane. Since the density of methane is 0.72 kg/m^3 and density of oxygen is 1.43 kg/m^3

$$\frac{\Delta m}{m} = \frac{35.6 \cdot 6.24 \cdot 10^6 \cdot 10^{18}}{(0.72 + 2 \cdot 1.43) \cdot 5.61 \cdot 10^{35}} = 1.1 \cdot 10^{-10} \ ,$$

where in the nominator eq.(35) and in denominator eq.(36) are used.

We can look at this result differently by starting from Avogadro number:

$$N_A = 6.022 \cdot 10^{23} \text{ mol}^{-1} \tag{39}$$

and molar volume (for ideal gas)

$$22.4 \cdot 10^{-3} \text{ m}^3 \cdot \text{mol}^{-1} \ .$$

This means that a cubic meter of methane contains $2.69 \cdot 10^{25}$ molecules. Thus, burning of one molecule of methane releases

$$\frac{35.6 \cdot 6.24 \cdot 10^{24}}{2.69 \cdot 10^{25}} = 8.3 \text{ eV}$$

Now we estimate the mass of one molecule of methane plus 2 molecules of oxygen: $16 \times 5 \cdot 0.94$ GeV $= 75$ GeV, and calculate $\Delta m/m$ at molecular level (we use 0.94 GeV as the mass of a nucleon):

$$8.3 \text{ eV}/75 \text{ GeV} = 1.1 \cdot 10^{-10} \ .$$

Thus, we see that the sum of masses of molecules on the right hand side of eq. (38) is by 8.3 eV smaller than that on the left-hand side. This mass difference is exploited in cooking.

Another example is the melting of ice. It takes $0.334 \cdot 10^6$ J to melt a kilogram of ice. That means that in this case the relative increase of mass $\Delta m/m$ is (see eqs. (35) and (36)):

$$\Delta m/m = 0.334 \cdot 10^6 \cdot 6.24 \cdot 10^{18} \cdot 1.78 \cdot 10^{-36} = 3.7 \cdot 10^{-12} \ .$$

If the temperature of a flat iron is increased by 200^0 its mass increases by $\Delta m/m = 10^{-12}$. This is readily estimated using the specific heat (25 J $\cdot$ mol^{-1} K^{-1} $= 450$ J kg^{-1} K^{-1}):

$$\frac{\Delta m}{m} = 450(\text{J kg}^{-1}\text{K}^{-1})200 \text{ K} = 10^{-12} \ .$$

All these mass differences are too tiny to be measured directly. Let us note that the defect of mass in a hydrogen atom 13.5 eV is also too small to be observed directly because the mass of the proton is known with large uncertainty ± 80 eV.

The tiny values of Δm in atomic transitions and chemical reactions were the basis for the statement that in non-relativistic physics mass is additive, and of the law of conservation of this additive mass.

However in nuclear and particle physics the defect of mass is much larger. For instance, in the case of deutron, which is a nuclear bound state of proton and neutron, the binding energy and hence the defect of mass is 2.2 MeV, so that $\Delta m/m \simeq 10^{-3}$.

Of special pedagogical interest is the reaction of annihilation of electron and positron into two photons (two γ-quanta). Photons are massless particles, which always move with velocity c. The latter statement follows from eqs. (31), (34):

$$|\frac{\mathbf{v}}{c}| = |\frac{\mathbf{p}c}{E}| = 1 \ , \text{ if } |\mathbf{p}c| = E \ . \tag{40}$$

Depending on their energy, photons are referred to as quanta of radio waves, visible and invisible light, X-rays, γ-quanta.

The reaction of annihilation is

$$e^+ + e^- \to \gamma + \gamma \ . \tag{41}$$

Let us consider the case when electron and positron annihilate at rest. Then their total energy is $E = E_0 = 2m_e c^2$, while the total momentum $\mathbf{p} = 0$. Due to conservation of energy and momentum the two photons will fly with opposite momenta, so that each of them will have energy equal to $m_e c^2$. The rest frame of $e^+ + e^-$ will be obviously the rest frame of two photons. Thus, the rest energy of the system of two photons will be $2m_e c^2$ and hence the mass of this system will be $2m_e$, in spite of the fact that each of the photons is massless. We see that mass in relativity theory is conserved, but not additive.

In general case the system of two free particles with energies and momenta E_1, $\mathbf{p}_1$ and E_2, $\mathbf{p}_2$ has total energy and momentum, respectively,

$$E = E_1 + E_2 \ , \quad \mathbf{p} = \mathbf{p}_1 + \mathbf{p}_2 \ . \tag{42}$$

These equations follow from additivity of energy and momentum. The mass of the system is defined as before by eq. (34). Hence

$$m^2 = (E_1 + E_2)^2 - (\mathbf{p}_1 + \mathbf{p}_2)^2 = m_1^2 + m_2^2 + 2E_1E_2(1 - \mathbf{v}_1\mathbf{v}_2) \ . \tag{43}$$

It follows from eq. (43) that the mass of a system of two particles depends not only on masses and energies of these particles, but also on the angle between their velocities. Thus for two photons m is maximal when this angle is π and vanishes when it is zero.

The mass of system has lost its Newtonian meaning of an amount of substance, its main characteristic being now rest energy (in units, where $c = 1$). Newtonian equations can be obtained from relativistic ones in the limiting case of low velocities ($v/c \ll 1$). In that case γ given by eq. (32) becomes

$$\gamma = (1 - v^2/c^2)^{-1/2} \simeq 1 + \frac{v^2}{2c^2} \ , \tag{44}$$

so that for one particle we get:

$$E = mc^2 + \frac{mv^2}{2} = E_0 + E_{kin} \ , \quad \mathbf{p} \simeq m\mathbf{v} \ . \tag{45}$$

For a system of two particles in the limit of vanishing v we get from eq. (43)

$$m^2 \simeq (m_1 + m_2)^2. \tag{46}$$

Thus the approximate additivity of mass is restored.

We started the description of Newtonian mechanics by consideration of static gravitational and electric interactions, in particular, their potentials (8) and (12). For particles at rest these potentials do not depend on time. The situation is drastically changed when the velocity of particles is not negligible. Let us start with electrodynamics. First, in addition to the scalar potential φ we have now vector potential $\mathbf{A}$, so that $\varphi, \mathbf{A}$ form a four vector. Second, because of finite velocity c of propagation of electromagnetic perturbations, both φ and $\mathbf{A}$ are retarded:

$$\varphi(t_2, \mathbf{r}_2) = \frac{e}{r - \frac{\mathbf{vr}}{c}} \ , \quad \mathbf{A}(t_2, \mathbf{r}_2) = \frac{e\mathbf{v}}{c(r - \frac{\mathbf{vr}}{c})} \ , \tag{47}$$

where

$$\mathbf{r} = \mathbf{r}_2(t_1) - \mathbf{r}_1(t_1) \ , \tag{48}$$

while

$$\mathbf{v} = \mathbf{v}(t_1, \mathbf{r}_1) \ . \tag{49}$$

Thus defined φ and $\mathbf{A}$ allow one to calculate the strengths of electric and magnetic fields:

$$\mathbf{E} = -\frac{1}{c}\frac{\partial \mathbf{A}}{\partial t} - \text{grad}\,\varphi \ , \quad \mathbf{H} = \text{rot}\,\mathbf{A} \ , \tag{50}$$

where the differentiation is performed with respect to t_2, $\mathbf{r}_2$.

In a four-dimensional form the six components of antisymmetric tensor of the strength of electromagnetic field F_{ik} are expressed in terms of derivatives of four-dimensional potential:

$$F_{ik} = \frac{\partial A_k}{\partial x^i} - \frac{\partial A_i}{\partial x^k} \ . \tag{51}$$

The lower indices are referred to as covariant, while the upper ones as contravariant.

$$A_i = (\varphi, -\mathbf{A}) \ , \ \ x^i = (ct, \mathbf{r}) \ , \ \ i = 0, 1, 2, 3 \tag{52}$$

The components of $\mathbf{E}$ and $\mathbf{H}$ are expressed in terms of components of F_{ik}:

$$E_x = F_{11} \ , \ \ E_y = F_{02} \ , \ \ E_z = F_{03} \ , \tag{53}$$

$$H_x = F_{23} \ , \ \ H_y = F_{31} \ , \ \ H_z = F_{12} \ . \tag{54}$$

The Lorentz invariant products of four-vectors are constructed by using the so-called metric tensor η_{ik}, which in an inertial reference frame is given by a diagonal 4×4 matrix:

$$\eta^{ik} = \eta_{ik} = \text{diag}(1, -1, -1, -1) \tag{55}$$

with vanishing non-diagonal elements. Multiplication of a covariant vector by η^{ik} gives a contravariant vector, e.g.:

$$p^i = \eta^{ik} p_k \ , \tag{56}$$

where summation over index k is presumed.

Up to now we considered point-like particles. If the charge is smeared over a finite volume with density ρ, the total charge of a particle is given by integral:

$$e = \int \rho(\mathbf{r}) d\mathbf{r} \ . \tag{57}$$

Similarly:

$$e\mathbf{v} = \int \mathbf{v}(\mathbf{r}) \rho(\mathbf{r}) d\mathbf{r} \ . \tag{58}$$

The four-vector $j^i = (c\rho, \mathbf{v}\rho)$ describes the density of the four-current. (In the case of a point-like particle $\rho = e\delta(\mathbf{r} - \mathbf{r}_1)$.)

The famous Maxwell's equations of classical electrodynamics have the form:

$$\frac{\partial F^{ik}}{\partial x^k} = -\frac{4\pi}{c} j^i \ , \quad \frac{\partial \tilde{F}^{ik}}{\partial x^k} = 0 \ , \tag{59}$$

Here

$$\tilde{F}^{ik} = \varepsilon^{iklm} F_{lm} \ , \tag{60}$$

where ε^{iklm} is Lorentz-invariant antisymmetric tensor. We see that current j^i is the source of electromagnetic field.

4 Mass in General Relativity

Let us now consider relativistic gravity. The role of gravitational potentials is played by 10 components of symmetric metric tensor $g_{ik}(x^i)$, four of them being diagonal, while six off-diagonal. What is very important is that in the case of gravity the ten components of g^{ik} are functions of space-time points x^i: they change from one point to another. The source of gravitational field, the analogue of vector j^i, is the density of energy-momentum tensor T^{ik}. T^{ik} is symmetric and conserved

$$\frac{\partial T^{ik}}{\partial x^k} = 0 \ . \tag{61}$$

The total 4-momentum of a system

$$p^i = \frac{1}{c}\int T^{i0}(\mathbf{r})d\mathbf{r} \tag{62}$$

Hence T^{00} is the density of energy, while T^{10}/c, T^{20}/c and T^{30}/c represent the density of momentum. For a point-like particle with mass m the density of mass μ is given by

$$\mu = m\delta(\mathbf{r} - \mathbf{r}_1) \ . \tag{63}$$

$$T^{ik} = \mu c \frac{dx^i}{ds} \cdot \frac{dx^k}{dt} = \mu c u^i u^k \frac{ds}{dt} \ , \tag{64}$$

where u^i is contravariant velocity, while ds is an invariant interval:

$$u^i = dx^i/ds \ , \ \ ds^2 = g_{ik}dx^i dx^k \ , \ \ ds = cd\tau = \sqrt{g_{00}}dx^0 \ . \tag{65}$$

Hence

$$\tau = \frac{1}{c}\int \sqrt{g_{00}}dx^0 \ , \tag{66}$$

where τ is the proper time for a given point in space.

The connection between u^i and ordinary 3-velocity $\mathbf{v}$ is

$$u^i = (\gamma, \frac{\mathbf{v}}{c}\gamma) \ . \tag{67}$$

Thus

$$u_i u^i = 1 \ . \tag{68}$$

The most important conclusion is that the source of gravitational field is proportional to the mass of a particle.

The equation for gravitational potential g_{ik}, derived by Einstein in 1915, is more complex than the Maxwell equation for A^i:

$$R_{ik} - \frac{1}{2}g_{ik}R = 8\pi G T_{ik} \ . \tag{69}$$

Here R_{ik} is the so-called Ricci tensor, while R is scalar curvature:

$$R = g^{ik}R_{ik} \ . \tag{70}$$

The role of electromagnetic field strength F_{ik} is played in gravity by the affine connection:

$$\Gamma^i_{kl} = \frac{1}{2}g^{im}\left(\frac{\partial g_{mk}}{\partial x^l} + \frac{\partial g_{ml}}{\partial x^k} - \frac{\partial g_{kl}}{\partial x^m}\right) \ , \tag{71}$$

while the role of derivative $\partial_k F^{ik}$ is played by the left-hand side of eq. (69), where the Ricci tensor is given by:

$$R_{ik} = g_{in} g^{rs} \left(\frac{\partial \Gamma^n_{ks}}{\partial x^r} - \frac{\partial \Gamma^n_{kr}}{\partial x^s} + \Gamma^n_{pr}\Gamma^p_{ks} - \Gamma^n_{ps}\Gamma^p_{kr} \right) \ . \tag{72}$$

The drastic difference of gravidynamics from electrodynamics is the nonlinearity of Einstein equation (69): it contains products of affine connections. This nonlinearity manifests itself at low values of v/c as a tiny effect in the precession of perihelia of planets (Mercury). However it is very important for strong gravitational fields in such phenomena as black holes.

From principal point of view of highest priority is the dual role of the tensor g_{ik}, which is both dynamical and geometrical. Dynamically g_{ik} represents the potential of gravitational field. On the other hand g_{ik} and its derivatives determine the geometry of space-time. Einstein gave to his theory of relativistic gravity the name of General Theory of Relativity (GTR).

It is clear from the above equations of GTR that in non-relativistic limit $v/c \ll 1$ the gravitational interaction is determined by only one mass m, while notations m_g and m_i are redundant and misleading. Even more useless are concepts of active and passive gravitational mass often considered by some authors.

5 The pedagogical virus of "relativistic mass"

The "famous formula $E = mc^2$" and the concept of "relativistic mass" increasing with velocity, which follows from it, are historical artifacts, contradicting the basic symmetry of Einstein's Special Relativity, the symmetry of 4-dimensional space-time. The relation discovered by Einstein is not $E = mc^2$, but $E_0 = mc^2$, where E_0 is the energy of a free body at rest introduced by Einstein in 1905. The source of the longevity of the "famous formula" is the irresponsible attitude of relativity theory experts to the task of explaining it to the non-experts.

The notion of "relativistic mass" presents a kind of pedagogical virus which very effectively infects new generations of students and professors and shows no signs of decline. Moreover in the Year of Physics it threatens to produce a real pandemia.

I published my first articles [11, 12] against the "Einstein famous equation $E = mc^2$" in 1989. The subject seemed important to me because it concerned the proper teaching of special relativity at high schools, colleges and universities and explaining its genuine meaning to a wide audience of non-physicists, the so-called "pedestrians" in popular science magazines and books.

The task looked also not absolutely formidable because a consistent presentation of relativity existed for a long time in the world-wide accepted textbook by Landau and Lifshitz [13], which was the basis of my own understanding, and in some other textbooks.

The existence of countless texts, in which the essence of relativity was mutilated (or semi-mutilated) had two sides. On one hand, it looked discouraging, especially because among the authors of these texts there were many famous physicists, the fathers and greatest authorities of modern physics. On the other hand, it was a challenge. So I tried to explain clearly to the readers the beauty of four-dimensional space-time approach and the ugliness and inconsistency of "relativistic mass", an illegitimate child of relativistic and non-relativistic equations.

My optimism had increased when in 1992 Taylor and Wheeler in the second edition of the influential and popular "Spacetime Physics" [14] included a "Dialog: Use and Abuse of the Concept of Mass", in which they supported my articles [11, 12]. A copy of this book is in my bookcase with a postcard sent to me in October 1991 by John Archibald Wheeler. The postcard has a photo of the famous Albert Einstein

Memorial in front of the building of the National Academy of Sciences, Washington, DC. The bronze sculpture of Einstein includes a copybook with $E = mc^2$ on an open page.

Since that time I received hundreds of letters from physicists (both professors and students) stating their adherence to the four-dimensional formulation of relativity and to the Lorentz invariant concept of mass. In a few cases I helped the authors to correct erroneous explanations of the concept of mass in preparing new editions of their textbooks. However the number of proponents of relativistic mass seemed not to decrease.

A leading role in promoting the relativistic mass have played the books by Max Jammer [15, 16]. Especially aggressive the proponents of relativistic mass became in connection with the World Year of Physics, which marks the 100th anniversary of fundamental articles published by Einstein in 1905.

The campaign started by the September 2004 issue of "Scientific American", full with "relativistic mass" equal to $m_0/\sqrt{1 - v^2/c^2}$, where m_0 is rest mass, and "the most famous equation $E = mc^2$". A letter to the editors, defending the four-dimentional approach and invariant mass had been rejected by the editor G. Collins who in April 2005 wrote: "Most important, we believe that tackling the issue head-on in the manner you and your coauthors want in the letters column of Sci. Am. would be very confusing to our general audience and it would make the subject seen all the more mysterious and impenetrable to them". Thus to avoid "head-on" collision of correct and false arguments the editors of Sci. Am. preferred to hide from the readers the correct viewpoint.

P. Rodgers – the Editor of European "Physics World" wrote in January 2005 in editorial [17]: "... $E = mc^2$ led to the remarkable conclusion that mass and energy are one and the same". Unlike G. Collins, P. Rodgers published a letter criticizing this statement and partly agreed with the criticism [18].

In September 2005 the bandwagon of relativistic mass was joined by "The New York Times", which published an article by B. Green [19].

The journalists were supported by renowned scientists, such as R. Penrose, who in a new thousand pages thick book had written [20]:

"In a clear sense mass and energy become completely equivalent to one another according to Einstein's most famous equation $E = mc^2$."

How many students, teachers and journalists will be infected by this sentence? How many readers had been infected by the famous book by S. Hawking [21], the second edition of which appeared in 2005? On the very first page of it Hawking wrote:

"Someone told me that each equation I included in the book would halve the sales. I therefore resolved not to have any equations at all. In the end, however, I did put in one equation, Einstein's famous equation $E = mc^2$. I hope that this will not scare off half of my potential readers."

I am sure that the usage of $E = mc^2$ had doubled the sales of his book, the buyers being attracted by the famous brand. But is it possible to estimate the damage done to their understanding of relativity theory and to the general level of the literature on relativity incurred by this case of spreading the virus?

Two recent preprints by Gary Oas [22, 23] written in the framework of Educational Program for Gifted Youth at Stanford University were devoted to the use of relativistic mass. The author "urged, once again, that the use of the concept at all levels to be abandoned" [22]. The manuscript has been submitted for publication to the "Americal Journal of Physics", but was rejected as being "too lengthy" (it contains 12 pages!). A lengthy bibliography (on 30 pages) of books referring to special and/or general relativity is provided in ref. [23] to give a background for discussions of the historical use of the concept of relativistic mass. It is easy to forecast the aggressive reaction of the virus infected community to this attempt to cure it.

Acknowledgments

This work is supported by the grant of Russian ministry of education and science No. 2328.2003.2.

References

[1] I. Newton, *Philosophiae Naturalis Principia Mathematica* [*Mathematical Principles of Natural Philosophy*], translated in English by A. Motte, revised and annotated by F. Cajori (University of California Press, 1966).
[2] G. Galilei, *Dialogue Concerning the Chief World Systems – Ptolematic and Copernican*, Engl. Transl. S. Drake, foreword A. Einstein, 1967 (Berkeley, CA: University of California Press) 2nd edn.
[3] Nikolas of Cusa, *On learned ignorance*: a translation and an appraisal of *De docta ignorantia* by Jasper Hopkins. 2nd ed., rev. Minneapolis: A.J. Bannning Press, 1990, [c1985].
[4] H. Lorentz, *Electromagnetic phenomena in a system moving with any velocity smaller than that of light.* Proc. Acad. Sci., Amsterdam **6**, 809 (1904).
[5] H. Poincare, *Sur la Dynamique de l'electron*, Comptes Rendues **140**, 1504 (1905).
[6] H. Poincare, *Sur la Dynamique de l'electron*, Rendiconti del Circolo Matematico di Palermo **XXI**, 129 (1906).
[7] A. Einstein, *Zur Elektrodynamik bewegter Körper*, Ann. d. Phys. **17**, 891 (1905).
[8] A. Einstein, *Ist die Trägheit eines Körpers von seinem Energiegehalt abhängig?* Ann. d. Phys., **18**, 639 (1905).
[9] M. Planck, *Das Prinzip der Relativität und die Grundgleichungen der Mechanik*, Verh. d. Deutsch. Phys. Ges. **4**, 136 (1906).
[10] H. Minkowski, *Raum und Zeit*, Phys. Z. **10**, 104 (1909).
[11] L.B. Okun, Soviet Physics-Uspekhi **32** (July 1989) 629 - 638.
[12] L.B. Okun, *Physics Today*, June 1989, pages 31-36; http://www.physicstoday.org/vol-42/iss-6/vol42no6p31_36.pdf;
L.B. Okun, *Physics Today*, May 1990, page 147.
[13] L.D. Landau, E.M. Lifshitz, *The Classical Theory of Fields*, translated from the second Russian ed., Pergamon, New York (1951).
[14] E.F. Taylor, J.A. Wheeler, *Spacetime Physics*, 2nd ed. Freeman and Company. New York, 1992, pages 246 - 252.
[15] M. Jammer, *Concepts of Mass in Classical and Modern Physics*, Harvard, 1961.
[16] M. Jammer, *Concepts of Mass in Contemporary Physics and Philosophy*, Princeton, 2000.
[17] P. Rodgers, *Physics World* **18** No.1, 13 (2005).
[18] P. Rodgers, *Physics World* **18** No.10, 20 (2005).
[19] B. Green, *That Famous Equation and You*, The New York Times, Sept. 30, 2005.
[20] Roger Penrose, *The road to reality. A complete guide to the laws of the universe*, A. Knopf, New York, 2004, p. 434.
[21] S. Hawking, *A Brief History of Time*, New York, 1988.
[22] G. Oas, *On the abuse and use of relativistic mass*, arXiv: physics/0504110.
[23] G. Oas, *On the Use of Relativistic Mass in Various Published Works*, arXiv: physics/0504111.

ACCELERATOR NEUTRINO EXPERIMENTS

Takashi Kobayashi [a]
Institute for Particle and Nuclear Studies, High Energy Accelerator Research Organization,
1-1 Oho, Tsukuba, Japan

Abstract. Recent progress in accelerator neutrino experiments is reviewed.

1 Introduction

Accelerator neutrino experiments have been playing important role in understanding nature of neutrinos. Recently, after Super-Kamiokande (SK) reported the evidence of neutrino oscillation [1], potential of the accelerator neutrino experiments has been attracting much attention.

In accelerator experiments, neutrino beam is produced from pion decay in flight which are produced by hitting proton beam on a target. In order to increase neutrino flux in forward direction, the parent pions are usually focused by single turn toroidal magnet called electromagnetic horns. The produced neutrino beam is almost purely ν_μ with the order of 1% contamination of ν_e. Anti-neutrino beam can be produced by flipping the polarity of the horns.

Prominent advantage of accelerator neutrino experiments over the observation of neutrino from nature is that condition and systematics can be under control. For example, the direction and flight distance, energy spectrum, initial flavor content can be known.

One major field of accelerator neutrino experiments is oscillation experiments. These experiment measures flavor content at certain distance from the production. The flavor change indicates finite masses and mixing between generations. Non oscillation experiments include studies of νN interaction in GeV region for future precision oscillation experiments and deep inelastic scattering experiments which probe nucleon structure and/or test electroweak standard model. In these proceedings, recent and near future accelerator neutrino experiments are reviewed.

2 Oscillation Experiments

2.1 Long baseline experiments

First evidence of neutrino oscillation was reported by SK in 1998 in the observation of atmospheric neutrino [1]. Several long baseline (LBL) oscillation experiments to confirm the evidence have been planned.

[a] e-mail: takashi.kobayashi@kek.jp

K2K

K2K (KEK-to-Kamioka) is the first LBL oscillation experiment [2] whose main goals are (1) to confirm the oscillation observed in atmospheric neutrinos by measuring ν_μ disappearance and (2) to search for ν_e appearance. The ν_μ beam is produced using 12-GeV proton synchrotron at KEK and detected by SK at 250 km from KEK. The mean energy of ν_μ beam is 1.3 GeV. Near neutrino detectors (ND) are placed at 300 m from the production target. The experiment was started in 1999 and terminated in 2004. Total accumulated number protons on target (POT) is 1.05×10^{20} during the whole experimental period and data corresponding to 0.922×10^{20} POT has been analyzed.

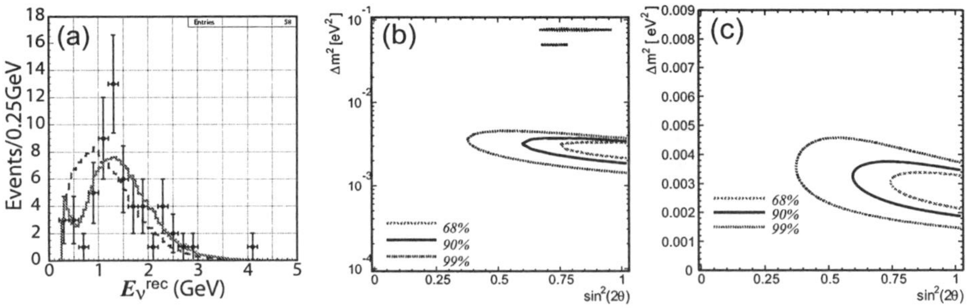

Figure 1: (a) $E_\nu^{\rm rec}$ distribution for single ring μ-like sample. Points with error bars are data. Dashed (blue) histogram is expected spectrum without oscillations and the solid (red) histogram is the best fit oscillated spectrum. These histograms are normalized to data by the number of events. (b) Allowed regions of oscillation parameters in log scale in Δm^2. Dashed, solid and dotted lines are 68%, 90% and 99% C.L. contours, respectively. (c) Same contour as (b) with linear scale.

Observed number of fully contained events in whole experimental period is 112 and the distribution of reconstructed neutrino energy ($E_\nu^{\rm rec}$) for 58 single ring μ-like events are plotted by points in Fig. 1(a). The expected number of events and energy spectrum at SK are predicted by extrapolating observed number of events and spectrum at ND using far/near flux ratio (F/N ratio). The F/N ratio is estimated using beam MC simulation and is validated by in-situ measurement of pion distribution just downstream of the horn system [3] and the HARP experiment [4]. Expected number of events at SK without oscillation is estimated to be $155.9^{+11.5}_{-10.2}$ events and expected spectrum is drawn in Fig. 1(a).

Combining the deficit of events and the spectrum distortion, probability of no oscillation is found to be less than 0.003% which corresponds to 4.2σ significance. Allowed regions for the oscillation parameters are drawn in Fig. 1 (b) and (c). At maximum mixing, $\sin^2 2\theta = 1$, the mass squared difference is constrained to be within $\Delta m^2 = 1.9 \sim 3.5 \times 10^{-3}$ eV^2 at 90% confidence level. The best fit point in physical region is at $(\sin^2 2\theta, \Delta m^2) = (1.0, 2.76 \times 10^{-3}\ \text{eV}^2)$.

Electron neutrino appearance is also searched for in K2K. Electron neutrino events are selected by requiring fully contained, single electron-like (showering) ring with visible energy greater than 100 MeV without any decay electron signal in 30 μs time window. In order to further reduce the background which is dominated by NC π^0 production, an additional selection is introduced. The π^0 background events remained since the standard ring finding algorithm failed to find one of 2 rings of two decay gammas. In the new algorithm, the missing ring is assumed to exist and forced to be reconstructed. Then a 2-dimensional cut is applied in the plane of invariant mass of the 2 rings and the energy of the primary ring (Fig. 2(a) and (b)).

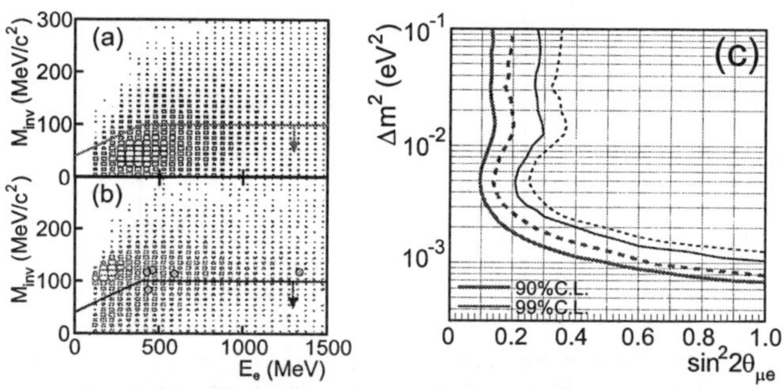

Figure 2: Expected distribution of invariant mass and electron energy for (a)ν_e signal and (b)ν_μ background events (box histogram). The region below the line is selected as signal. Observed events after all cuts except this π^0 cut is plotted in (b). (c) Excluded region of oscillation parameters by ν_e appearance search. Upper right area of the lines are excluded. The solid and dashed lines are obtained limit and sensitivity, respectively.

One event is selected as an electron candidate, while expected background is estimated to be 1.7 events. Since no significant signal events over the background is observed, upper bound on the mixing angle is set and plotted in Fig. 2(c). The region $\sin^2 2\theta_{\mu e} (\simeq \frac{1}{2}\sin^2 2\theta_{13}) > 0.13$ at $\Delta m^2_{\mu e} = 2.8 \times 10^{-3}\text{eV}^2$ is excluded at 90% C.L.

MINOS

MINOS is a LBL neutrino experiment in US where the ν_μ beam is produced by using 120 GeV Main Injector at FNAL and detected by a detector placed in Soudan mine at 735 km from FNAL [5]. The main purposes of the experiment is the precise measurement of ν_μ disappearance at Δm^2 region suggested by atmospheric neutrino observation. The experiment started taking data in March, 2005.

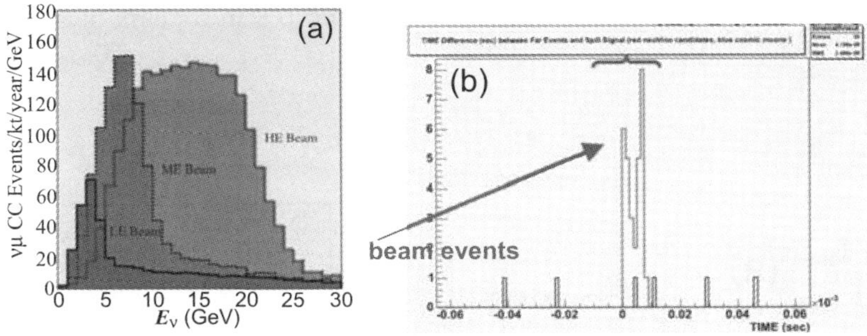

Figure 3: (a) Expected energy spectra of ν_μ charged current interactions. (b) Time difference between neutrino candidates and beam timing signal.

The beam is conventional horn-focused wide-band beam. By choosing different setting of the distance between target, 1st horn, and 2nd horn, neutrino energy regions can be selected as shown in Fig. 3(a). The far detector is a magnetized Iron and scintillator sandwich detector with Iron thickness of 2.5 cm. To measure properties of neutrinos just after production, near detector with the same configuration as the far detector is placed. The fiducial masses of the near and far detectors are 980 ton and 5,400 ton, respectively. Expected number of charged current interactions is about 2,500/yr in the fiducial volume of the far detector with the LE beam in Fig. 3(a).

After the commissioning in early 2005, MINOS has accumulated 1.9×10^{19} POT as of July 2005 and achieved up to 2.5×10^{13} protons/pulse and 125 kW beam power. Neutrino events from FNAL have already been observed at MINOS far detector with the HE beam option (Fig. 3(b)).

In 5 years of MINOS, precision of the oscillation parameters reaches about 5% and 2×10^{-3} eV^2 for $\sin^2 2\theta$ and Δm^2, respectively.

CNGS projects

The CERN neutrino beam to Gran Sasso (CNGS) is a project to produce high energy ν_μ beam using 400-GeV CERN-SPS and send toward Gran Sasso laboratory at 732 km from CERN [6]. The beam is the horn-focused wide-band ν_μ beam with mean energy of ~17 GeV. Expected number of ν_μ CC interactions is ~2900/kt/yr at Gran Sasso for 4.5×10^{19}POT/yr. Main goal of the project is the detection of ν_τ appearance.

OPERA [7] is a far detector at Gran Sasso. It consists of 206k bricks of Emulsion Cloud Chamber (ECC), which is a sandwich of 1-mm thick Pb plates and emulsion sheets. Total mass of the ECC is 1.7 kton. The ECC part is followed by a magnetic spectrometer with electronic tracking system. The τ events from CC ν_τ interactions are identified by the event topology, a kink in a track caused by $\tau\to\mu$ decay. Expected number of signal and background events for the ν_τ appearance search are 12.8 and 0.8 in 5 years of running at the $\Delta m^2 = 2.4 \times 10^{-3}$ eV^2.

T2K

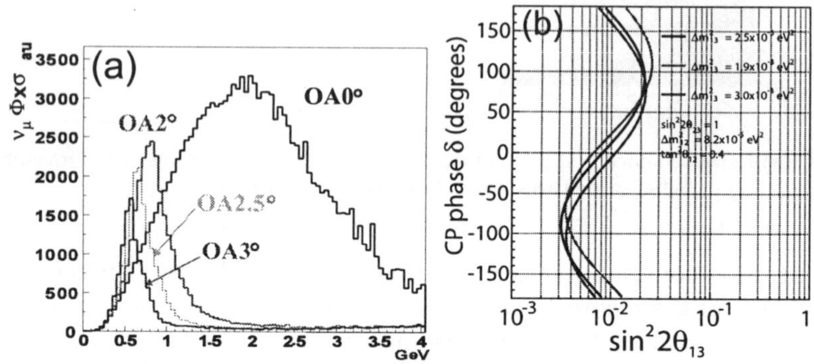

Figure 4: (a) Expected spectrum at SK. (b)Expected sensitivity for ν_e appearance search at 90% C.L.

The T2K (Tokai-to-Kamioka) [8] is a next generation LBL experiment in which ν_μ beam is produced using the 50 GeV proton synchrotron in J-PARC [9] and sent to SK at 295 km from J-PARC. The main goals are discovery of ν_e appearance and the precision measurements of ν_μ disappearance. At the beginning, the accelerator will be operated at 40 GeV and the goal of the beam power is 750 kW.

Off-axis beam [10] is adopted for the first time to produce high intensity low energy narrow band beam and maximize the sensitivity. The expected ν_μ spectrum at SK without oscillation is plotted in Fig. 4 (a). Expected number of CC interactions at SK with 2.5° off-axis is 1600 in fiducial volume of 22.5 kt in 1 year (10^{21}POT) at 40 GeV operation.

The expected sensitivities are plotted in Fig. 4 (b). The ν_e appearance can be searched down to $\sin^2 2\theta_{13} \sim 0.008$ at CP violating phase $\delta = 0$. The goal of precision in ν_μ disappearance measurement are $\delta(\sin^2 2\theta_{23}) = 0.01$ and $\delta(\Delta m^2_{23}) = 1 \times 10^{-4}$ eV^2. The construction of the neutrino beam line has started in April 2004 and will take 5 years. The first proton beam on target is expected in the beginning of 2009.

2.2 Short baseline experiment: MiniBooNE

MiniBooNE is the experiment to confirm (or refute) the evidence of $\bar{\nu}_\mu \rightarrow \bar{\nu}_e$ oscillation at typical oscillation parameters of $\Delta m^2 \sim 1$eV^2 and $\sin^2 2\theta \sim 10^{-2\sim-3}$ reported by LSND experiment [11]. Wide-band ν_μ beam is produced by the proton beam from FNAL 8-GeV Booster (Fig. 5 (a)). The length of the decay region is 50 m and the neutrino detector is located at 541m from the production target. The detector is 800-ton pure mineral oil filled in sphere vessel and surrounded by 1280 PMTs deployed on the inner wall to detect Čerenkov and scintillation light.

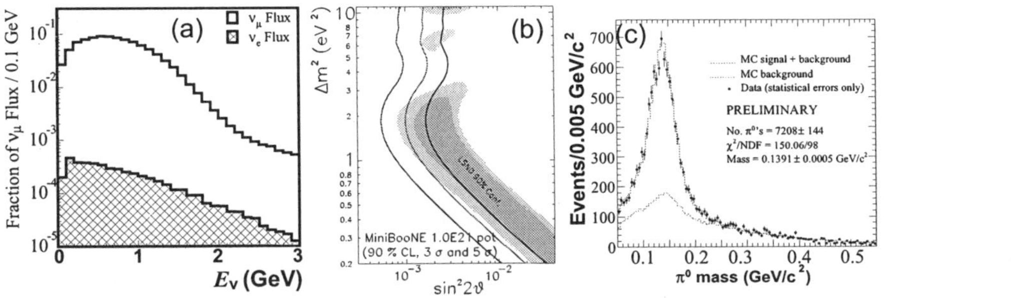

Figure 5: (a) Expected MiniBooNE neutrino flux vs. E_ν. (b) MiniBooNE $\nu_\mu \to \nu_e$ sensitivity at 90% C.L., 3σ, and 5σ coverage of the LSND region. (c) Reconstructed π^0 data for energy calibration.

The experiment started taking data in Aug. 2002. So far 5.78×10^{20} POT is accumulated and more than 600k neutrino interactions have been recorded as of Jun.20,2005. Expected sensitivities are drawn in Fig. 5 (b) and number of signal and background events at 10^{21}POT are expected to be 300 and 780 events, respectively.

3 Non-oscillation Experiments

3.1 Measurements of νN interactions at GeV region

For the next generation high precision oscillation experiments, detailed knowledge on the neutrino cross sections is essential to achieve their goals. Especially cross section data at GeV region is very poor since it mainly comes from low statistics bubble chamber experiments although interaction at this region is complicated as quasi-elastic, resonance pion production and deep inelastic multi-pion production are mixed. Recently, accelerator-based oscillation experiments start providing neutrino cross section data based on their high statistics data.

Precise knowledge of ν_μ NC π^0 production is important since it is one of the dominant sources of background in ν_e appearance search. In the K2K experiment, its cross section relative to total CC cross section has been measured using 1kt water Čerenkov detector in the near detector [12]. The final state particles in this process are only 2 gammas from π^0 decay. The selection criteria are fully contained, 2 electron-like rings with their invariant mass $85 < M_{\gamma\gamma} < 215$MeV. The result is $\sigma(\nu_\mu \mathrm{NC} 1\pi^0)/\sigma(\nu_\mu \mathrm{CCall}) =$

$0.064 \pm 0.001(stat) \pm 0.007(syst)$ where model prediction is 0.0065. Similar analysis is being done also in MiniBooNE.

K2K also tried to detect CC coherent pion production ($\nu_\mu + A \rightarrow \mu^- + \pi^+ + A$) using SciBar detector in the near detector [13]. Events satisfying the following criteria are selected; (only) 2 tracks, one is identified as muon and the other is pion like, not satisfy kinematics of quasi-elastic scattering, no large energy deposit around the vertex to reject short-range stopping proton, reconstructed q^2, $q^2_{\rm rec} < 0.1$ (GeV/c)^2. Selected number of candidate events is 113 while expected background coming from interactions other than CC coherent pion production is 111.4. No CC coherent pion production event is observed and upper limit of the cross section is set as $\sigma(\nu_\mu \mathrm{CCcoh}\pi)/\sigma(\nu_\mu \mathrm{CCall}) < 0.60 \times 10^{-2}$. The result is inconsistent with the prediction of the model by Rein and Sehgal of 2.67×10^{-2}.

An experiment dedicated to the measurements of neutrino interactions is proposed at FNAL. The experiment, MINERνA, will construct fully active scintillator tracker in front of the MINOS near detector [14]. Mass of the detector is 5.9 ton and ~15 million CC events are expected at 1.6×10^{21}POT in 4 years. The experiment got stage 1 approval in Apr.2004 and plan to start construction within 2006 and start taking data in 2008.

3.2 Deep inelastic scattering experiments

Deep inelastic scattering experiments have been providing precise data on nucleon structure and precision test of the electroweak theory. In 2001, the NuTeV experiment reported measurement result $\sin^2\theta_W = 0.2277 \pm 0.0013(stat.) \pm 0.0009(syst)$, which is 3σ away from the prediction of the electroweak theory, 0.2227 ± 0.0003 (Fig. 6) [15]. This "anomaly" is still open question. In order to confirm the result, data of the NOMAD experiment is being analyzed [16]. NOMAD was a high energy (~100 GeV) neutrino experiment which searched for $\nu_\mu \rightarrow \nu_\tau$ oscillation using CERN SPS from 1995 to 1998. The goal of the precision is $\delta(\sin^2\theta_W) \sim 1\%$ which is comparable to the NuTeV experiment.

4 Summary

Accelerator-neutrino experiments have been playing important role in understanding properties of neutrino, νN interaction and electroweak theory. Oscillation experiments study mass and mixing of neutrinos. The first long baseline experiment, K2K, has definitely confirmed the oscillation observed in atmospheric neutrino. MINOS experiment started in 2005 and OPERA will start in May 2006. The short baseline experiment, MiniBooNE, is going to check the LSND result.

High precision data of neutrino interaction cross sections are coming out from high statistics accelerator neutrino experiments such as K2K, MiniBooNE, NO-

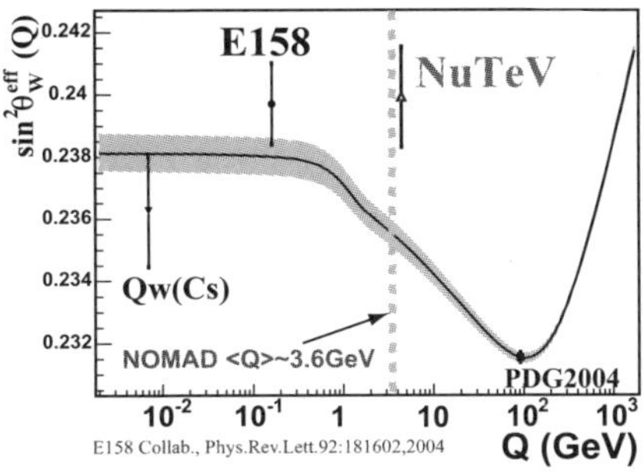

Figure 6: Theoretical prediction of $\sin^2\theta_W$ as a function of momentum transfer Q (solid line) with its error (shaded area). Measurement results are also plotted.

MAD and future experiment MINERνA. The "anomaly" in $\sin^2\theta_W$ reported by NuTeV experiment is still mystery and being investigated by NOMAD group.

The accelerator neutrino experiments have a potential to give breakthrough toward physics beyond the standard model for coming decades and many future projects are being constructed or planned.

References

[1] Y.Fukuda *et al* , Phys. Rev. Lett. **81**, 1562 (1998), Phys. Rev. Lett. **85**, 3999 (2000)

[2] E. Aliu *et al.* [K2K Collaboration], Phys. Rev. Lett. **94**, 081802 (2005).

[3] T.Maruyama, Ph.D. Thesis, Tohoku University (2000).

[4] HARP Collaboration (M.G. Catanesi *et.al.*), Nucl. Phys. B732, 1 (2006).

[5] http://www-numi.fnal.gov, A. Marchionni [MINOS Collaboration], FERMILAB-CONF-05-429-AD-E.

[6] G. Acquistapace *et al.*, CERN-98-02.

[7] A. Olshevski, in these proceedings.

[8] http://neutrino.kek.jp/jhfnu. Y. Itow, *et al* , hep-ex/0106019, Y. Hayato [T2K Collaboration], Nucl. Phys. Proc. Suppl. **143**, 269 (2005).

[9] http://j-parc.jp

[10] D. Beavis *et al* , Proposal of BNL AGS E-889 (1995).

[11] http://www-boone.fnal.gov/, S. J. Brice [MiniBooNE Collaboration], Nucl. Phys. Proc. Suppl. **143**, 115 (2005).

[12] S. Nakayama *et al.*, Phys. Lett. B **619**, 255 (2005).

[13] M. Hasegawa *et al.*, Phys. Rev. Lett. **95**, 252301 (2005).

[14] http://minerva.fnal.gov/.

[15] G. P. Zeller *et al.* [NuTeV Collaboration], Phys. Rev. Lett. **88**, 091802 (2002) [Erratum-ibid. **90**, 239902 (2003)].

[16] R.Petti, in these proceedings, R. Petti [NOMAD Collaboration], arXiv:hep-ex/0411032.

SEARCHING FOR NEUTRINO OSCILLATIONS WITH OPERA

N. Savvinov [a] (on behalf of the OPERA collaboration)
Laboratory for High Energy Physics, University of Bern, CH-3012, Switzerland

Abstract. The OPERA experiment will search for neutrino oscillations using a muon neutrino beam and a hybrid emulsion-scintillator detector. Basic principles, current status and expected performance of the experiment are discussed.

1 Motivation

Results from Super-Kamiokande [1] and K2K [2] together with those of CHOOZ [3] gained strong evidence in favor of the $\nu_\mu \to \nu_\tau$ scenario for the atmospheric neutrino anomaly. OPERA intends to confirm this result by observing the appearance of ν_τ in a ν_μ beam. As a byproduct of the experiment, a significant improvement of the current CHOOZ limit [3] on θ_{13} is expected.

2 Experimental strategy

OPERA will use the CNGS (CERN Neutrinos to Gran Sasso) ν_μ beam. The detector will be placed in the Gran Sasso underground laboratory in Italy, 735 km away from the source.

The detection of ν_τ will be based on direct observation of τ decay topologies. Since τ decay length under experimental conditions is very short (of order of 1 mm), a detector with a very high spatial resolution is required. This will be accomplished by using nuclear emulsions.

The OPERA emulsions films consist of two 40 μm emulsion layers separated by a 200 μm plastic base. The emulsions are sandwiched with 1 mm lead sheets, according to the so-called emulsion cloud chamber (ECC) technique. The basic unit of the OPERA detector, a brick, contains 57 emulsion films and 56 lead plates. The brick dimensions are 10.2 × 12.7 × 7.5 cm and the weight is 8.6 kg. The overall weight of the OPERA detector will be 1.7 kton.

In order to identify the location of the bricks containing the neutrino interaction point, brick layers are interspaced with electronic detectors. The Target Tracker (TT) planes consist of horizontal and vertical plastic scintillator strips read out by 64-channel PMTs. The DAQ of TT defines the trigger for the brick extraction for analysis.

The OPERA detector will be organized into two large sections ("supermodules") of 31 brick and TT planes. Each supermodule will be followed by a muon spectrometer for μ identification, momentum and charge measurement and charm background reduction. Each muon spectrometer is composed of a 1.55 T dipolar magnet, 22 RPC layers and 6 sections of drift tubes.

[a] e-mail: nikolay.savvinov@lhep.unibe.ch

Once the interaction candidate brick(s) is (are) identified, its (their) extraction will be handled by two designated robots called BMS ("brick manipulation system"). After extraction, the brick is exposed to cosmic rays for film alignment, and then disassembled. The emulsions are developed on-site and then sent to participating laboratories for scanning. After scanning, particle tracks are analyzed and the event is reconstructed using momentum information from multiple Coulomb scattering and readouts of the TT and the muon spectrometers.

3 Installation of the experiment

3.1 Neutrino Beam

The CNGS project has successfully completed the civil engineering stage as well as the installation of hadron shop, decay tube and general services. Currently, installation of the proton beamline, target, horn, deflector and shielding is underway. The comissioning is expected in Spring 2006.

3.2 Spectrometers

The magnets for both muon spectrometers are now in place. Installation of detectors for the spectrometers is in progress. The 4 downstream RPC planes of the first supermodule are already taking cosmics data.

3.3 Target modules

All required TT modules have been produced and are now delivered to Gran Sasso on a ready-to-install basis. As of September 2005, about 2/3 of TT and brick walls for the first supermodule are installed. It is expected that by November 2005 all TT installations for the first supermodule will be completed. The brick assembly machine (BAM) is expected to be fully commissioned by December 2005 and should start producing lead-emulsion bricks in January 2006 . Brick filling of the first supermodule should start in January 2006 and continue until the expected CNGS beam arrival in July 2006. After that, while taking data with one supermodule, the collaboration will continue to fill the second supermodule with bricks.

4 Emulsion scanning

The OPERA scanning system uses many ideas and approaches that originated during development of the automatic scanning system for CHORUS. The total surface of OPERA emulsions is, however, more than 170,000 m^2, as compared to 500 m^2 for CHORUS [4]. Such a high scanning load requires innovations in

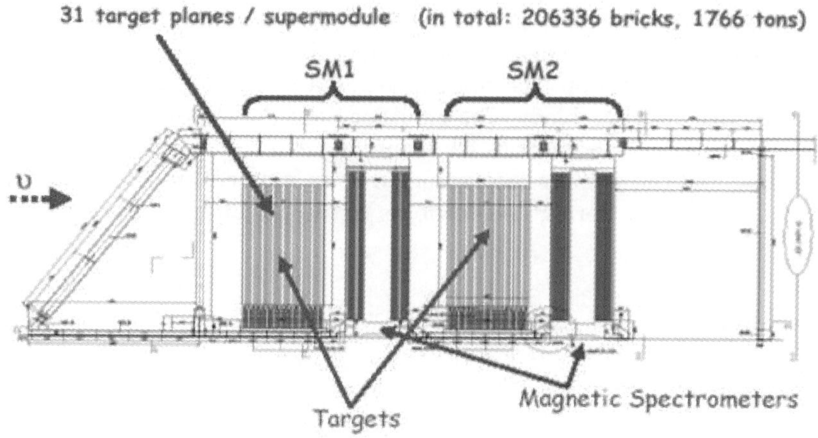

Figure 1: Schematic view of the OPERA detector.

Figure 2: Construction of the OPERA detector (June 2005). The neutrino beam will come from the left.

Figure 3: Few sample reconstructed events from a 10 GeV pion test beam.The beam is entering the picture from the right.

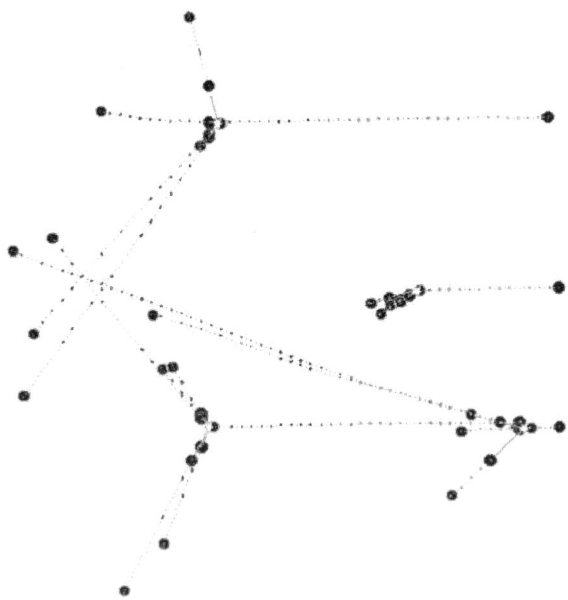

both scanning hardware and analysis algorithms, which have been successfully implemented in OPERA computer-controlled scanning stations.

Thanks to these innovations, the scanning speed has been increased more than by a factor of 20 with respect to the previous systems and reached over 20 cm^2 per hour. Online 3D reconstruction of particle tracks is performed with a sub-micron precision and efficiency better than 95%.

Scanning systems installed in Europe and Japan are now operational and are analyzing data from various test beams. A sample reconstructed event from a CERN 10 GeV pion test beam is shown in Figure 3.

The overall scanning capacity of the collaboration is estimated around 30 bricks per day, which fulfills the requirements of the experiment.

5 Expected event and background rates

The overall efficiency of τ-decay observation in OPERA was calculated to be 9.1%. This figure is based on both short and long decays of $\tau \to e$, $\tau \to \mu$ and $\tau \to h$ channels. A possibility of inclusion of the $\tau \to 3h$ channel with potential improvement in efficiency by about 1% is being investigated. Finally, a recent

Table 1: Expected number of observed signal and background events in 5 years of running for nominal calculations and envisaged improvements of efficiency and background reduction.

	Signal			BG
Δm^2 (10^{-3}eV^2)	1.9	2.4	3.0	
Nominal	6.6	10.5	16.4	0.7
Improved eff.	8.0	12.8	19.9	1.0
+ BG reduction	8.0	12.8	19.9	0.8

study shows a possibility of a 10 % improvement in brick finding efficiency.

The event rate of OPERA strongly depends on the oscillation parameters. Calculations for several values of Δm^2_{atm} (corresponding to the central value of Super-Kamiokande best fit and the limits of the 90% confidence interval) are presented in Table 1.

The main sources of background are charm decays, hadron re-interactions and large angle μ scattering. Algorithms employing dE/dx information for improved π/μ separation are currently being developed. It is expected that they would reduce charm background by about 40% (down to about 0.28 event). The result of recalculation of the large angle μ scattering background including nucleon form factors is 5 times lower than the upper limit from the CHORUS measurement used earlier.

6 Conclusions

The CNGS and the OPERA detector construction are progressing and the experiment should be ready to take data in August 2006. Based on Super-Kamiokande best fit for Δm^2_{atm}, OPERA should see in 5 years about 12.8 τ events. The background is expected on the level of 1 event. An ongoing effort to improve the efficiency and to further suppress the background may improve these figures.

References

[1] Super-Kamiokande Collaboration, *Phys. Rev.* **D** 71, 112005 (2005).
[2] C. Mariani, *hep-ex/0505019.*
[3] CHOOZ Collaboration, M. Appollonio et al., *Phys. Lett.* **B** 466, 415 (1999).
[4] E. Eskut, et al., *Nucl. Instr. and Meth.* **A** 401, 7 (1997).

REACTOR NEUTRINOS AND KAMLAND

J.Shirai [a]
(for the KamLAND Collaboration)
Research Center for Neutrino Science, Tohoku University, Sendai, 980-8578, Japan

Abstract. Great progress has been made in neutrino physics in recent years especially by series of experiments which have established neutrino oscillation and opened a new window to the physics beyond the standard model. Reactor neutrino experiments have made important contribution to the studies. In this talk experiments using reactor neutrinos are reviewed and the results of KamLAND experiment are presented, followed by planned reactor neutrino oscillation experiments and a very recent results made by KamLAND on the first detection of geologically produced antineutrinos.

1 Introduction

Neutrino mass is one of the most fundamental quantities to be studied in particle physics. In fact the finite mass of the neutrino directly means the physics beyond the standard model and studies on neutrinos have long been devoted to searching for the mass of the neutrinos. The extreme smallness of the neutrino masses compared to other quarks and leptons is believed to be the result of fundamental mechanisms in very high energy scales as speculated in many scenarios in grand unification theories. Also, the neutrinos are considered to have played important roles in the early Universe and its evolution with the neutrino massses as the critical parameters.

Experimental challenges to measure the neutrino mass have been made by searching for neutrino oscillation, neutrino-less double β decays and direct mass measurements, etc. Among them neutrino oscillation searches have played a central role with high sensitivities. This is because the neutrino oscillation is the results of the interference of the different mass eigenstates which, in a simple 2-neutrino flavor scheme (for example ν_e and ν_μ), is expressed for the survival probability of the original flavor, say ν_e as

$$P(\nu_e \to \nu_e) = 1- \sin^2 2\theta \sin^2(1.27\Delta m^2_{[eV^2]} L_{[m]}/E_{[MeV]}), \quad (1)$$

where E is the neutrino energy, L the distance from the source, and θ and Δm^2 are the mixing angle and the difference of the squared masses, respectively. In experiments using low energy neutrinos it is easy to get a large factor of L/E to be sensitive to small Δm^2. The neutrino oscillation would appear as a deficit of the flux and a characteristic change of the energy spectrum. Therefore, in this kind of experiment, so called a disapperance experiment, it is essential to have an intense neutrino flux of the known flavor, whose absolute intensity and the energy spectrum are well understood.

[a] e-mail: shirai@awa.tohoku.ac.jp

2 Reactor ν experiments

Nuclear reactor is an ideal tool for the disappearance-type neutrino oscillation experiment because it serves as a pure and a high intensity $\bar{\nu}_e$ source with low energies, less than ~8MeV. In a nuclear reactor fission products of the fuel elements are neutron rich and proceed via β^- decays to generate $\bar{\nu}_e$'s. So, reactor neutrinos are almost purely $\bar{\nu}_e$'s. Modern reactor experiments use power reactors where ^{235}U, ^{239}Pu, ^{238}U and ^{241}Pu dominantly generate the power with $\sim$ 200MeV/fission, and the numbers of generated $\bar{\nu}_e$'s is ~6/fission, which amounts to $6{\times}10^{20}\bar{\nu}_e$'s/s in a typical reactor with a thermal power of $3\mathrm{GW}_{th}$.

Reactor neutrino experiments have a long history since the first detection of neutrinos by F.Reines et al. in 1956 [1], by using the inverse β decay reaction,

$$\bar{\nu}_e + p \rightarrow e^+ + n, \quad (2)$$

where the reaction occurred in H_2O target containing $CdCl_2$ and the liquid scintillator (LS) beside the target was used to detect the prompt 2γ's from e^+e^- annihilation and a delayed γ ray from the neutron capture by Cd nucleus in $\sim 10\mu$s.

The detection technique of the delayed concidence of the reaction (2) is basically the same even in today's experiments. The process (2) occurs only by $\bar{\nu}_e$ and the backgrounds are significantly removed by taking the time and space correlation. Moreover, the prompt e^+ energy (ionization loss of e^+ and 2 γ's from the e^+e^- annihilation) is connected to the incident $\bar{\nu}_e$ energy by

$$E_{prompt}{=}E_{\bar{\nu}_e} - 0.8\mathrm{MeV}. \quad (3)$$

Also, it has a large cross section which is about hundred times that of the elastic scattering, $\bar{\nu}_e + e^- \rightarrow \bar{\nu}_e + e^-$, and the value is directly connected by the standard V-A theory to the precisely measured neutron lifetime τ_n by

$$\sigma = [(2\pi^2/m_e^5)/(f_{ps}\tau_n)]E_e p_e, \quad (4)$$

where f_{ps} is the phase space factor. Note that the threshold energy is 1.8MeV and expected E_{prompt} spectrum without $\bar{\nu}_e$-oscillation rises from 1.02MeV and makes a peak at ~3MeV due to a combination of the decreasing flux and increasing cross section with $\bar{\nu}_e$ energy.

Production rate of $\bar{\nu}_e$ is estimated from the initial isotopic composition of the fuel and the thermal output to obtain the change of the elements (burn-up) with an uncertaity less than 1%. The $\bar{\nu}_e$ sepectra for ^{235}U, ^{239}Pu and ^{241}Pu are evaluated based on the measured β^- spectra [2] for the fission products produced by thermal neutrons. The $\bar{\nu}_e$ spectrum for ^{238}U whose fission is made by only fast neturons is obtained by theoretical caluculation [3]. The uncertainty of the $\bar{\nu}_e$ spectra is ~2.5% in total.

Lots of reactor neutrino experiments were carried out in 1980's and 1990's

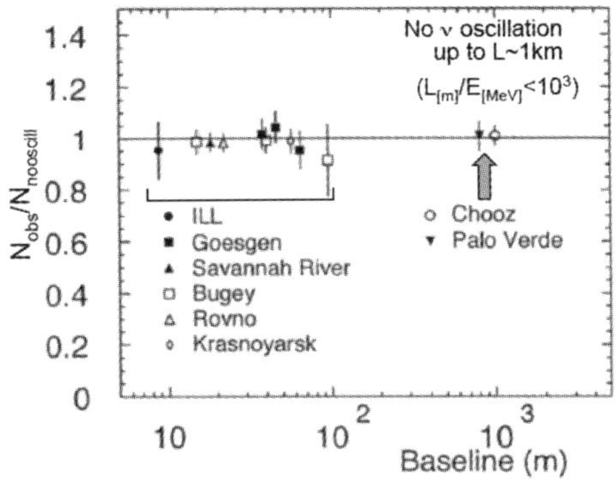

Figure 1: Ratios of observed vs. no-oschillation expectation of the reactor $\bar{\nu}_e$ flux with flight distances up to ~1km.

at Bugey, Rovno, Goesgen, Krasnoyarsk and Savannah River with the flight distances up to 100m [4]. The results showed an excellent agreement with the expected no-oscillation spectra, and the measured cross section of the reaction (2) by reactor $\bar{\nu}_e$'s agreed well with the prediction of the standard V-A theory at the 3% level [4].

In 1998 Super-Kamiokande [5] reported on the studies of atmospheric neutrinos that show a clear dependence of $\nu_\mu(\bar{\nu}_\mu)$ on azimuthal angle which strongly suggests $\nu_\mu \rightarrow \nu_\tau(\bar{\nu}_\mu \rightarrow \bar{\nu}_\tau)$ oscillation with Δm^2 around 10^{-2} ~10^{-3}eV^2. The Δm^2 region can be studied by reactor experiments by taking a baseline L ~1km. Two reactor experiments (Chooz and Palo Verde) [6] were carried out using O(10)ton LS detectors placed at ~1km from power reactors of ~10GW$_{th}$ to search for Δm^2 down to 10^{-3}eV^2 for large $\sin^2 2\theta$ ~1 region. No evidence for $\bar{\nu}_e$ oscillation found in these experiments confirmed that the $\nu_\mu \rightarrow \nu_e$ is not the major process of the atmospheric neutrino oscillation.

To summarize reactor neutrino experiments up to 1990's showed no indication of neutrino oscillation in Δm^2 down to around 10^{-3}eV^2. Figure 1 shows the ratios of the observed reactor $\bar{\nu}_e$ fluxes to the no-oscillation expectations. However, the reactor $\bar{\nu}_e$ flux and the spectrum have been well understood and the measured cross section of the reaction (2) is in good agreement with the standard V-A theory.

There was a long-standing problem called a solar neutrino problem (SNP) [7] since late 1960's when R.Davis observed ν_e's from the Sun and confirmed the burning mechanism of the Sun which is summarily expressed as $4p + 2e^- \rightarrow$ ^{4}He$+2\nu_e + 26.7$MeV$-E_\nu$. All the measured ν_e fluxes afterwards using C_2Cl_4, H_2O and Ga as targets but with different thresholds are only fractions between 1/3 and 1/2 of the standard solar model (SSM) prediction. The discrepancy remained even after great efforts on both experimental and theoretical sides and the neutrino oscillation was getting considered more seriously as the solution to the SNP. However, the experimentally allowed regions of oscillation parameters were not unique and there were four candidate solutions speculated by the matter-induced MSW [8] flavor transformation as well as the vacuum oscillation mechanisms.

In 2001, SNO [9] which uses 1000ton D_2O target reported not only the deficit of the solar ν_e flux measured by $\nu_e + d \rightarrow e^- + p + p$, but a good agreement with the SSM prediction of the total active neutrino flux measured by $\nu_x + d \rightarrow \nu_x + p + n$. This means the ν_e's generated in the Sun transform into another types of neutrinos on the way to the Earth. This strongly suggests the ν_e oscillation. Combined with other solar neutrino experiments oscillation parameters in a region of Δm^2 ~several $\times 10^{-5}$eV^2 and $\tan^2\theta$ ~1 called LMA is quite promising. However, it is not clear whether the oscillation is the true mechanism or not due to the limited knowledge of the solar interior. A decisive experiment using a neutrino flux with much less uncertainty was strongly awaited.

With the well-understood $\bar{\nu}_e$ flux and the spectrum, a reactor experiment with a sensitivity to Δm^2 ~10^{-5}eV^2 could check the LMA solution. However, it would require a long baseline of O(100)km from powerful reactors and a sensitive detector with a large target mass. This is really a big challenge!

3 KamLAND reactor experiment

KamLAND (*Kam*ioka *L*iquid scintillator *A*nti- *N*eutrino *D*etector) is the world's largest LS detector located in a rock cavern of the former Kamiokande detector. The site is 1000m underground (2,700m.w.e) in Kamioka mine in Japan and the cosmic-ray muon flux is 1/100,000 of the earth level. The main purpose of KamLAND is to carry out a reactor neutrino oscillation search for Δm^2 ~10^{-5}eV^2 region where the LMA solution to the SNP can be checked.

The central part of the detector [10] is a 1000ton ultrapure LS filled in a transparent balloon and viewed through non-scintillating oil by 1325 17-inch and 554 20-inch photomultiplier tubes (PMTs) to detect scintillation lights. It is contained in a 18m-diameter spherical tank and the outside is filled with purified water to make a radiation shield against the surrounding rocks and form a water Cherenkov detector with 225 20-inch PMTs to identify cosmic ray muons.

KamLAND is surrounded by many distant power reactors and 26 of them are locating in a narrow band of 180km from KamLAND and generate thermal power of ~75GW$_{th}$ in total, which is about 7% of the total reactor power in the world. The $\bar{\nu}_e$ flux at KamLAND are calculated from the operation data of every reactor in Japan and estimated for reactors in Korea and other countries. The $\bar{\nu}_e$ flux and spectrum uncertaity is estimated to be 3.4%.

$\bar{\nu}_e$ is detected through the reaction (2) in the LS by the two correlated events

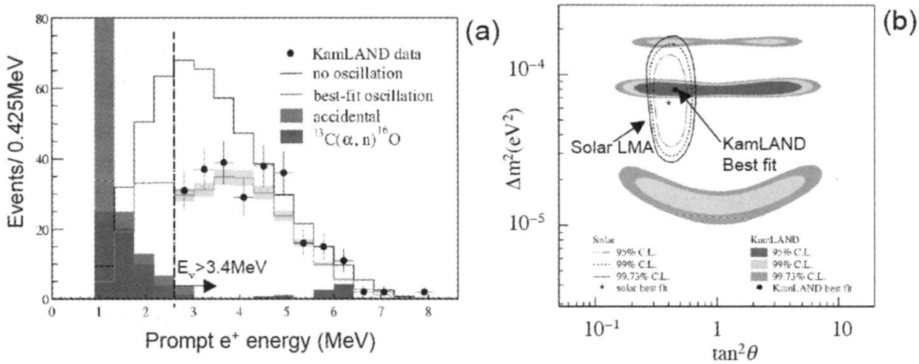

Figure 2: (a)Prompt energy spectrum ($E_{\bar{\nu}_e} - 0.8$MeV) of the observed reactor $\bar{\nu}_e$ candidates and the no-oscillation expectation, and the associated backgrounds. (b)Allowed region of the neutrino oscillation parameters from KamLAND $\bar{\nu}_e$ and solar ν_e experiments [12].

of a prompt e^+ and a delayed 2.2MeV γ-ray of the neutron capture in $\sim$200μs. Thanks to the high radiation shields and the very low concentrations of ^{238}U, ^{232}Th and ^{40}K in the LS confirmed by the data KamLAND announced the first reactor results [10] based on a 162ton·yr data sample in 2002. The observed $\bar{\nu}_e$ flux above 3.4MeV is only $0.611 \pm 0.085(stat) \pm 0.041(syst)$ times that of the no-oscillation expectation. This shows for the first time the reactor $\bar{\nu}_e$ disappearance at the 99.95%CL. If the neutrino oscillation is assumed along with CPT invariance the result excludes all solutions to the SNP except for the LMA and the analysis including the spectral shape constrained the allowed region of Δm^2 to narrow bands at around $10^{-4} \sim 10^{-5}$eV^2.

The results was further confirmed by the 2nd reactor results [11] with a larger data sample of 766 ton·yr exposure as shown in Figure 2. The observed 258 $\bar{\nu}_e$ candidate events with energies above 3.4MeV including 17.8±7.3 estimated background events are significantly less than 365±23.7 events of the no-oscillation expectation and the $\bar{\nu}_e$ disappearance is confirmed at 99.998%CL. The $\bar{\nu}_e$ energy spectrum shape highly preferred the oscillation to other assumptions such as neutrino decay and decoherence. The oscillation parameters are determined as $\Delta m^2 = (7.9^{+0.6}_{-0.5}) \times 10^{-5}$eV^2 and $\tan^2\theta = 0.46$, and the latter is $0.40^{+0.10}_{-0.07}$ when combined with the solar neutrino experiments under the assumption of the CPT invariance.

4 θ_{13} measurement

The mixing angle θ_{13} is the fundamental parameter like other mixing angles in the P-MNS matrix and should be determined experimentally. Currently an upper limit is given as $\sin^2 2\theta_{13} < 0.12$ [13] at $\Delta m^2_{13} \sim 3\times10^{-3}\mathrm{eV}^2$. The importance of the θ_{13} should be noted by the fact that it provides detectability of the CP-phase (δ_{CP}) in the future neutrino oscillation experiments as it appears in the following expression,

$$P(\nu_\mu \to \nu_e) - P(\bar{\nu}_\mu \to \bar{\nu}_e) = \cos^2\theta_{13}\sin 2\theta_{12}\sin 2\theta_{23}\sin\theta_{13}\ \sin\delta_{CP}\cdot(\sin\Delta_{31}+\sin\Delta_{23}+\sin\Delta_{12}), \tag{5}$$

where $\Delta_{ij} = (m_i^2 - m_j^2)L/(2E_\nu)$.

Reactor experiments can perform the pure θ_{13} measurements. By taking the L to O(1)km to maximize a known factor of $\sin^2(\Delta_{31}/2)$, surviving probability of $\bar{\nu}_e$ is expressed in a very good approximation as

$$P(\bar{\nu}_e \to \bar{\nu}_e) \approx 1-\sin^2 2\theta_{13}\sin^2(\Delta_{31}/2), \tag{6}$$

due to the fact of $\Delta m^2_{13} \approx \Delta m^2_{23} \approx 30\Delta m^2_{12}$.

Several reactor θ_{13} experiments have been proposed [14] with a goal to achieve the sensitivity of $\sin^2 2\theta_{13} = 0.01$ which is around 10 times better than the previous experiment. The basic idea is to use identical detectors (near and far) to cancel out the common uncertainties [15].

The value of $\sin^2 2\theta_{13}$ may resolve the uncertainty of $\sin^2\theta_{23}$ contained in the expression of $P(\nu_\mu \to \nu_e)$ which will be measured in the planned long baseline accelerator neutrino experiments. In this point of view the reactor θ_{13} measurement can play a complementary role.

5 Geoneutrino detection by KamLAND

This is the application of the KamLAND reactor $\bar{\nu}_e$ results to extract geologically produced $\bar{\nu}_e$'s called geoneutrinos. The study was made for the first time by KamLAND [16] with the ultra-low radioactivity environment.

The heat produced in the Earth is a fundamental quantity in understanding the internal dynamics and evolution of the Earth. It is generated by radioactive decays of ^{238}U, ^{232}Th and ^{40}K contained deep inside the Earth. However, the amount is not well known and indirectly evaluated from the studies of chondritic meteorites that most of which ($\sim$16TW) is produced by ^{238}U and ^{232}Th decays. If geoneutinos emitted from β^- decays of these elements can be detected, they are used as a direct means to measure the radiogenic heat and provide very important information about the inside of the Earth [17].

KamLAND can detect $\bar{\nu}_e$'s from 2 and 3 β^- decays in ^{238}U and ^{232}Th chains by using the reaction (2) with 1.8MeV threshold, which would appear in the spectrum between the threshold and 3.4MeV. The data sample of 749.1 days of the live-time is anlaysed by tighter cuts than the reactor $\bar{\nu}_e$ analysis. Figure 3(a) shows the results. In total 152 $\bar{\nu}_e$ candidates are selected with 127$\pm$13 estimated backgrounds from; (1) 80.4$\pm$7.2 reactor $\bar{\nu}_e$'s by an extrapolation of the spec-

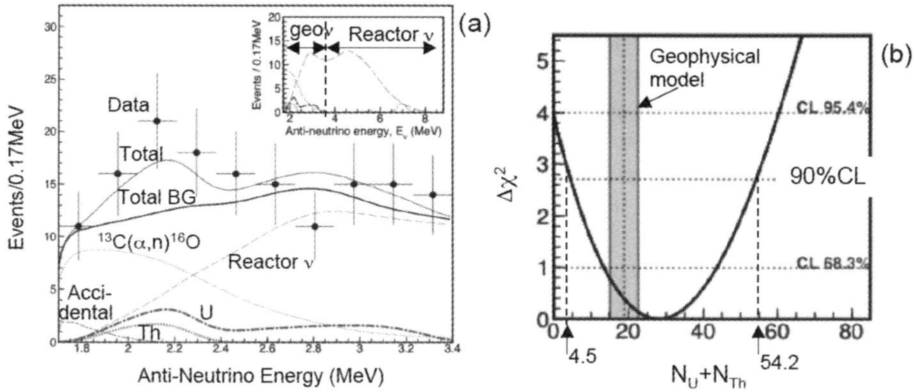

Figure 3: (a)Energy spectra of the observed $\bar{\nu}_e$ candidates and the expectations for ^{238}U+^{232}Th $\bar{\nu}_e$'s and backgrounds. (b)$\Delta\chi^2$ obtained by using the known spectra of the signal and backgrounds as a function of total number of $\bar{\nu}_e$ candidates from ^{238}U and ^{232}Th with a constraint of Th/U mass ratio of 3.9.

trum with oscillation parameters [11], (2) 42±11 events from $^{13}\mathrm{C}(\alpha,n)^{16}\mathrm{O}^{(*)}$ reaction with an α from ^{210}Po decay and the prompt signals from de-excitation of ^{16}O, the neutron scattering with a proton and/or ^{12}C, (3) 2.38±0.01 events from accidental coincidence, and (4) 1.9±0.2 events from $\bar{\nu}_e$'s from long lived reactor fission products, (5) 0.30±0.05 events from $^{9}\mathrm{Li}\rightarrow 2\alpha + n + e^- + \bar{\nu}_e$.

Geoneutrino candidates from ^{238}U+^{232}Th are obtained by a simple subtraction as 25^{+19}_{-18} events. Including known spectra of ^{238}U and ^{232}Th decay chains and the backgrounds an unbinned maximum likelihood analysis gives 3 ^{238}U and 18 ^{232}Th geoneutrinos. If the widely-accepted Th/U mass ratio of 3.9 is used as a constraint a 90%CL interval of the number of ^{238}U+^{232}Th geoneutrino candidates is obtained as 4.5-54.2 as shown in Figure 3(b). This is a finite value and the central value (28) is consistent with the rate analysis and a prediction of a geophysical model [16]. Also, an upper bound of 60TW at 99%CL for the radiogenic heat from ^{238}U and ^{232}Th is obtained based on the geophysical model. This is the first results constraining the radiogenic heat by direct measurement. Although statistical power is low, KamLAND has made a first step to use neutrinos to study the inside of the Earth.

6 Summary

• Since the discovery of the neutrino, reactor experiments have continued to play crucial role in the studies of neutrinos. Although no evidence for neutrino oscillation was found in baselines up to ~1km, understanding the reactor $\bar{\nu}_e$ flux and the spectrum was greatly improved and precise measurements of the process (2) have been made to confirm the V-A structure of the SM in MeV region. Also, it should be noted that reactor experiments with O(10)ton LS is technically established.

• Discovery of the reactor $\bar{\nu}_e$ disappearance and the confirmation of the $\bar{\nu}_e$ oscillation were made by KamLAND with an unprecedented sensitivity. The oscillation parameters are precisely determined and the long-standing solar neutrino problem has been finally resolved under the assumption of CPT invariance.

• θ_{13} is the fundamental parameter and provides the detectability of the CP-violation in the lepton sector in future accelerator-based neutrino experiments. Planned reactor θ_{13} experiments will play crucial roles in the studies.

• KamLAND has challenged for the first time the geoneutrino detection, and opened a new window to study the inside of the Earth.

• Lots of physics outputs have been provided by reactor neutrino experiments. Exciting results will come out near future. Reactors are highly cost effective and will continue to be a powerful tool for the studies of neutrinos.

References

[1] F.Reines, Rev. Mod. Phys. **68**, 317 (1996).

[2] ^{235}U: K.Schreckenbach *et al.*, Phys. Lett. B **160**, 325 (1985), ^{239,241}Pu: A.A.Hahn *et al.*, Phys. Lett. B **218**, 365 (1989).

[3] P.Vogel *et al.*, Phys. Rev. C **24**, 1543 (1981).

[4] Y.Declais *et al.*, Phys. Lett. B **338**, 383 (1994), and references therin.

[5] Super-Kamiokande Collab., Y.Fukuda *et al.*, Phys. Rev. Lett. **81**, 1562 (1998).

[6] M.Apollonio *et al.*, Phys. Lett. B **466**, 415 (1999); F.Boehm *et al.*, Phys. Rev. D **62**, 072002 (2000)

[7] K.Nakamura, Phys. Lett. B **592**, 459 (2004).

[8] L.Wolfenstein, Phys. Rev. D **17**, 2369 (1978); S. P. Mikheyev and A. Yu. Smirnov, Sov. J. Nucl. Phys. **42**, 913 (1986).

[9] SNO Collab., Q.R.Ahmad *et al.*, Phys. Rev. Lett. **87** 071301 (2001).

[10] KamLAND Collab., K.Eguchi *et al.,* Phys. Rev. Lett. **90**, 021802 (2003).

[11] KamLAND Collab., T.Araki *et al.,* Phys. Rev. Lett. **94**, 081801 (2005).

[12] SNO Collab., S.N. Ahmed et al., Phys. Rev. Lett. **92**, 181301 (2004).

[13] CHOOZ Collab., Eur. Phys. J. C **27**, 331 (2003).

[14] M.H.Shaevitz, Nucl. Phys. B (Proc. Suppl.) **145**, 154 (2005).

[15] L.Mikaelyan, Nucl. Phys. B (Proc. Suppl.) **91**, 120 (2001)

[16] KamLAND Collab., T.Araki *et al.,* Nature, **436**, 499 (2005).

[17] G.Eders, Nucl. Phys., **78**, 657 (1966); G.Marx, Czech. J. Phys. B **19**, 1471 (1969).

DOUBLE BETA DECAY EXPERIMENTS

A.S. Barabash [a]
Institute of Theoretical and Experimental Physics, B.Cheremushkinskaya 25, 117218 Moscow, Russia

Abstract. The present status of double beta decay experiments are reviewed. The results of the most sensitive experiments, NEMO-3 and CUORICINO, are discussed. Proposals for future double beta decay experiments are consided. In these experiments sensitivity for the effective neutrino mass will be on the level of (0.1-0.01).

1 Introduction

Interest in neutrinoless double beta decay has seen a significant rebirth in recent years after evidence for neutrino oscillations was obtained from the results of atmospheric and solar neutrino experiments (see, for example, the discussions in [1–3]).

This observation of oscillations was recently confirmed by the KamLAND experiment with reactor antineutrinos [4,5] and by the new SNO result [6]. These results are an impressive proof that neutrinos have a nonzero mass. However, the experiments studying neutrino oscillations are not sensitive to the nature of the neutrino mass (Dirac or Majorana?) and provide no information on the absolute scale of the neutrino masses, since such experiments are sensitive only to the difference of the masses, Δm^2. The detection and study of $2\beta 0\nu$ decay may clarify the following problems of neutrino physics (see discussions in [7–9]): (i) neutrino nature; is the neutrino a Dirac or a Majorana particle?, (ii) absolute neutrino mass scale (a measurement or a limit on m_1), (iii) the type of neutrino mass hierarchy (normal, inverted, or quasidegenerate), (iv) CP violation in the lepton sector (measurement of the Majorana CP-violating phases).

2 Results of experimental investigations

2.1 Two neutrino double beta decay

This decay was first recorded in 1949 in a geochemical experiment with ^{130}Te [10]; in 1967, it was also found for ^{82}Se [11]. Attempts to observe this decay in a direct experiment employing counters had been futile for a long time. Only in 1987 could M. Moe, who used a time-projection chamber (TPC), observe $2\beta(2\nu)$ decay in ^{82}Se for the first time [12]. Within the next few years, experiments employing counters were able to detect $2\beta(2\nu)$ decay in many nuclei. In ^{100}Mo [13–15], and ^{150}Nd [16] $2\beta(2\nu)$ decay to the 0^+ excited state of the daughter nucleus was recorded. Also, the $2\beta(2\nu)$ decay of ^{238}U was detected in a radiochemical experiment [17], and in a geochemical experiment for the first time the ECEC process was detected in ^{130}Ba [18]. Table 1 displays the present-day averaged and recommended values of $T_{1/2}(2\nu)$ from [19].

2.2 Neutrinoless double beta decay

In contrast to two-neutrino decay, neutrinoless double-beta decay has not yet been observed [b].

---

[a]e-mail: Alexander.Barabash@itep.ru

[b]The possible exception is the result with ^{76}Ge, published by a fraction of the Heidelberg-Moscow Collaboration, $T_{1/2} = 1.2 \cdot 10^{25}$ y [20] (see Table 2). The Moscow portion of the Collaboration does not agree with this conclusion [21] and there are others who are critical

Table 1: Average and recommended $T_{1/2}(2\nu)$ values (from [19]).

Isotope	$T_{1/2}(2\nu)$
^{48}Ca	$4.3^{+2.1}_{-1.0} \cdot 10^{19}$
^{76}Ge	$(1.5 \pm 0.1) \cdot 10^{21}$
^{82}Se	$(0.92 \pm 0.07) \cdot 10^{20}$
^{96}Zr	$(2.0 \pm 0.3) \cdot 10^{19}$
^{100}Mo	$(7.1 \pm 0.4) \cdot 10^{18}$
^{100}Mo-^{100}Ru(0_1^+)	$(6.8 \pm 1.2) \cdot 10^{20}$
^{116}Cd	$(3.0 \pm 0.2) \cdot 10^{19}$
^{128}Te	$(2.5 \pm 0.3) \cdot 10^{24}$
^{130}Te	$(0.9 \pm 0.1) \cdot 10^{21}$
^{150}Nd	$(7.8 \pm 0.7) \cdot 10^{18}$
^{150}Nd-^{150}Sm(0_1^+)	$1.4^{+0.5}_{-0.4} \cdot 10^{20}$
^{238}U	$(2.0 \pm 0.6) \cdot 10^{21}$
^{130}Ba; ECEC(2ν)	$(2.2 \pm 0.5) \cdot 10^{21}$

The present-day constraints on the existence of $2\beta(0\nu)$ decay are presented in Table 2 for the nuclei that are the most promising candidates. In calculating constraints on $\langle m_\nu \rangle$, the nuclear matrix elements from [23–25] were used (3-d column). It is advisable to employ the calculations from these studies, because the calculations are the most thorough and take into account the most recent theoretical achievements. In column four, limits on $\langle m_\nu \rangle$, which were obtained using the NMEs from a recent paper [26]. In this paper g_{pp} values (g_{pp} is parameter of QRPA theory) were fixed using experimental half-life values for 2ν decay and then NME(0ν) were calculated. Those authors analyse results of all existing QRPA calculations and demonstrate that their approach gives most accurate and reliable values for NMEs (but see, nevertheless, critics in [27]). The respective phase-space volumes were taken from [28].

Table 2: Best present results on $2\beta(0\nu)$ decay (limits at 90% C.L.).

Isotope	$T_{1/2}$, y	$\langle m_\nu \rangle$, eV [23–25]	$\langle m_\nu \rangle$, eV [26]	Experiment
^{76}Ge	$> 1.9 \cdot 10^{25}$	$< 0.33 - 0.84$	$< 0.53 - 0.59$	HM [29]
	$\simeq 1.2 \cdot 10^{25}$(?)	$\simeq 0.5 - 1.3$(?)	$\simeq 0.7$(?)	Part of HM [20]
	$> 1.6 \cdot 10^{25}$	$< 0.36 - 0.92$	$< 0.58 - 0.64$	IGEX [30]
^{130}Te	$> 1.8 \cdot 10^{24}$	$< 0.4 - 0.9$	$< 1 - 1.6$	CUORICINO [31]
^{100}Mo	$> 4.6 \cdot 10^{23}$	$< 0.65 - 1.0$	$< 2.4 - 3.0$	NEMO- 3 [32]
^{136}Xe	$> 4.5 \cdot 10^{23}$	$< 0.8 - 4.7$	$< 2.9 - 5.6$	DAMA [33]
^{116}Cd	$> 1.3 \cdot 10^{23}$	$< 1.4 - 2.5$	$< 3.7 - 4.3$	SOLOTVINO [34]
^{82}Se	$> 1 \cdot 10^{23}$	$< 1.7 - 3.7$	$< 3.8 - 4.7$	NEMO-3 [32]

of this result [22]. Thus at the present time this "positive" result is not accepted by the "2β decay community" and it has to be checked by new experiments.

2.3 Double beta decay with Majoron emission

Table 3 displays the best present-day constraints for an ordinary (triplet) Majoron (spectral index n = 1). For nonstandard models of the Majoron (spectral index n = 2, 3 and 7) the most strong limits were obtained for ^{100}Mo and ^{82}Se with NEMO-3 detector [35].

Table 3: Best present results on $2\beta(0\nu\chi^0)$ decay (ordinary Majoron) at 90% C.L. The NME from the following works were used, 3-d column: [23–25], 4-th column: [26].

Isotope	$T_{1/2}$, y	$\langle g_{ee} \rangle$, [23–25]	$\langle g_{ee} \rangle$, [26]
^{76}Ge	$> 6.4 \cdot 10^{22}$ [29]	$< (1.2 - 3.0) \cdot 10^{-4}$	$< (1.9 - 2.3) \cdot 10^{-4}$
^{82}Se	$> 1.5 \cdot 10^{22}$ [35]	$< (0.66 - 1.4) \cdot 10^{-4}$	$< (1.2 - 1.9) \cdot 10^{-4}$
^{100}Mo	$> 2.7 \cdot 10^{22}$ [35]	$< (0.4 - 0.7) \cdot 10^{-4}$	$< (1.6 - 1.8) \cdot 10^{-4}$
^{116}Cd	$> 8 \cdot 10^{21}$ [34]	$< (1.0 - 2.0) \cdot 10^{-4}$	$< (2.8 - 3.3) \cdot 10^{-4}$
^{128}Te	$> 2 \cdot 10^{24}$(geochem) [36]	$< (0.7 - 1.6) \cdot 10^{-4}$	$< (1.9 - 2.4) \cdot 10^{-4}$

3 Best present experiments

In this section the two large-scale experiments NEMO-3 and CUORICINO are discussed.

3.1 NEMO-3 experiment [32,41,42]

This is a tracking experiment that, in contrast to experiments with ^{76}Ge, detects not only the total energy deposition, but also other parameters of the process, including the energy of the individual electrons, angle between them, and the coordinates of the event in the source plane. Since June of 2002, the NEMO-3 detector has operated at the Frejus Underground Laboratory (France) located at a depth of 4800 m.w.e. The detector has a cylindrical structure and consists of 20 identical sectors. A thin (about 30-60 mg/cm^2) source containing beta-decaying nuclei and having a total area of 20 m^2 and a weight of up to 10 kg was placed in the detector. The energy of the electrons is measured by plastic scintillators (1940 individual counters), while the tracks are reconstructed on the basis of information obtained in the planes of Geiger cells (6180 cells) surrounding the source on both sides. In addition, a magnetic field of strength of about 25 G parallel to the detector axis is created by a solenoid surrounding the detector.

The main characteristics of the detector are the following: the energy resolution of the scintillation counters lies in the interval 14-17% FWHM for electrons of energy 1 MeV; the time resolution is 250 ps for an electron energy of 1 MeV; and the accuracy in reconstructing of the vertex of 2e$^-$ events is about 1 cm.

At the present time, the investigations are being performed for seven isotopes; these are ^{100}Mo (6.9 kg), ^{82}Se (0.93 kg), ^{116}Cd (0.4 kg), ^{150}Nd (37 g), ^{96}Zr (9.4 g), ^{130}Te (0.45 kg), and ^{48}Ca (7 g).

Fig. 1 and Fig. 2 display the spectrum of $2\beta(2\nu)$ events, in ^{100}Mo and ^{82}Se that were collected over 389 days [32]. For ^{100}Mo angular distribution (Fig. 1b) and single electron spectrum (Fig. 1c) are also shown. The total number of useful events is about 219,000, which is much greater than the total statistics of all of the preceding

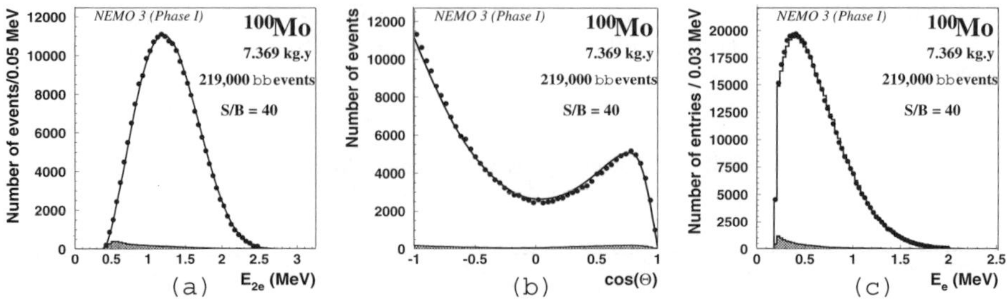

Figure 1: (a) Energy sum spectrum of the two electrons, (b) angular distribution of the two electrons and (c) single energy spectrum of the electrons, after background subtraction from ^{100}Mo with 7.369 kg·years exposure [32]. The signal contains 219,000 2β events and the signal-to-background ratio is 40. The solid line corresponds to the expected spectrum from $2\beta(2\nu)$ simulations and the shaded histogram is the subtracted background computed by Monte-Carlo simulations.

experiments with ^{100}Mo. By employing the calculated values of the detection efficiencies for $2\beta(2\nu)$ events, the following half-life values were obtained for ^{100}Mo and ^{82}Se:

$$T_{1/2}(^{100}Mo; 2\nu) = [7.11 \pm 0.02(stat) \pm 0.54(syst)] \cdot 10^{18}\ y \tag{1}$$

$$T_{1/2}(^{82}Se; 2\nu) = [9.6 \pm 0.3(stat) \pm 1.0(syst)] \cdot 10^{19}\ y \tag{2}$$

Using a maximal likelihood method the following limits on neutrinoless double beta decay of ^{100}Mo and ^{82}Se (mass mechanism; 90% C.L.) have been obtained:

$$T_{1/2}(^{100}Mo; 0\nu) > 4.6 \cdot 10^{23}\ y \tag{3}$$

$$T_{1/2}(^{82}Se; 0\nu) > 1 \cdot 10^{23}\ y \tag{4}$$

The corresponding limits on $\langle m_\nu \rangle$ are presented in Table 2. The NEMO-3 experiment is on going and new improved results will be obtained in the near future.

3.2 CUORICINO [31]

This program is the first stage of the larger CUORE experiment (see Subsection 4.1). The experiment is running at the Gran Sasso Underground Laboratory. The detector consists of low-temperature devices based on $^{nat}TeO_2$ crystals. The detector consists of 62 individual crystals, their total weight being 40.7 kg. The energy resolution is 7.5-9.6 keV at an energy of 2.6 MeV.

The experiment has been running since March of 2003. The summed spectra of all crystals in the region of the $2\beta(0\nu)$ energy of ^{130}Te is shown in Fig. 3. The total exposure is 3.09 $kg\cdot$ y (^{130}Te). The background at the energy of the $2\beta(0\nu)$ decay is 0.18 $\mathrm{keV}^{-1} \cdot kg^{-1} \cdot y^{-1}$. No peak is evident and the limit is $T_{1/2}(^{130}Te; 0\nu) > 1.8 \cdot 10^{24}$ y (90% CL). The corresponding limits on $\langle m_\nu \rangle$ are presented in Table 2.

The sensitivity of the experiment to $2\beta(0\nu)$ decay of ^{130}Te will be on the level of $9 \cdot 10^{24}$ for 5 y of measurement. This in turn means the sensitivity to $\langle m_\nu \rangle$ is on the level of 0.2-0.6 eV.

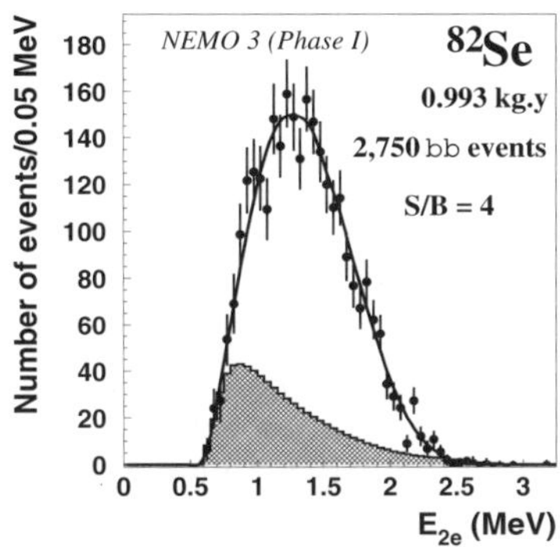

Figure 2: Energy sum spectrum of the two electrons after background subtraction from ^{82}Se with 0.993 kg·years exposure (same legend as Fig. 1) [32]. The signal contains 2,750 2β events and the signal-to-background ratio is 4.

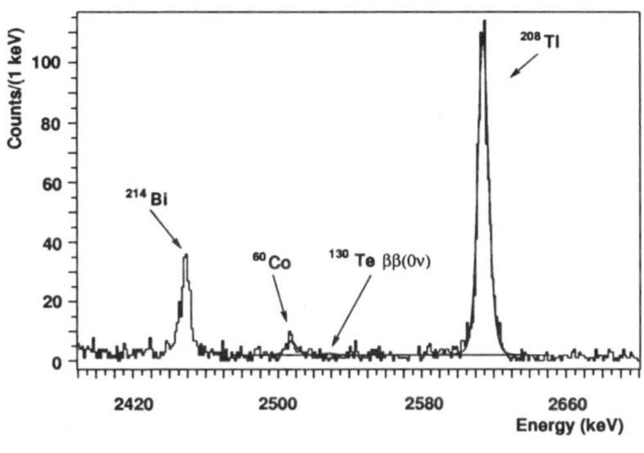

Figure 3: The sum spectra of all crystal in the region of the $2\beta(0\nu)$ energy of ^{130}Te [31].

4 Planned experiments

In this section, mention of five promising experiments which can be realized within the next five to ten years is discussed. The estimation of the sensitivity in all experiments is made using NMEs from [23–26].

4.1 CUORE [45]

This experiment is to be run at the Gran Sasso Underground Laboratory. The plan is to investigate 760 kg of $^{nat}TeO_2$, with a total of 206 kg of ^{130}Te. One thousand low-temperature ($\sim$ 8 mK) detectors, each having a weight of 760 g, will be manufactured and arranged in 25 towers (one tower is approximately equivalent to the CUORICINO detector, see Subsection 3.2). One of the main problems here is to reduce the background level by a factor of about 10 to 100 in relation to the background level achieved in the detector CUORICINO [31]. Upon reaching a background level of 0.001 $\text{keV}^{-1} \cdot kg^{-1} \cdot y^{-1}$, the sensitivity of the experiment to the 0ν decay of ^{130}Te will become approximately $3.5 \cdot 10^{26}$ y ($\langle m_\nu \rangle \sim$ 0.03-0.11 eV). The experiment has been approved and has financial support.

4.2 GERDA [46]

This is one of two (along with the MAJORANA experiment) planned experiments with ^{76}Ge. The experiment is to be located in Gran Sasso Underground Laboratory (3500 m v.e.). The proposal is based on ideas and approaches which were proposed for GENIUS [47] and the GEM [48] experiments. The plan is to place "naked" HPGe detectors in highly purified liquid nitrogen.

The proposal involves three phases. In the first phase, the existing HPGe detectors ($\sim$ 15 kg), which previously were used in Heidelberg-Moscow [29] and IGEX [30] experiments, will be utilized. In the second phase $\sim$ 40 kg of enriched Ge will be investigated. In the third phase the plan is to use $\sim$ 500 kg of ^{76}Ge.

The first phase, lasting one year, is to measure the sensitivity to $3 \cdot 10^{25}$ y, which gives a possibility to check the "positive" result of [20]. The sensitivity of the second phase (for three years of measurement) will be $\sim 2 \cdot 10^{26}$ y, which corresponds to a sensitivity for $\langle m_\nu \rangle$ on the level of $\sim$ 0.09-0.3 eV.

The first two phases are funded. Measurements will start in $\sim$ 2007(2008). The results of these two phases will play an important role in the decision to support the full scale experiment.

4.3 MAJORANA [49, 50]

The MAJORANA facility will consist of 210 sectioned HPGe detectors manufactured from enriched germanium (the degree of enrichment is about 86%). The total mass of enriched germanium will be 500 kg. The use of sectioned HPGe detectors, pulse shape analysis, anticoincidence, and low radioactivity structural materials will make it possible to reduce the background to a value below $3 \cdot 10^{-4}$ $\text{keV}^{-1} \cdot kg^{-1} \cdot y^{-1}$ and to reach a sensitivity of about $4 \cdot 10^{27}$ y within ten years of measurements. The corresponding sensitivity to the effective mass of the Majorana neutrino is about 0.02 to 0.07 eV.

In the first step $\sim$ 90-120 kg of ^{76}Ge will be investigated. It is anticipated that in the sensitivity to $2\beta(0\nu)$ decay for 3 years of measurement will be $3 \cdot 10^{26}$ y. It will reject or to confirm the "positive" result from [20]. Sensitivity to $\langle m_\nu \rangle$ will be $\sim$ 0.07-0.25 eV. During this time different methods and technical questions will be checked and possible background problems will be investigated.

4.4 EXO [51]

In this experiment the plan is to implement M. Moe's proposal of 1991 [52]. Specifically it is to record both ionization electrons and the Ba^+ ion originating from the double-beta-decay process ^{136}Xe - $^{136}Ba^{++} + 2\text{e}^-$. In reference [51], it is proposed to operate with 1t of ^{136}Xe. The actual technical implementation of the experiment has not yet been developed. One of the possible schemes is to fill a TPC with liquid enriched xenon. To avoid the background from the 2ν decay of ^{136}Xe, the energy resolution of the detector must not be poorer than 3.8% (FWHM) at an energy of 2.5 MeV (ionization and scintillation signals will be detected).

As the first stage of the experiment it is planned the EXO-200 will use 200 kg of ^{136}Xe without Ba ion registration. This experiment is currently under preparation and measurement will start probably in 2007(2008). The 200 kg of enriched Xe is a product of Russia. If the background will be 40 events, as estimated by the authors, then the sensitivity of the experiment will be $\sim 6 \cdot 10^{25}$ y (this corresponds to sensitivity for $\langle m_\nu \rangle$ at the level $\sim$ 0.07-0.4 eV).

4.5 SuperNEMO [53–55]

The NEMO Collaboration has studied and is pursing an experiment that will observe 100 kg of ^{82}Se with the aim of reaching a sensitivity to the 0ν decay mode at the level of $T_{1/2} \sim (1-2) \cdot 10^{26}$ y (the corresponding sensitivity to the neutrino mass is about 0.04 to 0.15 eV). In order to accomplish this goal, it is proposed to use the experimental procedures nearly identical to that in the NEMO-3 experiment (see Subsection 3.1). The new detector will have planar geometry and will have a modular structure. A ^{82}Se source having a thickness of about 40 mg/cm^2 and a very low content of radioactive admixtures is placed at the center of the modules. The detector will record all features of double- beta decay: the electron energy will be recorded by counters based on plastic scintillators ($\Delta E/E \sim 10-12\%$ at E =1 MeV), while tracks will be reconstructed with the aid of Geiger counters.

5 Conclusion

In conclusion, two-neutrino double-beta decay has so far been recorded for ten nuclei (^{48}Ca, ^{76}Ge, ^{82}Se, ^{96}Zr, ^{100}Mo, ^{116}Cd, ^{128}Te, ^{130}Te, ^{150}Nd, ^{238}U). In addition, the $2\beta(2\nu)$ decay of ^{100}Mo and ^{150}Nd to 0^+ excited state of the daughter nucleus has been observed and the ECEC(2ν) process in ^{130}Ba was recorded.

Neutrinoless double-beta decay has not yet been confirmed. There is a conservative limit on the effective value of the Majorana neutrino mass at the level of 0.9 eV. Within the next few years, the sensitivity to the neutrino mass in the CUORICINO and NEMO-3 experiments will be improved to become about 0.2 to 0.6 eV with measurements of ^{130}Te and ^{100}Mo.

The next-generation experiments, where the mass of the isotopes being studied will be as grand as 100 to 1000 kg, will have started within three to five years. In all probability, they will make it possible to reach the sensitivity to the neutrino mass at a level of 0.1 to 0.01 eV.

Acknowledgments

I would like to thank Prof. C.S. Sutton for his useful remarks. A portion of this work was supported by grants from INTAS (03051- 3431) and NATO (PST CLG 980022).

References

[1] P.C. de Holanda, A.Yu. Smirnov, *Phys. Rev.* **D** 66, 113005 (2002).
[2] J.N. Bahcall et al., *JHEP* 0207, 054 (2002); hep-ph/0204314.
[3] M. Maltoni et al., *Phys. Rev.* **D** 67, 013011 (2003).
[4] K. Eguchi et al., *Phys. Rev. Lett.* 90, 021802 (2003).
[5] T. Araki et al., *Phys. Rev. Lett.* 94, 081801 (2005).
[6] S.N. Ahmed et al., *Phys. Rev. Lett.* 92, 181301 (2004).
[7] S.M. Bilenky, hep-ph/0403245.
[8] S. Pascoli, S.T. Petcov, W. Rodejohann, *Phys. Lett.* **B** 558, 141 (2003).
[9] S. Pascoli, S.T. Petcov, *Phys. Lett.* **B** 580, 280 (2004).
[10] M.G. Inghram, J.H. Reynolds, *Phys. Rev.* 76, 1265 (1949); 78, 822 (1950).
[11] T. Kirsten, W. Gentner, O.A. Schaeffer, *Z. Phys.* 202, 273 (1967).
[12] S.R. Elliott, A.A. Hahn, M.K. Moe, *Phys. Rev. Lett.* 59, 2020 (1987).

[13] A.S. Barabash et al., *Phys. Lett.* **B** 345, 408 (1995).
[14] A.S. Barabash et al., *Phys. At. Nucl.* 62, 2039 (1999).
[15] L. De Braeckeleer et al., *Phys. Rev. Lett.* 86, 3510 (2001).
[16] A.S. Barabash et al., *JETP Lett.* 79, 10 (2004).
[17] A.L. Turkevich, T.E. Economou and G.A. Cowan, *Phys. Rev. Lett.* 67, 3211 (1991).
[18] A.P. Meshik et al., *Phys. Rev.* **C** 64, 035205 (2001).
[19] A.S. Barabash et al., *MEDEX'05*, Corfu, Greece, September 26-29, 2005.
[20] H.V. Klapdor-Kleingrothaus et al., *Phys. Lett.* **B** 586, 198 (2004).
[21] A.M. Bakalyarov et al., *Phys. Part. Nucl. Lett.* 2, 77 (2005); hep-ex/0309016.
[22] A. Strumia and F. Vissani, *Nucl. Phys.* **B** 726, 294 (2005).
[23] F. Simkovic et al., *Phys. Rev.* **C** 60, 055502 (1999).
[24] S. Stoica, H.V. Klapdor-Kleingrothaus, *Nucl. Phys.* **A** 694, 269 (2001).
[25] O. Civitarese and J. Suhonen, *Nucl. Phys.* **A** 729, 867 (2003),
[26] V.A. Rodin et al., nucl-th/0503063.
[27] J. Suhonen, *Nucl. Phys.* **A** 752, 53c (2005).
[28] J. Suhonen, O. Sivitarese, *Phys. Rep.* 300, 123 (1998).
[29] H.V. Klapdor-Kleingrothaus et al., *Eur. Phys.J.* **A** 12, 147 (2001).
[30] C.E. Aalseth et al., *Phys. Rev.* **C** 65, 09007 (2002).
[31] C. Arnaboldi et al., *Phys. Rev. Lett.* 95, 142501 (2005).
[32] R. Arnold et al., *Phys. Rev. Lett.* 95, 182302 (2005).
[33] R. Bernabei et al., *Phys. Lett.* **B** 546, 23 (2002).
[34] F.A. Danevich et al., *Phys. Rev.* **C** 67, 035501 (2003).
[35] R. Arnold et al., *Nucl. Phys.* **A** 765, 483 (2006).
[36] O.K. Manuel, *J. Phys.* **G** 17, 221 (1991).
[37] M. Gunther et al., *Phys. Rev.* **D** 55, 54 (1997).
[38] R. Arnold et al., *Nucl. Phys.* **A** 678, 341 (2000).
[39] J. Tanaka J., H. Ejiri, *Phys. Rev.* **D** 48, 5412 (1993).
[40] C. Arnaboldi et al., *Phys. Lett.* **B** 557, 167 (2003).
[41] R. Arnold et al., *Nucl. Instr. Meth.* **A** 536, 79 (2005).
[42] R. Arnold et al., *JETP Lett.* 80, 377 (2004).
[43] F. Simkovic, P. Domin and S. Semenov, *J. Phys.* **G** 27, 2233 (2001).
[44] P. Domin et al., *Nucl. Phys.* **A** 753, 337 (2005).
[45] C. Arnaboldi et al., *Nucl. Instr. Meth.* **A** 518, 775 (2004).
[46] I. Abt et al., hep-ex/0404039.
[47] H.V. Klapdor-Kleingrothaus, J. Hellmig and M. Hirsch, *J.Phys.* **G** 24, 483 (1998).
[48] Yu.G. Zdesenko, O.A. Ponkratenko, V.I. Tretyak, *J. Phys.* **G** 27, 2129 (2001).
[49] Majorana White Paper, nucl-ex/0311013.
[50] C.E. Aalseth et al., *Nucl. Phys.* **B** *(Proc. Suppl.)* 138, 217 (2005).
[51] M. Danilov et al., *Phys. Lett.* **B** 480, 12 (2000).
[52] M. Moe, *Phys. Rev.* **C** 44, R931 (1991).
[53] A.S. Barabash, *Czech. J. Phys.* 52, 575 (2002).
[54] A.S. Barabash, *Phys. At. Nucl.* 67, 1984 (2004).
[55] F. Piquemal, report on NANP05 (Dubna, June 20-25, 2005).

SPONTANEOUS AND INDUCED TWO-BETA PROCESSES

S.Semenov[1 a], Yu.Gaponov[1 b], F.Šimkovic[2 c], V.Khruschov[1 d]

[1] *Russian Research Centre "Kurchatov Institute", 123182 Kurchatov Square 1, Moscow, Russia;*

[2] *Department of Nuclear Physics, Comenius University, Mlynska Dolina, F1 84248 Bratislava, Slovakia*

Abstract. Two-beta processes, both spontaneous and induced by external neutrino or electron beam, have been reviewed. These prosesses are of great interest, for search of lepton number violating channel ($0\nu\beta\beta$-mode) is a path to study fundamental neutrino properties beyond Standard Model of electroweak interactions.

1 Introduction

Double beta decay in nuclei are being intensively examined now both theoretically and experimentally. These investigations are of a great importance for particle and nuclear physics, because, particulary, observation of neutrinoless $0\nu 2\beta$ decay, forbidden in SM, would indicate the Majorana nature of neutrino and give valuable information on neutrino mass matrix and lepton number violating parameters. The experimental and theoretical situation in two-beta decay study is considered in [1–5]. Double beta-decay can be considered as an simultaneus beta decay of two neutrons ($2\beta^-$) or two protons ($2\beta^+$) in even-even nucleus. The possible double beta-decay modes can be distinguished as double beta-decay with and without lepton number violation.

High-sensitivity $0\nu\beta\beta$ investigation is one of the main directions of weak interaction physics, because these experiments probe lepton number violation mechanisms beyond the Standard Model. Many international large-scale projects, aimed to study this problem are beeing developed now [5].

2 Two-neutrino double beta-decay

There are four types of lepton number conserving two-neutrino double beta-transitions, allowed by SM: 1) Emission of two electrons ($\beta^-\beta^-$); 2)Capture of double bound atomic electrons (EC/EC); 3)Capture of one bound atomic electron and emission of one positron (β^+/EC); 4)Emission of two positrons ($\beta^+\beta^+$). The intensity of the first type process, $2\nu\beta^-\beta^-$-decay, has been directly measured for the following nuclei: ^{48}Ca, ^{76}Ge, ^{82}Se, ^{96}Zr, ^{100}Mo, ^{116}Cd, ^{128}Te, ^{130}Te, ^{150}Nd, ^{238}U, including decays in two excited states (^{100}Mo and ^{150}Nd) [6]. This is a second order process in weak interaction, so inverse half

[a] e-mail: semenov@imp.kiae.ru

[b] e-mail: gaponov2@imp.kiae.ru

[c] e-mail: simkovic@fmph.uniba.sk

[d] e-mail: khru@imp.kiae.ru

decay time of two neutrino double beta decay involves summation over full set of states of intermediate nucleus [7].

Thus, $2\nu\beta^-\beta^-$-decay offers a useful test of nuclear structure models for nuclear matrix elements calculations. Theoretical values of nuclear matrix elements play essential role in determination of Majorana neutrino mass from the $0\nu\beta\beta$-decay. However calculation of $2\nu\beta\beta$ matrix elements is a complicated problem, for the nuclei under consideration are medium and heavy open shells ones and modern models of nuclear structure have certain difficulties to reproduce matrix elements values. It should be noted, that in nuclear models, like shell model and QRPA [3] the evaluation of the nuclear matrix element is restricted to the two subsequent Gamow-Teller transitions through a number of low-lying discrete states of the intermediate nucleus.

According to single state dominance hypothesis (SSD) [8,9] it is assumed that $2\nu\beta\beta$-decays with 1^+ ground state of intermediate nucleus are determined only by the two virtual beta decay transitions: the first one connecting the ground state of the initial nucleus with ${1^+}_{g.s.}$ intermediate state and the second one proceeding from the ${1^+}_{g.s.}$ state to the final ground state. The main advantage of this approach is that it is model independent since the values of of the single β-decay matrix elements can be deduced from the measured log*ft* values. So, the inverse $2\nu\beta^-\beta^-$ half decay time for $0^+ \rightarrow 0^+$ transitions takes the form:

$$[T_{1/2}^{2\nu}]^{-1} = \frac{G_\beta^4 g_A^4 m_e^9}{96 ln 2 \pi^7} |M_1^i M_1^f|^2 H(T), \tag{1}$$

where the dimensionless integral over lepton space H(T) is

$$H(T) = \int_1^{T+1} d\epsilon_1 \int_1^{T+2-\epsilon_1} d\epsilon_2 \int_0^{T+2-\epsilon_1-\epsilon_2} d\omega_1 \times$$

$$F(Z_f, \epsilon_1) F(Z_f, \epsilon_2) p_1 \epsilon_1 p_2 \epsilon_2 \omega_1^2 \omega_2^2 D(K, L) \tag{2}$$

Here ω_i, p_i, and ϵ_i (i=1,2) are energies of antineutrinos, momenta, and energies of electrons in the units of the mass of electron respectively, $\omega_2 = T + 2 - \epsilon_1 - \epsilon_2 - \omega_1$, T is the kinetic energy of leptons in the final state. F(Z_f, ϵ) denotes the relativistic Coulomb factor, K and L are built of the energy denominators of perturbation theory [7]. The beta transition amplitudes ${M^i}_1$ and ${M^f}_1$ can be deduced from log*ft* values of electron capture and single beta decay processes as follows:

$${M^i}_1 = \frac{1}{g_A}\sqrt{\frac{3D}{ft_{EC}}}, \qquad {M^f}_1 = \frac{1}{g_A}\sqrt{\frac{3D}{ft_{\beta^-}}} \tag{3}$$

In the previous SSD considerations of $2\nu\beta\beta$-decay the approximate treatment of the energy denominators has been applied [9]. The sum of lepton enrgies entering energy denominators were replaced with a half of energy release of this process. Recently it was shown, that this approximation may lead to significant inaccuracy in determining $\beta\beta$ half-decay times for transition to ground and excited states [10, 11]. It was demonstrated that exact treatment of energy denominators is especially important for $\beta\beta$-nuclear system with a small energy difference between the intermediate and ground states, such as ^{100}Mo and ^{116}Cd. The effect of energy denominators may strongly influence energy and angular distribution of outgoing electrons [11, 12].

It should be noted, that previously used approximation, when energy denominators are independent of lepton energies corresponds to the dominant transition through higher lying states of intermediate nuclei - higher states dominance hypothesis (HSD). Thus a comparison of the experimental data on the energy distribution of outgoing electrons with SSD and HSD predictions can give a valuable information about the structure of involved nuclear matrix elements. The high statistics measurements, produced in the NEMO 3 experiment with ^{100}Mo [13] are in favor of the SSD mechanism for transition to the ground state [11, 12].

3 Neutrinoless double beta-decay

Recent neutrino oscillation experiments [14] give strong evidence in favor of new physics beyond SM, such as neutrino masses and mixing. Complementary information is necessary to solve the problem of the nature of massive neutrinos. This is possible via investigation of the process in which the total lepton number is not conserved. The discovery of neutrinoless double beta decay $(A, Z) \to (A, Z+2) + 2e^-$ whould be the experimental proof, that neutrino is a Majorana particle [15]. The inverse half-decay time in assumption of the dominance of light neutrino mixing mechanism is given by [7]:

$$\frac{1}{T^{0\nu}_{1/2}}(A,Z) = |m_{\beta\beta}|^2 |M^{0\nu}(A,Z)|^2 G^{0\nu}(E_0, Z) \tag{4}$$

Here $G^{0\nu}(E_0, Z)$ and $M^{0\nu}(A, Z)$ are the known phase factor and theoretical nuclear matrix element respectively, E_0 is the energy release. Nowadays many large-scale experiments on the search of $0\nu\beta\beta$-decay are beeing carried out with the aim to determine effective Majorana neutrino mass $m_{\beta\beta}$. It should be noted, that estimation of $m_{\beta\beta}$ depends on nuclear matrix element value.

The most restrictive limit for $m_{\beta\beta}$ offers Heidelberg-Moscow experiment [16], were the best result $T^{0\nu}_{1/2} > 1.9 \cdot 10^{25}$ years have been achieved. The improved calculations of neutrinoless nuclear matrix elements have been produced in [17]. It was shown, that fixing the strength of particle-particle interaction in Quasiparticle Random Phase Approximation (QRPA)-approach, so that $2\nu\beta\beta$ half-decay time is correctly reproduced, allows to reduce theoretical uncertainty of $0\nu\beta\beta$ matrix elements. With the obtained $M^{0\nu}$ for ^{76}Ge [17]

and experimental result [16] the upper limit on the effective Majorana mass is $m_{\beta\beta} < 0.55eV$ [18]. The analysis of atmosheric neutrino oscillation data [14] gives the following best-fit values of two independent neutrino mass-squared differencies: $\delta m^2_{atm} = 2.0 \cdot 10^{-3} eV^2$. Thus about an order of magnitude improvement in sensitivity to the $m_{\beta\beta}$ should be achieved, and this is planned in future large-scale projects which include significant quantities of high enriched stable isotopes. These experiments will probe the inverted hierarchy of neutrino masses [19].

4 Induced 2β-transitions

The theoretical study of neutrino induced $\beta\beta$ transitions [20] enables one to extend the number of nuclei for experimental investigation of the $\beta\beta$ processes and to develope new methods for detection of solar, atmospheric and terrestrial neutrinos.

The neutrino induced $\beta\beta$ transition $\nu_e + (A, Z) \to (A, Z+2) + 2e^- + \bar{\nu}_e$ is a second-order process in weak interaction, allowed within the Standard Model. The amplitude of this process exhibits resonant behavior and strongly depends on the widths of intermediate nuclear states. It was pointed out that ^{100}Mo is a suitable isotope for registration of the low-energy solar neutrino [21] due to its low reaction threshold of 0.168 MeV. An analysis of the associated $\beta\beta$ transitions for ^{100}Mo was carried out in [22] and for $^{112,114,116}Cd$ and ^{18}O in [23]. A comlementary information on neutrino Majorana properties can give consideration of charge excange process without neutrino emission in electron beam [24]: $e^- + (A, Z) \to (A, Z-2) + e^+$.

The cross section of this reaction is proportional to ${m_{\beta\beta}}^2$. For sufficiently large electron energies the denominator of the process direct diagramm has a singularity, so width of intermediate states should be taken into account. Similar behavior is present in the $(\mu^- - e^+)$ conversion problem [25].

Acknowledgments

This work has been supported by the JINR-Slovak Republic common project for 2005, by the VEGA grant 1/3039/06 and by the RRC "Kurchatov Institute" grant N14 on fundamental research for 2005.

References

[1] K.Grotz, H.V.Klapdor-Kleingrothaus, *"The Weak Interaction in Nuclear Particle and Astrophysics"*, (Adam Hilger, New York) 1990.

[2] J.Suhonen, O.Civatarese, *Phys.Rep.* 300, 123 (1998).

[3] A.Faessler, F.Šimkovic, *J.Phys.*G 24, 2139 (1998).

[4] V.Tretyak, Yu.Zdesenko, *At.Data.Nucl.Data Tables* 80, 84 (2002).
[5] Yu.Gaponov, S.Semenov, V. Khruschev, A. Balysh in *"Proc.of the All-Russia (Intern) Conf. on Phys. and Chem. Processes on Selection of Atoms and Molecules, Zvenigorod, 2004"*, TZNIIATOMINFORM, Moscow, 2004, p.130 (in Russian); Yu.Gaponov, L.Inzhechik, S.Semenov in *"Isotopes: properties, production, enrichment"*, ed. by V. Baranov, Fizmatlit, Moscow, 2005, Vol.2, p.7 (in Russian).
[6] A.Barabash, et al, *Phys. Lett.* **B** 345, 408 (1995); A. Barabash, F.Hubert, Ph. Hubert, V.Umatov, *JETP* 79, 10 (2004).
[7] M.Doi, T.Kotani, E. Takasugi, *Progr.Theor.Phys.Suppl.* 83, 1 (1985).
[8] J.Abad, A.Morales, R.Nunez-Lagos, A.Pacheco*Ann.Phys* **A** 80, 9 (1980).
[9] O.Civitarese, J.Suhonen, *Phys.Rev.* **C** 58, 1535 (1998); O.Civitarese, J.Suhonen, *Nucl.Phys.* **A** 653,321 (1999).
[10] S.Semenov, F.Šimkovic, V.Khruschev, P.Domin, *Phys.At.Nucl.* 63, 1196 (2000); S.Semenov, F.Šimkovic, P.Domin, *Part.Nucl.Lett.* 109, 26 (2001).
[11] F.Šimkovic, P.Domin, S.Semenov, *J.Phys.* **G** 27, 2233 (2001).
[12] P.Domin, S.Kovalenko, F.Šimkovic, S.Semenov, *Nucl.Phys.* **A** 753, 337 (2005).
[13] R.Arnold et NEMO Collaboration, *Pis'ma Zh.Eks.Teor.Fiz.* 80, 429 (2004); A.Barabash et NEMO Collaboration, *Nucl.Phys.* **B** (Proc. Suppl.) 138, 207 (2005); X. Sarazin et NEMO Collaboration, *Nucl.Phys.* **B** (Proc. Suppl.) 143, 221 (2005).
[14] S.Fukuda et al, *Phys.Rev.Lett.* 81, 1562 (1998); Q.Ahmed et al, *Phys.Rev.Lett.* 87, 071301 (2001); T.Araki et al, *Phys.Rev.Lett.* 94, 081801 (2005).
[15] J.Schechter and J.Valle, *Phys.Rev.* **D** 25, 2951 (1982).
[16] H.Klapdor-Kleingrothaus et al, *Eur.Phys.J.* **A** 12, 147 (2001).
[17] V.Rodin, A.Faessler, F.Šimkovic, P.Vogel, *Phys.Rev.* **C** 68, 044302 (2003).
[18] S.Bilenky, A.Faessler, F.Šimkovic, *Phys.Rev.* **D** 70, 033003 (2004).
[19] S.Bilenky, A.Faessler, T.Gutsche and F.Šimkovic, *Phys.Rev.* **D** 72, 053015 (2005).
[20] S.Semenov, Yu.Gaponov, R.Khafizov, *Yad.Fiz* 61, 1379 (1998); L.Inzhechik, Yu.Gaponov, S.Semenov, *Yad.Fiz.* 61, 1384 (1998).
[21] H.Ejiri et al, *Phys.Rev.Lett.* 85, 2817 (2000).
[22] P.Domin, F.Šimkovic, S.Semenov, Yu.Gaponov, *Czech.J.Phys.* 52, 451 (2002).
[23] S.Semenov, Yu.Gaponov, F.Šimkovic, P.Domin, *Yad.Fiz.* 65, 2247 (2002).
[24] S.Semenov, Yu.Gaponov, L.Inzhechik,*Izv.RAN (ser.fiz.)* 64, 38 (2000).
[25] F.Šimkovic, P.Domin, S.Kovalenko, A.Faessler, Part.Nucl.Lett. 104, 40 (2001).

NEUTRINOLESS DOUBLE BETA DECAY IN THEORIES BEYOND THE STANDARD MODEL: ELECTRON ANGULAR DISTRIBUTIONS

A. Ali [a]
Deutsches Electronen-Synchrotron, DESY, 22607 Hamburg, Germany
A.V. Borisov [b], D.V. Zhuridov [c]
Faculty of Physics, Moscow State University, 119992 Moscow, Russia

Abstract. The angular distribution of the final electrons in the long range mechanism of the neutrinoless double beta decay is derived for the general Lorentz invariant effective Lagrangian. Possible theories beyond the SM are classified from their effects on the angular distribution.

1 Introduction

Neutrinoless double beta decay ($0\nu2\beta$) is forbidden in the Standard Model (SM) by lepton number (LN) conservation, which is a consequence of the renormalizability of the SM. However, being the low energy limit of a more general theory, an extended version of the SM could contain nonrenormalizable terms (tiny to be compatible with experiments), in particular, terms, that violate LN and allow the $0\nu2\beta$ decay. Probable sources of LN violation are: Majorana neutrinos ν_M [1,2] (one of the main candidates after the observation of neutrino oscillations [3]), SUSY Majorana particles [4, 5] (gluinos, neutralinos), scalar bilinears (SBs) [6], e.g. leptoquarks (LQs) [7] or the doubly charged dilepton (the component ξ^{--} of the $SU(2)_L$ triplet Higgs scalar etc.). All these particles should be much lighter than the electron (light ν_Ms) or much heavier than the proton (heavy ν_Ms, SUSY particles, SBs) [3], giving rise to two possible classes of mechanisms for the $0\nu2\beta$-decay, which we call the long range and short range mechanism, respectively. For both the classes, the separation of the lepton physics from the nuclear physics takes place [8] that simplifies calculations. Note that according to the Schechter–Valle theorem [9] any mechanism of the $0\nu2\beta$-decay produces an effective Majorana mass for the neutrino, which must therefore contribute to this decay in any case. Despite the lack of confirmation for the claimed observation of the $0\nu2\beta$-decay [10], the restrictions on the decay half-life [11] make it possible to get bounds on the parameters of the models with LN violation. Our aim is to examine the possibility of the determination of the decay mechanism from the angular correlation. The experimental facilities that can measure the electron angular distribution in the $0\nu2\beta$-decay are NEMO3 [12], ELEGANT V and others [13].

2 Angular distribution for the long range mechanism of $0\nu2\beta$ decay

For the decay mediated by light ν_Ms, the most general effective Lagrangian is the Lorentz invariant combination of the leptonic j_α and the hadronic J_α

[a] e-mail: ahmed.ali@desy.de
[b] e-mail: borisov@phys.msu.ru
[c] e-mail: jouridov@mail.ru

currents of definite tensor structure and helicity [14]

$$\mathcal{L} = \frac{G_F}{\sqrt{2}}[(U_{ei} + \epsilon^{V-A}_{V-A,i})j^{\mu i}_{V-A}J^{+}_{V-A,\mu} + \sum_{\alpha,\beta}{}' \epsilon^{\beta}_{\alpha i} j^{i}_{\beta} J^{+}_{\alpha} + \text{H.c.}]. \quad (1)$$

Here and thereafter, a summation over the repeated indices is assumed; $\alpha, \beta = V \mp A, S \mp P, T_{L,R}$ ($T_\rho = 2\sigma^{\mu\nu}P_\rho$, $\sigma^{\mu\nu} = \frac{i}{2}[\gamma^\mu, \gamma^\nu]$, P_ρ is the projector, $\rho = L, R$); the prime indicates the summation over all Lorentz invariant contributions, except for $\alpha = \beta = V - A$; the leptonic currents contain neutrino mass eigenstates and the index i runs over the light eigenstates. Parameters $\epsilon^{\beta}_{\alpha i}$ encode new physics, i.e., deviations of the Lagrangian from the standard $V - A$ current-current form and mixing of the non-SM neutrinos.

The coefficients $\epsilon^{\beta}_{\alpha i}$ are connected with the parameters of different SM extensions. In Ref. [2] only non-standard terms with $\epsilon^{V-A}_{V+A,i}$, $\epsilon^{V+A}_{V\mp A,i}$ have been considered. In the R-parity-violating (RPV) SUSY accompanying the neutrino exchange mechanism [4], the nonzero parameters are $\epsilon^{V-A}_{V-A,i}$, $\epsilon^{S\mp P}_{S+P,i}$ and $\epsilon^{T_R}_{T_R,i}$. For the mechanism with leptoquarks [7] $\epsilon^{S+P}_{S\mp P,i}$, $\epsilon^{V+A}_{V\mp A,i}$ are nonzero. The bounds on $\epsilon^{V+A}_{V\pm A}$, $\epsilon^{S+P}_{S\pm P}$, $\epsilon^{T_R}_{T_{R,L}}$ ($\epsilon^{\beta}_{\alpha} = U_{ei}\epsilon^{\beta}_{\alpha i}$) from the Heidelberg–Moscow experiment are presented in [14, 15].

We have calculated only the leading contribution of the parameters $\epsilon^{\beta}_{\alpha}$ to the decay width using the approximation of relativistic electrons and non-relativistic nucleons. Following Ref. [1] we describe the outgoing electrons by plane waves taking into account the effect of the nuclear Coulomb field by the Fermi factors $F(\varepsilon_s)$ [1, 2] with the s-th electron energy ε_s. The non-relativistic structure of the nuclear currents in the impulse approximation is taken from Ref. [16]. We neglect the nucleon recoil terms and the dipole dependence of the nuclear form factors on the momentum transfer. Considering only one nonzero parameter in the Lagrangian at a time, we obtain the differential width in the angle θ between the electron momenta for the $0^+ \to 0^+$-transitions

$$\frac{d\Gamma}{d\cos\theta} = C|M_{GT}|^2 I[(a+b)(1 - k\cos\theta)],$$
$$I[x] = \int d\varepsilon_1 \varepsilon_1^2 \varepsilon_2^2 F(\varepsilon_1)F(\varepsilon_2)x, \quad (2)$$

where $\varepsilon_2 = \Delta - \varepsilon_1$ and Δ is the energy release of the process. The constant

$$C = \frac{G_F^4 g_A^4 m_e^2}{64\pi^5 R_0^2}$$

contains the electron mass m_e and the nuclear radius R_0 for the a and b functions and the neutrino potentials to be dimensionless. The Gamow–Teller nuclear matrix element M_{GT} contains the neutrino potential $h(r, \omega) = R_0\phi_0/r$ with $\phi_0 = e^{i\omega r}$, where r is the distance between nucleons and ω is the neutrino average energy. The angular correlation coefficient in (2) is

$$k = \frac{a - b}{a + b}, \quad |k| \le 1. \quad (3)$$

The expressions for a and b for different choices of ϵ_α^β are shown in Table 1.

Table 1: Expressions for a and b in (2) and (3) for the stated choice of ϵ_α^β.

ϵ	a	b
ϵ_{V-A}^{V-A}	$\lvert[\langle m\rangle/m_e + 2\sum_i U_{ei}\epsilon_{V-A,i}^{V-A}(m_i/m_e)](1-\chi_F)\rvert^2$	0
ϵ_{V+A}^{V-A}	$\lvert(1-\chi_F)\langle m\rangle/m_e - 2\sum_i U_{ei}\epsilon_{V+A,i}^{V-A}(m_i/m_e)(1+\chi_F)\rvert^2$	0
$\epsilon_{S\mp P}^{S+P}$	$a_0 + \frac{1}{9}(\Delta/m_e)^2(F_S^{(3)}/g_V)^2\lvert\epsilon_{S\mp P}^{S+P}\chi_F'\rvert^2$	0
$\epsilon_{T_L}^{T_R}$	$a_0 + \frac{16}{81}(T_1^{(3)}/g_A)^2(\Delta/m_e)^2\lvert\epsilon_{T_L}^{T_R}(\chi_{GT}' + 3\chi_T')\rvert^2$	0
$\epsilon_{T_L}^{T_L}$, $\epsilon_{T_R}^{T_R}$	a_0	0
ϵ	b	a
ϵ_{V-A}^{V+A}	$\frac{1}{2}\lvert\epsilon_{V-A}^{V+A}\rvert^2[(\varepsilon_{12}/m_e)^2\lvert\chi_{2+}\rvert^2 + \frac{4}{9}(\Delta/m_e)^2\lvert\chi_P'\rvert^2]$	a_0
ϵ_{V+A}^{V+A}	$\frac{1}{2}(\varepsilon_{12}/m_e)^2\lvert\epsilon_{V+A}^{V+A}\chi_{2-}\rvert^2$	a_0
$\epsilon_{S\mp P}^{S-P}$	$2(F_S^{(3)}/g_V)^2\lvert\sum_i U_{ei}\epsilon_{S\mp P,i}^{S-P}(m_i/m_e)\chi_F\rvert^2$	a_0
$\epsilon_{T_R}^{T_L}$	$32(T_1^{(3)}/g_A)^2\lvert\sum_i U_{ei}\epsilon_{T_R,i}^{T_L}(m_i/m_e)\rvert^2$	a_0

Here $\varepsilon_{12} = \varepsilon_1 - \varepsilon_2$, $\langle m\rangle = \sum_i U_{ei}^2 m_i$ is the effective Majorana mass , $F_S^{(3)}$ and $T_1^{(3)}$ are the renormalization constants [17]. Some notations similar to ones of Ref. [2] are used:

$$\chi_{2\pm} = \chi_{GT\omega} \pm \chi_{F\omega} + \frac{1}{9}\chi_{1\mp}, \quad \chi_{1\pm} = \chi_{GT}' - 6\chi_T' \pm 3\chi_F',$$

$$\chi_F = \left(\frac{g_V}{g_A}\right)^2 \frac{M_F}{M_{GT}}, \quad \chi_P = \left(\frac{g_V}{g_A}\right)\frac{M_P}{M_{GT}}, \quad \chi_X = \frac{M_X}{M_{GT}}, \quad X = T, GT,$$

with Fermi M_F, pseudoscalar M_P and tensor M_T nuclear matrix elements. The prime and the index ω show that the matrix element in the numerator instead of h contains the potential $h' = h + \omega R_0 h_1$ or $h_\omega = h - \omega R_0 h_1$ correspondingly, with $h_1 = -d\phi_0/d(\omega r)$. The a for all zero ϵ_α^β is $a_0 = \lvert(1-\chi_F)\langle m\rangle/m_e\rvert^2$. For the $V \mp A$ part of (1) our result agrees with Ref. [2] for the relativistic electrons ($m_e/\Delta \to 0$) that weakly interact with the nucleus ($\alpha Z \to 0$).

If the effects of all interactions beyond the SM extended by the ν_Ms are zero (all $\epsilon_\alpha^\beta = 0$), than $k = 1$ and the distribution (2) is proportional to $1 - \cos\theta$; $k \neq 1$ only for $b \neq 0$ irrespective of the value of a. Therefore the "non-standard" effects of the *first* set of parameters, $\epsilon_{V\mp A}^{V-A}$, $\epsilon_{S\mp P}^{S+P}$, $\epsilon_{T_L}^{T_R}$, do not change the form of the distribution, but the effects of the *second* set (see lower part of Table 1), $\epsilon_{V\mp A}^{V+A}$, $\epsilon_{S\mp P}^{S-P}$, $\epsilon_{T_R}^{T_L}$, do change this distribution, again varying the parameters ϵ_β^α one at a time. The effects of $\epsilon_{T_L}^{T_L}$ and $\epsilon_{T_R}^{T_R}$ are zero. The coefficient k and the set $\{\epsilon\}$ of nonzero ϵ_α^βs that change the $1 - \cos\theta$ form of the distribution for the SM are given in Table 2. They correspond to the following extensions of the SM: only ν_M, ν_M plus SUSY [4], ν_M plus right-handed currents (RC) (connected with right-handed W bosons [2] or LQs [7]).

Table 2: The angular correlation coefficient k for various SM extensions.

SM extension	$\{\epsilon\}$	k
ν_M	-	1
ν_M + SUSY	ϵ_{S-P}^{S+P}	$-1 < k < 1$
ν_M+RC	$\epsilon_{V\mp A}^{V+A}$	$-1 < k < 1$

For a complete understanding of the influence of the decay mechanism on the angular distribution, it is necessary to take into account the nucleon recoil terms, the effect of the nuclear Coulomb field on the electrons in the terms with $\epsilon_{S\mp P}^{S\mp P}$ and $\epsilon_{T_{L,R}}^{T_{L,R}}$, and the short range mechanism of the $0\nu2\beta$-decay.

Acknowledgments

We thank Alexander Barabash for the helpful discussion.

References

[1] M.G. Shchepkin, *Sov. Phys. Usp.* **27**, 555 (1984).
[2] M. Doi, T. Kotani, E. Takasugi, *Prog. Theor. Phys. Suppl.* **83**, 1 (1985).
[3] Particle Data Group: S. Eidelman et al., *Phys. Lett.* **B592**, 1 (2004).
[4] M. Hirsch, H.V. Klapdor-Kleingrothaus, S.G. Kovalenko, *Phys. Lett.* **B372**, 181 (1996); **B381**, 488 (Erratum) (1996); H. Päs, M. Hirsch, H.V. Klapdor-Kleingrothaus, *Phys.Lett.* **B459**, 450 (1999).
[5] M. Hirsch, H.V. Klapdor-Kleingrothaus, S.G. Kovalenko, *Phys. Rev. Lett.* **75**, 17 (1995); *Phys. Lett.* **B352**, 1 (1995); *Phys. Lett.* **B403**, 291 (1997); *Nucl. Phys. Proc. Suppl.* **A52**, 257 (1997); *Phys. Rev.* **D57**, 1947 (1998).
[6] H.V. Klapdor-Kleingrothaus, U. Sarkar, *Phys. Lett.* **B554**, 45 (2003).
[7] M. Hirsch, H.V. Klapdor-Kleingrothaus, S.G. Kovalenko, *Phys. Rev.* **D54**, 4207 (1996).
[8] J.D. Vergados, hep-ph/0409319.
[9] J. Schechter, J.W. Valle, *Phys. Rev.* **D25**, 2951 (1982); E. Takasugi, *Phys. Lett.* **B149**, 372 (1984); J.F. Nieves, *Phys. Lett.* **B147**, 375 (1984).
[10] H.V. Klapdor-Kleingrothaus, A. Dietz, H.V. Harney, I.V. Krivosheina, *Mod. Phys. Lett.* **16**, 2409 (2001).
[11] A.S. Barabash, *Phys. At. Nucl.* **67**, 438 (2004); hep-ex/0602037.
[12] NEMO3 Collab.: R. Arnold et al., *JETP Lett.* **80**, 377 (2004).
[13] H. Ejiri et al., *Phys. Rev.* **C63**, 065501 (2001); DCBA Collab.: N. Ishihara et al., *Nucl. Phys.* **B** *(Proc. Suppl.)* **111**, 309 (2002); N. Ishihara, in *Proceedings of the 8th Accelerator and Particle Physics Institute (APPI2003), Appi, Iwate, Japan, 25–28 Feb. 2003*, Ed. by Y. Fujii, KEK Proceedings 2003–6, p. 192.
[14] H. Päs, M. Hirsch, H.V. Klapdor-Kleingrothaus, S.G. Kovalenko, *Phys. Lett.* **B453**, 194 (1999).
[15] H.V. Klapdor-Kleingrothaus, H. Päs, hep-ph/0002109.
[16] T. Tomoda, *Rep. Prog. Phys.* **54**, 53 (1991).
[17] S.L. Adler et al., *Phys. Rev.* **D11**, 3309 (1975).

SEARCH FOR PHENOMENA OUTSIDE THE STANDARD MODEL WITH PROTOTYPE OF THE BOREXINO DETECTOR

A.Derbin [a], O.Smirnov [b]

Abstract. Counting Test Facility (CTF) is a prototype of the Borexino detector built by collaboration to test the possibility of the deep purification of the liquid scintillator (LS) and shielding water from the natural radioactive isotops. After achieving the record purification of liquid scintillator on the level of 10^{-16} g/g of U/Th, the CTF became the world's biggest (about 4 tones of LS) ultrapure detector. The data collected with an upgraded version of the CTF were analyzed by Borexino collaboration in order to search for a number of possible manifestations of the non-standard physics. Here we present the main results on the rare physics obtained recently using the CTF data.

1 Introduction

The Standard Model (SM) agrees with experimental data, but it has a number of intrinsic problems which makes it theoretically unsatisfactory. It contains at least 19 arbitrary parameters, and needs even more parameters to incorporate recently discovered neutrino oscillations. Big number of free parameters of SM, absence of experimental evidence of the Higgs particles, the nonclear situation with the cold dark matter origin,- all these facts are challenging experimenters to search for the fenomena outside the SM frames. The search for rare processes in the low energy region, usually named "non-accelerating physics", is a method for reaching, maybe in an indirect way, the energies at which unification of particles and fields occurs, and which are non-reachable for the modern and future particle accelerators. Below we give a brief review of the main results on the rare physics obtained recently using the CTF data.

2 CTF detector

The CTF is a 4-ton unsegmented liquid scintillator (LS) detector located deep undeground (1,400 m of rock equivalent to ~3,800 m.w.e.) at the Gran Sasso laboratory, Italy. The active detector (formed by 4 tones of ultra pure LS contained inside a transparent, 2 m diameter, spherical nylon vessel, 500 μm thick) is placed inside a cylindrical tank, 10m in diameter and 11m height, filled with 1000 m^3 of high-purity water providing 4.5 meters water layer for shielding of active detector against neutrons from rocks and gamma rays from the photomultipliers. The scintillator is carefully purified to ensure the ^{238}U and ^{232}Th in it are less than some units 10^{-16} g/g. The active volume is viewed by 100

[a] e-mail: derbin@mail.pnpi.spb.ru;St.Petersburg Nuclear Physics Institute RAS, Gatchina, Russia

[b] e-mail: osmirnov@jinr.ru; Joint Institute for Nuclear Research, 141980, Dubna, Moscow region, Russia

photomultipliers mounted on the open stainless steel structure, providing 21% geometrical coverage. The PMT's are 8 inch ETL 9351 tubes made of low radioactivity glass and characterized by high quantum efficiency (26% at 420 nm). In addition, 16 PMTs are mounted on the bottom of the tank looking upward, forming active muon veto system. A detailed description of the CTF can be found in [1]. The CTF has been in operation since 1993. During the first campaign (CTF1) from 1993 to 1995 the detector was filled with the scintillator on the base of pseudocumene or PC (1,2,4-Trimetylbenzene, $C_6H_3(CH_3)_3$, $\rho = 0.88$ g/cm^3) with 1.5 g/l of PP0 as fluor (2,5-Diphenyloxazole, $C_{15}H_{11}NO$). In this period the data analysis was mainly focused on background studies. In 1999 the CTF was filled with PXE (1-Phenyl-1-xylylethane, $C_{16}H_{18}$, $\rho = 0.995$ g/cm^3) and equipped with an active muon-veto. Moreover, an outer nylon vessel designed to reduce Rn diffusion from the photomultiplers and the water toward the scintillator was added [2]. All the construction materials were carefully selected from the point of view of their natural radioactivity.

Most of the physical results reported here are obtained with this modification of CTF (CTF2). The CTF2 has good spatial (12 cm at 1 MeV) and energy resolution (6% at 1 MeV). The detector has a very low threshold, corresponding to ~140 keV with 99% efficiency, and very low background. Total counting rate is about 1 cps (mainly due to ^{14}C), the counting rate in the " neutrino window" (250-800 keV) is as low as 0.06 keV^{-1}kg^{-1}y^{-1}.

In 2002 a third campaign of the data taking with PC as a LS started. This last campaign is still in progress and aims to define the purification strategy for the scintillator to be used during the filling of Borexino [3]

3 CTF results

The expected response function of the CTF detector to the considered rare process was simulated with Monte Carlo method using EGS4 code. The energy calibration were performed with the end-point energy of β-spectrum of ^{14}C isotope contained in LS, and with 1.45 MeV γ from ^{40}K decay.

3.1 Neutrino magnetic moment and Solar antineutrino flux.

The interest in the huge (~$10^{-11}\mu_B$) for the SM neutrino magnetic moment was inspired by the solar neutrino problem. The model of spin-flavour precession has had better agreement with experimental data than the oscillation solution, favoured by the KamLand experiment. At the same time, the oscillation (LMA MSW) solution of the problem does not exclude the possibility of observing weaker effects caused by the neutrino magnetic moment. Anomalously big electromagnetic neutrino moments should lead to the neutrino scattering on the electron and to the radiative neutrino decay $\nu_H \rightarrow \nu_L + \gamma$. The lower limit on the mean lifetime of the *pp*– and ^{7}Be- neutrino relative radiative decay obtained from the CTF data is $\tau_{c.m.}(\nu_H \rightarrow \nu_L + +\gamma)/m_\nu \geq 4.2 \cdot 10^3\ s \cdot eV^{-1}$ ($\alpha = 0$) [4]. It is more than one order of magnitude stronger than that obtained in the previous experiments using nuclear reactors and accelerators. A new upper limit on the neutrino magnetic moment from the CTF data for the $pp-$ and ^{7}Be- solar neutrinos is $\mu_\nu \leq 5.5 \times 10^{-10}\ \mu_B$ (90% CL) in the assumption of SSM neutrino fluxes [5]. This value is only 3 times weaker than the ones obtained using reactor neutrinos and ^{8}B solar neutrinos.

Another possible manifestation of a large neutrino magnetic moment would be the presence of the antineutrino in the Solar neutrino spectrum. The 2 t year of the exposion of the CTF has been used to limit the antineutrino flux in the low energy region starting at the inverse beta-decay threshold $E > 1.81$ MeV. After the rejection of the muon induced background no candidate events have been observed at practically zero expected background. This fact limits the antineutrinos production at the Sun by no more than 5% of the Standard Solar Model ^{8}B flux [6]. The limit, being weaker than that of other experiments, is the first one valid in the low energy region.

3.2 Electric and barion charge conservation.

Many extensions of the SM include interactions violating the barion and lepton charge B and L, and are predicting processes with $\triangle B = 1,2$ and $\triangle(B-L) = 0$, 2 resulting in the decay of protons and neutrons in nuclei. The main direction of the searches is the attempts to observe the nucleon decay into strongly interacting or charged particles. At the same time, for the processes with nucleon disappearance or decay into weakly interacting particles (neutrinos, majorons etc.) the experimental limits are 5-6 orders of magnitude lower. A check on the electric and baryonic charge conservation can be performed by looking for the electron decay into the channel $e \rightarrow \nu + \gamma$, and by looking for the decay of nucleons and dinucleons into the "invisible" channel $N \rightarrow 3\nu$, $NN \rightarrow 2\nu$. A new lower limit on the mean lifetime for the decay e$\rightarrow\gamma + \nu$ established using CTF data is $\tau \geq 4.6 \times 10^{26}$ y (90% CL) [7]. New limits on the N and NN decays into the invisible channels (disappearance, decays to neutrinos, majorons, etc.) established using the CTF data for the nn and pp decays ($\tau(nn \rightarrow inv) > 4.9 \cdot 10^{25}$ y, $\tau(pp \rightarrow inv) > 5.0 \cdot 10^{25}$ y with 90% CL) are the best up-to-date limits set by any method, including radiochemical and geochemical experiments [8].

3.3 Pauli exclusion principle.

Experiments on the PEP test are either looking for the anomalous atoms or nuclei in a non-paulian state, or searching for the emission of radiation following the transitions violating the Pauli principle in nuclei, or searching for the non-Paulian β^-- and β^+-decays. The PEP was tested for nucleons (n, p) in ^{12}C and

^{16}O nuclei, using the results of background measurements with the CTF. The approach consisted of a search for γ, n, p and/or α's emitted in a non-Paulian transition of $1P$- shell nucleons to the filled $1S_{1/2}$ shell in nuclei. Similarly, the Pauli-forbidden $\beta^{\pm}$ decay processes were searched for. Due to the extremely low background and the large mass (4.2 tons) of the CTF detector, the following most stringent up-to-date experimental bounds on PEP violating transitions of nucleons were established: $\tau(^{12}C \rightarrow ^{12}\widetilde{C} + \gamma) > 2.1 \cdot 10^{27}$ y, $\tau(^{12}C \rightarrow ^{11}\widetilde{B} + p) > 5.0 \cdot 10^{26}$ y, $\tau(^{12}C(^{16}O) \rightarrow ^{11}\widetilde{C}(^{15}\widetilde{O}) + n) > 3.7 \cdot 10^{26}$ y, $\tau(^{12}C \rightarrow ^{8}\widetilde{Be} + \alpha) > 6.1 \cdot 10^{23}$ y, $\tau(^{12}C \rightarrow ^{12}\widetilde{N} + e^- + \widetilde{\nu_e}) > 7.6 \cdot 10^{27}$ y and $\tau(^{12}C \rightarrow ^{12}\widetilde{B} + e^+ + \nu_e) > 7.7 \cdot 10^{27}$ y, all at 90% CL [9].

3.4 Heavy neutrino.

Before the discovery of the neutrino oscillations with the mixing parameters $\Delta m_{12} \approx 6{\cdot}10^{-5}$ eV and $\Delta m_{23} \approx 3{\cdot}10^{-3}$ eV, the search for the mass eigenstates of m_2 and m_3 in the mass range from a few keV up to some MeV was an actual problem because the upper limits on the masses of ν_μ and ν_τ, obtained in the direct experiments are 170 keV and 18 MeV respectively. At present, the possibility of the existence of new mass eigenstates is supported by the models with a sterile neutrino. The last one in the common case can have an arbitrary mass and can be mixed with 3 active neutrinos.

If heavy neutrinos with mass $m_{\nu_H} \geq 2m_e$ are emitted in the decays of ^{8}B in the Sun, then decays $\nu_H \rightarrow \nu_L + e^+ + e^-$ should be observed. The search of the new mass eigenstates of neutrino was performed using the CTF data looking for the neutrino decay with a mass bigger than $2 \cdot m_e$. The obtained limits on the mixing parameter $|U_{eH}|^2 \leq 10^{-3}$–10^{-5} in the mass range 1.1 MeV-12 MeV [10] are stronger than ones obtained in previous experiments using nuclear reactors and accelerators.

4 Conclusions

Using the unique features of the Borexino Counting Test Facility- the extremely low background and low energy threshold- a number of new physical results have been obtained, despite of the comparatevely small detectors's volume. These results demonstrate the sensitivity level of the modern low-background low-treshold liquid scintillator detector to the rare processes. In a course of some years we expect to improve significantly the presented limits with the data of the full-scale 300 tones Borexino detector [3], approaching the limits of great theoretical importance for some of the cited processes.

[1] G.Alimonti et al., *Nucl. Instrum. Meth.* **A** 406, 411 (1998); G. Alimonti et al., Astropart. Phys. 8, 141 (1998); G.Alimonti et al., Phys. Lett. B 422, 349 (1998).

[2] Borexino coll., H.O. Back et al., arXiv:physics/0408032.
[3] G.Alimonti et al., *Astropart. Phys.* 16, 205 (2002).
[4] A.V.Derbin, O.Yu.Smirnov, *JETP Lett.* 76, 483 (2002); A.V.Derbin, V.N.Muratova, O.Yu.Smirnov, O.A.Zaimidoroga, preprint PNPI-2003, No.2504, (2003).
[5] H.O.Back et al., *Phys.Lett.* **B** 563, 35 (2003).
[6] A. Derbin, A. Ianni, O. Smirnov. *Nucl. Phys.* **B** (Proc.Suppl.), 143 (2005).
[7] H.O.Back et al., *Phys.Lett.* **B** 525, 29(2002).
[8] H.O.Back et al., *Phys.Lett.* **B** 563, 23 (2003).
[9] H.O.Back et al., *European Physics Journal* **C** 37, 421 (2004).
[10] H.O.Back et al., *JETP Lett.* 78, 261 (2003).

CROSS-SECTION MEASUREMENTS IN THE NOMAD EXPERIMENT

R.Petti [a] [b]
CERN, CH-1211 Genéve 23, Switzerland
University of South Carolina, Columbia SC 29208, USA

Abstract. The NOMAD experiment collected valuable neutrino data samples, matching both the large statistics of massive calorimeters and the reconstruction quality of bubble chambers. This paper describes the recent measurements of neutrino cross-sections on carbon target. The approach followed for cross-section modeling is also explained.

1 Introduction

The NOMAD experiment was designed to search for ν_τ appearance from neutrino oscillations in the CERN wide-band neutrino beam produced by the 450 GeV proton synchrotron. The single-particle reconstruction and lepton identification capability of the NOMAD detector allowed the search for ν_τ appearance in most of the leptonic and hadronic τ decay channels [1] and also to look for $\nu_\mu \to \nu_e$ oscillations [2]. No evidence for oscillations was found.

The high resolution and reconstruction quality makes the NOMAD data samples a valuable resource for the neutrino physics community. The recent measurements of cross-sections and particle production would help to clarify our understanding of neutrino interactions with nuclei at intermediate energies. In addition, it must be noted the precise tracking in a light target provides a powerful tool to tune and validate Monte Carlo simulation programs for future experiments. This analysis activity can benefit from the beam and detector studies performed for the oscillation searches.

2 Detector and data samples

The NOMAD detector is described in detail in Ref. [3]. Inside a 0.4 T magnetic field there is an active target consisting of drift chambers (DC) [4] with a fiducial mass of about 2.7 tons and a low average density (0.1 g/cm^3). The main target, 405 cm long and corresponding to about one radiation length, is followed by a transition radiation detector (TRD) [5] for electron identification, a preshower detector (PRS), and a high resolution lead-glass electromagnetic calorimeter (ECAL) [6]. A hadron calorimeter (HCAL) and two stations of drift chambers for muon detection are located just after the downstream part of the magnet coil. An iron-scintillator sampling calorimeter with a fiducial mass of about $17t$ (FCAL) is located upstream of the central part of the NOMAD target. The detector is designed to identify leptons and to measure muons, pions, electrons and photons with comparable resolutions. Momenta are measured in the DC with a resolution:

$$\frac{\sigma_p}{p} \simeq \frac{0.05}{\sqrt{L[m]}} \oplus \frac{0.008 \times p[GeV/c]}{\sqrt{L[m]^5}}$$

where L is the track length and p is the momentum. The energy of electromagnetic showers, E, is measured in the ECAL with a resolution:

$$\frac{\sigma_E}{E} = 0.01 \oplus \frac{0.032}{\sqrt{E[GeV]}}.$$

[a] e-mail: Roberto.Petti@cern.ch
[b] On behalf of the NOMAD collaboration

The relative composition of Charged Current (CC) events in NOMAD is estimated [7] to be ν_μ,: $\bar{\nu}_\mu$,: ν_e,: $\bar{\nu}_e = 1.00$: 0.0227: 0.0154: 0.0016, with average neutrino energies of 45.4, 40.8, 57.5, and 51.5 GeV, respectively. Neutrinos are produced at an average distance of 625 m from the detector.

The NOMAD experiment collected data from 1995 to 1998. Most of the running, for a total exposure of 5.1×10^{19} protons on target (pot), was in neutrino mode. This resulted in three distinct data samples, according to the different targets: $1.3 \times 10^6 \nu_\mu$ CC interactions from the drift chambers (mainly carbon), $1.5 \times 10^6 \nu_\mu$ CC interactions from the region of the magnet coil (mainly aluminium) located in front of the DC and $1.2 \times 10^7 \nu_\mu$ CC interactions from FCAL (iron).

3 Modeling of inelastic cross-sections

3.1 Structure functions at high Q^2

At large momentum transfer structure functions are described as series in Q^{-2} on the basis of the operator product expansion of the correlator of the weak current (twist expansion):

$$F_i(x,Q^2) \quad = \quad F_i^{\rm TMC}(x,Q^2) + \frac{H_i^{(4)}(x)}{Q^2} + \frac{H_i^{(6)}(x)}{Q^4} + \cdots,$$

where $i = T, 2, 3$ refers to the type of the structure function, $F_i^{\rm TMC}$ are the leading twist (LT) terms corrected for the target mass effects and $H_i^{(t)}$ are the higher twist (HT) terms of twist t. The target mass corrections are computed using the approach of Ref. [8] (see also [11] for the treatment of the thereshold problem at $x \to 1$). The calculation of the leading twist is performed in the NNLO approximation and we include additional phenomenological terms up to twist-6. We use PDFs and HT based on Ref. [9], obtained from dedicated fits optimized at low Q^2 [10] and including additional data from (anti)neutrinos CC (NOMAD, CHORUS [17] and NuTeV [16]) and charged lepton (JLab) Deep Inelastic Scattering (DIS), as well as from Drell-Yan production (E605 and E886).

3.2 Low Q^2 structure functions

In the low-Q region (anti)neutrino cross sections are dominated by the longitudinal structure function F_L and the latter is driven by the axial-current interactions. The structure function F_T vanishes as Q^2 at low Q^2. This behaviour is similar to the charged lepton case and holds for both the vector and the axial-vector contributions. However, in the longitudinal channel the low-Q behavior of the vector and axial-vector parts are different.

The conservation of the vector current (CVC) suggests $q_\mu W_{\mu\nu} = 0$ for the vector current part of hadronic tensor. From this condition we conclude $F_L^{\rm VC}$ vanishes faster than $F_T^{\rm VC}$ at low Q^2 and $F_L^{\rm VC}/F_T^{\rm VC} \sim Q^2$. This behavior is similar to the charged-lepton case.

In contrast to the vector current, the axial-vector current is not conserved since the divergence of the axial-vector current is proportional to the pion field (Partially Conserved Axial Current or PCAC). We introduce explicitely a PCAC contribution to F_L:

$$F_L^{\rm AC} = F_L^{\rm PCAC} + \widetilde{F}_L^{\rm A}$$

where $F_L^{\rm PCAC}$ is finite for $Q^2 = 0$ and the last term $\widetilde{F}_L^{\rm A}$ is similar to $F_L^{\rm VC}$ and vanishes as Q^4 [12]. The structure functions F_T and $\widetilde{F}_L = F_L^{\rm VC} + \widetilde{F}_L^{\rm A}$, which are vanishing for $Q^2 \to 0$ like in the charged lepton case, are parameterized as smooth interpolations between the high Q^2 regime calculated as described in Section 3.1 and the $Q^2 \to 0$ predictions derived from current conservation arguments [10]. We choose the value $Q_0^2 = 1\ GeV^2$ as matching point for the twist expansion. In the region $0 < Q^2 < 1\ GeV^2$ we use cubic splines calculated for fixed x values. The corresponding coefficients are fully determined by the condition both functions and derivatives match with the twist expansion at Q_0^2.

3.3 Nuclear and electroweak corrections

A detailed calculation of nuclear corrections to structure functions is performed [11] [13]. The model takes into account a number of different effects including nuclear shadowing, Fermi motion and binding, nuclear pion excess and off-shell correction to bound nucleon structure functions. The off-shell effect and the effective scattering amplitude describing nuclear shadowing are treated phenomenologically and are extracted from statistical analysis of data from charged lepton scattering on a wide range of nuclear targets. The main differences with respect to (anti)neutrino scattering are related to the axial-vector current on coherent nuclear effects and are more evident at low Q^2 [13].

One-loop electroweak effects are taken into account [14] as corrections to the parton distributions used in the structure function calculation. The initial quark mass singularities from QED corrections are subtracted within the $\overline{\rm MS}$ scheme since they are effectively incorporated in the quark density functions. Results are cross-checked with a second independent calculation [15].

4 Measurement of inelastic cross-section

The present knowledge of neutrino cross-sections is rather nonuniform. In the region $E_\nu > 30$ GeV, where data from the large massive calorimeters (CCFR, NuTeV) are available, the uncertainty is about 2%. This increases to about 20% at lower energies, due to the limited statistics of bubble chamber experiments. Measurements of both the total $\sigma_{CC}^{\rm tot}$ and the differential $d\sigma_{CC}^2/dxdy$ cross-sections for ν_μ on carbon are performed in NOMAD. One of the primary goals is to constrain the (anti)neutrino cross-section model and to reduce the corresponding systematic uncertainties. This also provides the first measurement of inclusive cross-section on carbon target, with $\langle Q^2 \rangle \sim 13$ GeV^2.

The absolute normalization is obtained from the world average cross-section on isoscalar target in the energy range $40 < E_\nu < 300$ GeV. The measurement is performed in the kinematic region $\nu > 3$ GeV and $Q^2 > 1$ GeV^2. The analysis of the lower Q^2 region down to about 0.3 GeV^2 is currently being finalized.

A comparison of the measured differential cross-section on carbon with the model predictions (Sec. 3) is illustrated in Figure 1 for $E = 85$ GeV, indicating a good agreement in the entire kinematic region.

The NOMAD results are also consistent with the measurements performed by CHORUS and NuTeV on different targets [13]. Only in the recent NuTeV data a slight excess is observed for $x > 0.5$ [13]. The low Q^2 data from CHORUS support our treatment of the PCAC contribution to neutrino structure functions.

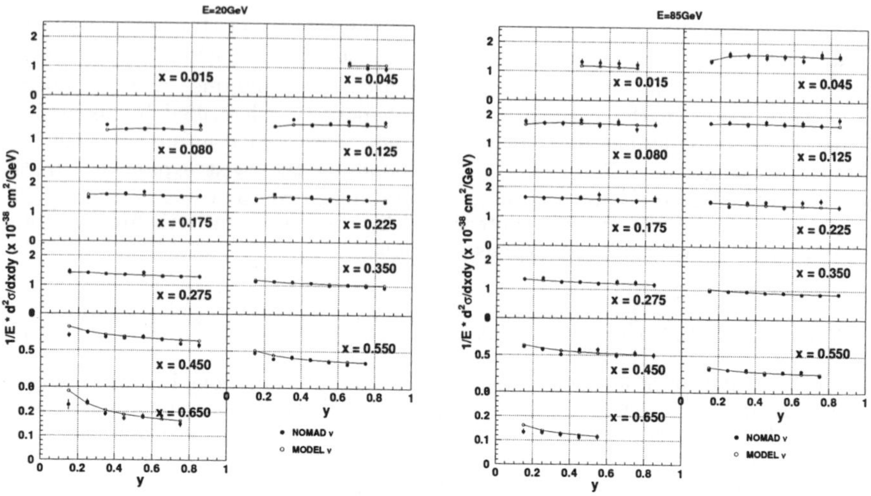

Figure 1: Differential neutrino CC cross-section on carbon at $E = 20$ GeV (left) $E = 85$ GeV (right). The full circles show the NOMAD measurement while the curve with open circles is the prediction from our cross-section model.

5 Measurement of quasi-elastic cross-section

The reconstruction and identification of the recoiling proton track allows a measurement of the quasi-elastic cross-section $\nu_\mu n \rightarrow \mu^- p$ on carbon in NOMAD. A kinematic analysis based upon a three-dimensional likelihood function is used to reject backgrounds from DIS and resonance production. Overall, about 8000 two track events are selected for the cross-section determination in the energy range $3 < E_\nu < 100$ GeV. A complementary sample of about 16000 single track events is used as a cross-check of the reconstruction efficiency.

In order to reduce systematic uncertainties, NOMAD is measuring the ratio of quasi-elastic cross-section with respect to DIS ($W^2 > 4$ GeV^2) processes. The absolute normalization is provided also in this case by the world average cross-section on isoscalar target.

The simulation of the quasi-elastic events is based upon the formalism by Llewellyn-Smith [18] and the axial form factor is parameterized in the conventional dipole form. Nuclear effects and the Pauli suppression factor are implemented following a simple Fermi gas model. Final state interactions are described by the DPMJET package [19]. The main systematic uncertainties are related to the understanding of nuclear corrections and are currently being finalized.

Acknowledgments

The author would like to thank the organizers of the conference for the invitation and for the pleasant stay.

References

[1] P. Astier et al. *Nucl. Phys.* **B** 611, 3 (2001).
[2] P. Astier et al. *Phys. Lett.* **B** 570, 19 (2003).
[3] J. Altegoer et al. *Nucl. Instr. and Meth.* **A** 404, 96 (1998).
[4] M. Anfreville et al. *Nucl. Instr. and Meth.* **A** 481, 339 (2002).
[5] G. Bassompierre et al. *Nucl. Instr. and Meth.* **A** 403, 363 (1998); G. Bassompierre et al. *Nucl. Instr. and Meth.* **A** 411, 63 (1998).
[6] D. Autiero et al. *Nucl. Instr. and Meth.* **A** 372, 556 (1996); D. Autiero et al. *Nucl. Instr. and Meth.* **A** 373, 358 (1996); D. Autiero et al. *Nucl. Instr. and Meth.* **A** 387, 352 (1997); D. Autiero et al. *Nucl. Instr. and Meth.* **A** 411, 285 (1998); D. Autiero et al. *Nucl. Instr. and Meth.* **A** 425, 188 (1999).
[7] P. Astier et al. *Nucl. Instr. and Meth.* **A** 515, 800 (2003).
[8] H. Georgi and H. D. Politzer, Phys. Rev. D **14**, 1829 (1976).
[9] S.I. Alekhin, Phys. Rev. D **68** (2003) 014002, [arXiv:hep-ph/0211096].
[10] S. Alekhin, S. Kulagin and R. Petti, paper in preparation.
[11] S. Kulagin and R. Petti, Nucl. Phys. A 765, (2006) 126, [arXiv:hep-ph/0412425].
[12] S. Kulagin and R. Petti, paper in preparation.
[13] S. Kulagin and R. Petti, arXiv:hep-ex/0602090.
[14] A. Arbuzov, D. Bardin and L. Kalinovskaya, arXiv:hep-ph/0407203.
[15] K. Diener, S. Dittmaier and W. Hollik, arXiv:hep-ph/0310364.
[16] M. Tzanov [NuTeV Collaboration], arXiv:hep-ex/0507040.
[17] G. Önengüt *et al.* [CHORUS Coll.], Phys. Lett. B. 632 (2006) 65.
[18] C.L. Llewellyn-Smith, Phys. Rep. 3, (1972) 261.
[19] J. Ranft, Phys. Rev. D51, (1995) 64; J. Ranft, arXiv:hep-ph/9911213.

SOLAR NEUTRINOS: SPIN FLAVOUR PRECESSION AND LMA

João M. Pulido [a], R. S. Raghavan [b], B. C. Chauhan [c]

Abstract. The time dependence that appears to be hinted by the data from the first 13 years of the solar neutrino Gallium experiments is viewed as resulting from a partial conversion of active neutrinos to light sterile ones through the resonant interaction between the magnetic moment of the neutrino and a varying solar field. A summary of the model and its predictions are presented for the forthcoming experiments Borexino and LENS.

1. After LMA has been ascertained as the dominant solution to the solar neutrino problem, the next step in solar neutrino experiments will be the search for time variations of the active flux. Such an effect remains much of an open question because so far most analyses have relied on time averaged solar neutrino data. Moreover, its investigation is of extreme importance because it may reveal the existence of electromagnetic properties of neutrinos and provide a significant addition to our understanding of solar dynamics.

Although the idea of a sizeable neutrino magnetic moment interacting with the solar field [1] lay dormant for some time, it was reintroduced in 1986 [2] to explain the claimed periodicity of the Chlorine event rate, but the effect remains inconclusive. In summary it works as follows: active neutrinos can be converted to sterile ones owing to the interaction of μ_ν with $B_\odot$. At times of intense solar activity

$$\textit{Strong } B_\odot \rightarrow \textit{large } \mu_\nu B_\odot \rightarrow \textit{large conversion} \tag{1}$$

with little or no conversion otherwise. Hence a neutrino flux anticorrelated to solar activity.

Period	1991-97	1998-03
SAGE & Ga-GNO	77.8 ± 5.0	63.3 ± 3.6
Ga-GNO only	77.5 ± 7.7	62.9 ± 6.0
no. of suspots	52	100

Table I - Event rates in solar neutrino units for Ga experiments in two different periods of solar activity and number of sunspots in the same periods.

In table I the neutrino event rate in Gallium experiments (SAGE and Gallex-GNO [3]) is shown as a function of time with the number of sunspots in the same period. It is seen from this table that there is a 2.4σ discrepancy in the combined results over the two periods. Since the sunspot number is correlated to solar activity, this is suggestive of an anticorrelation of the Ga event rate

[a]CFTP, Instituto Superior Técnico P-1049 Lisboa Portugal, e-mail: pulido@cftp.ist.utl.pt
[b]Virginia Pol. Inst. and State Univ. Blacksburg VA 24060, e-mail: raghavan@vt.edu
[c]CFTP, Instituto Superior Técnico P-1049 Lisboa, e-mail: chauhan@cftp.ist.utl.pt

with the 11-year solar cycle. No other experiments show such a variational effect, therefore since Ga are the only ones with a significant contribution of pp, ^7Be neutrinos ($\simeq$ 80%), the time dependence of these fluxes becomes an open possibility.

2. The bottom of the solar convective zone is a region where, owing to the relatively strong rotation gradient, the solar field is expected to have a maximum which can be as high as 300 kG and may be connected to the magnetic surface activity. The typical density profile of the sun requires the mass square difference associated to resonant conversion from active to nonactive neutrinos to be of the order of $10^{-8}eV^2$ [4], [5]. This excludes the solar and atmospheric mass square differences, Δm^2_{21} and Δm^2_{32}, and therefore conversion to $\bar{\nu}_\mu$ or $\bar{\nu}_\tau$. Since only three neutrino families are known to exist, we must introduce sterile neutrinos. We consider the simplest possible assumption whereby they mix with active ones through the magnetic moment only, so that in the vacuum

$$\begin{pmatrix} \nu_s \\ \nu_e \\ \nu_x \end{pmatrix} = \begin{pmatrix} 1 & 0 & 0 \\ 0 & c_\theta & s_\theta \\ 0 & -s_\theta & c_\theta \end{pmatrix} \begin{pmatrix} \nu_0 \\ \nu_1 \\ \nu_2 \end{pmatrix} \tag{2}$$

while in matter with a magnetic field

$$\mathcal{H}_{\rm M} = \begin{pmatrix} \frac{-\Delta m^2_{10}}{2E} & \mu_\nu B & 0 \\ \mu_\nu B & \frac{\Delta m^2_{21}}{2E} s^2_\theta + V_e & \frac{\Delta m^2_{21}}{4E} s_{2\theta} \\ 0 & \frac{\Delta m^2_{21}}{4E} s_{2\theta} & \frac{\Delta m^2_{21}}{2E} c^2_\theta + V_x \end{pmatrix} \tag{3}$$

with standard notation [d]. The 'new' mass square difference $\Delta m^2_{10} = m^2_1 - m^2_0$ fixes the location of the active $\rightarrow$ sterile transition for each neutrino energy.

For typical parameter values (peak field $B_0 = 2.2 \times 10^5\ G$, $\Delta m^2_{10} = -6.0 \times 10^{-9}eV^2$, $E = 0.33MeV$) the eigenvalue evolution within the sun is shown in fig.1. All pp neutrinos resonate around the bottom of the convective zone in this case. It is seen that the strongly adiabatic LMA transition, whose location is determined for each energy by Δm^2_{21}, takes place in the solar core. The active $\rightarrow$ sterile transition, whose adiabaticity depends on the magnetic field strength and its extension, takes place around the upper radiative/lower convective zone (i. e. the tachocline). The large order of magnitude difference between Δm^2_{21} $(7 - 8 \times 10^{-5}eV^2)$ and Δm^2_{10} $(-6 \times 10^{-9}eV^2)$ ensures the two resonances to be located far apart so they do not interfere. In fig.1 the vanishing field case is denoted by the two dotted lines crossing at the place where λ_1 and λ_0 nearly meet (critical density).

The two field profiles chosen are peaked at the bottom of the convective zone and can be seen in fig.1 of ref. [5]. If these are time dependent (possibly

[d] For details see ref. [4].

connected to solar activity), the effect provides a modulation of the pp neutrino flux mainly. For larger Δm^2_{10} (eV^2) $(= -1.0 \times 10^{-8}\ eV^2)$ the spin flavour precession resonance moves in the direction of the solar surface and both pp and ^7Be fluxes become modulated ($pp+^7Be$ modulation).

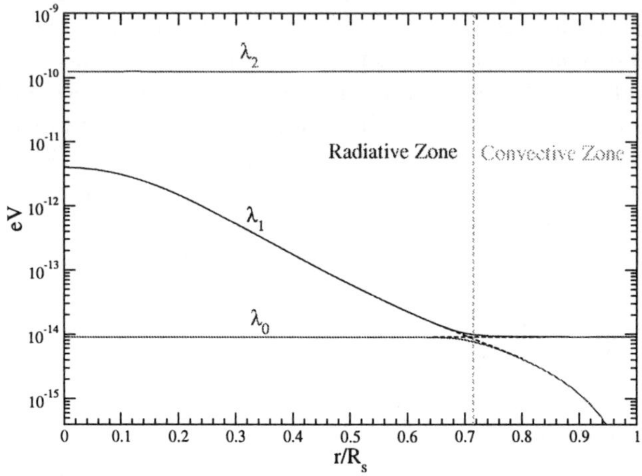

Figure 1: *The evolution of the mass matter eigenvalues along the solar neutrino trajectory.*

3. Our knowledge of the solar neutrino spectrum is very limited. ^8B, the best well known flux, accounts for just 10^{-4} of the total, while pp and ^7Be accounting for more than 98%, have only been observed in the radiochemical Ga experiments in accumulation with all other fluxes. In view of a hint from Ga experiments that low energy fluxes might be time dependent, we need real time dedicated low energy experiments to ascertain this possibility. We next consider the predictions for two of the forthcoming ones: Borexino and LENS [4].

(I) Borexino

This is a liquid scintillator detector installed in Gran Sasso Laboratory using $\nu\ e^-$ scattering, aiming at starting physics in late 2006. The kinetic energy threshold of 250 keV and maximum of 664 keV ensures that this experiment is directed mainly at ^7Be neutrino flux ($E_{Be} = 0.86 MeV$). The reduced rate for Borexino is, in standard notation

$$R_{Bor} = \frac{\int_{T_{min}}^{T_{max}} dT \int_{E_{min}}^{E_{max}} dE \phi(E)[P_{ee}(E)\frac{d\sigma_{\nu_e}}{dT} + (1-P_{ee}(E))\frac{d\sigma_{\nu_x}}{dT}]}{\int_{T_{min}}^{T_{max}} dT \int_{E_{min}}^{E_{max}} dE \phi(E)\frac{d\sigma_{\nu_e}}{dT}} \tag{4}$$

and is shown in fig.2 as a function of the peak field B_0. It is seen that in the $pp+^7Be$ dominated modulation the rate decreases faster for increasing B_0, thus

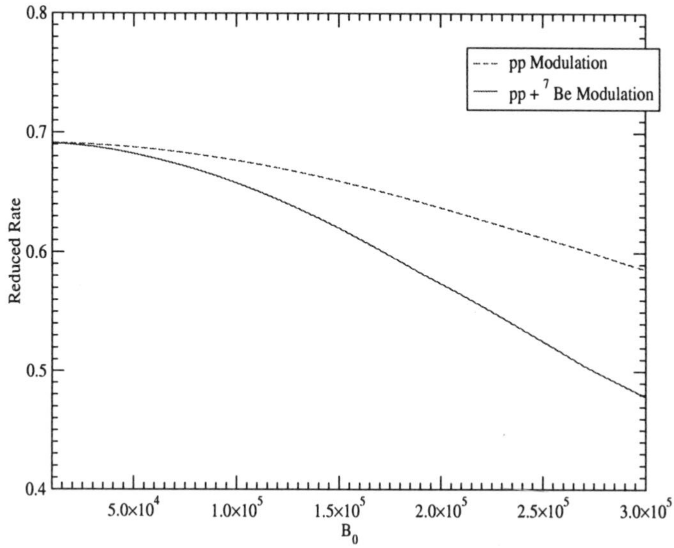

Figure 2: *Reduced Borexino event rate as a function of the peak field value.*

exhibiting more sensitivity to solar activity than in the *pp* case: the more sensitive ^7Be flux is to the peak field, the more sensitive will the event rate be. As the Borexino collaboration expects a combined error (statistical+systematic) of 10% and 5% after 1 and 3 years run, both scenarios can be perceptible after 1 year.

(II) LENS (Low Energy Neutrino Spectroscopy)

LENS is a real time detector in late stages of development measuring solar neutrinos through the charged current reaction

$$\nu_e +^{115} In \rightarrow^{115} Sn + e^- \tag{5}$$

with the lowest threshold yet: Q=114 keV. The signal energy is directly and uniquely related to the neutrino energy, hence a resolved spectrum of all low energy components (*pp*, ^7Be, *pep*, *CNO*) can be obtained that qualitatively shows how the sun shines. The LENS event rate is

$$R_{LENS}(E_e) = \int_Q^{E_{max}} P_{ee}(E) f(E_e^{'}, E_e) \phi(E) dE \tag{6}$$

where E_e' is the physical electron energy, $E_e' = E - Q$. For $pp + {}^7Be$ modulation ($\Delta m_{10}^2 = -1.0 \times 10^{-8}\ eV^2$) $R_{LENS}(E_e)$ is shown in fig.3.

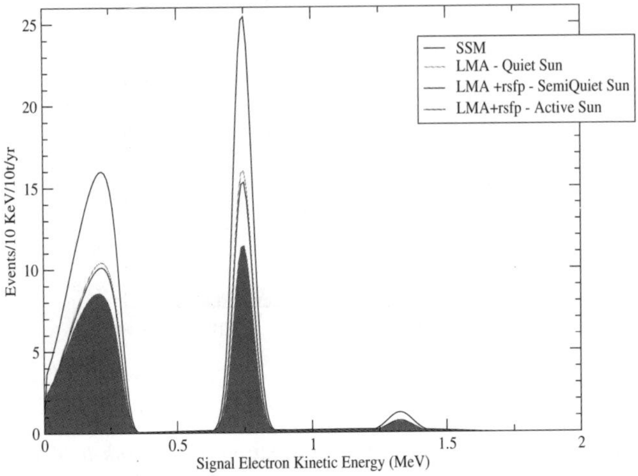

Figure 3: *LENS rate for pp+^{7}Be modulation and a peak field 250 kG for the lower line.*

4. To conclude: low energy solar neutrino experiments should at present be regarded as a major objective in the solar neutrino program, as time modulation is hinted by the Gallium experiments. Running Borexino and LENS during a significant fraction of a solar cycle will certainly test the possible modulation effect in the low energy sector, thus providing evidence for μ_ν and new physics.

References

[1] A. Cisneros, Astrophys. Space Sci. **10** (1971) 87.

[2] L. B. Okun, M. B. Voloshin and M. I. Vysotsky, Sov. J. Nucl. Phys. **44** (1986) 440 [Yad. Fiz. **44** (1986) 677].

[3] C. M. Cattadori, talk at XXI International Conference on Neutrino Physics and Astrophysics, Neutrino 2004, http://neutrino2004.in2p3.fr/.

[4] B. C. Chauhan and J. Pulido, JHEP **0406** (2004) 008 [arXiv:hep-ph/0402194].

[5] B. C. Chauhan, J. Pulido and R. S. Raghavan, JHEP **0507** (2005) 051 [arXiv:hep-ph/0504069].

NEUTRINO SPIN-FLAVOR OSCILLATIONS IN RAPIDLY VARYING MAGNETIC FIELDS

M.Dvornikov [a]

Institute of Terrestrial Magnetism, Ionosphere and Radiowave Propagation, 142190 Troitsk, Moscow region, Russia

Abstract. The general formalism for the description of neutrino oscillations in arbitrary rapidly varying external fields is elaborated. We obtain the new effective Hamiltonian which determines the evolution of the averaged neutrino wave function. The general technique is applied to the neutrino oscillations in rapidly varying magnetic fields. We evaluate the transition probabilities of the neutrino spin-flavor oscillations in magnetic fields of the Sun and compare them with the numerical solutions of the Schrödinger equation with the exact Hamiltonian.

One of the most interesting problems in the neutrino physics is the solar neutrino deficit. Nowadays it is experimentally established that the disappearance of solar electron neutrinos can be accounted for by the LMA-MSW solution. Apart from the LMA-MSW solution other theoretical models of the solar neutrino oscillations like spin-flavor precession (see work [1] and references therein) are also considered. We mention that neutrino oscillations in electromagnetic fields of various configurations were examined in our previous works (see Refs. [2]).

In this paper we study neutrino oscillations in presence of general rapidly varying fields. We derive the new effective Hamiltonian for the time evolution of the averaged neutrino wave function. Then we apply the general technique to neutrino oscillations in rapidly varying magnetic fields and discuss the neutrino conversion in solar magnetic fields. We present the numerical solutions of Schrödinger equation for the neutrino system interacting with the constant transversal and twisting magnetic fields and compare them with the approximate analytical solutions found in this work.

Let us consider the evolution of the two neutrinos $\nu = (\nu_1, \nu_2)$, which can belong to different flavors and helicity states. The evolution of the system is described by the Schrödinger type differential equation,

$$i\frac{d\nu}{dt} = H\nu. \tag{1}$$

Here we do not specify the explicit form of the Hamiltonian but just suppose that it is decomposed into two terms,

$$H = H_0 + \mathcal{H}, \quad \mathcal{H}(t+T) = \mathcal{H}(t). \tag{2}$$

The Hamiltonian H_0 corresponds to the neutrino interaction with constant or slowly varying external fields. In presence of only this term the solution of

[a] e-mail: maxdvo@izmiran.ru

Eq. (1) can be easily found. It is known to be periodical with the typical frequency $\Omega_0 \sim 1/L_{\text{eff}}$, where L_{eff} is the oscillations length. The function $\mathcal{H}(t)$ corresponds to rapidly varying external fields. The frequency $\omega = 2\pi/T$ should be much greater than Ω_0.

We will seek the solution of Eqs. (1) and (2) in the form (see also Ref. [3]),

$$\nu(t) = \nu_0(t) + \xi(t). \tag{3}$$

In Eq. (3) the function $\xi(t)$ is the small rapidly oscillating one with zero mean value and the function $\nu_0(t)$ is the slowly varying one during T. The mean value of a function is the time averaging over the period T. After substituting Eq. (3) in Eq. (1) we obtain two groups of rapidly and slowly varying terms in Eq. (1). In order for this expression to be an identity the terms of each group in left-handed side of the equation should be equal to the terms of the same group in right-handed side. With help of this fact one finds the time dependence of ξ in the explicit form,

$$\xi(t) = -i\left(\int \mathcal{H}(t)\ \mathrm{d}t\right)\nu_0(t). \tag{4}$$

Averaging residual terms in Eq. (1) over the period T and accounting for Eq. (4) we obtain the equation for the function ν_0,

$$i\frac{d\nu_0}{dt} = H_{\text{eff}}\nu_0, \quad H_{\text{eff}} = H_0 - i\overline{\mathcal{H}\left(\int \mathcal{H}\mathrm{d}t\right)}. \tag{5}$$

In derivation of Eq. (5) we do not make any assumptions about the smallness of the interaction described by the Hamiltonian $\mathcal{H}$.

On the basis of the general technique we discuss the neutrino evolution in matter under the influence of a combination of the constant transversal $\mathbf{B}_0$ and twisting $\mathbf{B}(\mathbf{r})$ magnetic fields. In our case the Hamiltonians H_0 and $\mathcal{H}$ have the form

$$H_0 = \begin{pmatrix} V/2 & \mu B_0 \\ \mu B_0 & -V/2 \end{pmatrix}, \quad \mathcal{H} = \begin{pmatrix} 0 & \mu B e^{-i\omega t} \\ \mu B e^{i\omega t} & 0 \end{pmatrix}, \tag{6}$$

where μ is the neutrino magnetic moment. In Eq. (6) we introduce the quantity $V/2 = (\Delta m^2/4E)\Theta - G_F n_{\text{eff}}/\sqrt{2}$, where Θ is the function of the vacuum mixing angle θ_{vac} (see Ref. [4]), Δm^2 is the difference of the neutrino mass squared, E is the neutrino energy, n_{eff} is the effective matter density, ω is the frequency of the transversal magnetic field variation, G_F is the Fermi constant.

Using Eqs. (5)-(6) it is possible to derive the expression for H_{eff},

$$H_{\text{eff}} = \begin{pmatrix} V/2 - (\mu B)^2/\omega & \mu B_0 \\ \mu B_0 & -V/2 + (\mu B)^2/\omega \end{pmatrix}. \tag{7}$$

Since B_0 and B do not depend on time in Eq. (7), we can solve Eq. (5) for the effective Hamiltonian given in Eq. (7). The transition probability is expressed in the following way

$$P(t) = A\sin^2\left(\frac{\pi t}{L}\right), \tag{8}$$

where

$$A = \frac{(\mu B_0)^2}{[V/2 - (\mu B)^2/\omega]^2 + (\mu B_0)^2}, \quad \frac{\pi}{L} = \sqrt{[V/2 - (\mu B)^2/\omega]^2 + (\mu B_0)^2}. \tag{9}$$

As it results from Eqs. (8) and (9), if the following condition is satisfied,

$$\frac{V}{2} \simeq \frac{(\mu B)^2}{\omega}, \tag{10}$$

then $A \simeq 1$ and the transition probability can achieve great values. This phenomenon is analogous to resonance amplification of spin-flavor oscillations.

It is interesting to discuss the situation when $B \gg B_0$. In this case the value of $A_0 = A(B = 0)$ is much less than unity. Hence the transition probability at the absence of the additional twisting magnetic field is small. In this case, if we choose the parameters according to Eq. (10), the amplitude of the transition probability becomes great. From Eqs. (9) and (10) one can derive the restriction imposed on B and B_0, $(B/B_0)^2 \gg 1$.

Now we consider one of the possible applications of the developed technique to neutrino spin-flavor oscillations in magnetic fields of the Sun. The solar magnetic field is unlikely to be only either constant transversal or twisting. Therefore we can apply the method elaborated in this paper to the description of the solar neutrino conversion. We examine one of the possible channels of neutrino oscillations, namely $\nu_{eL} \leftrightarrow \nu_{\mu R}$ conversion. First one should estimate the parameter V in Eq. (6). In our case $\Theta = (1 + \cos 2\theta_{\text{vac}})/2$ and the effective matter density $n_{\text{eff}} = (n_e - n_n/2)$. Let us discuss the neutrino with the following properties, $\Delta m^2 \sim 10^{-5}$ eV^2, $\theta_{\text{vac}} \sim \pi/4$. We take the neutrino energy $E \sim$ 10 MeV. Matter is supposed to consist mainly of hydrogen and has the density $d \sim 1.4$ g/cm^3. For these parameters we obtain that $V/2 \sim 10^{-15}$ eV.

We can evaluate the strength of the magnetic fields necessary for the 10% conversion of the initial ν_{eL} beam. We assume that neutrino conversion occurs along the distance $D \simeq 3.2 \times 10^{14}$ eV^{-1} $\sim R_{\odot}/10$, where $R_{\odot}$ is the solar radius. Neutrinos are supposed to have the transitional magnetic moment $\mu = 10^{-11}\mu_{\text{B}}$. Setting $P(t) = 0.1$ in Eq. (8) we obtain $B_0 \simeq 17.6$ kG. Then we can derive the strength of the twisting magnetic field: $B \simeq 3.2 B_0 \simeq 56.3$ kG. The resonance condition (10) is satisfied if $\omega \simeq 1.0 \times 10^{-14}$ eV.

In order to substantiate the correctness of the approach developed in our paper we obtain the numerical solutions of Eq. (1) with the Hamiltonian given

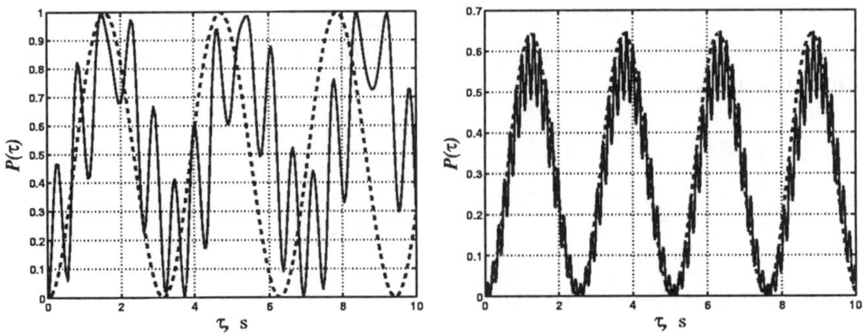

(a) (b)

Figure 1: Neutrino transition probability for $\nu_{eL} \leftrightarrow \nu_{\mu R}$ oscillations; (a) $\omega = 1.0 \times 10^{-14}$ eV; (b) $\omega = 4.0 \times 10^{-14}$ eV.

in Eq. (6). The probabilities of neutrino oscillations are shown on Fig. 1 versus $\tau = 1.5t$. We also compare them with the approximate analytical formula (8). The solid line corresponds to the transition probability computed with help of the numerical solution of Eq. (1) for the parameters taken above. The dashed line represents the transition probability given in Eq. (8). It can be seen in Fig. 1 that the numerical expression for the transition probability approaches to the approximate formula given in Eq. (8) at great frequencies of the twisting magnetic field. This comparison proves the validity of the elaborated technique.

Acknowledgments

The author is indebted to the organizers of the 12th Lomonosov Conference on Elementary Particle Physics for inviting him to participate in this activity.

References

[1] E.Kh.Akhmedov, *et al.*, *Phys.Rev.* **D** 48, 2167 (1993).
[2] M.S.Dvornikov, A.I.Studenikin, *Phys.At.Nucl.* 64, 1624 (2001); 67, 719 (2004).
[3] P.L.Kapitza, *Usp.Fiz.Nauk* 44, 7 (1951).
[4] G.G.Likhachev, A.I.Studenikin, *JETP* 81, 419 (1995).

SPIN LIGHT OF ELECTRON IN MATTER

Alexander Grigoriev [a], Sergei Shinkevich [b], Alexander Studenikin [b,a], Alexei Ternov [c], Ilya Trofimov [b]

Abstract. We further generalize the powerful method, which we have recently developed for description of the background matter influence on neutrinos, for the case of an electron moving in matter. On the basis of the modified Dirac equation for the electron, accounting for the standard model interaction with particles of the background, we predict and investigate in some detail a new mechanism of the electromagnetic radiation which is due to the electron charge in matter. We have termed this radiation the "spin light of electron" in matter and predicted that this radiation can have consequences accessible for experimental observations in astrophysical and cosmological settings.

1 Introduction

In a series of our papers [1–4] we have developed a rather powerful method of investigation of different phenomena that can appear when neutrinos and electrons move in the background matter. The method discussed is based on the use of the modified Dirac equations for the particles wave functions, in which the correspondent effective potentials, that account for the matter influence on particles, are included. It is similar to the Furry representation [5] in quantum electrodynamics, widely used for description of particles interactions in the presence of external electromagnetic fields. In [1–4] we apply the discussed method for elaboration of the quantum theory of the "spin light of neutrino" in matter. The spin light of neutrino in matter ($SL\nu$), one of the four new phenomena studied in our recent papers [6], is a new type of electromagnetic radiation that can be emitted by a massive neutrino (due to its magnetic moment) when the particle moves in the background matter. Within quasi-classical treatment the existence of this radiation was proposed and studied in [7], while the quantum theory of this phenomenon was developed in [1–4, 8].

In [1], it has been shown how the approach, developed at first for description of a neutrino motion in the background matter, can be spread for the case of an electron propagating in matter. The modified Dirac equation for an electron in matter has been derived [1] and on this basis we have considered the electromagnetic radiation that can be emitted by the electron (due to its electric charge) in the background matter. We have termed this radiation as the "spin light of electron" in matter. It should be noted here that the term "spin light" was introduced in [9] for designation of the particular spin-dependent contribution to the electron synchrotron radiation power.

[a]Skobeltsyn Institute of Nuclear Physics of Moscow State University, 119992 Moscow, Russia

[b]Physics Faculty of Moscow State University, 119992 Moscow, Russia

[c]Department of Theoretical Physics, Moscow Institute for Physics and Technology, 141700 Dolgoprudny, Russia

2 Modified Dirac equation for electron in matter

Let us consider an electron having the standard model interactions with particles of electrically neutral matter composed of neutrons, electrons and protons. This can be used for modelling a real situation exhisted, for instance, when electrons move in matter of a neutron star. We suppose that there is a macroscopic amount of the background particles in the scale of an electron de Broglie wave length. Then the addition to the electron effective interaction Lagrangian is

$$\Delta L_{eff}^{(e)} = \widetilde{f}^{\mu}\Big(\bar{e}\gamma_{\mu}\frac{1-4\sin^{2}\theta_{W}+\gamma^{5}}{2}e\Big), \tag{1}$$

where the explicit form of f^{μ} depends on the background particles densities, speeds and polarizations. The modified Dirac equation for the electron wave function in matter is [1]

$$\Big\{i\gamma_{\mu}\partial^{\mu}-\frac{1}{2}\gamma_{\mu}(1-4\sin^{2}\theta_{W}+\gamma_{5})\widetilde{f}^{\mu}-m_{e}\Big\}\Psi_{e}(x)=0, \tag{2}$$

where for the case of electron moving in the background of neutrons (that can be used as an abrupt model of a nuclear matter of a neutron star, see, for instance, [10])

$$\widetilde{f}^{\mu}=\frac{G_{F}}{\sqrt{2}}(n_{n},n_{n}\mathbf{v}). \tag{3}$$

Here n_n is the neutrons number density and $\mathbf{v}$ is the speed of the reference frame in which the mean momentum of the neutrons is zero.

The solution of equation (2) can be found in analogy with the case of neutrino (for details see [2,3]). The complete set of the electron wave functions in the matter is:

$$\Psi_{\varepsilon,\mathbf{p},s}(\mathbf{r},t)=\frac{e^{-i(E_{\varepsilon}t-\mathbf{pr})}}{2L^{3/2}}\begin{pmatrix}\sqrt{1+\frac{m_{e}}{E_{\varepsilon}-c\alpha_{n}m_{e}}}\ \sqrt{1+s\frac{p_{3}}{p}}\\ s\sqrt{1+\frac{m_{e}}{E_{\varepsilon}-c\alpha_{n}m_{e}}}\ \sqrt{1-s\frac{p_{3}}{p}}\ e^{i\delta}\\ s\varepsilon\sqrt{1-\frac{m_{e}}{E_{\varepsilon}-c\alpha_{n}m_{e}}}\ \sqrt{1+s\frac{p_{3}}{p}}\\ \varepsilon\sqrt{1-\frac{m_{e}}{E_{\varepsilon}-c\alpha_{n}m_{e}}}\ \sqrt{1-s\frac{p_{3}}{p}}\ e^{i\delta}\end{pmatrix}, \tag{4}$$

where L is the normalization length, $\delta=\arctan(p_2/p_1)$ and the so called matter density parameter α_n is defined as

$$\alpha_{n}=\frac{G_{F}}{2\sqrt{2}}\frac{n_{n}}{m_{e}}. \tag{5}$$

The electron energy in matter

$$E_{\varepsilon}^{(e)}=\varepsilon\sqrt{\mathbf{p}^{2}\Big(1-s\alpha_{n}\frac{m_{e}}{p}\Big)^{2}+m_{e}{}^{2}}+c\alpha_{n}m_{e} \tag{6}$$

depends on the electron helicity $s = \pm 1$ and the quantity $\varepsilon = \pm 1$ which splits the solutions into positive- and negative-frequency branches.

3 Spin light of electron in matter

To the lowest order of the perturbation theory the corresponding quantum process is described by the one-photon emission diagram with the initial ψ_i and final ψ_f electron states and with the vertex corresponding to the standard electromagnetic interaction with the photon due to the electron charge e. Thus, the transition amplitude is giving by

$$S_{fi} = -ie\sqrt{4\pi}\int d^4x\bar{\psi}_f(x)(\gamma^\mu e^*_\mu)\frac{e^{ikx}}{\sqrt{2\omega L^3}}\psi_i(x), \tag{7}$$

where $k^\mu = (\omega, \mathbf{k})$ and e^μ are the photon momentum and polarization vector, respectively. Choosing the three-dimensional transversal gauge and performing integration over the time variable, we get

$$S_{fi} = ie\sqrt{\frac{2\pi}{\omega L^3}}2\pi\delta(E - E' - \omega)\int d^3x\bar{\psi}_f(\mathbf{r})(\gamma\mathbf{e}^*)e^{i\mathbf{kr}}\psi_i(\mathbf{r}), \tag{8}$$

The δ-function in the last expression stands for the energy conservation law with E and E' being energies of the initial and final neutrino states. Performing further integration over the spatial coordinates we get also the momentum conservation law, which together with the energy conservation law,

$$E = E' + \omega, \quad \mathbf{p} = \mathbf{p}' + \mathbf{k}, \tag{9}$$

leads to the only possible transition of the electron when its chirality changes from $s_i = -1$ to $s_f = 1$. The corresponding photon energy then is given by the expression [1]:

$$\omega = \frac{2\alpha_n m_e p[\tilde{E} - (p + \alpha_n m_e)\cos\theta]}{(\tilde{E} - p\cos\theta)^2 - (\alpha_n m_e)^2}, \tag{10}$$

where θ is the angle between the directions of initial neutrino and emitted photon propagation. We also use the following notations

$$\tilde{E} = E - c\alpha_n m_e, \quad \tilde{E}' = E' - c\alpha_n m_e. \tag{11}$$

In the case of relativistic electrons and small values of the matter density parameter α_n the photon energy is

$$\omega_{SLe} = \frac{\beta_e}{1 - \beta_e\cos\theta_{SLe}}\omega_0, \quad \omega_0 = \frac{G_F}{\sqrt{2}}n_n\beta_e, \tag{12}$$

here β_e is the electron speed in vacuum. From this expressions we conclude that for the relativistic electron the energy range of the SLe may even extend up to energies peculiar to the spectrum of gamma-rays. We also predict the existence of the electron-spin polarization effect in this process.

4 Rate and power of the radiation

With the expressions for the amplitude (8) and the emitted photon energy (10) we arrive to general formulas for the total rate and power of the radiation:

$$\Gamma = \frac{e^2}{2}\int_0^{\pi} \frac{\omega}{1+\tilde{\beta}'_e y}\left(1-\tilde{\beta}'_e\tilde{\beta}_e - \frac{{m_e}^2}{\tilde{E}'\tilde{E}}\right)(1-y\cos\theta)\sin\theta\, d\theta, \qquad (13)$$

$$\mathrm{I} = \frac{e^2}{2}\int_0^{\pi} \frac{\omega^2}{1+\tilde{\beta}'_e y}\left(1-\tilde{\beta}'_e\tilde{\beta}_e - \frac{{m_e}^2}{\tilde{E}'\tilde{E}}\right)(1-y\cos\theta)\sin\theta\, d\theta, \qquad (14)$$

where

$$\tilde{\beta}_e = \frac{p+\alpha_n m_e}{\tilde{E}}, \ \tilde{\beta}'_e = \frac{p'-\alpha_n m_e}{\tilde{E}'}, \qquad (15)$$

are the quantities, describing the initial and final electron group velocities, and

$$E' = E-\omega, \quad p' = K_e\,\omega - p, \qquad (16)$$
$$K_e = \frac{\tilde{E}-p\cos\theta}{\alpha_n m_e}, \quad y = \frac{\omega - p\cos\theta}{p'}.$$

Taking integration in (13) and (14) we obtain:

$$\Gamma = e^2 \frac{m_e^2\left[(1+2\alpha_n^2)m_e^2+2p^2\right]}{4p^2(4\alpha_n p+m_e)^2\sqrt{(p+m_e\alpha_n)^2+m_e^2}}$$
$$\times\left[(4\alpha_n p+m_e)^2\ln\left(1+\frac{4\alpha_n p}{m_e}\right) - 4\alpha_n p\,(m_e+6\alpha_n p)\right] \qquad (17)$$

and

$$\mathrm{I} = e^2\frac{m_e^2}{2p^2(4\alpha_n p+m_e)^3}\left\{\left((1+\alpha_n^2)m_e^2+p^2\right)(4\alpha_n p+m_e)^3\ln\left(1+\frac{4\alpha_n p}{m_e}\right)\right.$$
$$-\frac{4}{3}\alpha_n p\left[88\alpha_n^2p^4+3(1+\alpha_n^2)m_e^4+30\alpha_n(1+\alpha_n^2)m_e^3p\right.$$
$$\left.\left.+p^2m_e^2(3+88\alpha_n^2(1+\alpha_n^2))+2\alpha_n(15+16\alpha_n^2)m_ep^3\right]\right\}. \qquad (18)$$

Let us estimate the total rate Γ of the radiation, and also the corresponding life-time T_{SLe} of the electron in respect to the considered process using expression (17). Consider the electron with momentum $p = 1\ MeV$ moving in matter characterized by the number density $n_n \sim 10^{37}\ \mathrm{cm}^{-3}$, in this case the matter density parameter is $\alpha_n = 0.6\times10^{-6}$. Then, for the rate of the process we get $\Gamma \sim 3.2\times10^{-10}MeV$ which corresponds to the characteristic life-time of the electron $T_{SLe} \sim 2\times10^{-2}s$.

Finally, we have developed the approach to description of the matter influence on an electron which is based on the exact solutions of the correspondent modified Dirac equation for the particle wave function. The approach developed (we have previously used it for the case of neutrino) is similar to the Furry representation in quantum electrodynamics. Note that our focus has been on the standard model interactions of electrons with the background matter. A similar approach, which implies the use of the exact solutions of the correspondent modified Dirac equations, can be developed in the case when electrons interact with different external fields predicted within various extensions of the standard model (see, for instance, [11, 12] and the paper of V.Zhukovsky et al in this book). We have predicted and investigated in some detail a new type of electromagnetic radiation (the spin light of electron in matter, SLe) that can be emitted by the electron due its electric charge within the standard model of interaction with the background matter. The obtained SLe energy spectrum shows that for ultra-relativistic electrons it can even extend up to energy range peculiar to the spectrum of gamma-rays. Comparing the rates of the spin light of neutrino and spin light of electron in matter, we have predicted (see also [1]) that the latter is more effective then the former. We have predicted also that the SLe emitted by ultra-relativistic electrons moving in dense astrophysical and cosmological media can have consequences accessible for experimental observations.

References

[1] A.Studenikin, J.Phys.A: Math. Gen. 39 (2006) 6769.

[2] A.Studenikin, A.Ternov, Phys.Lett.B 608 (2005) 107.

[3] A.Grigoriev, A.Studenikin, A.Ternov, Phys.Lett.B 622 (2005) 199; Grav.Cosm. 11 (2005) 132.

[4] A.Grigoriev, A.Studenikin, A.Ternov, Phys.Atom.Nucl. 69 (2006), in press.

[5] W.Furry, Phys.Rev. 81 (1951) 115.

[6] A.Studenikin, Nucl.Phys.B (Proc.Suppl.) 143 (2005) 570.

[7] A.Lobanov, A.Studenikin, Phys.Lett.B 564 (2003) 27; ibid. 601 (2004) 171; M.Dvornikov, A.Grigoriev, A.Studenikin, Int.J Mod.Phys.D 14 (2005) 309.

[8] A.Lobanov, Phys.Lett.B 619 (2005) 136; Dokl.Phys. 50 (2005) 286.

[9] I.M.Ternov, Sov.Phys.Usp. 38 (1995) 405; V.A.Bordovitsyn, I.M.Ternov, V.G.Bagrov, Sov.Phys.Usp. 38 (1995) 1037.

[10] A.Kusenko, M.Postma, Phys.Lett.B 545 (2002) 238.

[11] D.Colladay, V.A.Kostelecky, Phys.Rev.D 55 (1997) 6760.

[12] V.Ch.Zhukovsky, A.E.Lobanov, E.M.Murchikova, Phys.Rev.D 73 (2006) 065016.

NEUTRINO PHYSICS, BBN, LSS AND CMBR

A.D. Dolgov [a],
INFN, Ferrara 44100, Italy
ITEP, 117218, Moscow, Russia

Abstract. Cosmological bounds on neutrino mass, magnetic moment, and right-handed currents are presented and critically analyzed. Possible cosmological manifestations of violation of spin-statistics theorem for neutrinos are discussed.

1 Introduction

In this talk the impact of neutrinos on cosmology as well as the impact of cosmos on neutrinos are briefly reviewed. Though it is difficult to separate between the two but conventionally we include into a short list of the first subject, COSMOS-TO-NEUTRINOS, the following:
1. Measurement of neutrino mass.
2. Right-handed neutrinos and masses of right handed intermediate bosons.
3. Magnetic moments of neutrinos
4. Oscillations and mixing with sterile partners.
A short list of subjects of the other way around, NEUTRINOS-TO-COSMOS, is somewhat longer:
1. Big Bang Nucleosynthesis (BBN).
2. Large scale structure of the universe (LSS).
3. Cosmic microwave background radiation (CMBR).
4. Cosmological lepton asymmetry, baryo-thru-lepto-genesis.
5. Leptonic charge inhomogeneities, turbulence, primordial magnetic fields and gravitational waves.
6. Neutrino related dark energy (maybe???).
Some of these issues are discussed in what follows with very different level of details. The others are omitted because of lack of space, though they were mentioned in the talk. Detailed discussion with a long list of references can be found in review [1].

2 Thermal history of neutrinos and Gershtein-Zeldovich (GZ) bound

Neutrinos were in thermal equilibrium with $e^{\pm}$ and γ at $T > 1$ MeV. Correspondingly their number density at this stage could be expressed through the number density of photons in the primeval plasma as:

$$n_\nu = (3/8) n_\gamma \tag{1}$$

for each *left-handed* neutrino flavor, ν_e, ν_μ, and ν_τ, under assumption of vanishing charge asymmetry between neutrinos and antineutrinos, $n_\nu = n_{\bar\nu}$. At lower T neutrinos decoupled from the plasma and when at $T \leq m_e$ photons are heated by e^+e^--annihilation, the number density of neutrinos remained practically untouched. As a result the neutrino to photon ratio dropped down and today we expect

$$n_\nu + n_{\bar\nu} = (3/11) n_\gamma = 112/\mathrm{cm}^3 \tag{2}$$

Hence [2]:

$$\sum_a m_{\nu_a} < 94\,\Omega_m h^2 \text{ eV}, \tag{3}$$

[a] e-mail: dolgov@fe.infn.it

where $h = H/100km/sec/Mpc$ is the dimensionless Hubble parameter and $\Omega_m = \rho_m/\rho_c$ is the fraction of cosmological energy density of matter with respect to critical energy density, $\rho_c = 3H^2 m_{Pl}^2/8\pi$, and $m_{Pl} = 1.2 \cdot 10^{19}$ GeV is the Planck mass.

In particular, for $\Omega_m < 0.3$ and $h = 0.7$ the sum of neutrino masses is bounded by $\sum_a m_{\nu_a} < 14$ eV. This bound, derived by Gershtein and Zeldovich in 1966 [2] is sometimes called the Cowsik and McClelland bound [3]. However, the latter was derived 6 years after and the derivation contained a couple of serious errors: effects of photon heating are not accounted for and 2 spin states of neutrinos are assumed to be equally populated. Of course, if neutrino mass is non-vanishing, both helicity states are possible but in the standard electroweak theory, only left-handed neutrinos are present in cosmic plasma in noticeable amount [4].

Since it is known from neutrino oscillation experiments that the masses of neutrinos are almost equal at such scale, we can conclude that $m_{\nu_a} < 4.7\,\text{eV}$. This bound can be easily improved by factor ~3 because *hot dark matter* cannot contribute more than ~30% to the matter mass density:

$$m_{\nu_a} < 1.6\text{eV} \tag{4}$$

This is a very simple and robust bound, practically model independent, see however the discussion below.

For comparison, the best direct experimental bounds experiments by Troitsk and Meinz groups, found from tritium decay and thus valid for electronic type neutrinos (or better to say for mass eigenstates produced in this decay), are somewhat less accurate [5]:

$$m_{\nu_e} < (2-3)\text{ eV} \tag{5}$$

The neutrino oscillations experiments are much more precise but they measure only mass differences but not the absolute values of masses:

$$\delta m_{solar}^2 = (5.4 - 9.5) \cdot 10^{-5}\text{ eV}^2 \tag{6}$$
$$\delta m_{atm}^2 = (1.2 - 4.8) \cdot 10^{-3}\text{ eV}^2 \tag{7}$$

As we see in what follows, presently telescopes allow to weight neutrinos more accurately than direct experiments. Astronomy is sensitive to $m_\nu \sim$ (a few)$\times$0.1 eV if combined data on LSS and CMBR are used, but the arguments involve, though natural and testable, but still additional cosmological assumptions. It should be stressed that all these astronomical bounds are heavily based on the Gershtein-Zeldovich result.

In this connection a natural question arises if it is possible to avoid the GZ-limit? So let us analyze the underlying assumptions and see if they could be relaxed:

1. Thermal equilibrium in the early universe at $T > 1$ MeV. It is perfectly well fulfilled in the standard model. Equilibrium might be only broken if the universe was never hotter than MeV and number density of neutrinos would be smaller than the standard one [6]. However, if this were the case, BBN would encounter serious problems.
2. Vanishing or negligible lepton asymmetry. If there exists a noticeable excess of neutrinos over antineutrinos (or vice versa) the total energy density stored in cosmological neutrinos would be only larger and the GZ bound would be stronger.
3. Extra heating of photon plasma below MeV would result in a decrease of the ratio n_ν/n_γ. Care should be taken of potentially strong distortion of BBN and frequency spectrum of CMBR. It looks quite difficult but not impossible.
4. Neutrino stability at the cosmological time scale: $\tau_\nu \geq t_u \sim 10^{10}$ years. It is difficult to make smaller life-time and new interactions to achieve this goal are necessary.

Is not interesting for mass degenerate neutrinos, if a decay of heavier ν produces a lighter ν because in this case the total number of neutrinos is conserved. However, it there exists a new neutrino-like fermion, e.g. a light sterile ν_s, into which the usual active neutrinos decay, the GZ bound can be relaxed.

5. New interactions which could lead to strong annihilation of $\bar{\nu}\nu$ [7]. To this end a very light or massless boson coupled to neutrinos is needed. The corresponding new long range forces generated e.g. by scalar coupling, would lead to a very strong neutrino clumping - ν balls.

6. A possible existence of right-handed neutrinos would lead to a larger cosmological number and energy density of neutrinos and, as a result, to a stronger bound.

3 Cosmological large scale structure and neutrino mass

According to modern understanding, neutrinos play a destructive role in structure formation. After the temperature dropped below 1 MeV, neutrinos stopped interacting with the cosmic plasma and free streamed till their velocity red-shifted down to nonrelativistic value. Their free streaming length depends upon the mass and is astronomically huge:

$$l_{FS} \approx 250\,\mathrm{Mpc}\,(\mathrm{eV}/m_\nu) \tag{8}$$

If the free streaming length of particles is much larger than the galactic scale they are called hot dark matter (HDM). In neutrino dominated universe the structures below free streaming scale (8) are erased. Sub-dominant neutrinos would lead to suppression of power at small scales and to delayed formation of smaller structures. This is because in HDM cosmology larger structures formed first and smaller ones formed later by fragmentation of big ones. Thus an observation of small and early formed structures would be especially sensitive to contribution of neutrinos into the cosmological energy density and would allow to deduce a sensitive bound on their mass. In particular, Ly-α clouds observed at high redshifts, $z \geq 1$, are very sensitive to light massive neutrinos.

In fact in derivation of astronomical upper bounds on neutrino mass all available pieces of data are used: 1) WMAP data; 2) other CMB data; 3) 2dF data; 4) constraint on σ_8; 5) SDSS data; 6) constraint on H_0; 7) constraint from Lyman-α forest. Results of different groups are the following: 1) WMAP [8]: $\Sigma m_\nu <0.69$ eV, from 1-4,6, 7; 2) Hannestad [9]: $\Sigma m_\nu <1.01$ eV, from 1-3,6; 3) Allen et al [10]: $\Sigma m_\nu = 0.56^{+0.3}_{-0.26}$ eV, from 1-4,6; 4) SDSS [11]: $\Sigma m_\nu <1.8$ eV, from 1,5; 5) Barger et al. [12]: $\Sigma m_\nu < 0.75$ eV, from 1-3,5,6; 6) Crotty et al [13]: $\Sigma m_\nu < 1.0$ eV, from 1-3,5 (6); 7) Seljak et al. [14]: $\Sigma m_\nu <0.42$ eV, from 1,2,4-7; 8) Fogli et al. [15]: $\Sigma m_\nu <0.5$ eV, from 1-7; 9) Hannestad [16]: $\Sigma m_\nu <0.65$ eV, from 1,5-7.

These bounds are based on comparison of the theory of large scale structure formation with available astronomical data. To see how robust are these bounds one need to know the accuracy and systematic errors of observations and how reliable are the theoretical input. The latter is the following:

1. Spectrum of primordial fluctuations. Usually it is assumed to be flat, Harrison-Zeldovich type. It is predicted by inflation and confirmed by CMBR at large scales $\geq 10 Mpc$.
2. Properties of dark matter. Usually it is assumed to be non-interacting cold dark matter plus vacuum energy, (CDM+Lambda). It is interesting to check how the obtained bounds will be modified if there is a self-interacting dark matter, e.g. mirror matter.
3. Analytical calculations are done at linear regime, when $\delta\rho/\rho \ll 1$. The standard physics i.e. general relativity and hydrodynamics, is applied. The results are not distorted by non-linear processes at large scales accessible to CMBR and thus can be

verified by the data.
4. Numerical simulations at nonlinear regime, when $\delta\rho/\rho \geq 1$. The latter are necessary at smaller scales, ~ 10 Mpc when analytical methods are impossible. Some essential physics may be omitted in numerical simulations because of necessity of simplifications of the system.

If any of the above is not true, the bounds on m_ν may be very much relaxed. One should keep in mind that all above is confirmed but, strictly speaking, not proven by the data. In particular, an *ad hoc* modification of perturbation spectrum at intermediate scales could mimic the presence of massive neutrinos and no restrictive mass bound would be obtained.

"Massive" conclusions:
1. All bounds quoted above are based on the calculations of the present day number density of neutrinos by Gershtein and Zeldovich [2].
2. More restrictive results, i.e. $\Sigma m_\nu < 1 - 0.42$ eV are based on analysis of LSS at relatively small scales plus CMBR to shift degeneracy of parameters.
3. With almost equal neutrino masses (from oscillation experiments) one can conclude that $m_\nu < 0.33 - 0.14$ eV, lower value is probably too strong.
4. Direct experiment project, KATRIN, may approach this accuracy. Do we need it?
5. Possible ways to violate astronomical limits:
a) From particle physics: new particles or stronger ν-interactions;
b) From cosmology: unusual spectrum of perturbations at small scales.
We need KATRIN.

4 Neutrinos and BBN

Big bang nucleosynthesis proceeded when the universe was quite young, the universe age was in the interval $t = 1 - 300$ sec and the temperature was in the range $T = 1 - 0.07$ MeV. Essential physical processes started from neutron-proton transformations:

$$n + e^+ \leftrightarrow p + \bar\nu_e, \text{ and } n + \nu_e \leftrightarrow p + e^- \tag{9}$$

These processes became frozen at $T_f \approx 0.7$ MeV, and this determines the starting value of n/p-ratio:

$$n/p = \exp(-\Delta m/T_f) \tag{10}$$

where $\Delta m = 1.3$ MeV is the neutron-proton mass difference.

When T dropped down to $T_{BBN} = 60 - 70$ keV, practically all neutrons, which survived decay to this moment, quickly formed ^4He (about 25% by mass) and a little ^2H (3×10^{-5} by number), ^3He (similar to ^2H) and very little ^7Li ($10^{-9} - 10^{-10}$). For a review see, e.g. the article in Particle Data Group [5]. An interesting fine-tuning is worth mentioning here: if the Fermi coupling constant, G_F, were different by, say, a factor of two, the world would drastically change. For a larger G_F reactions (9) would be longer in equilibrium and less neutrons would survive. Hence there would be practically no cosmological helium and first starts would consist only of hydrogen. In the opposite case of smaller G_F the n/p-ratio can be as large as 1 and all baryonic matter in the universe would consist of ^4He without any hydrogen.

The role of neutrinos in BBN is very important and this allows to put restrictive bounds on neutrino properties by demanding that successive results of BBN are not destroyed and, another way around, to cure some potential problems in comparison of the theory with observations by new physics of neutrinos.

The cooling rate of the universe, i.e. the dependence of temperature upon time depends upon all particle species present in the cosmological plasma. At BBN epoch the dominant species are photons, e^+e^--pairs, and three left-handed neutrinos. If some other new light particles are present they would change the cooling rate. Their effects are parametrized as additional number of neutrino species, ΔN, such that $\Delta N = 1$ corresponds to additional energy density equal to that of one additional neutrino flavor in thermal equilibrium.

On one hand side, the energy density of the universe is equal to the critical one. On the other hand, it is equal to energy density of thermally equilibrium plasma with the effective number of species g_* - each bosonic degree of freedom contributes 1 into g_*, while each fermionic one gives 7/8. Equalizing the two expressions we obtain:

$$\rho = \frac{3m_{Pl}^2}{32\pi t^2} = \frac{\pi^2}{30} g_* T^4 \tag{11}$$

where

$$g_* = 10.75 + 1.75\Delta N_\nu \tag{12}$$

ΔN_ν includes any form of energy present at BBN, in particular, possible increase of ρ due to neutrino degeneracy:

$$\Delta N_\nu = \frac{15}{7}\left[\left(\frac{\xi}{\pi}\right)^4 + 2\left(\frac{\xi}{\pi}\right)^2\right]$$

where $\xi = \mu/T$ and μ is the neutrino chemical potential.

Increasing ΔN_ν leads:
1) to earlier n/p-freezing and higher n/p-ratio,
2) to faster reaching of T_{BBN} and thus also to higher n/p, because less neutrons could decay.

The net result of $\Delta N = 1$ is an increase of ^4He by 5%. Depending upon the analysis of the observational data the allowed upper limit on the number of additional neutrino species is $\Delta N_\nu < 0.3 - 1.0$. An indication of some disagreement between the observed abundances of ^4He and ^2H leads to very strange most favorable value $N_\nu \approx 2.5$ [17]. This claim however, disagree with the point of view expressed in review [5].

The impact of electronic neutrino on BBN is much stronger than that of ν_μ and ν_τ, because ν_e directly participates in reactions (9) and even a small deviation of ν_e from thermal equilibrium or a small degeneracy strongly shifts the n/p-ratio:

$$(n/p)_{eq} = \exp\left(-\frac{\delta m}{T} - \xi_e\right) \tag{13}$$

Hence the bounds on lepton asymmetry are very different for electronic and μ or τ flavors:

$$|\xi_{\mu,\tau}| < 2.5, \text{ and } |\xi_e| < 0.1, \tag{14}$$

if a compensation between $\xi_{\mu,\tau}$ and ξ_e is allowed; in absence of compensation the bounds are somewhat stronger.

These bounds on leptonic chemical potentials are obtained in absence of neutrino oscillations. The large mixing angle solution to Solar neutrino deficit allows to impose restrictive bound for all ν-flavors [18]:

$$|\xi_a| < 0.07 \tag{15}$$

because all asymmetries are equalized prior to BBN. If the bound (15) is valid, the neutrino degeneracy cannot have a noticeable cosmological impact, in particular, on

LSS and CMBR. The bound can be relaxed if ν have new interaction with light Majorons. The induced potential inhibits early oscillations [19].

If sterile neutrinos exist and are mixed with the known active ones, the effects of that could be observed through BBN. At the present time this allows to put more stringent bounds on the oscillation parameters than is possible by direct experiment. There are three possible phenomena which could influence BBN:
1. A larger N_ν due to excitation of sterile degrees of freedom
2. A modification of spectrum of ν_e. due to energy dependence of the oscillation probability.
3. A generation of lepton asymmetry in ν_e sector compensated by the opposite asymmetry in sterile sector.
The original calculations have been done for the case of negligible mixing between active neutrinos and lead to the bounds:

$$(\delta m^2_{\nu_e\nu_s}/\mathrm{eV}^2)\sin^4 2\theta^{\nu_e\nu_s} = 3.16\cdot 10^{-5}\ln^2(1-\Delta N_\nu), \tag{16}$$

$$(\delta m^2_{\nu_\mu\nu_s}/\mathrm{eV}^2)\sin^4 2\theta^{\nu_\mu\nu_s} = 1.74\cdot 10^{-5}\ln^2(1-\Delta N_\nu) \tag{17}$$

in non-resonance case and to much stronger ones in the resonance case.

Since we know now that active neutrinos are strongly mixed the calculations have been redone with with realistic mixing between active neutrinos [20]. The bounds depend upon all mixing angles and mass differences (there could be between three and null resonances) and it is probably impossible to present them in analytical form. The excluded regions can be found in figures of ref. [20].

5 Cosmological manifestations of statistics violation for neutrinos

What happens if neutrinos violate the spin-statistics theorem, and obey Bose statistics or mixed statistics, despite having non-integer spin? It is very difficult, maybe even impossible, to formulate a consistent quantum field theory with wrong spin-statistics relation but a study of phenomenological manifestation of this drastic assumption is surely of interest. In particular a modification of neutrino statistics would lead to pronounced cosmological consequences [21,22].

In this case neutrinos might form cosmological Bose condensate with the distribution function given by:

$$f_{\nu_b} = \frac{1}{\exp[(E-\mu_\nu)/T-1]} + C\delta(k), \tag{18}$$

where the constant C is the amplitude of the condensate. Because of condensation such light particles can make cosmological cold dark matter in addition to the usual hot dark matter. Thus dark matter could consist of old well known massive particles at expense of very new physics, instead of usual physics but new particles.

Evidently neutrinos with wrong statistics would lead to modified abundances of light elements produced at BBN. If neutrinos obey Bose statistics:
1) their energy density would be 8/7 of normal fermionic ν, giving an excess of neutrino species by $\Delta N_\nu = 3/7$ for all three neutrino families;
2) larger density of ν_e would lead to smaller temperature of n/p-freezing.
The last effect dominates and the net result in terms of ΔN happens to be negative:

$$N_\nu^{(eff)} = 2.43 \tag{19}$$

The equilibrium distribution for mixed statistics has the form:

$$f_\nu^{(eq)} = \left[\exp(E/T) + \kappa\right]^{-1}. \tag{20}$$

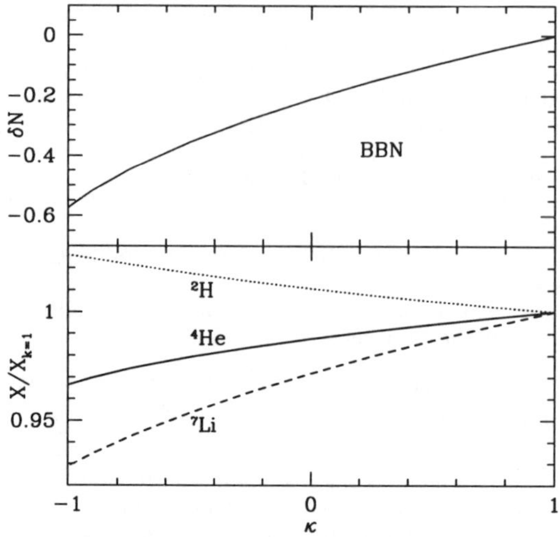

Figure 1: Upper panel: the change in the effective number of degrees of freedom which corresponds to the change of the ^{4}He abundance as a function of the effective Fermi-Bose parameter κ. Lower panel: the relative change of the primordial abundances of deuterium, helium-4, and lithium-7, as functions of κ. We take $\eta = n_B/n_\gamma = 6.5 \cdot 10^{-10}$

where the constant parameter κ interpolates between Fermi ($\kappa = 1$) and Bose ($\kappa = -1$) statistics.

In figs. 1 and 2 the results of calculations of ref. [22] are presented. Surprisingly the BBN with bosonic neutrinos is in a better agreement with the data than the BBN with normal fermionic neutrinos.

6 BBN and right-handed ν, W_R, Z_R, ν-magnetic moments, *etc*

A consideration of BBN allows also to put restrictive bounds on possible right-handed currents, masses of right-handed intermediate bosons, and neutrino magnetic moments. The results are based on the condition that there are not too many relativistic species present in the cosmic plasma at BBN, namely it is assumed that

$$\Delta N_\nu < 0.5 - 1.0 \tag{21}$$

If neutrino mass is non-zero then right-handed components of neutrinos are always produced kinematically. The production is most efficient when the temperature is comparable to the masses of the intermediate bosons and the production rate can be estimated as:

$$\Gamma_R = \frac{\dot{n}_{\nu_R}}{n_{\nu_R}} = 10 \left(\frac{m_\nu}{T}\right)^2 \frac{\Gamma_W^\nu n_W + \Gamma_Z^\nu n_Z}{T^3}, \tag{22}$$

where Γ_W and Γ_Z are the widths of decays of W and Z into neutrinos respectively.

One can check that with the existing upper bound on neutrino mass the right-handed components are never produced abundantly.

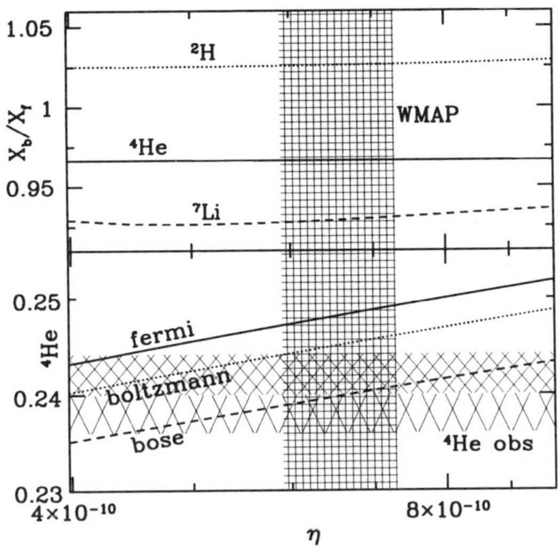

Figure 2: Upper: ratios of nuclear abundances for pure bosonic neutrinos with respect to fermionic ones as functions of $\eta = N_B/N_\gamma$. Vertical cyan region is WMAP 2σ determination of η. Lower: ^{4}He mass fraction for the purely bosonic, Boltzmann, and fermionic neutrinos. ($\kappa = -1, 0, +1$ respectively). The two skew hatched regions show the observation of primordial ^{4}He helium from Fields-Olive (in ref. [5]) (lower, yellow) and Izotov-Thuan [23] (upper, magenta),

Direct production of ν_R through possible right-handed currents would be allowed if ν_r decoupled before the QCD phase transition. In this case, even if ν_R were in equilibrium prior to the phase transition, their number density would be sufficiently diluted by the entropy release. This condition leads to:

$$\frac{m_{W_R}}{m_W} > \left(\frac{T_{QCD}}{3\,\text{eV}}\right)^{4/3} \tag{23}$$

Thus $m_{W_R} >$ (a few) TeV. It is an order of magnitude better than direct experimental limits.

If neutrinos are massive they may have a non-vanishing magnetic moment, μ_ν. They do have it in the standard model but it is extremely small, about $3{\cdot}10^{-19}\mu_B(m_\nu/\text{eV})$, where μ_B is the Born magneton of electron [24]. If there are some new interactions, then μ_ν may be much larger and experimental or cosmological/astrophysical bounds on μ_ν are of interest. With nonzero μ_ν right-handed ν_R would be produced in e.m. interactions:

$$e^\pm + \nu_L \to e^\pm + \nu_R \tag{24}$$

Demanding $\Delta N_\nu < 0.5$ we obtain:

$$\mu_\nu < 3 \times 10^{-10}\mu_B \tag{25}$$

There is another mechanism of production of right-handed neutrinos, namely spin rotation in magnetic field. The result depends upon the magnitude of primordial magnetic field:

$$\mu_\nu < 10^{-6}\mu_B\,(B_{primord}/\text{Gauss})^{-1} \tag{26}$$

If intergalactic magnetic fields, $B_{int-gal} \sim 10^{-6}$ Gauss were generated in the early universe and evolved adiabatically, then:

$$\mu_\nu < 10^{-19}\mu_B \tag{27}$$

This is an extremely strong limit by 8-9 orders of magnitude stronger than any other one, but it depends upon the hypothesis on the cosmological evolution of the primordial magnetic fields.

References

[1] A.D. Dolgov, *Phys. Repts.* **370**, 333 (2002).
[2] S.S. Gershtein, Ya.B. Zeldovich, *Pis'ma Zh. Eksp. Teor. Fiz.* **4**, 174 (1966); English translation: *JETP Lett.* **4**, 120 (1966).
[3] R. Cowsik, J. McClelland, *Phys. Rev. Lett.* **29**, 669 (1972).
[4] A. D. Dolgov, *Yad. Fiz.* **33** 1309 (1981); English translation: *Sov. J. Nucl. Phys.* **33** 700 (1981);
A.D. Dolgov, Ya.B. Zeldovich, *Rev. Mod. Phys.* **53** 41 (1981).
[5] S. Eidelman *et al.*, *Physics Letters* **B592**, 1 (2004).
[6] G.F. Giudice, E.W. Kolb, A. Riotto, *Phys. Rev.* **D64** 023508 (2001).
[7] J.F. Beacom, N.F. Bell, S. Dodelson, *Phys.Rev.Lett.* **93** 121302 (2004).
[8] D.N. Spergel *et al.*, *Astrophys. J. Suppl.* **148**, 175 (2003).
[9] S. Hannestad, *JCAP*, **0305**, 004 (2003).
[10] S.W. Allen, R.W. Schmidt, S.L. Bridle, *Mon. Not. Roy. Astron. Soc.* **346** 593 (2003).
[11] M. Tegmark *et al.* [SDSS Collaboration], *Phys. Rev.* **D69** 103501 (2004).
[12] V. Barger, D. Marfatia, A. Tregre, *Phys. Lett.* **B595** 55 (2004).
[13] P. Crotty, J. Lesgourgues, S. Pastor, *Phys.Rev.* **D69** 123007(2004).
[14] U. Seljak it et al [SDSS Collaboration], *Phys.Rev.* **D71** 103515 (2005).
[15] G.L. Fogli, E. Lisi, A. Marrone, A. Melchiorri, A. Palazzo, P. Serra J. Silk, *Phys.Rev.* **D70** 113003 (2004).
[16] S. Hannestad, hep-ph/0412181.
[17] G. Steigman, *Int. J. Mod. Phys.* **E15**, 1 (2006).
[18] A.D. Dolgov, S.H. Hansen, S. Pastor, S.T. Petcov, G.G. Raffelt D.V. Semikoz, *Nucl. Phys.* **B632**, 363 (2002).
[19] A.D. Dolgov, F. Takahashi, *Nucl. Phys.* **B688** 189 (2004).
[20] A.D. Dolgov, F.L. Villante, *Nucl. Phys.* **B679** 261 (2004).
[21] A.D. Dolgov, A.Yu. Smirnov, *Phys. Lett.* **B621**, 1 (2005);
A.D. Dolgov, hep-ph/0504238.
[22] A.D. Dolgov, S.H. Hansen, A.Yu. Smirnov, *JCAP* **0506** 004 (2005).
[23] Y.I. Izotov,T.X. Thuan, *Astrophys. J.* **602**, 200 (2004).
[24] B.W. Lee, R.E. Shrock, *Phys. Rev.* **D16** 1444 (1977). W.J. Marciano, A.I. Sanda, *Phys. Lett.* **B67** 303 (1977).

DIP IN UHECR SPECTRUM AS SIGNATURE OF PROTON INTERACTIONS WITH CMB

Veniamin Berezinsky,
INFN, Laboratori Nazionali del Gran Sasso, I-67010 Assergi (AQ) Italy and Institute for Nuclear Research of RAN, Moscow

Abstract. The dip is a feature in the diffuse spectrum of UHE protons in energy range $1 \times 10^{18} - 4 \times 10^{19}$ eV, which is caused by electron-positron pair production on CMB photons. Calculated for power-law generation spectrum with index $\gamma_g = 2.7$, the shape of the dip is confirmed with high accuracy by data of Akeno - AGASA, HiRes, Yakutsk and Fly's Eye detectors. The predicted shape of the dip is robust: it is valid for the rectilinear and diffusive propagation, for different discretenesses in the source distribution, for local source overdensity and deficit, for source inhomogeneities on scale $\ell \lesssim 100$ Mpc etc. Below the characteristic energy $E_b \approx 1 \times 10^{18}$ eV the spectrum of the dip flattens for both diffusive and rectilinear propagation, and more steep galactic spectrum becomes dominant at $E < E_b$. The energy of transition $E_{\rm tr} < E_b$ approximately coincides with the position of the second knee E_{2kn} observed in the cosmic ray spectrum. The critical energy E_b is determined by the energy $E_{\rm eq} = 2.3 \times 10^{18}$ eV, where adiabatic and pair-production energy losses are equal. Thus, position of the second knee is explained by energy losses of extragalactic protons on CMB photons.

1 Introduction

The most natural primary particles for UHECR are extragalactic protons. Due to interaction with the CMB radiation the UHE protons from extragalactic sources are predicted to have a sharp steepening of energy spectrum, so called GZK cutoff [1].

There are two other signatures of extragalactic protons in the spectrum: dip and bump [2, 3]. The dip is produced due to $p + \gamma_{\rm CMB} \to p + e^+ + e^-$ interaction. The bump is produced by pile-up protons which loose energy in the GZK cutoff. As was demonstrated in [3], the bump is clearly seen in the calculations for a single source at large redshift z, but it practically disappears in the diffuse spectrum, because individual peaks are located at different energies.

As it is discussed in this paper, the dip is a reliable feature in the UHE proton spectrum (see also [4] - [6]). Being relatively faint feature, it is however clearly seen in the spectra observed by AGASA, Fly's Eye, HiRes and Yakutsk arrays (see [7] - [12] for the data). We argue that it can be considered as the confirmed signature of interaction of extragalactic UHE protons with CMB.

The dip has low-energy and high-energy flattenings (see Fig 1). The high-energy flattening at $E_a \approx 1 \times 10^{19}$ eV reproduces the *ankle*, the feature well seen in the observational data. The low-energy flattening at $E_b \approx 1 \times 10^{18}$ eV, seen for both rectilinear [4] and diffusive [13–15] propagation, is a natural place for transition from extragalactic to galactic component of cosmic rays. Somewhere below E_b the more steep galactic component must appear. Thus E_b describes the beginning of transition if one follows it from high energy side. We shall demonstrate here (see also [15]) that E_b is connected with $E_{\rm eq} = 2.3 \times 10^{18}$ eV, where pair-production and adiabatic energy losses are equal. The visible transition from galactic to extragalactic cosmic rays occurs at $E_{\rm tr} < E_b$, and this energy coincides well with position of the *second knee* (Akeno - 6×10^{17} eV, Fly's Eye - 4×10^{17} eV, HiRes - 7×10^{17} eV and Yakutsk - 8×10^{17} eV).

The transition from galactic to extragalactic cosmic rays at $E_b \approx 1 \times 10^{18}$ implies the dominance of the proton composition of the observed cosmic rays at $E \gtrsim 1 \times 10^{18}$ eV. While HiRes [16] HiRes-MIA [17] and Yakutsk [18] data support this prediction and Haverah-Park [11] data do not contradict it at $E \gtrsim (1-2) \times 10^{18}$ eV, the Akeno [19] and Fly's Eye [9] data favour the mixed composition, dominated by heavy nuclei (for a review see [12] and [20]).

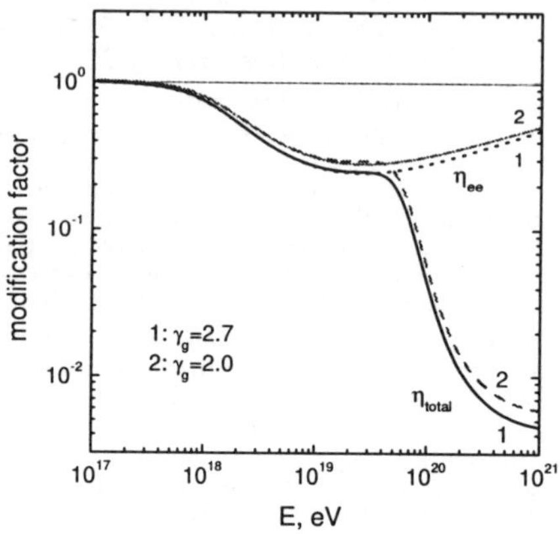

Figure 1: Modification factor for the power-law generation spectra with γ_g in the range 2.0 - 2.7. Curve $\eta = 1$ corresponds to adiabatic energy losses, curves η_{ee} - to adiabatic and pair production energy losses and curves η_{tot} - to total energy losses.

Below we shall analyze the features in UHE proton spectrum using basically two assumptions: the uniform distribution of the sources in the universe and the power-law generation spectrum. We do not consider the possible speculations, such as cosmological evolution of sources.

2 The dip

The analysis of the dip is convenient to perform in terms of the *modification factor*. Modification factor is defined as a ratio of the spectrum $J_p(E)$ with all energy losses taken into account, and unmodified spectrum $J_p^{\rm unm}$, where only adiabatic energy losses (red shift) are included.

$$\eta(E) = \frac{J_p(E)}{J_p^{\rm unm}(E)}. \tag{1}$$

Modification factor is less model-dependent quantity than the spectrum. In particular, it depends weakly on γ_g, because both numerator and denominator in Eq. (1) include $E^{-\gamma_g}$. In this paper we shall consider the non-evolutionary case $m = 0$. In Fig 1 the modification factors are shown as a function of energy for two spectrum indices $\gamma_g = 2.0$ and $\gamma_g = 2.7$. They do not differ much.

Comparison of the predicted dip (η_{ee} curve) with observational data are shown in Fig 2 for $\gamma_g = 2.7$. One can observe the excellent agreement with AGASA, HiRes and Yakutsk data (and also with Fly's Eye data not presented here), while the Auger spectrum, at this preliminary stage, does not contradict the dip.

2.1 Energy calibration of the detectors by dip

The systematic errors in energy determination of existing detectors exceed 20%. The dip can be used for energy calibration of detectors. Assuming the energy-independent

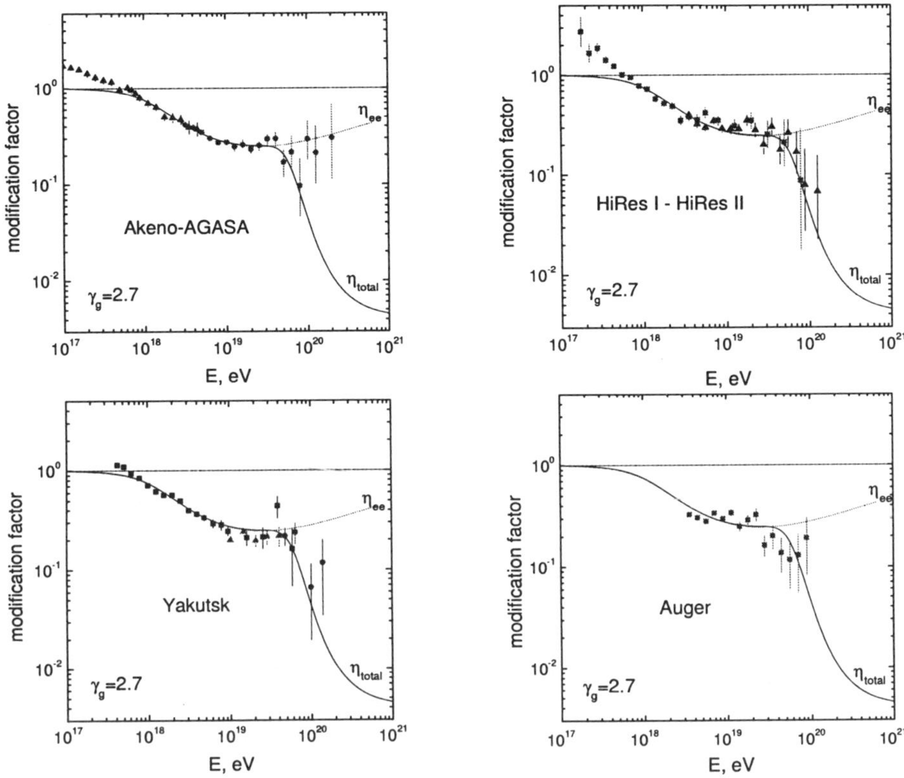

Figure 2: Predicted dip in comparison with AGASA, HiRes, Yakutsk and Auger [21] data.

systematic errors one can shift measured energy by factor λ to obtain the minimum χ^2 in comparison with theoretical curve. This procedure has been performed for each detector and it results in energy shift $\lambda_{\rm Ag} = 0.9$, $\lambda_{Ya} = 0.75$ and $\lambda_{\rm Hi} = 1.2$, for AGASA, Yakutsk and HiRes detectors, respectively. It is very interesting that after this energy calibration the measured fluxes for all detectors coincide. In Fig 3 the spectra of Akeno-AGASA, Yakutsk and HiRes are shown before and after this energy calibration. In particular one can see the good agreement of AGASA and HiRes data at $E < 1 \times 10^{20}$ eV and their consistency at $E > 1 \times 10^{20}$ eV. This result should be considered together with calculations [22], where it was demonstrated that 11 superGZK AGASA events can be simulated by the spectrum with GZK cutoff in case of 30% error in energy determination. We may tentatively conclude that existing discrepancy between AGASA, HiRes and Yakutsk fluxes and spectra at all energies may have the statistical origin. The noticeable difference between energy shift in AGASA and Yakutsk (0.75 - 0.9) on one side and HiRes (1.2) on the other, may be the evidence for difference in energy measurement by on-ground and fluorescent methods.

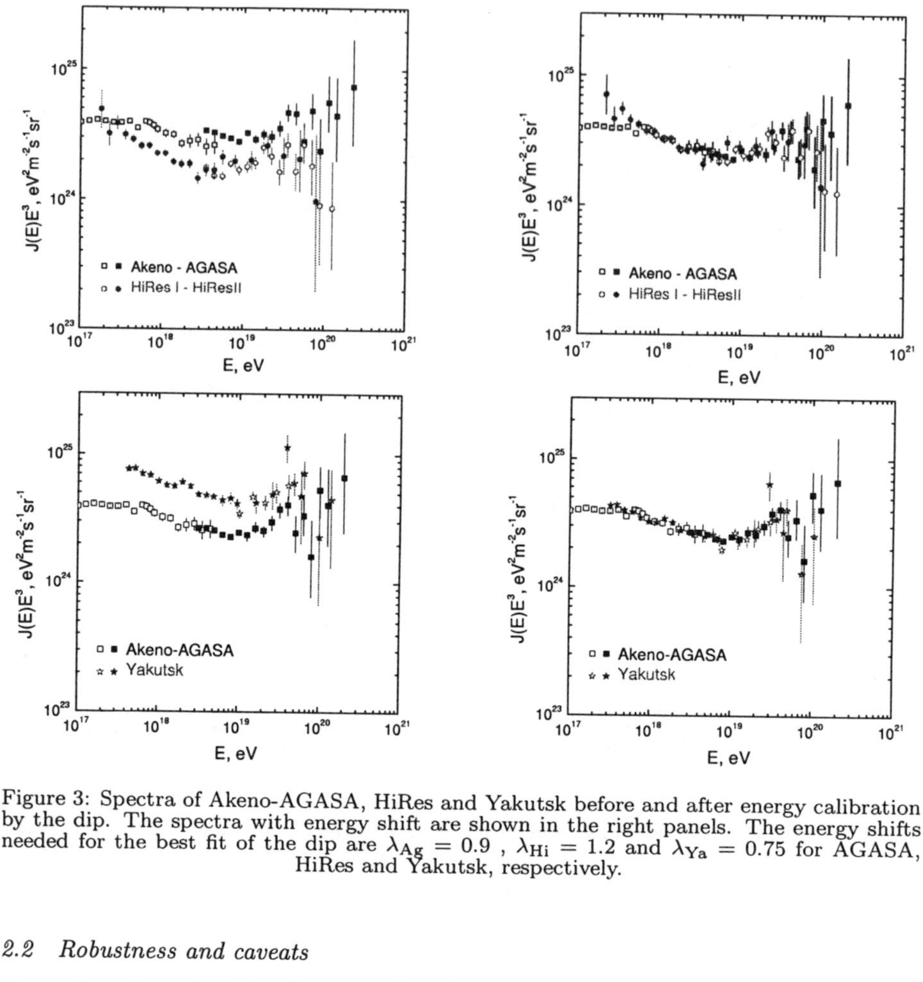

Figure 3: Spectra of Akeno-AGASA, HiRes and Yakutsk before and after energy calibration by the dip. The spectra with energy shift are shown in the right panels. The energy shifts needed for the best fit of the dip are $\lambda_{\rm Ag} = 0.9$, $\lambda_{\rm Hi} = 1.2$ and $\lambda_{\rm Ya} = 0.75$ for AGASA, HiRes and Yakutsk, respectively.

2.2 Robustness and caveats

How robust is prediction of the dip and what are uncertainties?

The shape of the calculated proton dip is stable relative to many phenomena included in calculations, namely, discreteness in the source distribution, mode of propagation (diffusive and rectilinear), inhomogeneities in source distribution on the scale $\ell \lesssim 100$ Mpc, local source overdensity and deficit, acceleration $E_{\rm max}$ and fluctuations in energy losses. However, it is modified by the assumption of source evolution [4] and by presence of UHE nuclei in primary flux [6] (see also [23]). In Fig 4 the dips for helium and iron nuclei are presented in comparison with the proton dip. One can see that presence of 15 - 20 % of nuclei in the primary flux upsets the good agreement of proton dip with observations. The presence of primary nuclei is unavoidable feature of acceleration origin of UHECR. Any acceleration mechanism operates in the gas where at least helium with cosmological concentration $n_{\rm He}/n_{\rm H} = 0.08$ is present. However, the nuclei abundance in the accelerated flux is very much model-dependent: the nuclei fraction depends on the model of acceleration, most notably through injection mechanism, on ionization state of the gas etc. The non-ionized He-nuclei are not involved in acceleration process. Accelerated He nuclei can destroyed in a source.

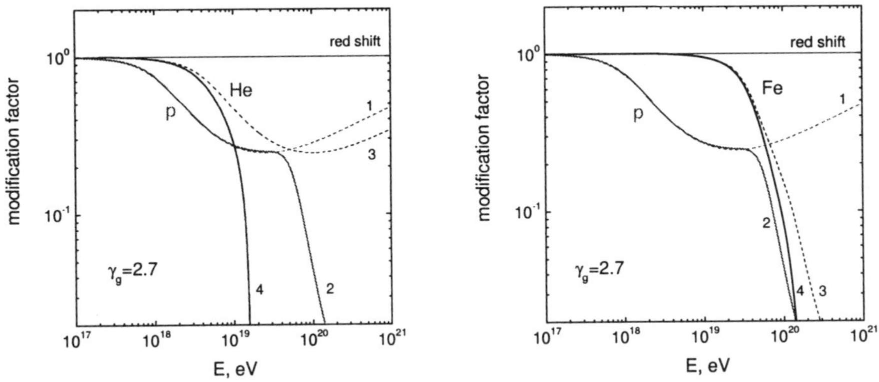

Figure 4: Modification factors for helium and iron nuclei in comparison with that for protons. Proton modification factors are given by curves 1 and 2. Nuclei modification factors are given by curves 3 (adiabatic and pair production energy losses) and by curves 4 (with photodisintegration included).

If the ratio of generation rates $Q_{\rm He}(E)/Q_p(E) = \lambda$, the modification factor for the mixed composition is given by

$$\eta(E) = \frac{\eta_p(E) + \lambda\eta_{\rm He}(E)}{1+\lambda}, \qquad (2)$$

In Fig 5 the modification factors for mixed composition with $\lambda = 0.1$ and $\lambda = 0.2$ are shown in comparison with the AGASA data. The case of $\lambda = 0.1$ (c.f. with $n_{\rm He}/n_{\rm H} = 0.08$) agrees well with observations.

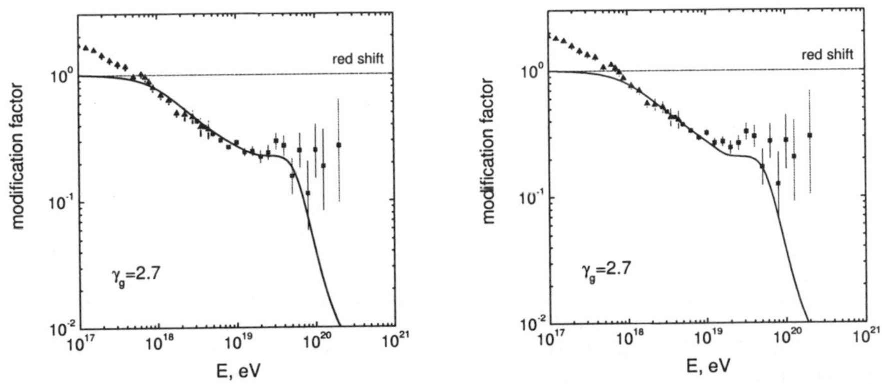

Figure 5: Modification factors for the mixed composition of protons and helium nuclei in comparison with AGASA data. The left panel corresponds to mixing parameter $\lambda = 0.1$, and the right panel to $\lambda = 0.2$.

If agreement of the proton dip with observations is not accidental (the probability of this is small according to small χ^2/d.o.f.), Fig 5 should be interpreted as indication to possible acceleration mechanism. It cannot be non-relativistic shock acceleration, which results in large λ, but it can be ultra-relativistic shock acceleration. Correlation with BL Lacs indicates the jet acceleration.

Another option is given by photodisintegration of UHE nuclei by radiation.

3 Transition from extragalactic to galactic cosmic rays

We obtained several indications that transition from extragalactic to galactic cosmic rays occurs at $E \approx 1 \times 10^{18}$ eV:
The dip predicted above 1×10^{18} eV describes perfectly well the observed spectra (see the modification factors in Fig. 2). However, one can observe that at $E \lesssim 1 \times 10^{18}$ experimental modification factor becomes $\eta > 1$, in contrast to definition $\eta \leq 1$. It signals for appearance of a new component, which can be nothing but the galactic cosmic rays.

The spectra calculated for rectilinear [4] and diffusive [14,15] propagation also show the flattening at $E \lesssim 1 \times 10^{18}$ eV.

What is the reason of this universality? We study the transition, moving from high towards low energies, denoting by E_b the beginning of transition. It is determined by energy $E_{\rm eq} = 2.3 \times 10^{18}$ eV, where adiabatic and pair-production energy losses become equal. The quantitative analysis of this connection is given in [15]. We shall give here the semi-quantitative explanation.

The flattening of the spectrum occurs at energies $E \leq E_b$, where $E_b = E_{\rm eq}/(1+z_{\rm eff})^2$ and $z_{\rm eff}$ should be estimated as redshift up to which the main contribution to unmodified spectrum occurs. The simplified analytic estimate for $\gamma_g = 2.6 - 2.8$ gives $1 + z_{\rm eff} \approx 1.5$ and hence $E_b \approx 1 \times 10^{18}$ eV.

In experimental data the transition is searched for as a feature started at some low energy $E_{\rm 2kn}$ - the second knee. Its determination depends on experimental procedure, and all we can predict is $E_{\rm 2kn} < E_b$. Determined in different experiments $E_{2kn} \sim (0.4 - 0.8) \times 10^{18}$ eV.

Being thought of as purely galactic feature, the second knee in our analysis appears as consequence of extragalactic proton energy energy losses.

The transition at the second knee is illustrated by Fig. 6. The clue to understanding of this transition is given by the KASCADE data. They confirm the rigidity model, according to which position of a knee for nuclei with charge Z is connected with the position of the proton knee E_p as $E_Z = ZE_p$. The heaviest nuclei (iron) start to disappear at $E > E_{\rm Fe} = 6.5 \times 10^{16}$ eV, if the proton knee is located at $E_p \approx 2.5 \times 10^{15}$ eV. The shape of the spectrum above the iron knee ($E > E_{\rm Fe}$) is model dependent, with two reliably predicted features: it must be steeper than the spectrum below the iron knee, shown by the dash-dot curve, and iron nuclei must be the dominant component there. We calculated this spectrum (curve gal.Fe in Fig. 6) subtracing the extragalactic proton spectrum (see below) from observational data shown in Fig. 6. The extragalactic proton spectrum has a characteristic energy E_b, below which the spectrum becomes more flat, i.e. drops down when multiplied to $E^{2.5}$ (see Fig. 6). This part of the spectrum is presented in Fig. 6 for the diffusive propagation according to calculations in [15]. The two falling parts of the spectrum in Fig. 6 (galactic and extragalactic) inevitably intersect at some energy $E_{\rm tr}$, which can be defined as transition energy. The end of galactic cosmic rays $E_{\rm Fe} = 6.5 \times 10^{16}$ eV and the beginning of full dominance of extragalactic component $E_b \approx 1 \times 10^{18}$ eV differ by an order of magnitude.

The second-knee transition gives an alternative possibility in comparison with ankle-transition hypothesis known from end of 70s. It is inspired by flattening of the spectrum at $E_a \approx 1 \times 10^{19}$ eV seen in the AGASA and Yakutsk data (left panels

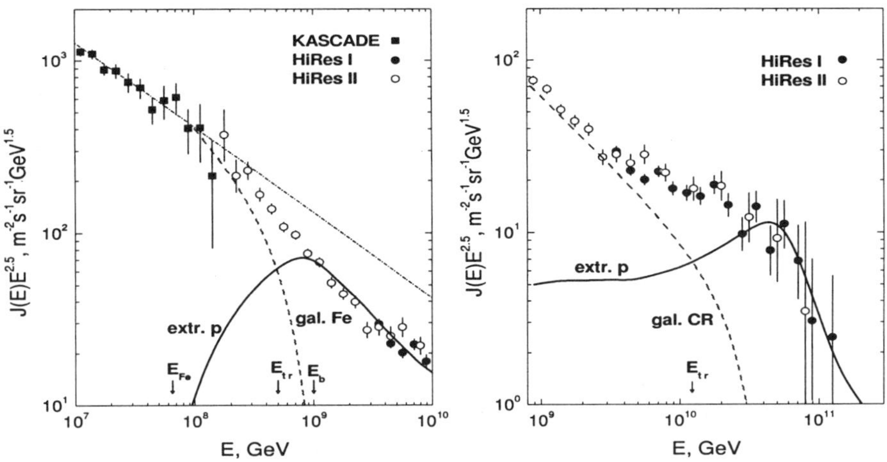

Figure 6: Transition from extragalactic to galactic cosmic rays at the second knee (left panel) and ankle (right panel) models. In the left panel are shown: KASCADE total spectrum, which above the iron knee $E_{\rm Fe}$ is composed mostly by iron nuclei (gal.Fe curve), below E_b the extragalactic proton spectrum (extr.p curve) is calculated for diffusive propagation [15] and $E_{\rm tr}$ is the energy of transition from galactic cosmic rays to extragalactic protons. The dot-dash line shows power-law extrapolation of low-energy KASCADE spectrum. In the right panel extragalacic proton spectrum is calculated for generation spectrum $\propto E^{-2}$, while galactic spectrum (curve gal.CR) is taken as difference between the observed total spectrum and calculated spectrum of extragalactic protons.

in Fig. 2). Being multiplied to factor $E^{2.5}$, as in Fig. 6, the ankle transition looks very similar to that at the second knee. Note that in the latter case the ankle is just an automatic part of the dip.

In the ankle model it is assumed that galactic cosmic ray spectrum has a power-law shape $\propto E^{-\gamma}$ from the proton knee $E_p \approx 2.5 \times 10^{15}$ eV to about $E_a \sim 1 \times 10^{19}$ eV where it becomes steeper and crosses the more flat extragalactic spectrum. The ankle transition in Fig. 6 is shown for extragalactic proton spectrum with generation index $\gamma_g = 2$, while galactic spectrum, given by curve "gal.CR" is calculated as difference of the observed total spectrum and calculated extragalactic proton spectrum.

The ankle model has the problems with galactic component of cosmic rays. In the rigidity models the heaviest nuclei (iron) start to disappear at $E > E_{\rm Fe} = 6.5 \times 10^{16}$ eV. How the gap between 1×10^{17} eV and 1×10^{19} eV is filled?

Galactic protons start to disappear at $E > 2.5 \times 10^{15}$ eV. Where they came from at $E > 1 \times 10^{17}$ eV to be seen e.g. in the Akeno detector with fraction 10%?

The ankle model needs acceleration by galactic sources up to 1×10^{19} eV, which is difficult to afford. The second knee model ameliorates this requirement by one order of magnitude.

The second-knee model predicts the spectrum shape above 1×10^{18} eV with extremely good accuracy (χ^2/d.o.f.= 1.12 for Akeno-AGASA and χ^2/d.o.f.= 0.908 for HiRes). In the ankle model one has to consider this agreement as accidental, though such hypothesis has very small probability, determined by χ^2 cited above.

4 Conclusions

I have presented the model-independent analysis of spectra of extragalactic protons interacting with CMB. Basically only two assumptions are involved in this analysis: the generation spectrum at $E \gtrsim 1 \times 10^{18}$ is power-law, $\propto E^{-\gamma_g}$, and sources are distributed uniformly in the space, with some distance d between them. We found *dip*, produced by $p + \gamma_{\rm CMB} \to e^+ + e^-$ at energies $1 \times 10^{18} \lesssim E \lesssim 4 \times 10^{19}$ eV, as most interesting and reliable feature. The predicted shape of the dip is robust: it is valid for the rectilinear and diffusive propagation, for different discretenesses in the source distribution, for local source overdensity or deficit, for source inhomogeneities on scale $\ell \lesssim 100$ Mpc, and it is not affected by fluctuations in proton interaction with CMB photons. For comparison with experimental data only two free parameters are needed: the spectral index γ_g and normalization of the flux. The number of energy bins in the data of Akeno-AGASA, HiRes and Yakutsk (independently) is 21 - 23. The agreement is characterized by χ^2/d.o.f. ≈ 1. The agreement with Fly's Eye is also very good. Being the faint spectrum feature, dip thus is reliably seen in experimental data.

It implies confirmation of the proton dip by observations and discovery of UHE proton interaction with CMB.

The dip gives a possibility of energy calibration for the detectors. Shifting the measured energy by factor λ for each detector, one may obtain the minimum χ^2 for agreement the data and predicted spectrum. This factor is found to be 0.75, 0.9 and 1.2 for Yakutsk, AGASA and HiRes. *After this energy shift (calibration) the fluxes and spectra of all three detectors remarkably coincide (see Fig. 3).* The difference between shift for on-ground measurements of Yakutsk and AGASA (0.75 - 0.9) and fluorescent method of HiRes (1.2) may signal about discrepancy of these methods for energy determination. At energy $E > 1 \times 10^{20}$ eV the AGASA data show the excess of events in comparison with Yakutsk and HiRes. After energy calibration this excess is not statistically significant (about 2.5σ).

The presence of $\gtrsim 20\%$ of UHE nuclei in primary radiation upsets the good agreement of the dip with observations. This can be considered as indication to mechanism of acceleration. The different shapes of the dip for protons and nuclei (see Fig. 4) can be used for determination of the chemical composition.

Presence of the confirmed dip implies the transition from extragalactic to galactic cosmic rays at the second knee. This transition is explained by energy losses of extragalactic protons, whose spectrum becomes flatter at $E < 1 \times 10^{18}$ eV, and thus more steep galactic spectrum starts to dominate (Fig. 6).

References

[1] K. Greisen, *Phys.Rev.Lett.* **16**, 748 (1966);
G. T. Zatsepin, V.A. Kuzmin, *Pisma Zh.Experim.Theor.Phys.* **4**, 114 (1966).
[2] C. T. Hill and D. N. Schramm, *Phys.Rev.* **D 31**, 564 (1985).
[3] V. S. Berezinsky and S. I. Grigorieva, *Astron.Astroph.* **199**, 1 (1988).
[4] V. Berezinsky, A. Z. Gazizov, S. I. Grigorieva, hep-ph/0204357.
[5] V. Berezinsky, A. Z. Gazizov, S. I. Grigorieva, astro-ph/0210095.
[6] V. Berezinsky, A. Z. Gazizov, S. I. Grigorieva, *Phys. Lett.* **B 612**, 147 (2005) [astro-ph/0502550].
[7] K. Shinozaki and M. Teshima, *Nucl. Phys. B (Proc. Suppl.)* **136**, 18 (2004).
[8] HiRes collaboration, *Phys. Rev. Lett.* **92**, 151101 (2004).
[9] D. J. Bird *et al.* [Fly's Eye collaboration],*Ap.J.* **424**, 491 (1994).
[10] A. V. Glushkov et al. [Yakutsk collaboration] Proc. of 28th Int. Cosmic Ray Conf. (Tsukuba, Japan), **1**, 389 (2003);

[11] M. Ave et al. [Haverah Park collaboration], *Astrop.Phys.* **19**, 61 (2003).
[12] M. Nagano and A. Watson, *Rev.Mod.Phys.* **72**, 689 (2000).
[13] R. Aloisio and V. Berezinsky, *Astroph.J.* **612**, 900 (2004).
[14] M. Lemoine, *Phys.Rev.* **D 71**, 083007 (2005).
[15] R. Aloisio and V. Berezinsky, *Astroph.J.* **625**, 249 (2005).
[16] G. Archbold and P. V. Sokolsky, Proc. of 28th ICRC, 405 (2003).
[17] T.Abu-Zayyad et al.,*Phys.Rev.Lett.* **84**, 4276 (2003).
[18] A. V. Glushkov *et al.* (Yakutsk collaboration) *JETP Lett.* **71**, 97 (2000).
[19] M. Honda et al [Akeno collaboration], *Phys.Rev.Lett.* **70**, 525 (1993).
[20] A. Watson, *Nucl.Phys. B (Proc.Suppl.)* **136**, 290 (2004).
[21] The Pierre Auger Collaboration, astro-ph/0507150 (2005).
[22] D. De Marco, P. Blasi, A. Olinto, astro-ph/0507324 (2005).
[23] D. Allard et al, astro-ph/0505566 (2005).

THE SCIENCE OF PAMELA SPACE MISSION

P.Picozza [a], A.Morselli [b]
University and INFN of Roma "Tor Vergata", Roma, Italy

Abstract. Many different mechanisms can contribute to antiprotons and positrons production, ranging from conventional reactions up to exotic processes like neutralino annihilation. The open problems are so fundamental (i.e.: is the universe symmetric in matter and antimatter ?) that experiments in this field will probably be of the greatest interest in the next years. Here we will summarize the present situation, showing the different hypothesis and models and the experimental measurements needed to lead to a more established scenario.

1 The main questions in Cosmic Ray Research

There are some fundamental questions in Cosmic Ray Research that still need to be answered:

- Where do the particles come from?
- How and where do they get accelerated?
- How do they propagate through the interstellar medium and what kind of interactions do they encounter?
- What role do they play in the energy budget of the interstellar medium?
- Do we find hints of the existing of "exotic" particles?

The study of antimatter and antiparticle content in cosmic rays is a unique tool to investigate several physics and astrophysical phenomena. The idea of exploiting cosmic antiprotons measurements to probe unconventional particle physics and astrophysics scenarios has a long history (see references in [1]) and moved the cosmologists for several decades. The search of antimatter is strictly connected with the baryon antibaryon asymmetry in the Universe. The present observational limit in the search of antihelium is in the order of 10^{-6} in the $\bar{\text{He}}/\text{He}$ ratio. PAMELA will extend this limit to the $\sim 10^{-8}$ level.

The majority of the antiproton and positron components in cosmic rays are of secondary origin, but part of them could come from other primary sources. The search and the identification of such possible sources is one of the major challenges in cosmic rays studies. In fact several observations give us a coherent picture of a Universe dominated by dark matter and dark energy: the detection of an anomalous content of antiprotons or positrons in cosmic rays could be a clear signature of a source connected with the presence of dark matter, e.g. WIMPs or Kaluza-Klein particles annihilation. Furthermore, the precise measurement of the secondary antiproton flux is a decisive factor to well understand the propagation of cosmic rays in the galaxy, while secondary to primary CR ratios are very sensitive to the change of the propagation parameters. The

---

[a] e-mail: piergiorgio.picozza@roma2.infn.it
[b] e-mail: aldo.morselli@roma2.infn.it

knowledge of secondary antiproton flux gives complementary information with respect to that obtained from nuclei like Boron or sub-iron nuclei [2].

2 The PAMELA experiment

Figure 1 shows a picture of the apparatus that consists of: a time of flight system, a magnetic spectrometer, an anticoincidence system, an electromagnetic imaging calorimeter, a shower tail catcher scintillator and a neutron detector. PAMELA will be launched into space by a Soyuz TM2 rocket in the 2006. The position in the satellite is shown in figure 2. The orbit is elliptical and semi-polar, with an inclination of 70^0 and an altitude varying between 350 km and 600 km; the mission will last at least three years. The geometric acceptance in the standard trigger configuration is $\simeq$ 20.5 cm^{2}sr. Moreover PAMELA is equipped with another trigger based only on the calorimeter signals: it permits to increase the acceptance of the $e^+ + e^-$ measurement at high energy up to about 600 cm^{2}sr.

The main PAMELA scientific primary goals and objectives are:

• Search for evidence of exotic matter as heavy antinuclei and nonbaryonic particles outside the Standard Model

• Understanding formation and evolution of our Galaxy and the Universe

• Exploring the cycles of matter and energy in the Universe.

These objectives can be persued with the search for structure in antiparticle spectra; the search for antinuclei and antimatter; the study of cosmic rays propagation by means of light nuclei and light isotopes spectra; the study of electron spectrum and investigation of the contribution of local sources; the study of solar physics and solar modulation.

The capabilities of PAMELA are illustrated in table 1 [1,4].

Particles	**Numbers (3 years)**	**Energy range**
positrons	3×10^5	50 MeV – 270 GeV
antiprotons	3×10^4	80 MeV – 190 GeV
limit on antinuclei		$\sim 10^{-8}$ $\bar{\text{He}}$/He
electrons	6×10^6	50 MeV – 2 TeV
protons	3×10^8	80 MeV – 700 GeV
electrons+positrons		50 MeV – 10 TeV
light nuclei (up to Z=6)		100 MeV/n – 200 GeV/n
light isotopes (^{2}H, ^{3}He)		100 MeV/n – 1 GeV/n

Table 1: PAMELA Capabilities

The expected performance of PAMELA in the case of a pure secondary positrons and antiprotons are shown with red boxes, respectively in figure 3

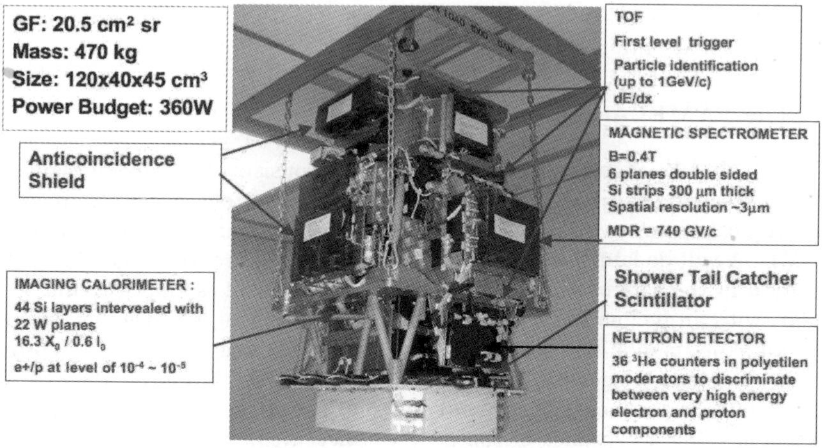

Figure 1: the PAMELA telescope in a photo taken just prior delivery to Russia. The detector is approximately 120 cm tall, has a mass of $\sim$ 470 kg and the power consumption is $\sim$ 360 W.

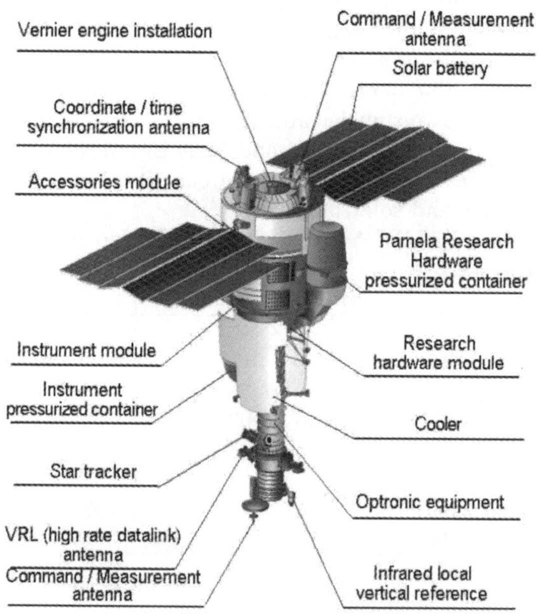

Figure 2: A sketch of the Resurs DK1 Satellite. PAMELA is mounted inside a Pressurized Container, attached to the Satellite.

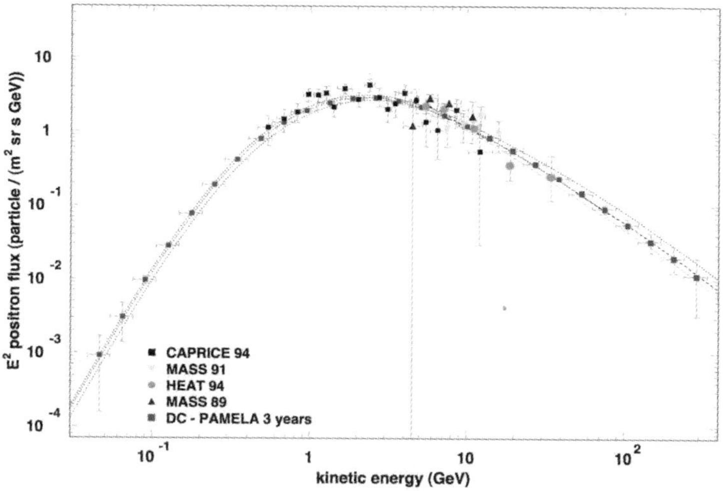

Figure 3: Experimental data ([1]) compared with PAMELA's expectations for positrons for diffuse and convention (DC) model background [2]. The central line corresponds to the parameters of the best B/C fit while the others show the uncertainties band.

and 4 in the case of a propagation model that includes diffuse and convention [2]. The errors of the expected PAMELA data points include only statistical uncertainties. An average PAMELA orbit has been used to estimate the vertical geomagnetic cutoffs and the consequent expected number of antiproton and positron events.

The primary contribution to the $\bar{p}$ flux has been computed using the public code DarkSUSY [3]. The SUSY contribution to the $\bar{p}$ flux is shown in figure 5 for a neutralino mass of 1 TeV (obtained from a particular choice of mSUGRA parameters) and a clumpiness factor fd of 5 10^4. For different values of the mSUGRA parameters one can found the minimal values of the clumpiness factors fd needed to disentangle a neutralino induced component in the antiproton flux with PAMELA. This factor can be computed as a function of the mSUGRA parameters. Fixing the less sensitive parameters A_0, $\tan\beta$ and sign(μ), the clumpiness factor become a function of m_0 and $m_{1/2}$.

For each model the minimal value of the clumpiness factor fd needed to satisfy both conditions has been found. As the clumpiness factor is a function of m_0 and $m_{1/2}$ parameters the contour plots can be made calculating equi-clumpiness factors lines. The results is shown in figure 5 for $\tan\beta = 55$. Black color represents the regions in the parameter space that are excluded either by accelerator bounds or because electroweak symmetry breaking is not achieved

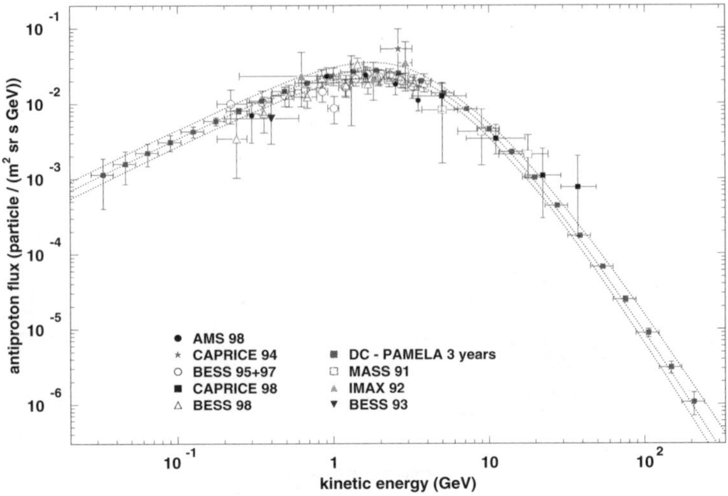

Figure 4: Same as figure 3 but for the antiproton flux.

or because the neutralino is not the lightest supersymmetric particle. Red (dark shaded) are domains with Ωh^2 in the WMAP region $0.09 < \Omega h^2 < 0.13$, while green (light shaded) are the parameter space domains with $0.13 < \Omega h^2 < 0.3$. The equi-neutralino mass contours is in blue dashed lines [2]. Other example are in [2]. It can be see that PAMELA will be able to disentangle a neutralino induced component for halo models that has fd as low as ~ 10.

PAMELA will give an fundamental contribution to the definition of the propagation parameters with the measure of different cosmic ray ratio and of Hydrogen and Helium spectra. Some examples are shown in figure 7.

Solar physics is another relevant items that PAMELA will deeply study. In particular PAMELA will measure:

- Solar Modulation effects
- High energy component of electron and positron in Solar Proton Events (from 50 MeV to 100 GeV)
- Nuclear composition of gradual and impulsive events
- ^{3}He and ^{4}He isotopic composition.

There are osservational evidence that the negative charge component of galactic cosmic rays is modulated in the heliosphere differently than the positive one. The modification and modulation of Galactic Cosmic Ray spectra

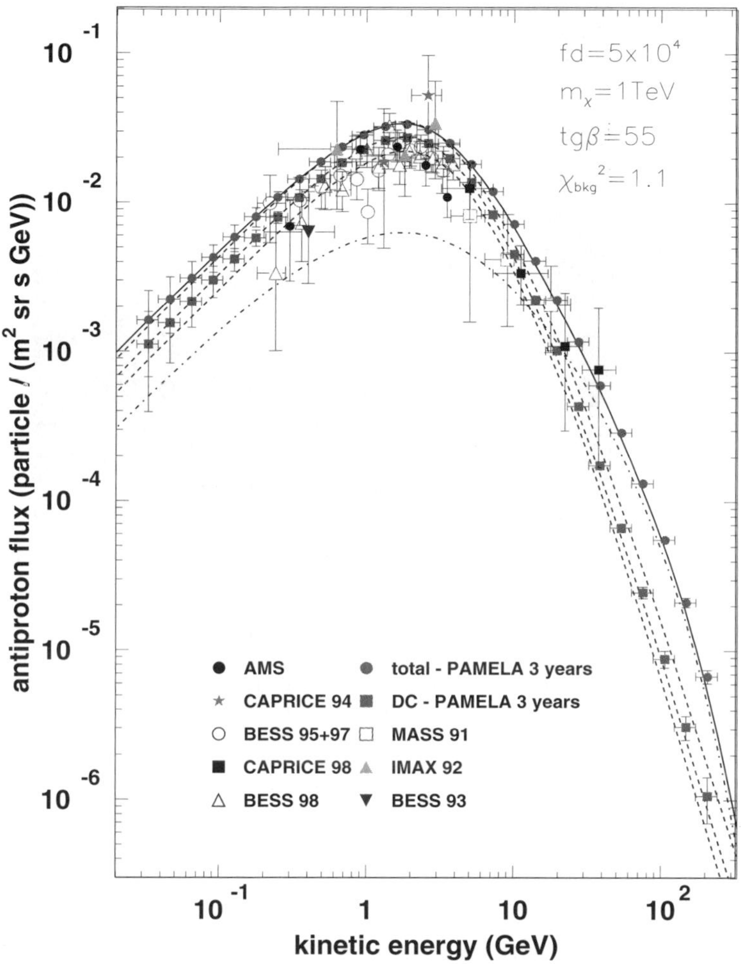

Figure 5: Antiproton absolute flux: theoretical predictions for total uncertainty and best B/C fit for DC model (dashed lines). Experimental data are from [1]. The PAMELA expectations points (red squares) for DC background are for three years of data taking. The dash-dotted line is a neutralino induced contribution for a neutralino mass of 1 TeV (see text) and a clumpiness factor fd of 5 10^4 while the solid line is the total contribution calculated with the addition of the DC background and the red circles are the corresponding PAMELA points [2].

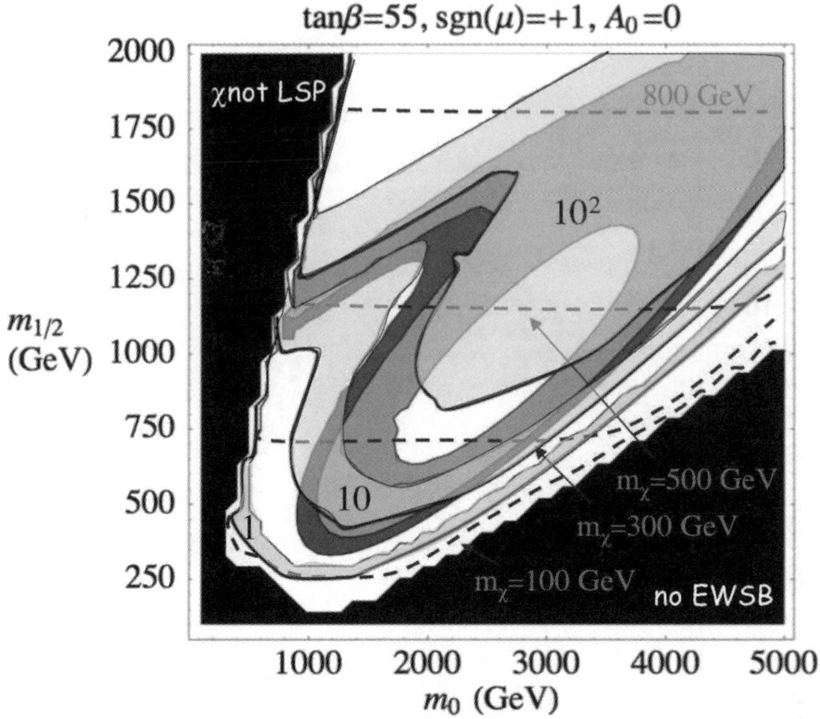

Figure 6: Contour plots for the minimum fd needed for a PAMELA disentanglement (upper bounds of the translucent bands) and for the maximum fd allowed by current experimental data (lower bounds of the translucent bands). The translucent regions denote the domains that correspond to models detectable by PAMELA. .

in the heliosphere complicate the interpretation of the exotic matter results at low energy. PAMELA will monitor the solar modulation for a significative part of the solar cycle.

3 Conclusion

PAMELA is the first space experiment which will measure the antiprotons and positrons to high energies, $\sim$ 200 GeV, with an unprecedented statistical precision. It will set a new lower limit of finding antihelium. It will look for dark matter candidates. It will provide measurements on elemental spectra and low mass isotopes with an unprecedented statistical precision and will help to improve the understanding of particle propagation in the interstellar medium.

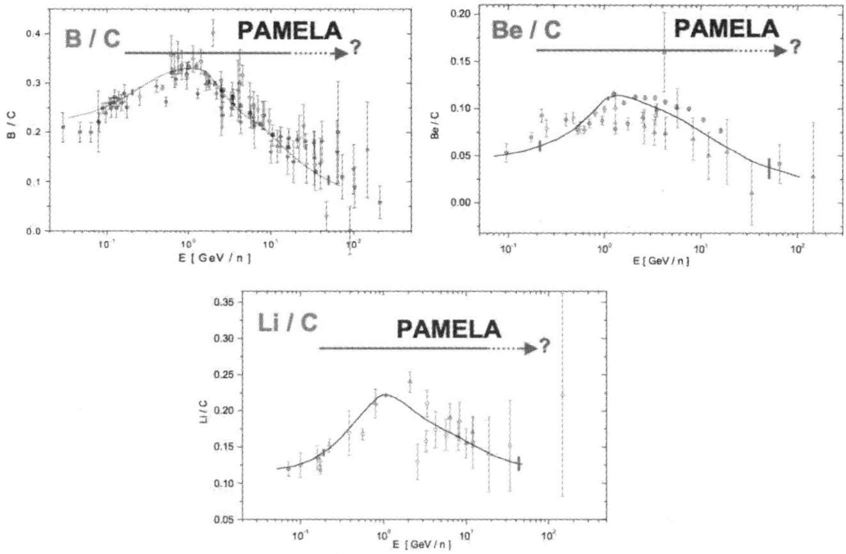

Figure 7: Present status of secondary to primary ratio measurements with the PAMELA energy range capability and some example of the PAMELA statistical errors.

It will be able to measure the high energy tail of solar spectra and for the first time solar positrons. It will be able to measure electrons at very high energy to discover sources near the solar system.

References

[1] P.Picozza, A.Morselli, (2003) J. Phys. G: Nucl. Part. Phys., 29, 903-911

[2] A.Lionetto, A.Morselli, V.Zdravkovic (2005), J. Cosmol. Astropart. Phys. JCAP09, 010 [astro-ph/0502406]

[3] Gondolo P, Edsjö J, Ullio P Bergström L, Schelke M and Baltz E A 2004 *JCAP* **07** 008 *Preprint* astro-ph/0211238

[4] Y. I. Stozhkov et al, the PAMELA Collaboration, International Journal of Modern Physics A Vol. 20, No. 29, 6745, 2005
M. Bongi et al. IEEE Transactions on Nuclear Science 51 (3): 854-859 Part 3, 2004

CURRENT STATUS AND PROSPECTS OF THE AMS EXPERIMENT

D.Rapin [a] , (on behalf of the AMS collaboration)
DPNC, Université de Genève, CH 1211 GENEVA, Switzerland

Abstract. The Alpha Magnetic Spectrometer is presently being built in prevision of a more than 3 years operation on the International Space Station ISS. With high statistics measurements of primary cosmic ray spectra from 0.5 GeV to a few TeV, its purpose is to provide a sensitive search for cosmic anti-matter and dark matter as well as cosmic ray properties. AMS is an international collaboration of ~500 persons from 56 institutes in 3 continents. The detector, its current status, its physics goals and capabilities are described.

1 Introduction

AMS addresses unsolved questions of particle physics and cosmology:

- The apparent matter-antimatter asymmetry in the Universe deserves attention. Until now, all last 20 years of cosmic ray searches for antinuclei have given negative results. High sensitivity experimental input is essential, either with positive or negative outcome. This aspect is described elsewhere in this conference [1].
- Observations indicate that the Universe should contain a large amount of unknown dark matter (DM). It may be composed of Weakly Interacting Massive Particles (WIMPs). The lightest SUSY particle, the χ neutralino, may be a candidate. It can be searched in various annihilation channels:
  $\chi + \chi \rightarrow e^+ + ...$
  $\chi + \chi \rightarrow \bar{p} + ...$
  $\chi + \chi \rightarrow \bar{d} + ...$
  $\chi + \chi \rightarrow \gamma + ...$
  Presence of such processes causes distortion of spectra and of the e^+/e^-, $\bar{p}/p$ and $\bar{d}/d$ ratios. Simultaneous measurements of all channels will add confidence in the results.
- Astrophysical models can benefit from a precise measurement of spectra for various nuclear species in cosmic rays. For instance, light nuclei like Li, Be and B are products of CNO spallation by the interstellar medium. Radio-active secondaries like ^{10}Be ($\tau = 2 \times 10^6\, y$) are *cosmic clocks*. Chemical composition and isotope ratios depend on cosmic ray confinement time and density of interstellar medium.
- High energy (GeV) γ directional detection and measurements are relevant to the knowledge and understanding of astronomical objects like *blasars*, *pulsars* and *AGNs*.

[a] e-mail: Divic.Rapin@physics.unige.ch

From an experimentalist's point of vue, direct observation of cosmic ray on a Low Earth Orbit (LEO) is challenging for what concerns the technical constraints of operation in space: reduction of weight together with mechanical resistance to vibrations during the launch, large temperature variations ($\sim 100°$), difficulty to cool electronics in vacuum and the quasi-impossibility to repair any failure. On the other hand, the quality of measurements benefits from the absence of atmosphere and from low background from trapped particles. The Space Shuttle allows transportation of large volumes and weights, and the International Space Station (ISS) provides long exposure time and high power.

2 AMS-01 prototype qualification mission

A prototype reduced version, AMS-01, was built and had the opportunity to fly and collect data for 10 days on board of the *Discovery* orbiter during the STS-91 space shuttle flight in 1998. The mean altitude was 370 km and the orbit inclination was 51.4°. This flight was the last NASA mission to dock with the MIR space station.

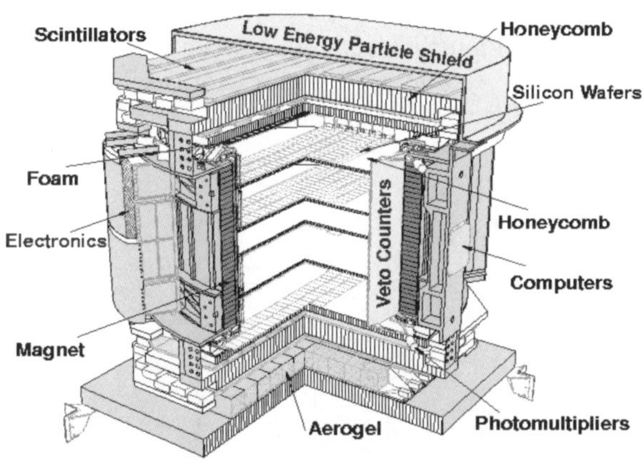

Figure 1: AMS-01 detector

Figure 1 shows the AMS-01 detector. Its heart was a magnetic spectrometer. The 0.15 Tesla field was provided by a permanent magnet arranged in a *magic ring* structure in order to have a zero total magnetic moment. Trajectories were measured by six layers of double sided Silicon microstrip detectors. The detector was not totally equipped. The total Silicon surface of the tracker reached 2 m^2. The set-up was completed by a scintillator time-of-flight system with two double layers, an Aerogel threshold Cerenkov detector and veto counters located on the magnet inner wall.

The detector thus measured the specific energy loss dE/dx, the velocity β, the charge Z and the rigidity $R \equiv pc/Ze$ of particles traversing its aperture.

The primary goal of the STS-91 mission in June 1998 was docking to and resupplying the MIR space station. For AMS, as secondary payload, the mission was to test the equipment under the stringent conditions of launch and space environment, as well as to familiarize the AMS team with space operations. After MIR undocking, AMS had control of the orbiter orientation for ninety hours. Data were taken with variables attitudes, at altitudes between 320 and 370 km. The orbit covered latitudes between $\pm 51.7°$ and all longitudes. A total of 10^8 events were recorded. Results have been obtained on spectra of electrons, positrons, protons, antiprotons, deuterons and helium. Searches have been made for antihelium and heavy anti-ions. Secondary spectra (albedo resulting from interactions in high atmosphere in relation with the geomagnetic field) have also been extensively studied. [2]

3 Cosmic ray spectra and particle ID

As shown in figure 2, large differences (several orders of magitude) between abundances of the various species of particles or nuclei, gives paramount importance to the particle identification. In particular, the detector must have very good capability of e^+/p and $\bar{p}/e^-$ separation. A large dynamic range and a good resolution in dE/dx measurements are needed for good identification of atomic species (Z).

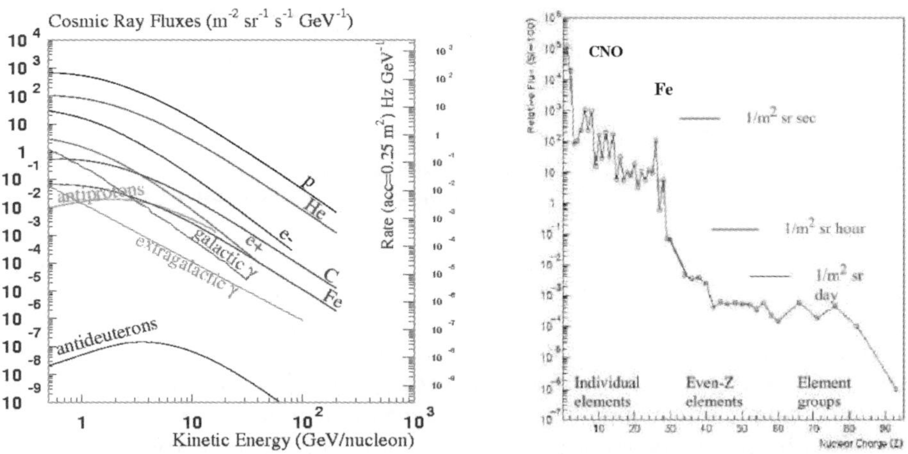

Figure 2: Left: Cosmic Ray spectra. Right: Relative abundances of nuclei in CR

4 AMS-02 Apparatus

Following the above requirements and in view of detection of rare phenomena together with high statistics measurement of cosmic ray spectra during 3 years, the AMS collaboration is constructing a large superconducting magnetic spectrometer (figure 3). Its main components are:

- Transition Radiation Detector (TRD) with capability to distinguish $e^{\pm}$ from haevier particles with a factor better than 10^2 up to 300 GeV/c.
- The central spectrometer, with magnet and silicon tracker. It allows rigidity (pc/Ze), charge (Z) and sign measurement.
- Time of Flight scintillation counter (TOF) for measurement of particle velocity and charge ($|Z|$) through dE/dx deposition in scintillators.
- Ring Image Cherenkov Counter (RICH) measuring independantly velocity and charge. It will allow isotope identification of relativistic ions.
- Electromagnetic Calorimeter (ECAL) with 3D sampling to measure electrons and photons.
- Veto scintillating counters to reject particles going sideways.

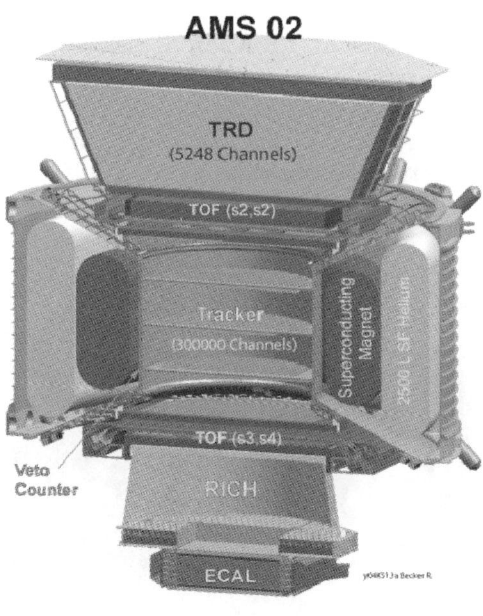

Figure 3: Cut-through of the AMS-02 detector

5 Detector Components

5.1 Transition Radiation Detector (TRD)

Transition radiation is emitted (with less than 1% probability) when a particle crosses boundaries of media with different dielectric properties. The radiation energy is proportional to the Lorenz-γ factor of the particle. For 10-300 GeV electrons, it corresponds to X-rays.

The detector is made of 20 layers of 22 mm thick fleece radiator alternating with 6 mm diameter straw tubes containing a Xe/CO_2 gaz mixture for X-ray detection. The detector is also sensitive to ionization (dE/dx) and gives larger signals for electrons than for protons, at same momentum, due to transition radiation. A detailed description can be found in [3].

5.2 Superconducting Magnet

As shown on figure 4, the magnet consists of two "dipole" coils providing 70% of the field with two sets of 6 "racetrack" coils concentrating the return flux. This arrangement limits the stray field to 300 Gauss and ensures a zero total dipole magnetic moment in order to avoid mechanical torque by interaction with the Earth magnetic field. The field is perpendicular to the axis of the detector and reaches 0.86 Tesla. For 3 years of operation, 2500 liters of superfluid He will be sufficient to keep the coils at 1.8 K. He-II has several advantages for use in space under zero gravity: it has zero viscosity, larger density and specific latent heat than He-I. The magnet coils and the He reservoir are enclosed in a vacuum tank shown in figure 4. The inner and outer diameters are 1.1 and 2.7 m respectively. Reference [4] provides more details. The coils are presently manufactured, tested and assembled.

Figure 4: AMS-02 magnet

5.3 Microstrip Silicon Tracker

Trajectories of charged particles are localised by double sided silicon sensors, arranged in eight layers of $\sim$1 m^2 each, on five planes of an ultra light support structure. Close to 2500 sensors provide a tracking resolution of $\sim 10\mu$m in the bending direction, $\sim 30\mu$m orthogonal to it.

Together with the magnet, particle rigidity can be measured with an accuracy better than 2% at 20 GV, the maximum detectable rigidity is around 1 TV. Nuclei can also be identified by their energy loss ($dE/dx \sim Z^2$) in the silicon. High energy γ's, converted into $e^{\pm}$ pairs in the material above the tracker, can also be detected with excellent directional resolution and good energy resolution [5].

5.4 Time of Flight Counters (TOF)

Two double planes of scintillation counters, separated by 1 m, provide a fast trigger and time of flight measurement. Each plane is divided in 8 or 10 paddles, 12 cm wide and 1 cm thick, which are disposed crosswise along x or y directions. Light guides have been shaped such that each of the 144 PMTs (Hamamatsu R5946) is aligned along the local stray field [6]. The expected time resolution is 130 ps. Upward particles can mimic downward going anti-particles. They can be distinguish at the level of 10^9.

The TOF counter also measures the energy loss in scintillator, allowing ion identification up to Z$\simeq$20.

5.5 Ring Imaging Cerenkov Detector (RICH)

This detector provides further accurate velocity and charge measurements. Light emitted from the combined Aerogel (n=1.05) and NaF (n=1.33) radiators is detected, as a pattern of photons, by an array of 680 position sensitive photomultipliers (4×4 pixels), occasionally after reflection off a conical mirror [7].

5.6 Electromagnetic Calorimeter (ECAL)

Situated at the very bottom of the AMS detector, the 640 kg ECAL [8] is a fine grained lead-scintillating fiber sampling calorimeter allowing precise 3D imaging of the lateral and longitudinal shower development, and proton suppression by a factor 10^4 up to several hundred GeV. Its expected energy resolution is $\delta E/E \simeq 10.1\%\sqrt{E(GeV)} \oplus 2.6\%$. With its 10 radiation length, it will also be used as an independant photon detector with an angular resolution of $\sim 1°$.

6 Performance

Performance of all subdetectors were confirmed with particle beams at accelerators. As an example figure 5 shows the good agreement between tracker [9] and RICH [10] for the identification of nuclei resulting from fragmentation of a Pb ion beam at CERN. Figure 6 shows the expected quality of measurement of the ratio of abundance of Be isotopes in cosmic ray.

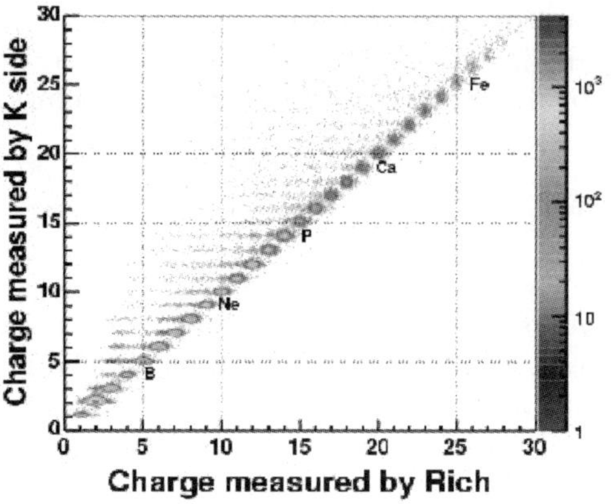

Figure 5: Event by event correlation of nuclei identification by Tracker and RICH

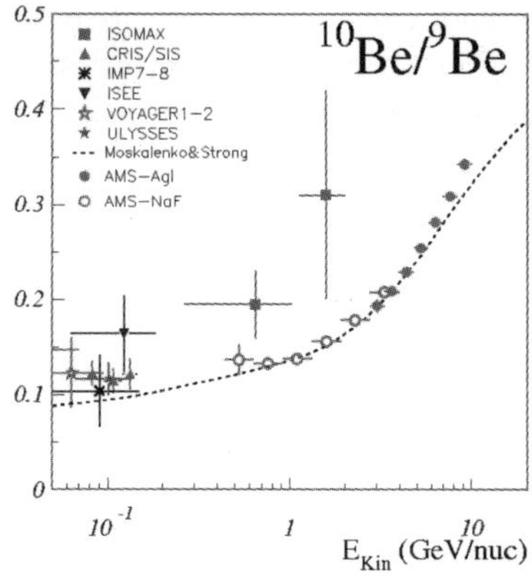

Figure 6: Expected accuracy of Be isotope ratio. Curve from [11].

7 Conclusion

AMS-02 is a complete and sophisticated detector which should be completed in 2006. Thanks to its large acceptance, momentum range, its particle identification capabilities and its long duration exposure (at least 3 years), it will collect orders of magnitude more statistics than AMS-01, under much better conditions. Its observation of cosmic ray and high energy photons in space will be complementary to other projects. A rich and diversified physics program will result from its data taking on the ISS.

Acknowledgments

The construction of AMS-02 is an undertaking of many individuals and organizations. The support of NASA and the U.S. Dept. of Energy has been vital in the inception, development and fabrication of the experiment. The interest and support of Mr. Daniel S. Goldin, former NASA Administrator, is gratefully acknowledged. The dedication of Dr. John O'Fallon, Dr. Peter Rosen and

Dr. P.K. Williams of U.S. DOE, our Mission Management team, Dr. Douglas P. Blanchard, Mr. Mark J. Sistilli and Mr. James R. Bates, NASA, Dr. Susan Breon, GSFC-NASA, Mr. Kenneth Bollweg and Mr. T. Martin, Lockheed-Martin, the support of the space agencies from Germany (DLR), Italy (ASI), France (CNES), Spain (CDTI) and China (CALT) and the support of CSIST, Taiwan, have made the construction possible.

The support of GSI–Darmstadt, particularly of Dr. Reinhard Simon made it possible for us to test electronics components for radiation effects.

The support of INFN, Italy, IN2P3, Region Rhône-Alpes and Haute Savoie, France, CIEMAT and CICYT, Spain, LIP, Portugal, CHEP, Korea, the Chinese Academy of Sciences, the National Natural Science Foundation and the Ministry of Science and Technology of China, Academia Sinica, Taiwan, the U.S. NSF, M.I.T., ETH-Zürich, the University of Geneva, National Central University and National Cheng Kung University, Taiwan, Moscow State University, SEU, Nanjing, Shanghai Jiao Tong University, RWTH-Aachen, the University of Turku and the University of Technology of Helsinki, is gratefully acknowledged.

We are also grateful for the strong support and interest shown from the private sector, including Dr. E. Ettlinger, Linde, Dr. R. Herzog, ILK, Dresden, Mr. J. Ross, Mr. S. Milward and Mr. S. Harrision of SCL, Culham, UK, Mr. M. Molina, CGS, Milan, Mr. F. Petroni, CAEN, Viareggio, Mr. A. Poseda, CRISA (Astrium), Madrid, Ing. A. Pontetti, G&A Engineering, Italy, Dr. E.A. Werner and Dr. J. Krieger, ISATEC, Aachen, and Dr. H. Bieri, Bieri Engineeering, Switzerland.

We thank Professors A. De Rujula, J. Ellis, A. Guth, L. Maiani, for many interesting discussions and support.

References

[1] C.Sbarra, *"Astroparticle Physics with AMS-02: The quest of antimatter"* (Proceedings of this conference).
[2] M.Aguilar et al. [AMS Collaboration], *Phys.Rep.* 366/6,331 (2002).
[3] Th.Kirn, Th.Siedenburg, *Nucl.Instr.Meth.* **A** 535, 165 (2004).
[4] B.Blau et al., *Nucl.Instr.Meth.* **A** 518, 139 (2004).
[5] R.Battiston et al., *Astroparticle Physics* 13, 51 (2000).
[6] D.Casadei et al., *Nuclear Physics* **B** *(Proc. Suppl.)* 113, 133 (2002).
[7] J.Casaus et al., *Nuclear Physics* **B** *(Proc. Suppl.)* 113, 147 (2002).
[8] F.Cadoux et al., *Nuclear Physics* **B** *(Proc. Suppl.)* 113, 159 (2002).
[9] B.Alpat et al., *Nucl.Instr.Meth.* **A** 540, 121 (2005).
[10] M.Buénerd et al., *Nucl.Instr.Meth.* **A** 553, 264 (2005).
[11] A.W.Strong and I.V.Moskalenko, *Astrophys. J.* 509, 212 (1998).

ASTROPARTICLE PHYSICS WITH AMS-02: THE QUEST OF ANTIMATTER

C.Sbarra [a]
on behalf of the AMS collaboration
INFN of Bologna, Viale B.Pichat 6/2 I-40127 Bologna, Italy

*Abstract.*The Alpha Magnetic Spectrometer, to be placed on the International Space Station, will provide data on cosmic radiations in a large range of energy from 0.5 GeV to 3 TeV. Its main physics goals in the astrophysical domain are the anti-matter and the dark matter searches. The potential discovery of primordial antimatter by AMS-02 is presented. The observation of antinuclei or even of high energy antiprotons can be a signal to the existence of antistars in our universe. The anti-He^3 and anti-He^4 signals can probe the universe up to a redshift of about 1000. Possible investigated scenarios of baryogenesis are briefly discussed. Antiparticles, like positrons and antideuterons, as well as low energy antiprotons, can also reveal some exotic physics like dark matter.

1 The AMS experiment

The Alpha Magnetic Spectrometer (AMS) [1] in its prototype version (AMS-01) was flown for ten days on board the space shuttle Discovery (1998) . It performed environmental and background studies and fulfilled the constraints imposed by space. A limit on antimatter/matter in our universe was established, for the energy spanned by the detector [1]:

$$\frac{\overline{He}}{He} < 1.1 \times 10^{-6} \quad @ \; 99 \; \% \; C.L. \tag{1}$$

The knowledge obtained with the precursor AMS-01 was used to redesign and improve the detector for the Space Station and for a longer mission. The International Space Station (ISS), at about 400 km of altitude, has good geomagnetic and galactic coverage. The main orbit characteristics are:

- sixteen revolutions/day
- 51.7° of inclination (degrees parallel to latitude)
- 5° precession/day

Table 1 compares the typical resolutions and maximum fluxes of the spectrometers AMS-01 and AMS-02.

The new AMS-02 detector improves largely the analyzing power and momentum resolution thanks to, respectively, the use of a superconducting magnet and the improved tracker performance: 8 layers of double sided silicon tracker with an accuracy of $\sim 10\mu m$ in the bending plane and $\sim 30\mu m$ in the non bending

[a] e-mail: sbarra@bo.infn.it

Value	AMS-01	AMS-02
MDR	150 GV	$1 \div 2$ TV
Statistics	2.86×10^6	10^9
$E_{Max}(e^-)$	~ 30 GeV	1.4 TeV
$E_{Max}(e^+)$	~ 3 GeV	350 GeV
$E_{Max}(\bar{p})$	~ 3 GeV	450 GeV
BL^2	$\sim 0.15\ Tm^2$	$\sim 0.9\ Tm^2$
Acceptance	$0.15\ m^2$sr	$\sim 0.45\ m^2$sr
$\frac{\Delta p}{p}(1GeV)$	$\sim 10\%$	$1 \div 2\%$

Table 1: Comparison of the overall statistics, Maximum Detectable Rigidity (MDR), analizing power (BL^2) , acceptance and momentum resolution (at 1 GeV) and maximum energy for different kind of particles, obtained with AMS-01 and expected for the next AMS-02 mission.

plane [2]. The basic idea in constructing AMS-02 detector has been: low matter traversed by the cosmic rays and many repeated measurements of particle velocity, momentum and charge (necessary redundancy to operate in space for at least three years). All the AMS-02 subdetectors [3] are designed to collect a huge statistics of all type of cosmic rays from 1 GeV/n to 1 TeV/n.

The subdetectors are, from top to bottom shown in figure 1:

- TRD: 20 layers for electrons/hadrons (e/h) separation ($1.5 \div 300$ GeV)
- TOF: 4 planes of counters for trigger, time, velocity and charge |Q| measurement ($\frac{\Delta\beta}{\beta} \sim 3 \div 4\%$)
- TRACKER: 8 layers of double sided Silicon sensors for momentum and charge $\pm$ Q measurement ($\frac{\Delta p}{p} \sim 2\%$ at 1GeV).
- VETO: counters sorrounding the magnet
- RICH: ring imaging cerenkov with pixellized PMTs, for β ($\frac{\Delta\beta}{\beta} \sim 0.1\%$) and |Q| measurements.
- ECAL: brick of lead and plastic fibers (15 X_0) read in 3-D by PMTs, for e/h separation (1.5 GeV $\div$ 1 TeV) with an energy resolution < 3 % for E>10 GeV

The spectrometer is designed in such a way to have a proton rigidity resolution of 20% at 0.5 TV and a Helium resolution of 20% at 1 TV, as you can see in figure 2. The RICH can allow good isotope separation, in fact: $\frac{\Delta m}{m} = \frac{\Delta p}{p} \oplus \gamma^2 \frac{\Delta\beta}{\beta}$ with p measured by the tracker. Clearly the error on the velocity is the dominant term as the momentum increases. For the Cerenkov angle, measured by the RICH, the relation is: $\frac{\Delta\beta}{\beta} = tan\theta_c \Delta\theta_c$ and the uncertainty on β scales as $1/\sqrt{N}$, where N is the number of detected photons.

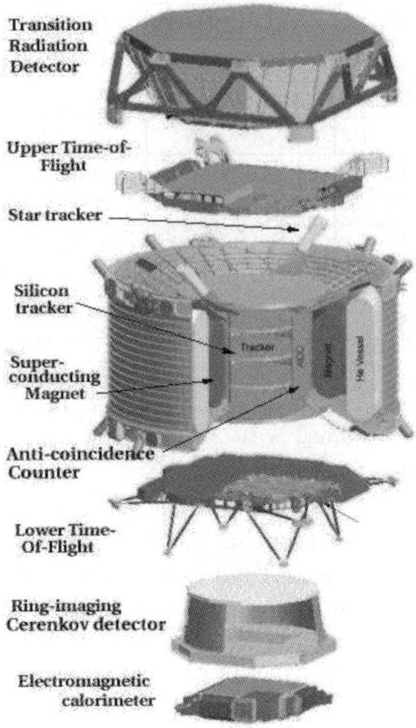

Figure 1: Exploded view of the AMS-02 sub-detectors.

2 Antimatter in our universe

The Physical laws are symmetric between baryons and anti-baryons at microscopic level. The physical universe, as we know, seems to be asymmetric because we have seen so far only matter.

The light elements (from proton up to ^7Li) are synthesised via nuclear reactions in the first $10^2 - 10^4$ second from the Big Bang (BB). In the standard theory of BB, the primordial abundances of light elements depend only on one free parameter: the nucleon energy density or, equivalentely, the ratio of the number of baryons (minus antibaryons) to photons: $\eta = \frac{n_B}{n_\gamma} = \frac{n_b - n_{\bar{b}}}{n_\gamma}$. This number is important because it remained constant in most of the evolution of the universe. By comparing the primordial abundances predicted by the BB with those inferred from observations, η can be constrained: $\eta \sim 10^{-10}$. Although the matter-antimatter asymmetry appears to be large today ($n_B \sim n_b \gg n_{\bar{b}}$) the fact that η is small implies that at very early times the asymmetry was small ($n_B \ll n_b \approx n_{\bar{b}} \approx n_\gamma$) [4].

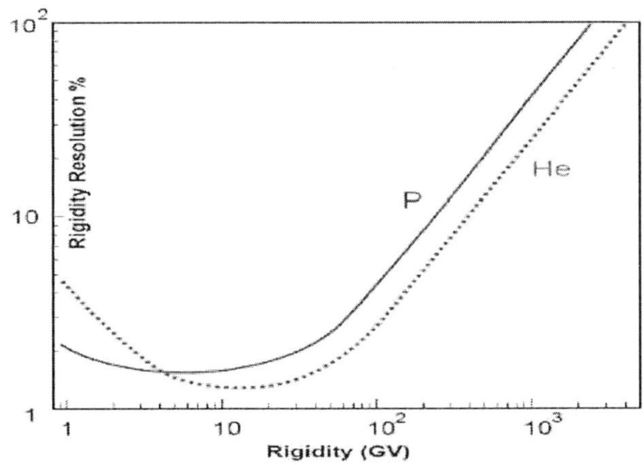

Figure 2: Simulated rigidity resolution of p and He [1].

The question that rises is: how an asymmetric universe can evolve (Baryogenesis), totally or even only locally, from a primordial Big Bang and with symmetric physical laws? Andrei Shakarov put the necessary conditions for baryogenesis [5]:

- Baryon number non conservation
- C and CP non conservation
- Out of thermal equilibrium decay

The inflation is a natural scenario where a baryogenesis can take place: it allows the out of equilibrium condition and can avoid the a-priori hypothesis of initial symmetric condition [6]. In a totally symmetric universe, antimatter nuclei could escape from their antimatter region and reach our galaxy, but, from γ rays observations, as γs would arise from domain borders annihilation, our matter dominated region is of the size of, at least, the cluster of galaxies ($\sim 100 Mpc$). Nevertheless, it is worth while investigating since there could be also place for antistars in a matter dominated universe, as described in the following. In our universe there is the possibility of small insertions of antimatter regions [7]. In an inhomogeneus baryogenesis scenario, in fact, well above the electroweak energies, the quantum fluctuations of a complex, baryonic charged scalar field caused by inflation can generate antimatter regions that can survive annihilation. As a consequence, antistar global clusters can exist in our galaxy and the expected signature is a flux of $^3\overline{He}$ and $^4\overline{He}$ accessible by AMS-02, but not by AMS-01 [8].

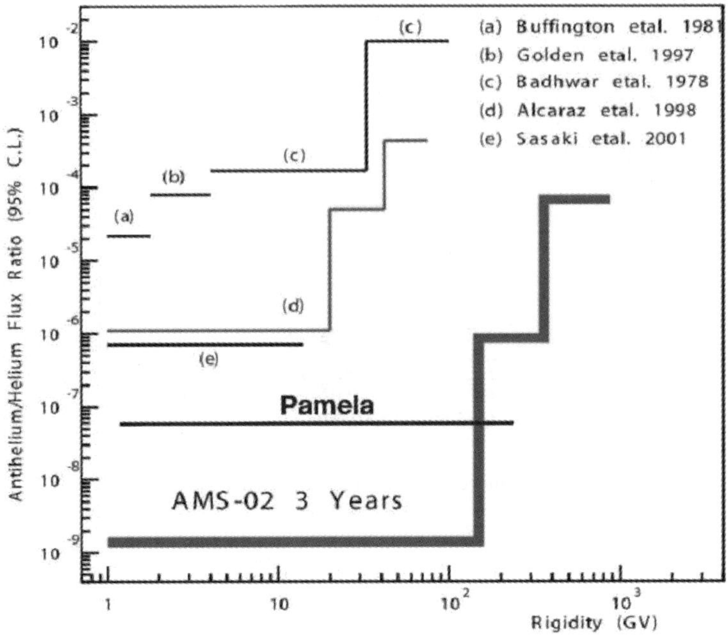

Figure 3: Present experimental limits for the antihelium/helium ratio and the sensitivities foreseen for the Pamela and AMS-02 experiments.

3 Antimatter and exotic physics search with AMS-02

Antinuclei can be found in cosmic rays with a spectrometer on top of the atmosphere (balloons) or on satellites (Pamela, AMS-02). In figure 3 are shown the limits reached so far for the ratio $\overline{He}/He$.

AMS-02 will have the possibility to detect an antinucleus or, at least, to lower the limit on antimatter/matter by a factor of 10^3. The detection of antimatter from an antimatter domain can also be done observing the high energy antideuterons or antiprotons ($E \geq 100GeV$), as described in 3.1 and 3.3.

Antiparticles can be produced by cosmic rays interactions with the interstellar medium (ISM) through inelastic collisions ("spallation" processes). Such secondaries fluxes are of the order: $\overline{He}$ O(10^{-12}), $\overline{D}$ O(10^{-8}), $\overline{p}$ O(10^{-4}), e^+ O(10^{-3}). Exotic sources of antimatter or dark matter annihilation, can produce additional fluxes of particles, as for instance the neutralino annihilation $\chi\chi \rightarrow WW \rightarrow \bar{p}, \bar{D}, e^+, ..$ leading to an excess of events, and idirect search in several channels can be made [9] [10].

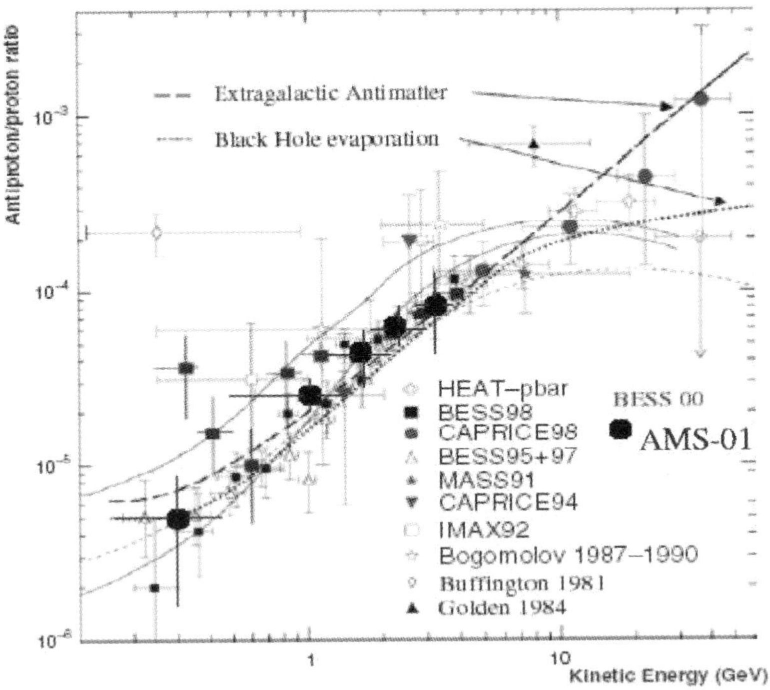

Figure 4: Collected data on $\frac{\overline{p}}{p}$ and various model superimposed [12] as discussed in 3.1. As you can notice, more data above $\sim 100GeV$ would discriminate between the various hypothesis.

3.1 antiprotons

Measurements of the ratio of secondary to primary nuclei suggest that the probability of a CR to escape from the galaxy (mean lifetime) falls with energy as $\sim E^{-\delta}$ with $\delta \sim 0.7$. Then the expected ratio $\frac{\bar{p}}{p}$, in the hypothesis of the existence of an anti-galaxy, should increase with the energy as $\sim E^{\delta}$, due to the fact that $\bar{p}$ would be mainly extragalactic while p mainly galactic (in the hypothesis of the same acceleration mechanism both in galaxy than in the anti-galaxy) [11]. In figure 4 you can see the most recent data on the ratio $\overline{p}/p$ and superimposed is the dashed line from an antigalaxy, that could be from an anti-super cluster of galaxies and also from an antistars global cluster in our own galaxy (as in 2). Superimposed on figure 4 are also other models: pure secondary production with and without solar modulation [12] and more exotic black hole evaporation model: from primordial black holes

$\overline{D}$ and $\overline{n}$, as also other particles, can evaporate through Hawking radiation [13]. The antiprotons are particularly sensitive to the physics details of cosmic ray propagation (controlled by B/C ratio), particularly at low momentum, so detailed measurements of their spectrum can discriminate between the various models of CRs propagation [9].

3.2 positrons

The ratio $\frac{e^+}{e^+ + e^-}$ can reveal neutralino annihilation, provided there is a sort of enhancement of the signal that could arise if we live in a clumpy halo. The HEAT collaboration found a possible excess of the ratio, around 10 GeV [14]. A "boost factor"can be estimated comparing the data with various Minimal Supersymmetric Standard Models and how fluxes are modified consequently. AMS-01 collected data up to $\sim$ 3 GeV and measured such a ratio [1]. AMS-02 can measure up to 350 GeV a statistics of $\approx 10^6$ positrons. Due to the large p background ($p/e^+ \sim 10^3$), a strong proton rejection factor is needed (ECAL + TRD together rejection: $R \geq 10^6$).

3.3 antideuterons

Antideuterons are still to be observed in cosmic rays; the current limit on antideuterons has been given by BESS collaboration [15]. The secondary production of $\overline{D}$ has a probability of $\sim 10^{-8}$ and the reaction: $p + \overline{p} \rightarrow \overline{D} + X$ has a proton threshold momentum of 17 GeV giving antideuterons above $\sim$ 2 GeV. As a consequence, an observation of antideuteron is more likely to be of "primary" origin, especially in the low energy region ($\sim$ 0.1 GeV) where secondary production is suppressed. The exotic component of $\overline{D}$ could come either from an antimatter domain or from neutralino annihilation in our galaxy halo [16]. The flux of $\overline{D}$ is O($\sim 10^{-4}$) the flux of $\overline{p}$, thus, for AMS-01, which observed $\sim 100\overline{p}$ the antideuteron was not observable. Small statistics $\bar{D}$ could be detected by AMS-02.

4 Conclusion

The AMS-02 detector on the International Space Station will measure cosmic ray fluxes with high precision for more than three years. The collected data on antinuclei or on high energy antiprotons will help to discover antimatter regions, both extragalactic and galactic as a relic from the inflationary period. AMS-02 will have also the sensitivity to detect the products of dark matter annihilation in our galaxy by analysing e^+, $\bar{D}$ and $\bar{p}$ data in a large energy range.

Acknowledgments

We wish to thank the many organizations and individuals listed in the aknowledgments of [3]. I would like also to thank Prof. G.Diambrini-Palazzi and Piergiorgio Picozza for the interesting discussions on antimatter search from space. Finally I would like to thank Alexander Studenikin and the 12th Lomonosov committee for inviting me and for their hospitality.

References

[1] M.Aguilar *et al. Physics Reports* **A** 366, (2002) 331.
[2] M.Paniccia *Frascati Physics Series* Vol. XXXVII, (2004) 83.
[3] D.Rapin "Current status and prospects of the AMS experiment" (*Proceedings of this conference*).
[4] W.Kolb and S.Turner *Ann. Rev. Nucl. Part. Sci.* 33, (1983) 645.
[5] Andrei Sakharov *JETP Lett.* **B** 91, (1967) 24.
[6] L.F.Abbott *et al. Physics Lett.* **B** 117, (1982) 29.
[7] M.Yu.Klophov *et al. hep-ph/0210012* (2002).
[8] M.Yu.Klophov *et al. Phys. Rev* **D** 62, (2000) 083505.
[9] D.Maurin *et al. astro-ph/0212111* (2002).
[10] M.Sapinski, *Eur.Phys.J.* **C** 33 (2004) 975.
[11] F.W.Stecker and A.W.Wolfendale, *"Proceedings of the 19th ICRC (La Jolla)"* (1985) 354.
[12] P.Picozza and A.Morselli, *J.Phys.* **G**: Nucl Part. Phys. 29 (2003) 903.
[13] D.Maurin *et al. A&A*, 398 (2003) 403.
[14] E.A. Baltz *et al. Physics Review* **D** 65/6, (2002) 331.
[15] H.Fuke *et al.*, *"Proceedings of the 28th ICRC"* (Tsukuba) (2003) 1797.
[16] P.Salati *et al.*, *"Proceedings of the 26th ICRC"* (1999) HE.5.1.06 .

DARK MATTER INVESTIGATIONS

R. Bernabei[1a], P. Belli[1], F. Montecchia[1], F. Nozzoli[1], F. Cappella[2], A. Incicchitti[2], D. Prosperi[2], R. Cerulli[3], C.J. Dai[4], H.L. He[4], H.H. Kuang[4], J.M. Ma[4] and Z.P. Ye[4,f]

[1]*Dip. di Fisica, Università di Roma "Tor Vergata" and INFN, sez. Roma Tor Vergata, I-00133 Rome, Italy*

[2]*Dip. di Fisica, Università di Roma "La Sapienza" and INFN, sez. Roma, I-00185 Rome, Italy*

[3]*INFN, Laboratori Nazionali del Gran Sasso, I-67010 Assergi (AQ), Italy*

[4]*IHEP, Chinese Academy, P.O. Box 918/3, Beijing 100039, China*

Abstract. Experimental observations and theoretical considerations have shown that a large part of the Universe is made of Cold Dark Matter particles, relics from the Big Bang; this has currently motivated a large experimental effort to investigate them both by direct and indirect detection methods. In particular, the DAMA/NaI experiment ($\simeq$ 100 kg highly radiopure NaI(Tl)) had the unique feature to effectively investigate the presence of a Dark Matter particle component in the galactic halo in a model independent way by exploiting the so-called annual modulation signature. With a total exposure of more than 10^5 kg $\cdot$ day, collected over seven annual cycles deep underground at the Gran Sasso National Laboratory of the I.N.F.N., it has pointed out – at 6.3 σ C.L. – an effect which satisfies all the many peculiarities of the signature. Several (but still few with respect to the possibilities) corollary model dependent quests for the candidate particle have been carried out and some of the obtained results are summarized here.

1 Introduction

The problem of the existence of Dark Matter in our Universe dates back to the astrophysical observations at the beginning of past century, but the presence of Dark Matter in our Universe has been definitively accepted by the scientific community only about 40 years later, when two groups performed systematic measurements of the rotational velocities of celestial bodies in spiral galaxies. After the '70 many other observations have further confirmed the presence of Dark Matter in the Universe and, at present, the measurements are mainly devoted to the investigation of the quantity, of the distribution (from the cosmological scale down to the galactic one) and of the nature of the Dark Matter in the Universe. Recent measurements of the CMB temperature anisotropy by WMAP, analysed in the framework of the *Big Bang* cosmological scenario, support a density of the Universe: $\Omega = 1$, further crediting that most of the Universe is dark. Recently, observations on the supernovae Ia at high red-shift as standard candles have suggested that about 73% of Ω might be in form of a *dark energy*; the argument is still under investigation and presents some problems on possible theoretical interpretations. However, even in this scenario – where the matter density in the Universe would be $\Omega_m \sim 0.3$ – large space for Dark Matter particles in the Universe exists. In fact, the luminous matter can only account for a density $\simeq 0.005$ and the baryonic Dark Matter for $\simeq 0.04$. On the other hand, the contribution of Dark Matter particles relativistic at the decoupling time (such as light neutrinos) is also restricted to be $\lesssim 0.01$ by considerations on large scale structure formations. Thus, most of the Dark Matter particles in the Universe, relics from the Big Bang, were either produced at rest or non relativistic at decoupling time; they are named Cold Dark Matter particles (CDM). The CDM candidates have to be neutral, stable or quasi-stable (e.g. with a time decay of order of the age of the Universe) and have to be elusive in the interaction with ordinary matter. These features are

[a]e-mail: rita.bernabei@roma2.infn.it

[f]also: University of Jing Gangshan, Jiangxi, China

respected by the axions (solar axions were also investigated by DAMA/NaI), by the axion-like candidates with mass in the keV range, by a class of candidates named WIMPs (Weakly Interacting Massive Particles) and by some other candidates which cannot properly be classified as WIMP, but which offer similar phenomenology.

Since in the Standard Model of particle physics no suitable particle as CDM candidate exists, a window beyond the Standard Model is investigated. At present, a widely considered candidate for CDM is the lighest supersymmetric particle named neutralino. In fact, in the supersymmetric theories where the R-parity is conserved, the lighest supersymmetric particle must be stable and can interact neither by electromagnetic nor by strong interactions, otherwise it would be detected in the galactic halo as the ordinary matter is. However, other candidates are also considered, such as e.g. a heavy neutrino of the 4th family or a sneutrino (the spin-0 supersymmetric partner of the neutrino) in a multi-component Dark Matter scenario. Moreover, a dominant contribution of a sneutrino candidate remains still possible in supersymmetric models with violation of lepton number, where two mass states and a small energy splitting is present, as reported in ref. [1]; a similar sneutrino can only inelastically scatter off nuclei, after its excitation to the low lying energy level. Other possible candidates are the particles from multi-dimensional Kaluza-Klein-like theories, the mirror Dark Matter particles [2], the self-interacting dark matter particles [3], etc. Moreover, in principle even whatever particle with suitable characteristics, not yet foreseen by theories, can be a good candidate as CDM.

Several observations have pointed out that our Galaxy should also be embedded in a dark halo with mass at least 10 times larger than that of the luminous matter. The presence of CDM particles of this dark halo can be investigated either by direct (deep underground) or by indirect (in space or underwater or underground) approaches. In the first case, their interaction on target nuclei can be detected by means of the signal induced by the recoiling nucleus or, in case of inelastic scattering, also by the successive de-excitation gamma's; moreover, other possibilities which directly involve ionization/excitation phenomena in the detector remain still open (see e.g. later). In the other case, the flux of secondary particles - mainly neutrinos, positrons, antiprotons, gamma's - produced by possible annihilation of some of the CDM candidates either in the Sun or in the Earth or in the halo may be detected. Most of the approaches are limited by the possibility to give only a model dependent result; related discussions can be found e.g. in refs. [4, 5] and references therein.

2 A direct detection experiment exploiting a model independent signature: DAMA/NaI

The DAMA/NaI experiment was proposed in 1990 [6], designed and realized having the main aim to investigate in a model independent way the presence of a Dark Matter particle component in the galactic halo [4, 5, 7–15]. For this purpose, we planned to exploit the effect of the Earth revolution around the Sun on the Dark Matter particles interactions on a suitable underground set-up. In fact, as a consequence of its annual revolution, the Earth should be crossed by a larger flux of Dark Matter particles in June (when its rotational velocity is summed to the one of the solar system with respect to the Galaxy) and by a smaller one in December (when the two velocities are subtracted).

In fact, the expected differential rate as a function of the recoil energy, dR/dE_R (see ref. [4] for detailed discussion), depends on the CDM velocity distribution and on the Earth's velocity in the galactic frame, $\vec{v}_e(t)$. Projecting $\vec{v}_e(t)$ on the galactic plane, one can write: $v_e(t) = v_\odot + v_\oplus cos\gamma cos\omega(t - t_0)$. Here $v_\odot$ is the Sun's velocity with the respect to the galactic halo ($v_\odot \simeq v_0 + 12$ km/s and v_0 is the local velocity whose value is in the range 170-270 km/s (90% C.L.) [9, 16]); $v_\oplus = 30$ km/s is the Earth's orbital velocity around the Sun on a plane with inclination $\gamma = 60^o$ with the respect to the galactic plane. Furthermore, $\omega = 2\pi/T$ with T=1 year and roughly $t_0 \simeq 2^{nd}$ June (when the Earth's speed is at maximum). The Earth's velocity can be conveniently expressed in unit of v_0: $\eta(t) = v_e(t)/v_0 = \eta_0 + \Delta\eta cos\omega(t - t_0)$, where – depending on the assumed value of the local velocity – η_0=1.04-1.07 is the yearly average of η and $\Delta\eta$ = 0.05-0.09. Since $\Delta\eta \ll \eta_0$, the expected counting rate in a k-th energy interval can be expressed by the first order Taylor approximation:

$$S_k[\eta(t)] = S_k[\eta_0] + [\frac{\partial S_k}{\partial \eta}]_{\eta_0} \Delta\eta cos\omega(t - t_0) = S_{0,k} + S_{m,k} cos\omega(t - t_0), \qquad (1)$$

with the contribution from the highest order terms less than 0.1%. This annual modulation signature is very distinctive since a seasonal effect induced by CDM particles must simultaneously satisfy all the following requirements: (i) the rate must contain a component modulated according to a cosine function; (ii) with one year period; (iii) a phase that peaks roughly around $\simeq 2^{nd}$ June; (iv) this modulation must only be found in a well-defined low energy range, where CDM particles can induce signals; (v) it must apply to those events in which just one detector of many actually "fires", since the probability that CDM particles would have multiple interactions is negligible; (vi) the modulation amplitude in the region of maximal sensitivity must be $\lesssim$7% for usually adopted halo distributions, but it can be significantly larger in case of some possible scenarios such as e.g. those in refs. [1,17]. Only systematic effects able to fulfil these requirements and to account for the whole observed modulation amplitude could mimic this signature; thus, no other effect investigated so far in the field of rare processes offers a so stringent and unambiguous signature. With the present technology, the annual modulation signature, originally suggested in the middle of '80 by [18], remains the main signature of a CDM signal in the galactic halo, offering an efficient model independent signature, able to test a large interval of cross sections and of halo densities;

The DAMA/NaI experiment was located deep underground in the Gran Sasso National Laboratory of I.N.F.N. [19–22]; it has been part of the DAMA project, which is also composed by several other low background set-ups, such as: i) DAMA/LXe ($\simeq$ 6.5 kg pure liquid Xenon scintillator, using Kr-free enriched Xenon) [23]; ii) DAMA/R&D, set-up devoted to tests on prototypes and small scale experiments [24]; iii) the new second generation larger mass NaI(Tl) radiopure set-up DAMA/LIBRA (see later); iv) DAMA/Ge detector for sample measurements.

In particular, the NaI(Tl) scintillator was chosen as the best target material to investigate the process since it offers several competing advantages with the respect to every other technique considered so far, such as: well known technology; large mass; high duty cycle; feasible well controlled operational conditions and monitoring; routine calibrations feasible down to keV range in the same conditions as the production runs; high light response, that is keV threshold reachable; absence of the necessity of re-purification or cooling down/warming up procedures (implying high reproducibility, high stability, etc.); absence of microphonic noise and an effective noise rejection at threshold (time decay of NaI(Tl) pulses is hundreds ns, while that of noise pulses is tens ns); reachable high radiopurity by material selections and protocols, by chemical/physical purifications, etc.; sensitivity to spin-independent (SI), spin-dependent (SD), mixed (SI&SD) couplings as well as to many other existing scenarios; sensitivity to both high (by Iodine target) and low (by Sodium target) mass candidates; possibility to effectively investigate the annual modulation signature in all the needed aspects; pulse shape discrimination feasible at reasonable level; possibility to achieve significant results on several other rare processes; no safety problems; necessity of a relatively small underground space; the lowest cost with the respect to every other considered technique; etc.

Considering its main goal, DAMA/NaI has been designed by employing and further developing all the necessary low-background techniques and procedures. A detailed description of the set-up, of its radiopurity, of its performance, of the used hardware procedures, of the determination of the experimental quantities and of the data reduction has been given in refs. [4,5,10,11,25].

The following part of this paper will shortly summarize the final model independent result, some of the corollary quests for the candidate particle, comparisons, implications and perspectives.

However, it is worth to remind that - thanks to its radiopurity and features - DAMA/NaI has also investigated other approaches for Dark Matter particles and several other rare processes such as: possible processes violating the Pauli exclusion principle, CNC processes in ^{23}Na and ^{127}I, electron stability, searches for neutral SIMPs, for neutral nuclearites, for Q-balls and for solar axions; moreover, DAMA/NaI allowed also the study of nuclear transition to a possible superdense nuclear state and possible two-body cluster decay of ^{127}I [26–35].

3 The model-independent result of DAMA/NaI

The evidence of the presence of a particle Dark Matter component at level of our Galaxy in the data of DAMA/NaI has firstly been reported by the DAMA collaboration at the TAUP conference in 1997 [36] and published also in [7], confirmed in [8,9], further confirmed in [10–14] and conclusively confirmed, at end of experiment in 2003 (see ref. [4,5,15]).

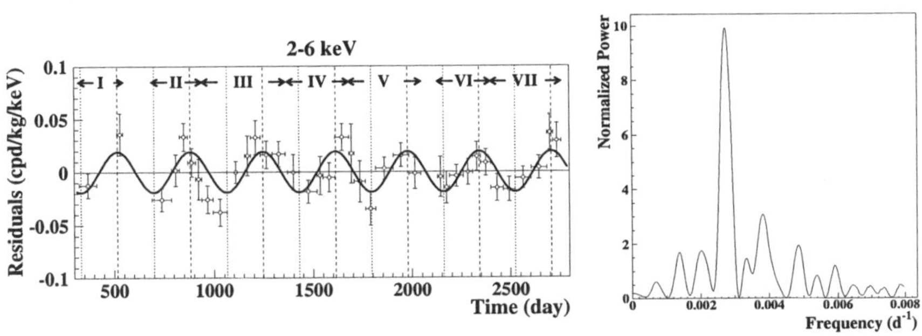

Figure 1: *On the left*: experimental residual rate for *single-hit* events in the cumulative (2–6) keV energy interval as a function of the time over 7 annual cycles, end of data taking July 2002. The experimental points present the errors as vertical bars and the associated time bin width as horizontal bars. The superimposed curve represents the cosinusoidal function behaviour expected for a Dark Matter particle signal with a period equal to 1 year and phase exactly at 2nd June; the modulation amplitude has been obtained by best fit [4]. *On the right*: power spectrum of the measured *single-hit* residuals for the cumulative (2–6) keV energy interval calculated including also the treatment of the experimental errors and of the time binning. As it can be seen, the principal mode corresponds to a frequency of $2.737 \cdot 10^{-3}$ d^{-1}, that is to a period of $\simeq$ 1 year.

In fact, during seven independent experiments each one of one annual cycle (total exposure: 107731 kg $\times$ day), DAMA/NaI has pointed out the presence of an annual modulation in the *single-hit* residual rate in the lowest energy interval (2 – 6) keV, at 6.3 σ C.L., (Fig.1 and Fig.2) satisfying all the features expected for a Dark Matter particle component in the galactic halo. No systematic effect or side reaction able to account for the observed effect has been found. This result is model-independent. It represents the first experimental evidence of the presence of DM particles in the halo independently on their nature.

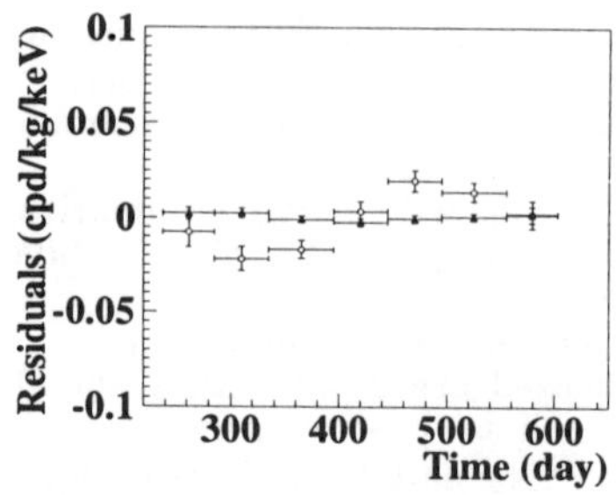

Figure 2: Experimental residual rates over seven annual cycles for *single-hit* events (open circles) – class of events to which DM events belong – and over the last two annual cycles for *multiple-hits* events (filled triangles) – class of events to which DM events do not belong – in the (2–6) keV cumulative energy interval. They have been obtained by considering for each class of events the data as collected in a single annual cycle and using in both cases the same identical hardware and the same identical software procedures. The initial time is taken on August 7th.

At present, apart from the now-running DAMA/LIBRA, no other experiment is sensitive, for mass, radiopurity and control of the stability, to such a model independent signature. For details see e.g. ref. [4, 5].

4 Some corollary model-dependent quests for a candidate

On the basis of the obtained 6.3 σ model-independent result, corollary investigations can also be pursued on the nature of the Dark Matter particle candidate. This latter investigation is instead model-dependent and – considering the large uncertainties which exist on the astrophysical, nuclear and particle physics assumptions and on the parameters needed in the calculations – has no general meaning (as it is also the case of exclusion plots and of the Dark Matter particle parameters evaluated in indirect detection experiments). Thus, it should be handled in the most general way, as pointed out with time passing [4, 5, 15].

The corollary quests for a Dark Matter candidate particle will be discussed considering either some candidates belonging to the WIMP class particles (i.e. particle that interact via scattering off to target nuclei) or light bosonic candidates (i.e. particles that interact via the conversion of the particle mass in electromagnetic radiation).

For simplicity, the results of these analyses are presented in terms of allowed volumes/regions obtained as superposition of the configurations at given C.L. for the considered model frameworks. Details on the followed procedure can be found e.g. in ref. [4, 5, 15].

4.1 Some corollary quests for the candidates of the WIMP class

For this class of Dark Matter particles, it has been considered so far low (of order of few GeV) and high mass (up to many hundreds of GeV) candidates interacting with ordinary matter via: i) mixed SI&SD coupling; ii) dominant SI coupling; iii) dominant SD coupling; iv) preferred SI inelastic scattering. A detailed discussion on the allowed volumes/regions for these candidates can be found e.g. in ref. [4, 5].

As an example (see our references for other cases), here we have only reported in Fig. 3-*left* the results obtained in the general case of particle belonging to the WIMP class with mixed SI&SD. The coloured regions are example of slices (of the 4-dimensional allowed volume) in the plane $\xi\sigma_{SI}$ vs $\xi\sigma_{SD}$ for some of the possible particle mass, m_W, and θ values. Here, ξ is the fractional amount of local density of Dark Matter particles, σ_{SI} and σ_{SD} are the point-like SI and SD DM particle-nucleon cross sections and $tg\theta$ is the ratio between the effective SD coupling strengths of the

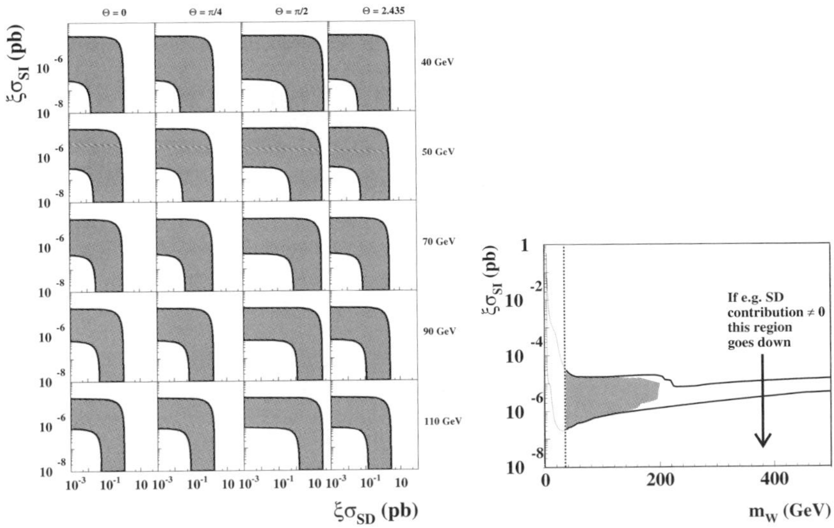

Figure 3: *On the left:* example of slices (of the 4-dimensional allowed volume) in the plane $\xi\sigma_{SI}$ vs $\xi\sigma_{SD}$ in the case of a Dark Matter particle with mixed SI&SD interaction for the model frameworks of ref. [4, 5]. *On the right:* allowed region in the plane (m_W, $\xi\sigma_{SI}$) obtained for the case of a Dark Matter particle with dominant SI interaction for some model frameworks [4,5]. The vertical dotted line represents a bound in case of a neutralino candidate when supersymmetric schemes based on GUT assumptions are adopted to analyse the LEP data; the low mass region is allowed for neutralino when other schemes are considered and for every other WIMP candidate [4, 5]. While the area at WIMP masses above 200 GeV is allowed only for few configurations, the lower one is allowed by most configurations (the colored region gathers only those above the vertical line) [4, 5].

Dark Matter particle with neutron and proton (θ is defined in the $[0,\pi)$ interval). In Fig. 3-*right*, instead, the allowed region obtained for the case of a particle belonging to the WIMP class with dominant SI interaction for some model frameworks is reported [4, 5]. Obviously, larger sensitivities than those reported in these figures would be reached when including the effect of other existing uncertainties on the astrophysical, nuclear and particle physics assumptions and related parameters; similarly, the set of the best fit values would also be enlarged as well.

It is worth to note that the DAMA allowed region of Fig. 3-*right* is already compatible with theoretical expectations in the particular case of a purely SI coupled neutralino candidate in MSSM with gaugino mass unification at GUT scale released [37, 38].

4.2 Some corollary quests for light bosonic candidates

Recently, DAMA collaboration has also pursued a corollary quest for the candidate particle considering a light ($\sim$ keV mass) bosonic candidate, either with pseudoscalar or with scalar couplings, as Dark Matter component in the galactic halo [15]. For these candidates, the direct detection process is based on the total conversion in NaI(Tl) crystal of the mass of the absorbed bosonic particle into electromagnetic radiation. Thus, in these processes the target nuclei recoil is negligible and is not involved in the detection process; therefore, signals from these light bosonic Dark Matter candidates are lost in experiments based on rejection procedures of the electromagnetic contribution to the counting rate.

The main processes involved in the detection of a DM light bosonic particle both

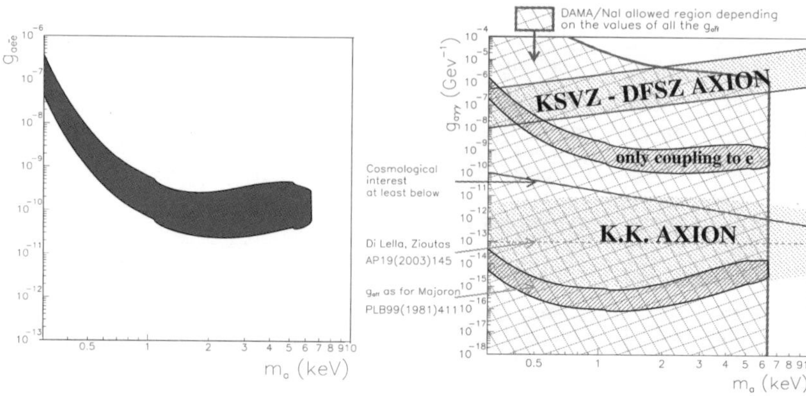

Figure 4: Case of a pseudoscalar boson candidate (here named a). *Left:* region allowed at 3σ C.L. in the plane $g_{a\bar{e}e}$ vs m_a by the DAMA/NaI annual modulation data in the considered model framework. *Right:* allowed region in the plane $g_{a\gamma\gamma}$ vs m_a (*crossed hatched region*). All the configurations in this region can be allowed depending on the values of all the $g_{a\bar{f}f}$. In the plot examples of the expectation for some models have been reported. The solid line corresponds to a particle with lifetime equal to the age of the Universe; at least all the $g_{a\gamma\gamma}$'s below this line are of cosmological interest. See ref. [15].

in the pseudoscalar and in the scalar interactions are "Compton - like" effect, "axio-electric" or "photoelectric-like" effect and Primakoff effect. In all these processes the total (including the secondary processes: X-rays and Auger electrons) energy release, E_{rel}, in the detector (providing that its detection efficiency is $\simeq 1$ for low-energy electrons and low-energy photons) matches the total energy of the a particle, $E_a \simeq m_a$ since the a velocity is of the order of $10^{-3}c$.

For detailed discussion about this topic see ref. [15].

In the case of a pseudoscalar boson (here named a), the results can be presented in terms of $g_{a\bar{e}e}$, coupling constant of the particle to the electrons, and m_a, the particle mass (see Fig. 4-*left*). This allowed region is almost independent on the adopted coupling constants of the axion with the *up* and *down* quarks ($g_{a\bar{u}u}$ and $g_{a\bar{d}d}$, respectively) since the axioelectric effect largely dominates on the other detection processes [15]. The pseudoscalar boson particle can decay into two photons but its lifetime can be also of cosmological interest as shown in Fig. 4-*right*, where the allowed region in the plane $g_{a\gamma\gamma}$ vs m_a is reported, $g_{a\gamma\gamma}$ being the effective axion coupling constant to photons.

Furthermore, in the case of a scalar boson (here named h), only a coupling to hadronic matter is considered in order to account for the DAMA observations [15] and the result is an allowed multi-dimensional volume in the space defined by m_h (scalar boson mass) and by all the $g_{h\bar{f}f}$ coupling constants to quarks. For semplicity, here we report only a slice obtained considering the $g_{h\bar{N}N}$ coupling constants to nucleons (see Fig. 5-*left*). This region does not allow a direct information about the effective coupling constant to photons, $g_{h\gamma\gamma}$, and, therefore, about h lifetime; however, the large number of free coupling constants allows the existence of a large number of h configurations of cosmological interest that can be calculated in some model frameworks (see Fig. 5-*right*). For details see ref. [15].

4.3 Few arguments on comparisons with some model dependent results

As regards some claimed model-dependent comparisons by experiments insensitive to the annual modulation signature, often an incorrect/partial/not updated quotation of the DAMA/NaI result is performed. Moreover, these experiments have exposures

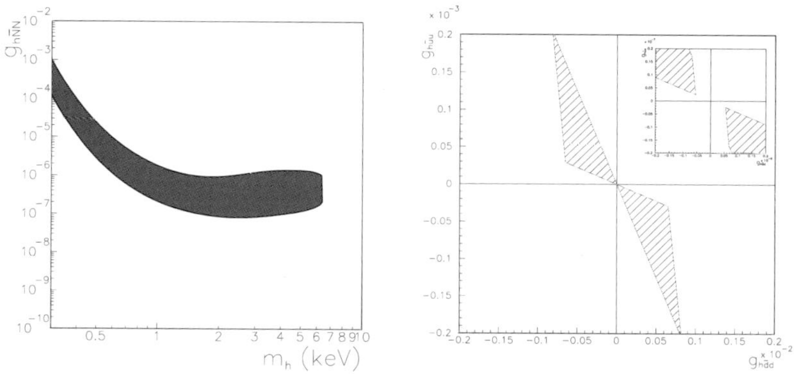

Figure 5: Case of a scalar boson candidate (here named h). *Left:* region allowed at 3σ C.L. in the plane $g_{h\bar{N}N}$ vs m_h for a scalar light boson Dark Matter candidate coupled only to the hadronic matter in the given frameworks [15]. Several configurations correspond to particles of cosmological interest (see ref. [15]). *Right:* DAMA/NaI allowed configurations of cosmological interest in the $g_{h\bar{u}u}$ vs $g_{h\bar{d}d}$ plane. In the inset: a magnification of the regions around (0,0).

orders of magnitude lower than the one by DAMA/NaI, use different target nuclei, etc.; some discussions can be found e.g. in ref. [4,5].

Generally they neglect for example the uncertainties on: i) the nature of the candidate particle(s); ii) the real coupling with ordinary matter; iii) the spin-dependent and spin-independent form factors and related parameters for each nucleus; iv) the spin factor used for each nucleus; v) the real scaling laws for nuclear cross sections among different target materials; vi) experimental and theoretical parameters; vii) the effect of different halo models and related parameters on different target materials, etc. Moreover, large differences are expected in the counting rate for instance among nuclei fully sensitive to the SD interaction (as ^{23}Na and ^{127}I) with the respect to nuclei largely insensitive to such a coupling (as e.g ^{nat}Ge, ^{nat}Si, ^{nat}Ar, ^{nat}Ca, ^{nat}W, ^{nat}O) and also when nuclei in principle all sensitive to this coupling but having different unpaired nucleon (e.g. neutron in case of the odd spin nuclei, such as ^{129}Xe, ^{131}Xe, ^{125}Te, ^{73}Ge, ^{29}Si, ^{183}W and proton in the ^{23}Na and ^{127}I). Moreover, in case the detection of the CDM particles would involve electromagnetic signals (see, for example, the case of the light bosons discussed above), all the experiments, such as e.g. CDMS, EDELWEISS, CREST, ZEPLIN, WARP and their extensions, do lose this signal in the rejection procedures of the e.m. contribution to the counting rate. For completeness, it is also worth to note that no results obtained with different target nuclei can intrinsecally be directly compared even for the same kind of coupling. In fact, this requires – among others – the knowledge of the scaling laws among the DM particle-nuclear cross sections (situation even worse than that in the field of double beta decay experiments when different materials are used). As regards the indirect searches, a comparison would always require the calculation and the consideration of all the possible CDM particle configurations in the given particle model, since it does not exist a biunivocal correspondence between the observables in the two kinds of experiments: CDM particle-nucleus elastic scattering cross section (direct detection case) and flux of either muons from secondary neutrinos or antimatter or photons following CDM annihilations (indirect detection cases).

Finally, we remind that recently it has been suggested [39] that the positive hints from the indirect detection – namely the excess of positrons and of gamma rays in the space – and the model independent effect observed by DAMA/NaI can also be described in a scenario with multi-component Dark Matter in the galactic halo, made of a subdominant component of heavy neutrinos of the 4th family and of a sterile dominant component.

5 The new DAMA/LIBRA set-up and beyond

In 1996 DAMA proposed to realize a ton set-up [40] and a new R&D project for highly radiopure NaI(Tl) detectors was funded at that time and carried out for several years in order to realize as an intermediate step the second generation DAMA/LIBRA experiment (successor of DAMA/NaI) with an exposed mass of about 250 kg. Thus, as a consequence of the results of this second generation R&D, the new experimental set-up DAMA/LIBRA (Large sodium Iodide Bulk for RAre processes), $\simeq$250 kg highly radiopure NaI(Tl) crystal scintillators (matrix of twenty-five $\simeq$ 9.70 kg NaI(Tl) crystals), was funded at the end of 1999 and realised. In fact, after the completion of the DAMA/NaI data taking on July 2002, the dismounting of DAMA/NaI occurred and the installation of DAMA/LIBRA started. The experimental site as well as many components of the installation itself have also been implemented (environment, shield of the photomultipliers, wiring, High-Purity Nitrogen system, cooling water of air conditioner, electronics and data acquisition system, etc.).

DAMA/LIBRA is taking data since March 2003 and the first data release will, most probably, occur at end 2008. The highly radiopure DAMA/LIBRA set-up is a powerful tool for further investigation on the Dark Matter particle component in the galactic halo [5] having many intrinsic merits and a larger exposed mass, an higher overall radiopurity and improved performances and increased sensitivity with the respect to DAMA/NaI.

At present a third generation R&D effort towards a possible NaI(Tl) ton set-up has been funded and related works are in progress.

6 Conclusion

DAMA/NaI has pointed out at 6.3 σ C.L. the presence of Dark Matter particle component in the galactic halo by investigating the model independent annual modulation signature in seven annual cycles. No systematic effect or side reaction able to account for the observed effect has been found. Several of the possible corollary quests for the investigation on the nature of the Dark Matter candidates have also been pursued considering the present poor knowledge on the many astrophysical, nuclear and particle physics related aspects. Here, the new corollary quest for light bosonic candidate has shortly been summarized. At present after a devoted R&D effort, the second generation DAMA/LIBRA (a $\simeq$250 kg more radiopure NaI(Tl) set-up) has been realised and put in operation since March 2003. A third generation R&D toward a possible ton NaI(Tl) set-up, we proposed in 1996 [40], is also in progress.

References

[1] D. Smith and N. Weiner, Phys. Rev. D 64 (2001) 043502.
[2] R. Foot, *hep-ph/0308254*.
[3] S. Mitra, Phys. Rev. D 71 (2005) 121302.
[4] R. Bernabei et al., La Rivista del Nuovo Cimento 26 (2003) 1.
[5] R. Bernabei et al., Int. J. Mod. Phys. D 13 (2004) 2127.
[6] P. Belli, R. Bernabei, C. Bacci, A. Incicchitti, R. Marcovaldi, D. Prosperi, DAMA proposal to INFN Scientific Committee II, April 24th 1990.
[7] R. Bernabei et al., Phys. Lett. B 424 (1998) 195.
[8] R. Bernabei et al., Phys. Lett. B 450 (1999) 448.
[9] P. Belli et al., Phys. Rev. D 61 (2000) 023512.

[10] R. Bernabei et al., Phys. Lett. B 480 (2000) 23.
[11] R. Bernabei et al., Eur. Phys. J. C 18 (2000) 283.
[12] R. Bernabei et al., Phys. Lett. B 509 (2001) 197.
[13] R. Bernabei et al., Eur. Phys. J. C 23 (2002) 61.
[14] P. Belli et al., Phys. Rev. D 66 (2002) 043503.
[15] R. Bernabei et al., Int. J. Mod. Phys. A in press (ROM2F/2005/19 & astro-ph/0511262).
[16] P.J.T. Leonard and S. Tremaine, Astrophys. J. 353 (1990) 486; C.S. Kochanek, Astrophys. J. 457 (1996) 228; K.M. Cudworth, Astron. J. 99 (1990) 590.
[17] K. Freese et al., Phys. Rev. Lett. 92 (2004) 11301.
[18] K.A. Drukier et al., Phys. Rev. D 33 (1986) 3495; K. Freese et al., Phys. Rev. D 37 (1988) 3388.
[19] M. Ambrosio et al., Astropart. Phys. 7 (1997) 109.
[20] P. Belli et al., Il Nuovo Cim. A 101 (1989) 959.
[21] M. Cribier et al., Astropart. Phys. 4 (1995) 23.
[22] C. Arpesella et al., Health Phys. 72 (1997) 629.
[23] P. Belli et al., Astropart. Phys. 5 (1996) 217; P. Belli et al., Il Nuovo Cim. 19 (1996) 537; P. Belli et al., Phys. Lett. B 387 (1996) 222; Phys. Lett. B 389 (1996) 783 (err.); P. Belli et al., Phys. Lett. B 465 (1999) 315; P. Belli et al., Phys. Rev. D 61 (2000) 117301; R. Bernabei et al., New Journal of Physics 2 (2000) 15.1; R. Bernabei et al., Phys. Lett. B 493 (2000) 12; R. Bernabei et al., Nucl. Instrum & Meth. A 482 (2002) 728; R. Bernabei et al., Eur. Phys. J. direct C 11 (2001) 1; R. Bernabei et al., Phys. Lett. B 527 (2002) 182; R. Bernabei et al., Phys. Lett. B 546 (2002) 23; R. Bernabei et al., Phys. Lett. B 436 (1998) 379.
[24] R. Bernabei et al., Astropart. Phys. 7 (1997) 73; R. Bernabei et al., Il Nuovo Cim. A 110 (1997) 189; P. Belli et al., Astropart. Phys. 10 (1999) 115; P. Belli et al., Nucl. Phys. B 563 (1999) 97; R. Bernabei et al., Nucl. Phys. A 705 (2002) 29; P. Belli et al., Nucl. Instrum & Meth. A 498 (2003) 352; R. Cerulli et al., Nucl. Instrum & Meth. A 525 (2004) 535.
[25] R. Bernabei et al., Il Nuovo Cim. A 112 (1999) 545.
[26] R. Bernabei et al., Phys. Lett. B 389 (1996) 757.
[27] R. Bernabei et al., Phys. Lett. B 408 (1997) 439.
[28] R. Bernabei et al., Il Nuovo Cim. A 112 (1999) 1541.
[29] P. Belli et al., Phys. Rev. C 60, 065501 (1999).
[30] P. Belli et al., Phys. Lett. B 460 236 (1999).
[31] R. Bernabei et al., Phys. Rev. Lett. 83 (1999) 4918.
[32] F. Cappella et al., Eur. Phys. J. direct C 14 1 (2002).
[33] R. Bernabei et al., Phys. Lett. B 515 (2001) 6.
[34] R. Bernabei et al., Eur. Phys. J. A 23 7 (2005).
[35] R. Bernabei et al., Eur. Phys. J. A 24 51 (2005).
[36] R. Bernabei et al., Nucl. Phys. B (Proc. Suppl.) 70, 79 (1999).
[37] A. Bottino et al., Phys. Rev. D 67 (2003) 063519; A. Bottino et al, Phys. Rev. D 69 (2003) 043506; D. Hooper and T. Plehn, Phys. Lett. B 562 (2003) 18. G. Bélanger, F. Boudjema, A. Pukhov and S. Rosier-Lees, hep-ph/0212227.
[38] A. Bottino et al., Phys. Rev. D 69 (2004) 037302, Phys. Rev. D 68 (2003) 043506.
[39] K. Belotsky, D. Fargion, M. Khlopov and R.V. Konoplich, *hep-ph/0411093*.
[40] R. Bernabei et al., Astrop. Phys. 4 45 (1995); R. Bernabei, in The identification of Dark Matter, World Sc. Pub. (1997) 574.

MIRROR DARK MATTER

Raymond R. Volkas [a]
School of Physics, Research Centre for High Energy Physics, The University of Melbourne, Victoria, 3010, Australia

Abstract. We review the mirror matter model and its application to the dark matter problem.

1 Mirror matter models

The mirror matter idea goes back to Lee and Yang in their famous paper on parity violation [1]. They pointed out that parity may still be a symmetry of the universe even if the weak interactions considered in isolation seem to violate it. To achieve overall parity invariance, one must double all degrees of freedom (except the graviton's) by introducing a mirror matter sector (it was actually not clear in the late 1950's that all particles needed to be doubled, but it is clear now). The idea was then occasionally revisited, including by Kobzarev, Okun and Pomeranchuk [2] in the late 1960's, and by Blinnikov and Khlopov [3] in the early 1980's. The latter looked at the cosmological implications of mirror matter, especially on galaxy formation. The idea that there is a large amount of mirror matter in the universe is still a tenable hypothesis, though the scenario is necessarily different in detail from that considered in the 1980's. The purpose of this talk is to review the possibility that dark matter might be mirror matter.

Foot, Lew and the author [4] independently reinvented the mirror matter idea in 1990, unaware of this history, purely for aesthetic reasons: we wanted to construct a phenomenolgically viable model where the full Lorentz group, including improper transformations, was a symmetry of the vacuum as well as the Lagrangian. Because we had the standard model to draw on, we wrote down an explicit gauge theoretic realisation of the mirror matter idea for the first time.

Left-right symmetry is maximally violated by the weak interactions, so it appears at first sight that nature must violate parity. However, this conclusion rests on the assumption that the left- and right-handed electrons are parity transforms of each other (and so on for the other fermions). In a mirror matter model (MMM), a new "hidden sector" is instead postulated to be the parity transform of the ordinary sector. This possibility is experimentally allowed. Hence one cannot yet conclude that the improper Lorentz transformations are broken symmetries.

Any MMM has gauge group $G \otimes G$ and a discrete symmetry such that

$$x^\mu \leftrightarrow x_\mu, \qquad \psi \sim (R, R') \leftrightarrow \psi' \sim (R', R), \tag{1}$$

[a] e-mail: r.volkas@physics.unimelb.edu.au

with

$$\mathcal{L}_{\text{total}} = \mathcal{L}(\psi) + \mathcal{L}(\psi') + \Delta\mathcal{L}(\psi, \psi'). \tag{2}$$

The idea of doubling all fields might seem uneconomic, but there is a strong precedent for particle doubling driven by Lorentz group considerations: antiparticles. It could be that the improper Lorentz transformations are also associated with a twinning (the supersymmetric example is another instance of a speculative Lorentz-related pairing). The mirror doubling does *not* imply an explosion in independent parameters, because the discrete parity symmetry keeps things under control. The limited number of new parameters is contained only in $\Delta\mathcal{L}$.

The key predictions of a MMM are: (i) Everything except for gravity is doubled: fermions, gauge bosons and Higgs bosons. (ii) The two sectors, ordinary and mirror, couple gravitationally. (iii) They can couple non-gravitationally via $\Delta\mathcal{L}$. These interactions are governed by unknown constants that can be small. (iv) Mirror protons and mirror electrons are as stable as their ordinary cousins.

Any gauge model can be made mirror symmetric. We shall focus on the mirror-symmetric standard model here. The gauge group is

$$[\,SU(3)_c \otimes SU(2) \otimes U(1)_Y\,] \otimes [\,SU(3)'_c \otimes SU(2)' \otimes U(1)'_Y\,] \tag{3}$$

and the fermion multiplet assignments are

$$q_L \sim \left(3, 2, \frac{1}{3}\,;\, 1, 1, 0\right) \leftrightarrow q'_R \sim \left(1, 1, 0\,;\, 3, 2, \frac{1}{3}\right) \tag{4}$$

and so on for the other fermions. For gauge bosons,

$$G \sim (\,8, 1, 0\,;\, 1, 1, 0\,) \leftrightarrow G' \sim (\,1, 1, 0\,;\, 8, 1, 0\,) \tag{5}$$

and similarly for the weak and hypercharge gauge bosons. The Higgs doublet also has a mirror partner:

$$\phi \sim (\,1, 2, 1\,;\, 1, 1, 0\,) \leftrightarrow \phi' \sim (\,1, 1, 0\,;\, 1, 2, 1\,). \tag{6}$$

Gravity is common to the two sectors because the action is simply

$$S = \int d^4x\,\sqrt{-g}\,\left(\,\kappa R + \rho_0 + \mathcal{L} + \mathcal{L}' + \Delta\mathcal{L}\,\right), \tag{7}$$

where R is the Ricci scalar, and κ and ρ_0 are related to Newton's constant and the cosmological constant, respectively. As an aside, an amusing variant was recently proposed by Kaplan and Sundrum [5]:

$$S = \int d^4x\,\sqrt{-g}\,\left(\,\kappa R + \rho_0 + \mathcal{L} - \mathcal{L}' + \Delta\mathcal{L}\,\right). \tag{8}$$

The mirror sector becomes a ghost sector, and any dynamically generated vacuum energy cancels between the ordinary and mirror sectors, alleviating the cosmological constant problem.

The non-gravitational ordinary-mirror couplings are very few in number. One is

$$\epsilon\, F^{\mu\nu}\, F'_{\mu\nu} \tag{9}$$

for the hypercharge field-strength tensors, leading to the kinetic mixing of photons with mirror-photons. This term is absent at tree-level for grand-unified MMMs, but can be generated radiatively after GUT symmetry breaking. Mirror particles gain an ordinary electric charge proportional to ϵ. The smoking gun for mirror matter is milli-charged particles with the same charge-ratios and masses as their ordinary counterparts.

A second term is

$$\lambda\, \phi^{\dagger}\, \phi\, \phi'^{\dagger}\, \phi' \tag{10}$$

in the Higgs potential. There are two physical neutral Higgs boson mass eigenstates, each a maximal superposition of the ordinary and mirror Higgses. The mass splitting is a free parameter governed by λ. If large enough, then the mass eigenstates will be resolved, and each will have a 50% invisible width. If so, a low reheating temperature for the universe is needed to prevent the above interaction from thermally equilibrating the ordinary and mirror cosmological plasmas [6].

Finally, if right-handed neutrino singlets and their mirror partners exist, then also [7]

$$\overline{\nu}_R\, \nu'_L, \qquad \overline{\ell}_L\, \phi\, (\nu'_L)^c, \tag{11}$$

and their mirror transforms appear in $\Delta\mathcal{L}$, where ℓ_L denotes lepton doublets. Mirror neutrinos are interesting candidates for sterile neutrinos: they are light for the same reason ordinary neutrinos are light, whatever that reason may be [7].

2 Large-scale structure (LSS) with mirror dark matter

An important general point about mirror-matter cosmology is that initial conditions, or inflation, must supply a lower temperature T' for the mirror-matter plasma compared to the ordinary plasma: $T' < T$. Big bang nucleosynthesis (BBN) requires $T' \lesssim 0.5T$ to suppress the mirror contribution to relativistic stress-energy. We require that the microscopic mirror symmetry of the MMM *not* imply that the universe's contents are 50 : 50 ordinary and mirror states. This can be arranged by exploiting fluctuations, such as in inflationary models [8, 9]. By their very nature, fluctuations will not be duplicated event-by-event in the two sectors. The initial temperature difference will be maintained provided $\Delta\mathcal{L}$ is small enough.

The temperature difference means that while the microphysics is the same for mirror-baryogensis, mirror-BBN and mirror-LSS as for their ordinary analogues, the outcomes will be different.

Berezhiani, Comelli and Villante [9] examined mirror-baryogensis for the cases of (i) out-of-equilibrium decays of massive particles, and (ii) electroweak baryogenesis. In both cases, one can get $\Omega'_B > \Omega_B$ because $x \equiv T'/T < 1$, where Ω_B is the baryon density in units of the critical density.

For mirror-BBN, $x < 1$ implies that the relative abundance of He' to H' is higher than that of He to H. Mirror stars have more helium when they are formed, so they evolve more quickly and produce more supernovae [10].

We now come to the main topic of LSS with mirror dark matter (MDM) [9, 11, 12]. MDM consists of mirror-protons, mirror-helium, mirror-electrons and mirror-photons at a temperature $T' < T$ (there are also mirror-neutrinos, but they will not play an important role here). The MDM mass is dominated by the baryons, so in that respect it is "cold". The mirror-photons make MDM not identical to collisionless cold dark matter (CDM): there is a mirror Silk length below which structure is damped. As $x \to 0$, MDM $\to$ CDM in the linear regime. The value of x is important.

There are two possible sequences of events:

1. Mirror-photon decoupling, then matter-radiation equality, then photon decoupling. This transpires if $x < x_{eq} \simeq 0.33$.

2. Matter-radiation equality, then mirror-photon decoupling, then photon decoupling: $x > x_{eq}$ required.

The relevent red-shifts are $z_{dec} \simeq 1100$, $z'_{dec} \simeq 1100/x$ and $z_{eq} \simeq 3300$.

MDM has a Jeans length and Silk length. The microphysics is the same as ordinary baryonic matter, so one just has to scale temperature from T to T'. There are two conditions for MDM perturbation growth to begin: matter domination (as for any DM), and mirror-photon decoupling (so that the mirror Jeans length is small). For the former, perturbation growth begins at matter-radiation equality, while for the latter it is slightly delayed until mirror-photons decouple. Subsequently, after ordinary photons have themselves decoupled, ordinary baryons condense into the MDM potential wells.

Perturbations at length scales below the mirror Silk length

$$\lambda'_S(x > 0,33) \simeq 11 x^{5/4}\ \mathrm{Mpc}, \quad \lambda'_S(x < 0,33) \simeq 14 x^{3/2}\ \mathrm{Mpc} \qquad (12)$$

are damped. Between λ'_S and the scale

$$\lambda_{max} \sim 370 \frac{x^2}{\sqrt{1 + \frac{0.33}{x}}}\ \mathrm{Mpc}, \qquad (13)$$

at which the processed power spectrum peaks, the perturbations oscillate. Galactic scale in this linear regime context is about 3.7 Mpc, so for λ_{max} to be subgalactic we need $x \lesssim 0.2$. Phenomenologically, the damping on small scales is similar to warm DM. Adapting a warm DM analysis by Narayanan et al. [13], we [11] deduced that $x \lesssim 0.06$ is favoured.

Ciarcelluti et al have examined cosmic microwave background anisotropies for MDM [12]. They found that for $x < x_{eq}$, MDM gives practically the same outcome as CDM.

In summary: in the linear regime, MDM behaves like CDM provided x is small. Damping on small scales is the place to look for observational evidence to distinguish MDM from collisionless CDM.

3 Why is $\Omega_B \simeq 0.2\Omega_{\rm dark}$?

The apparent coincidence that prompts the above question suggests that the microphysics of ordinary and dark matter might be similar [14]. This is, of course, the main characteristic of MDM. Is there a MDM cosmology that would explain the 0.2 ratio? Interestingly, the answer is yes [14].

The main stages in this cosmology are as follows:

1. After reheating, $T' \gg T$.

2. At some $T' = T_1$, mirror-baryon and/or mirror-lepton asymmetries are created by some mechanism, e.g. mirror-leptogenesis. No significant ordinary asymmetries are generated.

3. The initial asymmetries are chemically reprocessed. Ordinary asymmetries are created from mirror asymmetries via effective operators of the form
$$\overline{\ell}_{iL}\, \phi^c\, \ell'_{jR}\, \phi'. \tag{14}$$
Such operators are of course related to neutrino–mirror-neutrino mass mixing after electroweak symmetry breakdown. As well contributing to chemical reprocessing, these operators also establish kinetic equilibrium between the mirror and ordinary plasmas: $T' = T$ now.

4. A second, relatively brief, period of inflation (or some alternative process) must then happen to create a mild hierarchy $T'/T \lesssim 0.2$.

The outcomes depend on the flavour structure of the effective operator. Assume one of the following combinations dominates:

- Case 1: $i = j = 2$.

- Case 2: $i = j = 1$.

- Case 4: $i = 2$ and $j = 3$ plus $i = 3$ and $j = 2$.

(The numbering follows that of Ref. [14]; some cases omitted in the present review.) We also need an assumption about the initial asymmetries. We take the natural expectation that $B' \neq 0$ and $L' \neq 0$ are equally spread amongst the flavours. The outcomes are then

$$\text{Cases 1 and 2}: \quad \frac{\Omega_B}{\Omega'_B} = \frac{3795}{15487} \simeq 0.25 \tag{15}$$

$$\text{Case 4}: \quad \frac{\Omega_B}{\Omega'_B} = \frac{20754}{43415} \simeq 0.48. \tag{16}$$

The first is observationally allowed, the second is not. We conclude that successful outcomes can be achieved for certain flavour structures.

4 Other issues and conclusions

4.1 Why spheroidal haloes?

Dark matter must be spheroidally distributed around spiral galaxies in order to explain the flat galactic rotation curves. But if MDM has the same microphysics as ordinary matter, why does the latter condense into a disk while the former remains spheroidal? One way or another, the MDM component of a galaxy must be sufficiently heated to prevent the condensation. Two possibilities have been examined [15]:

- Mirror-sector supernovae (SN) can heat the mirror DM gas. One needs about one mirror SN per year for a typical spiral galaxy. This is 30-100 times the ordinary SN rate. But, as pointed out above, mirror stars are expected to evolve more quickly, and in addition the existence of more MDM in a galaxy than ordinary matter also increases the mirror-SN rate.
- Via kinetic mixing, Eq. 9, *ordinary* SN can heat the mirror gas. It is possible to convert a fraction of the SN energy emitted in the form of neutrinos into mirror photons. Since the energy emitted in neutrinos is about a hundred times higher than the kinetic energy released, this is a significant energy resource.

4.2 Compact mirror objects as MACHOs

If DM is mirror matter, then compact mirror objects should be abundant in the universe. This is one of the main ways to test the MDM hypothesis. In particular, the Milky Way halo should contain such compact objects, though much of the MDM should also remain as gas. Perhaps the Galactic halo MACHOs are mirror stellar objects [16]? (Mirror white dwarfs are favoured on mass grounds).

4.3 Direct detection of dark matter

Foot [17] has shown that the positive DAMA result for direct DM detection [18] is consistent with all the negative DM search results if DM is mirror matter. Recently, he has argued [19] that CDMS [20] should be on the verge of detecting MDM events near their low energy threshold.

4.4 Conclusions

Provided $T' \lesssim 0.1\,T$, mirror dark matter yields an acceptable early universe cosmology. Perhaps MDM can also explain why the observed baryon and dark matter densities are so similar. The non-linear regime of large-scale structure formation for MDM has not been well analysed, and maybe it cannot realistically be done. Some aspects of it, however, are well established, such as that mirror stars must evolve more rapidly than ordinary stars because of the former's helium-rich initial condition. In any case, we should look for observational evidence for MDM through MACHO searches, and direct DM detection experiments.

Acknowledgments

This work was supported by the Australian Research Council.

References

[1] T.D.Lee and C.N.Yang, *Phys.Rev.* 104, 256 (1956).
[2] I.Kobzarev, L.Okun and I.Pomeranchuk, *Sov.J.Nucl.Phys.* 3, 837 (1966).
[3] S.I.Blinnikov and M.Yu.Khlopov. *Sov.J.Nucl.Phys.* 36, 472 (1982); *Sov.Astron.* 27, 371 (1983).
[4] R.Foot, H.Lew and R.R.Volkas, *Phys.Lett.* **B** 272, 67 (1991).
[5] D.E.Kaplan and R.Sundrum, hep-th/0505265.
[6] A.Yu.Ignatiev and R.R.Volkas, *Phys.Lett.* **B** 487, 294 (2000).
[7] R.Foot, H.Lew and R.R.Volkas, *Mod.Phys.Lett.* **A** 7, 2567 (1992); R.Foot, *Mod.Phys.Lett.* **A** 9, 169 (1994); R.Foot and R.R.Volkas, *Phys.Rev.* **D** 52, 6595 (1995). See also, Z.G.Berezhiani and R.N.Mohapatra *Phys.Rev.* **D** 52, 6607 (1995).
[8] E.W.Kolb, D.Seckel and M.S.Turner, *Nature(London)* 314, 415 (1985); H.Hodges, *Phys.Rev.* **D** 47, 456 (1993); Z.G.Berezhiani, A.D.Dolgov and R.N.Mohapatra, *Phys.Lett.* **B** 375, 26 (1996); V.Berezinsky and A.Vilenkin, *Phys.Rev.* **D** 62, 083512 (2000).
[9] Z.Berezhiani, D.Comelli and F.L.Villante, *Phys.Lett.* **B** 503, 362 (2001).
[10] Z.Berezhiani, S.Cassisi, P.Ciarcelluti and A.Pietrinferni *Ap.J.* 24, 495 (2006).

[11] A.Yu.Ignatiev and R.R.Volkas, *Phys.Rev.* **D** 68, 023518 (2003).
[12] Z.Berezhiani, P.Ciarcelluti, D.Comelli and F.L.Villante, *Int.J.Mod.Phys.* **D** 14, 107 (2005); P.Ciarcelluti, astro-ph/0312607; *Int.J.Mod.Phys.* **D** 14, 223 (2005).
[13] V.K.Narayanan et al, *Ap.J.Lett.* 543, L103 (2000).
[14] R.Foot and R.R.Volkas, *Phys.Rev.* **D** 68, 021304 (2003); 69, 123510 (2004).
[15] R.Foot and R.R.Volkas, *Phys.Rev.* **D** 70, 123508 (2004).
[16] R.Foot, *Phys.Lett.* **B** 452, 83 (1999).
[17] R.Foot, *Phys.Rev.* **D** 036001 (2004); astro-ph/0403043; *Mod.Phys.Lett.* **A** 19, 1841 (2004).
[18] R.Bernabei et al (DAMA Collaboration), *Phys.Lett.* **B** 480, 23 (2000); *Riv.Nuovo Cimento* 26, 1 (2003).
[19] R.Foot, astro-ph/0510705.
[20] D.S.Akerib et al (CDMS Collaboration), *Phys.Rev.Lett.* 93, 211 301 (2004).

INDIRECT DARK MATTER SEARCH

V.Zhukov [a]

Institut für Experimentelle Kernphysik, Universität Karlsruhe (TH),P.O. Box 6980,76128 Karlsruhe, Germany and

Institute of Nuclear Physics, Moscow State University, 119992 Moscow, Russia

Abstract. Indirect Dark Matter (DM) search can be used to constrain the mass and distribution of the DM in our Galaxy. We illustrate this with the existing experimental data and discuss possible uncertainties.

1 Introduction

Evidence for Dark Matter (DM) comes from observations of anisotropies in the cosmic microwave radiation, structures hierarchy, distant supernovae and the galactic rotation curves. The gravity mediated supersymmetric model (mSUGRA) identifies the DM as a lightest SUSY particle - neutralino, which can self annihilate. In the Indirect DM search (IDM) the signal from DM annihilation (DMA) in the galactic halo can be observed as an excess of gamma, positrons and antiprotons fluxes on top of the Cosmic Rays (CR) background, which is relatively small for these components. The flux of i-component from DMA can be written as: $F_i(E) \sim \frac{1}{m_\chi^2} \int \rho^2(r) B(r) G(E,\epsilon,r) \sum < \sigma_k v > A_i^k(\epsilon) dr d\epsilon$, where $< \sigma v >$ is the thermally averaged annihilation cross section, $A_i(\epsilon)$- partons hadronization into i-component, $\rho(r)$ is the DM density distribution in the Galactic halo, $B(r)$ is the local clumpiness of the DM or boost factor, m_χ is the neutralino mass and the $G_i(E,e,r)$ is the propagation term (G_γ=1). The annihilation cross section is related to the mSUGRA parameters. The spatial distribution of DM and the clumpiness depend upon cosmological model. The propagation term for the charged components can be obtained from the galactic model, which also is used to calculate the CR backgrounds.

In this study the DMA and propagation of the products were included in the GALPROP galactic model [1]. A possible procedure for the IDM search is demonstarted for the existing experimental data.

2 Ingredients of the Indirect Dark Matter search

Annihilation cross section.

The mass of the neutralino χ_1^0 and the DMA cross section depends upon mSUGRA parameters (m_0,$m_{1/2}$,tanβ,A_0,sgnμ); $m_{\chi_1^0} \sim 0.4 m_{1/2}$, $\sigma \propto \frac{tan^2\beta}{m_{1/2}^4}$ [2]. Taking into account the existing experimental limits ($m_h < 114 GeV$, $bs\gamma$, α_μ) and the relic density(RD) constraint ($\Omega h^2 = 0.113 \pm 0.01$), the $m_{1/2}$, m_0 plane shrinks to a narrow band where a measurable DMA signal, assuming $B < 1000$,

[a] e-mail: zhukov@ekp.physik.uni-karlsruhe.de

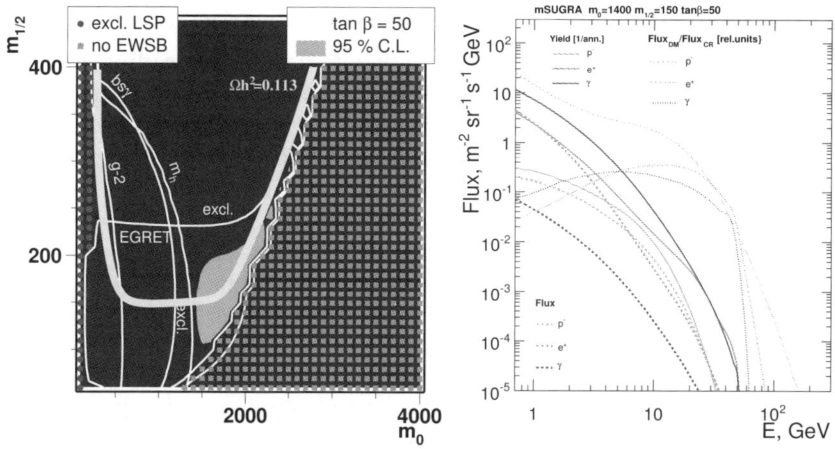

Figure 1: Left: The mSUGRA m_0, $m_{1/2}$ plane at tanβ=50 with existing experimental constraints and the limit from EGRET. Right: The yields per annihilation for different components, the fluxes after propagation(Φ = 0) and the ratio of the DM and CR fluxes calculated for the isothermal DM halo profile(B=1) at mSIGRA point (m_0=1400, $m_{1/2}$=150, tanβ=50).

is expected via an exchange of the pseudoscalar Higgs boson with a subsequent decay into $b\bar{b}$ quarks, see Fig.1(left) [6]. The hadronization of b quarks is well known and the annihilation yields for each component can be calculated, see Fig. 1 (right).

Gamma rays and the neutralino mass

The starting point of the IDM search is the gamma rays since they do not suffer from energy losses and can be used for reconstruction of the neutralino mass. The conventional GALPROP model (CM) underestimates the diffusive gamma in the energy range of E_γ=1-10 GeV in respect to the EGRET experimental data [3]. This excess is difficult to explain by the model uncertainties unless the locally measured CR fluxes are not the galactic averaged ones [5]. The excess is observed in all sky directions and can be interpreted as the DMA in the galactic halo [4]. The Fig.2 shows gamma fluxes for the CM without and with DMA for $m_{\chi_1^0} = 55 GeV$. The DM contribution is scaled with some 'boost' factor depending on the DM distribution. The neutralino mass constrains the $m_{1/2}$ and, together with the RD constraint, limits the tanβ in the range of 45-55 as shown in Fig. 1(left). For other tanβ the RD band is moving out of the $m_{1/2} = 100 - 250 GeV$ region. The $m_{1/2}$ and tanβ define the annihilation cross section and the yields.

DM bulk profile.

The general shape of the bulk DM distribution in the galaxy can be obtained from the N-body cosmological simulation and parametrized as a spherical profile [7]: $\rho_b(r) = \rho_0 \cdot (\frac{r}{a})^{-\gamma} \left[1 + (\frac{r}{a})^{\alpha}\right]^{\frac{\gamma-\beta}{\alpha}}$, where a is a scale radius and α, β, γ are the slopes at different r. The parameters can very in a big range and the spherical shape of the profile can be distorted by the infall of other DM clusters or by interaction with the baryionic matter although the axial symmetry is usually retained [4]. The normalization of the profile is done to the galactic rotation curve at the sun location $r_{\odot} = 8.5kpc$ ($\rho_{\odot} \sim 0.7GeV/cm^3$). Experimentally the DM profile can be reconstructed from the angular distributions of the gamma excess. The Fig.2(right) shows the longitudinal distribution of the EGRET and the simulations with different DM profiles. The cusped profiles NFW(1,3,1) or Moore(1.5,0,1.5) will be seen in a sky as a point like source ($\Delta\phi \sim 5^o$) in the galactic center. The best fit to the experimental data would require an isothermal bulk profile (α,β,γ=2,2,0) with a small ellipticity in the galactic plane($\sim$ 0.8) and ring of DM at $\sim$14 kpc [4]. Such profile reproduces the observed rotation curve and the outer ring can explain the kink in the curve at $r \sim 10kpc$ [8]. The averaged boost factor for such profile for $m_{\chi_1^0} = 55GeV$ is $\sim$ 35.

Clumpiness of DM and the boost factor.

In the inflation cosmology the smallest DM structures (clumps) are created first, which corresponds to the spectrum of primary fluctuations $P(k) = k^{np}$, where $np > 0.95$. Part of these clumps can survive the tidal destruction during evolution and contributes to the annihilation signal. Thus the DMA signal and the independent measurement of the rotation curve can be used to constrain the cosmological model. The DMA signal from the clump $I_c \sim \frac{1}{m_\chi^2} \int \rho_c^2(r)dr$ can be larger then the yield from the 'bulk' profile I_b by the boost factor $B = \frac{I_c + I_b}{I_b}$. In this case, the angular distribution of the observed DMA signal depends on the $n_c(r)$ distribution of the clumps in the Galaxy instead of the DM mass distribution $\rho_b(r)$. The $n_c(r)$ distribution can be affected by the tidal interactions with the galactic potential or stars in the bulge, decreasing the local boost factor $B(r)$ in the center [9] where most of the signal will produced by the bulk DM component. This will allow more cuspy profile. The Fig. 2(right) shows the angular distributions for $n_c(r)$ following the (1,3,1) profile truncated at r=0.5 kpc with the $B(r)$ modified according to the tidal destruction. The number of clumps and therefore the absolute value of the boost factor depends on the np. For the obtained boost factor, the np is estimated in the range of 1.0-1.1 if $\rho_c(r) \sim \frac{1}{r^{1.8-2.0}}$ [9].

Positrons and antiprotons from DMA

The propagation of the charged particles from DMA, $\bar{p}$ and e^+, is similar to the CR particles. The source function for the DMA products is the DM distribution and the injection energy spectrum is defined by the DMA, see Fig.1. In the GALPROP the propagation is described as a diffusion in the leaky box taking into account interactions with interstellar media (gas and light), convection and

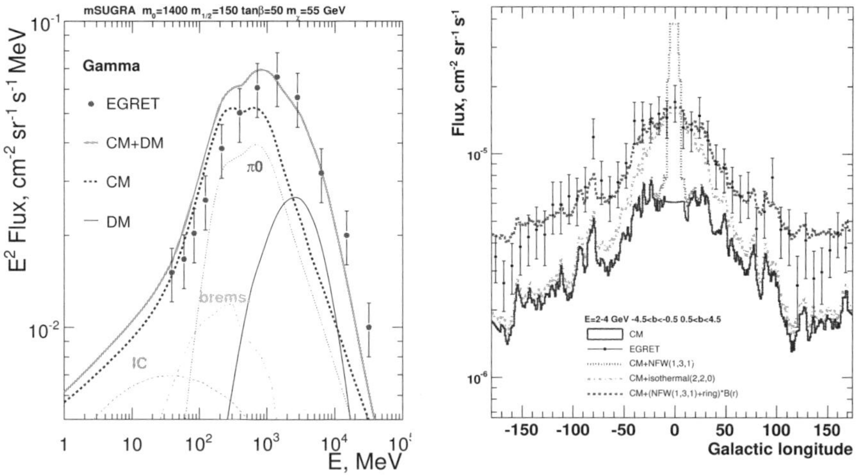

Figure 2: Left: The EGRET diffusive gamma spectrum and the CR fluxes calculated by GALPROP CM and with DM contribution(m_χ=55 GeV) from the isothermal DM profile. Right: The EGRET longitudinal distribution at E=2-4 GeV and the calculations with the CM and different DM profiles.

reacceleration. The CM parameters are tuned to reproduce most of the existing experimental data on CR fluxes. After propagation the energy spectrum is modified by the reaccelareation and energy losses; for electrons ($E > 10GeV$) this is mostly inverse Compton, synchrotron and bremsstrahlung, for antiprotons - the ionization($E < 1GeV$) and fragmentation. The fluxes from DMA in isothermal profile after propagation are shown in Fig.1(right) together with the signal significance in respect to CR backgrounds. The shape of the $\bar{p}$ and e^+ spectrums is model dependent and difficult to use for the m_χ estimation, but the distortions in the spectrums can be used as a signature for the DMA. The largest significance is expected for antiprotons at low energy $E < 10GeV$. However the flux at this energy is strongly affected by the uncertainties of the solar modulation and would require a special study. The experimental data, CR fluxes and the calculated fluxes after propagation and solar modulation are shown in Figure 3 for the DM profile and the boost factor defined from the gamma rays. The positron fraction demonstrates a relatively good agreement with the data but the antiprotons fluxes are overestimated at $E < 10GeV$.

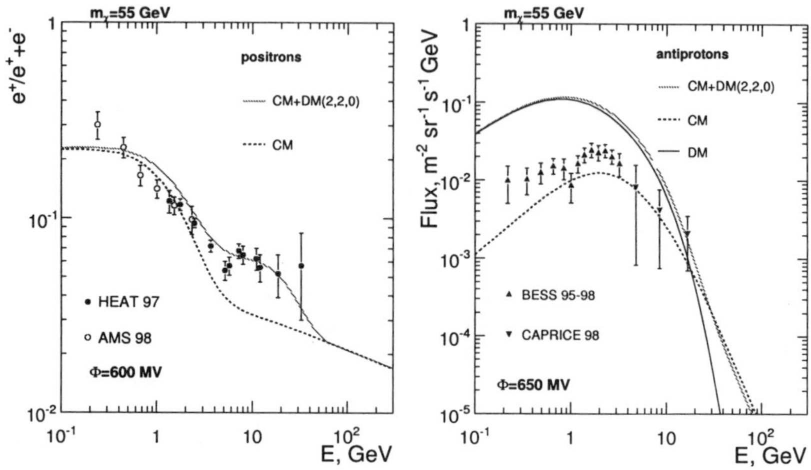

Figure 3: The positron fraction(left) and the antiproton flux(right) without and with DMA.

3 Conclusion

The existing experimental data on the diffusive gamma rays and positron fraction do not contradict the DMA hypothesis. In the mSUGRA model the mass of DM particle can be constrained in the range of m_χ=40-100 GeV and the DM distribution follows the isothermal profile with an external ring at $\sim 14kpc$. The *np* of primary fluctuations is estimated in the range of 1.0-1.1.

Acknowledgments

I thank W. de Boer, C.Sander, V.Dokuchaev, A.Strong, I.Moskalenko and D.Kazakov for useful discussions.

References

[1] A. Strong, I.Moskalenko *ApJ.* **509** 212 (1998).
[2] R. Barbieri, *Riv. Nuovo Cim.* **11** 1(1988);
[3] S.D. Hunter et al., *ApJ.* **481** 205 (1997).
[4] W.de Boer, C.Sander, V.Zhukov et al.,*AA*,**444**, 51(2005).
[5] A. Strong, I.Moskalenko, O.Reimer *ApJ.* **613** 962 (2004).
[6] W.de Boer, C.Sander, V.Zhukov et al., *hep-ph/0511154*.
[7] J.Navarro et al. *Astroph.J.* **490** 493, (1997).
[8] M.Honma, Y.Sofue, *Publ.Astr.Soc.Japan* **49** 453 (1997)
[9] V.Berezinsky, V.Dokuchaev, Y.Eroshenko, *astro-ph/0511494* .

DARK ENERGY AND BLACK HOLES

E. Babichev [a], V. Dokuchaev [b] and Yu. Eroshenko [c]
Institute for Nuclear Research of the Russian Academy of Sciences
60th October Anniversary prospect 7a, 117312 Moscow, Russia

*Abstract.*We describe the accretion of dark energy onto a black hole. Solution for a stationary spherically symmetric accretion of the relativistic perfect fluid with an arbitrary equation of state onto the Schwarzschild black hole is presented. It is shown that accretion of phantom energy is accompanied by a gradual decrease of the black hole mass. Masses of all black holes tend to zero in the phantom energy universe approaching to the Big Rip.

1 Introduction

In recent years the strong observational evidences of the accelerating expansion of the Universe has been obtained. In the framework of General Relativity this acceleration is explained by the dominance in the Universe of dark energy with a negative pressure [1–3]. There are several theoretical models of dark energy: the vacuum energy (the cosmological constant Λ) or such dynamical components as quintessence [4–7] and k-essence [8,9]. The other discussed possibilities are the extensions of General Relativity and extra-dimensional dark energy.

A peculiar property of cosmological models with dark energy is the possibility of a Big Rip [10, 11]: an infinite increase of the scale factor of the Universe in a finite time. The Big Rip scenario is realized in the extreme case of dark energy, the so-called phantom energy, when the sum of energy density and pressure is negative, $\rho + p < 0$. In the Big Rip scenario, the cosmological phantom energy density tends to infinity, and all bound objects in the Universe are finally torn apart up to the subnuclear scales. It should be noted, however, that condition $\rho + p < 0$ alone is not enough for the realization of the Big Rip scenario [12]. The nowadays data on distant supernovas analyzed in a model-independent way [13] showed that the presence of phantom energy with $-1.2 < w < -1$ in the Universe is highly likely.

The accretion of a scalar field with the specially chosen potential $V(\phi)$ onto a black hole was considered in [14–18]. We will describe here the accretion of dark energy onto a black hole by modelling it as perfect relativistic fluid with a negative pressure. In our recent paper [19] (see also [20, 21]), we showed that the masses of all black holes in the Universe with phantom energy gradually decrease, and the black holes disappear completely by the Big Rip. The masses of black holes can decrease during accretion in the case of phantom energy due to the violation of the weak energy dominance condition ($\rho + p \geq 0$) that underlies the theorem on the nondecreasing area of the event horizon of a classical black hole [23].

2 Accretion of dark energy onto black hole

Let us consider the stationary, spherically symmetric accretion onto the black hole of a perfect fluid with an arbitrary equation-of-state $p = p(\rho)$, in particular, with a negative pressure, $p < 0$. The dark energy density is assumed to be low enough (i. e. a test fluid) for metric to be the Schwarzschild one. In this case the flux of energy onto a black hole is [19]:

$$\dot{M} = 4\pi A M^2 [\rho_\infty + p(\rho_\infty)], \tag{1}$$

[a]e-mail: babichev@inr.npd.ac.ru
[b]e-mail: dokuchaev@inr.npd.ac.ru
[c]e-mail: erosh@inr.npd.ac.ru

where M is a black hole mass, A is a numerical factor depending on the equation-of-state and index ∞ corresponds to the values far enough from the black hole (i. e. at the Newtonian infinity). It follows from (1) that the black hole mass increases as it accretes the gas of particles when $p > 0$, but decreases as it accretes the phantom energy when $p + \rho < 0$. This result is rather general and does not depend on the specific form of the equation of state $p = p(\rho)$; only the condition $p + \rho < 0$ is important. The physical cause of the decrease in the black hole mass is as follows: the phantom energy flows into the black hole, but the energy flux of the phantom energy is directed opposite to the flow of dark energy. If we ignore the cosmological evolution of the density ρ_∞, then we find the law of change of the black hole mass due to accretion

$$M = M_i \left(1 - \frac{t}{\tau}\right)^{-1}, \quad \tau = \frac{1}{4\pi A M_i [\rho_\infty + p(\rho_\infty)]}, \tag{2}$$

where M_i is the initial mass of the black hole, and τ is the evolution time-scale. In a particular case of linear equation-of-state, $p = \alpha(\rho - \rho_0)$, where α and ρ_0 are constants, the numerical factor

$$A = \frac{(1+3\alpha)^{(1+3\alpha)/2\alpha}}{4\alpha^{3/2}}. \tag{3}$$

Among the other cases, this model describes an ultrarelativistic gas ($p = \rho/3$), a gas with ultra-hard equation of state ($p = \rho$) and the simplest model of dark energy ($\rho_0 = 0$ and $\alpha < 0$). This linear model describes the phantom energy, $\rho + p < 0$, when $\rho_\infty/\rho_0 < \alpha/(1+\alpha)$. However, the requirement that fluid energy density ρ be nonnegative should be taken into account. For some of the specific values of α (e. g. for $\alpha = 1/3$, $1/2$, $2/3$, and 1), the accretion is described by the simple analytical formulas [19, 20].

Using the method suggested in [24], we obtain the mass evolution law for the rotating Kerr black hole moving with a 4-velocity u in the fluid with an ultra-relativistic equation of state $p = \rho - \rho_0$:

$$\dot{M} = 4\pi(r_+^2 + a^2)[\rho_\infty + p(\rho_\infty)]u^0_{\rm BH}, \tag{4}$$

where $r_+ = M + (M^2 - a^2)^{1/2}$ is a radius of event horizon, $a = J/M$ is a specific angular momentum of the black hole (rotation parameter), and u^0_∞ is the zeroth 4-velocity component of the black hole relative to the fluid.

The other analytically solvable example is the Chaplygin gas with an equation-of-state $p = -\alpha/\rho$ where $\alpha = const > 0$. In this model a numerical factor in (1) is $A = 4\left(\rho_\infty^2/\alpha\right)^{3/2}$.

3 The fate of black holes in the Big Rip

Let us now consider the evolution of black holes in the cosmological Big Rip scenario, where the scale factor $a(t)$ increases to infinity in a finite time [10, 11]. For simplicity, we take into account only the dark energy and disregard the other forms of energy. In the described linear model for the equation-of state, the Big Rip takes place at $\rho + p < 0$ and $\alpha < -1$. The following relation can be derived from the cosmological Friedmann equations in the case of a linear equation of state:

$$|\rho + p| \propto a^{-3(1+\alpha)}.$$

Setting, for simplicity, $\rho_0 = 0$, we find the evolution law of the phantom energy density in this Universe:

$$\rho_\infty = \rho_{\infty,i}\left(1-\frac{t}{\tau}\right)^{-2}, \quad \tau^{-1} = -\frac{3(1+\alpha)}{2}\left(\frac{8\pi}{3}\rho_{\infty,i}\right)^{1/2}. \tag{5}$$

Here, $\rho_{\infty,i}$ is the initial cosmological phantom energy, and the initial time was chosen in such a way that the Big Rip occurs at time τ. It is easily seen that the Big Rip takes place at $\alpha \equiv \partial p/\partial\rho < -1$. Using (1) and (5), we find the evolution law of the black hole mass in the cosmological Big Rip scenario:

$$M = M_i\left(1+\frac{M_i}{\dot{M}_0\,\tau}\,\frac{t}{\tau-t}\right)^{-1}, \tag{6}$$

where constant

$$\dot{M}_0 = (3/2)\,A^{-1}|1+\alpha|, \tag{7}$$

and M_i is the initial mass of the black hole. In the particular case of phantom energy with $\alpha = -2$ and $A = 4$ (which corresponds to the fluid radial 4-velocity at the black hole horizon $u_{\rm H} = -1$), $\dot{M}_0 = 3/8$. In the limit $t \to \tau$ (i. e. near the Big Rip), the t dependence of the black hole mass becomes linear, $M \simeq \dot{M}_0\,(\tau - t)$. When t approaches to τ, the rate of decrease in the black hole mass ceases to depend on the initial black hole mass and the phantom energy density. In other words, the masses of all black holes near the Big Rip are approximately equal and approach zero. The accretion of phantom energy dominates over the Hawking evaporation until the black hole mass decreases to the Planck mass. Formally, however, all black holes in the Universe completely evaporate during the Hawking radiation in the Planck time before the Big Rip occurs.

4 Accretion of scalar fields

Jacobson [14] found a solution for the scalar field with a zero potential $V = 0$ in the Schwarzschild metric, $\phi = \dot{\phi}_\infty[t + 2M\ln(1-2M/r)]$, where ϕ_∞ is the scalar field value at infinity. For this solution $\dot{M} = 4\pi(2M)^2\dot{\phi}_\infty^2$. Frolov and Kofman [16] showed that this solution is also valid for other models of scalar fields with a nonzero potential $V(\phi)$ under the certain conditions. The energy-momentum tensor constructed using the Jacobson solution is identical to one for a perfect fluid with an extremely hard equation of state, $p = \rho$, after the substitution $p_\infty \to \dot{\phi}_\infty^2/2$ and $\rho_\infty \to \dot{\phi}_\infty^2/2$. This is not surprising, since a scalar field theory with the zero potential, $V(\phi) = 0$, is identical to the corresponding model of a perfect fluid [25]. In view of this correspondence, we easily see the agreement between our result (1) for $\dot{M}$ in the case of $p = \rho$ and the similar ones from [14, 16].

The Lagrangian of a scalar field that describes the phantom energy must have a negative kinetic term [10, 11], for example, $K = -\phi_{;\mu}\phi^{;\mu}/2$, (see e. g. [26] for more general cases). In this case, the phantom energy flux onto the black hole has the negative sign, $T_{0r} = -\phi_{,t}\phi_{,r}$, where ϕ is the solution of the same Klein-Gordon equation as that for the standard scalar field, but with the substitution $V \to -V$. For a zero potential, this solution is identical to the Jacobson solution [14] obtained for a scalar field with the positive kinetic term. However, the Lagrangian with a negative kinetic term and $V(\phi) = 0$ does not describe the phantom energy. At the same time, the solution for a scalar field with $V(\phi) = 0$ is identical to the solution for a positive constant potential, $V_0 = const$, which can be chosen in such a way that $\rho = -\dot{\phi}^2/2 + V_0 > 0$. In this case, the scalar field describes the phantom energy with $\rho > 0$ and $p < -\rho$, which leads to a decrease of the black hole mass at the rate $\dot{M} = -4\pi(2M)^2\dot{\phi}_\infty^2$.

Acknowledgments

This work has been supported in part by the Russian Foundation for Basic Research, projects 02-02-16762a, 03-02-16436a, and 04-02-16757a and by the Ministry of Science of the Russian Federation, grants 1782.2003.2 and 2063.2003.2.

References

[1] A. Riess, A. V. Filippenko, P. Challis, et al., *Astron.J.* 116, 1009 (1998).
[2] N.Bahcall, J.P.Ostriker, S.Perlmutter, P.J.Steinhardt, *Science* 284, 1481 (1999).
[3] S.J.Perlmutter, G.Aldering, G.Goldhaber, et al., *Astrophys.J.* 517, 565 (1999).
[4] C. Wetterich, *Nucl.Phys.* **B** 302, 668 (1988).
[5] B.Ratra, P.J.E.Peebles, *Phys.Rev.* **D** 37, 3406 (1988).
[6] J.A.Frieman, C.T.Hill, A.Stebbins, I.Waga, *Phys.Rev.Lett.* 75, 2077 (1995).
[7] R.R.Caldwell, R.Dave, P.J.Steinhardt, *Phys.Rev.Lett.* 80, 1582 (1998).
[8] C.Armendariz-Picon, T.Damour, V. Mukhanov, *Phys.Lett.* **B** 458, 209 (1999).
[9] C.Armendariz-Picon, V.Mukhanov, P.J.Steinhardt, *Phys.Rev.Lett.* 85, 4438 (2000).
[10] R.R.Caldwell, *Phys.Lett.* **B** 545, 23 (2002).
[11] R.R.Caldwell, M.Kamionkowski, N.N.Weinberg, *Phys.Rev.Lett.* 91, 071301 (2003).
[12] B.McInnes, *J. High Energy Phys.* 0208, 029 (2002); M.Bouhmadi-Lopez, J.A.J.Madrid, `arXiv:astro-ph/0404540`.
[13] U.Alam, V.Sahni, T.D.Saini, A.A.Starobinsky, `arXiv: astro-ph/ 0311364`.
[14] T.Jacobson, *Phys.Rev.Lett.* 83, 2699 (1999).
[15] R.Bean, J.Magueijo, *Phys.Rev.* **D** 66, 063505 (2002).
[16] A.Frolov, L.Kofman, *JCAP* 5, 9 (2003).
[17] L.A.Urena-Lopez, A.R.Liddle, *Phys.Rev.* **D** 66, 083005 (2002).
[18] M.Yu.Kuchiev, V.V.Flambaum, `arXiv:gr-qc/0312065`.
[19] E.Babichev, V.Dokuchaev, Yu.Eroshenko, *Phys.Rev.Lett.* 93, 021102 (2004).
[20] E.Babichev, V.Dokuchaev, Yu.Eroshenko. *JETP* 100, 528 (2005).
[21] V.Berezin, V.Dokuchaev, Yu.Eroshenko and A.Smirnov. *Class. Quant. Grav.* 22, 4443 (2005).
[22] E.Babichev, V.Dokuchaev, Yu.Eroshenko, *Class.Quant.Grav.* 22, 143 (2005).
[23] S.W.Hawking, G.F.R.Ellis, *"The Large Scale Structure of Space-Time"* (Cambridge Univ. Press, Cambridge), Chaps. 4.3, 1973.
[24] L.I.Petrich, S.L.Shapiro, S.A.Teukolsky, *Phys.Rev.Lett.* 60, 1781 (1988).
[25] V.N.Lukash, *Sov.Phys.—JETP* 52, 807 (1980).
[26] P.F.Gonzalez-Diaz, *Phys.Lett.* **B** 586, 1 (2004).

MICROLENSING WITH THE RADIOASTRON SPACE TELESCOPE

A.F. Zakharov[a]

National Astronomical Observatories of CAS, Beijing 100012, China;
Institute of Theoretical and Experimental Physics, 117259, Moscow, Russia;
Astro Space Centre of Lebedev Physics Institute, Moscow, Russia;
Bogoliubov Laboratory of Theoretical Physics, JINR, Dubna, Russia

Abstract. It is well-known that gravitational lensing is a powerful tool to investigate matter distributions including DM. Typical angular distances between images and typical time scales depend on gravitational lens masses. For microlensing case angular distances between images or typical astrometric shifts due to microlensing are about $10^{-5} - 10^{-6}$ *as*. Such an angular resolution will be reached with the space space–ground interferometer Radioastron. The basic targets for microlensing searches should be bright point-like radio sources at cosmological distances. In this case, an analysis of their variability and a solid determination of microlensing could lead to an estimation of their cosmological mass density, moreover, in this case one could not exclude a possibility that non-baryonic dark matter also form microlenses if the corresponding optical depth will be high enough. To search for microlensing the most perspective objects are gravitational lensed systems as usually, like CLASS gravitational lens B1600+434, for instance. However, for direct resolving these images and direct detection of apparent motion of the knots, a Radioastron sensitivity probably have to be improved, otherwise, it is necessary to increase an integration time (assuming that a radio source has a typical core – jet structure and microlensing phenomenon is caused superluminal apparent motion of knots). In the case of a confirmation (or a disproval) of claims about microlensing in gravitational lens systems one can speculate about a microlens contribution into the gravitational lens mass. Astrometric microlensing due Galactic Macho's actions is not very important because of low optical depths and long typical time scales. Therefore, a launch of space interferometer Radioastron will give new excellent facilities to investigate microlensing in radio band, since in this case there is a possibility not only to resolve microimages but also observe astrometric microlensing.

1 Parameters of the ground–space interferometer Radioastron

According to the schedule of the Russian Space Agency the space radio telescope Radioastron will be launched in 2007. This international project was initiated by Astro Space Center (ASC) of the Lebedev Physical Institute of the Russian Academy of Sciences (RAS) in collaboration with other institutions of RAS and Russian Space Agency. This space based 10-meter radio telescope will be used for space – ground VLBI measurements. For observations four wavelength bands will be used corresponding to $\lambda = 1.35$ cm, $\lambda = 6.2$ cm, $\lambda = 18$ cm, $\lambda = 92$ cm [1].

Since the apogee height for Radioastron should about $3.5 \cdot 10^5$ km and as a result the fringe size for the minimal wavelength will be smaller than 10 μas (in this case angular resolution could be about 1 μas). The minimal correlated

[a] e-mail:zakharov@itep.ru

flux for space-ground VLBI should be about 1 mJy for the 1.35 cm wavelength, therefore source fluxes should be higher than the threshold.

Thus, there are non-negligible chances to observe mirages (shadows) around the black hole at the Galactic Center and in nearby AGNs in the radio-band using Radioastron facilities [13, 14, 15, 16, 17].

2 Resolving microimages for distant quasars

Basics of microlensing and optical depth calculations for different location of microlenses are presented in a number of papers (see, for instance, [8, 9, 10, 11, 12]).

To prove the microlensing hypothesis for a distant quasar, the source have to have the following properties from a list of the most perspective targets of VSOP or Radioastron missions (or from its extended version):
a) A source should demonstrate signatures of microlensing which are different from typical features for scintillations at time scales $<$ 3–5 years (that is an estimated time of Radioastron mission);
b) A compact core for the source should have size $\lesssim 40\ \mu as$ and flux density should be higher than Radioastron thresholds $\gtrsim 20$ mJy at 6 cm wavelength and $\gtrsim 100$ mJy at at 1.35 cm wavelength.

About $14\% \pm 6\%$ of extragalactic sources have core size $\lesssim 40\ \mu as$ and the angle corresponds to the fringe size at the 6 cm wavelength. This part of sources could be used for photometric monitoring and for a further analysis of a preferable explanation of variability. If the analysis would indicate that microlensing is a preferable cause of variability the candidate could be selected as the first order one. But even in the case, if a source would demonstrate variability that could be explained by another cause (but not by microlensing), the source should be checked to search for image splitting or (and) astrometric image displacement since models for alternative explanation of variabilities could be not quit correct.

From theoretical point of view there is a possibility to detect microlensing for both core and bright knots. In this case the two situations will be characterized by different time scales.

3 Microlensing for gravitational lensed systems

Few years ago, Koopmans et al. [2, 3] claimed that the most realistic explanation of short-term variability of a gravitational lens CLASS B1600+434 at 5 GHz and 8.5 GHz. The authors considered different cases of variability such as scintillation due to scattering and microlensing. As a result they concluded, that microlensing phenomenon in radio band gives the natural fit for observational data.

Treyer & Wambsganss [5] claimed that for photometric fluctuations ~ 0.5 mag typical astrometric displacement should be about $20 - 40\ \mu as$ (to evaluate photometric and astrometric microlensing one could use numerical approaches and analytical asymptotical expansions near fold and cusp singularities [7, 8]. In principle such a displacement could be observed with Radioastron space interferometer at 6 cm and 1.35 cm wavelengths if flux densities for the object is high enough. For example, in the B1600+434 case the density flux is suitable for the core (at least, at 6 cm wavelength), but if the superluminal motion of knots is responsible for microlensing (as Koopmans et al. [2] claimed) the sensitivity of Radioastron should be improved in 10 times at 6 cm wavelength to observe such a displacement of knots. At the 1.35 cm wavelength the Radioastron flux density threshold is probably too high to detect the displacement.

4 Conclusions

First, one could point out that gravitational lensed systems are the most perspective objects to search for microlensing. Astrometric microlensing could be detected in the gravitational lens system such as B1600+434 in the case if a proper motion of source, lens and an observer are generated mostly by a superluminal motion of knots in jet (superluminal motion in jet was found with HALCA in the quasar PKS 1622-297 [6]). But in this case, based on density flux estimates [2], one could say that probably a required sensitivity of the Radioastron interferometer should be improved.

In the case if there is microlensing of core in the B1600+434 system for example, then astrometric microlensing in the system could be about should be about $20 - 40\ \mu as$ [5] and the Radioastron interferometer will have enough sensitivity to detect such an astrometric displacement.

Second, in principle microlensing for distant sources could be the only tool to evaluate Ω_L from microlensing event rate. To solve this problem with the Radioastron interferometer one should analyze variabilities of compact sources with a core size $\lesssim 40 \mu as$ and with high enough flux densities about $\gtrsim 20$ mJy at 6 cm wavelength and about $\gtrsim 100$ mJy at at 1.35 cm wavelength To fit the most reliable model for variabilities of the sources such as scintillations, microlensing etc. A fraction of the sources in the list of extragalactic targets for VSOP and Radioastron about 13% – 14 % [4]. In the case, if the analysis would indicate that other explanations (such as scintillations) are preferable and future observations with Radioastron interferometer would show that the are no features for astrometric microlensing, one could conclude that Hawkins hypothesis should be ruled out. But if an essential fraction of variability could be fitted by microlensing, the sources could be as the first order candidate to search for astrometric microlensing.

Therefore, one could say that astrometric microlensing (or direct image resolution with Radioastron interferometer) is the crucial test to confirm (or rule out) microlens hypothesis for gravitational lensed systems and for point like distant objects.

Astrometric microlensing due to Macho's action in our Galaxy is not very important for observations with the space interferometer Radioastron, since first, probabilities are not high; second, typical time scales are longer than estimated life time of the Radioastron space mission [12].

Therefore, just after the Radioastron launch it will be the first chance to detect microlensing by a direct way. So, the main goal of the paper to attract an attention to such a challenging possibility because, preflight time is very short now and perspective targets should be analyzed carefully by observational and theoretical ways in advance. A number of point like bright sources at cosmological distances and gravitational lensed systems with point like components demonstrating microlens signatures is not very high and the sources should be analyzed by the careful way to search for candidates where microlens model is preferable in comparison with alternative explanations of variabilities.

Acknowledgements

The author thanks N.S. Kardashev for fruitful discussions and A.I. Studenikin for his attention to this contribution. The author is also grateful to the National Natural Science Foundation of China (NNSFC) (Grant # 10233050) and National Key Basic Research Foundation (Grant # TG 2000078404) for a partial financial support of the work.

References

[1] N.S. Kardashev, *Experimental Astronomy*, **7**, 329 (1997).

[2] L.V.E. Koopmans and A.G. de Bruyn, *A&A*, **358**, 793 (2000).

[3] L.V.E. Koopmans, A.G. de Bruyn, J. Wambsganss, C.D. Fassnacht, astro-ph/0004285.

[4] Y.Y. Kovalev, K.I. Kellermann, M.L. Lister et al., astro-ph/0505536.

[5] M. Treyer, J. Wambsganss, *A & A*, **416**, 19 (2004).

[6] K. Wajima, H. Bignall, H. Kobayashi et al., astro-ph/0511063.

[7] A.F. Zakharov, *A & A*, **293**, 1 (2005).

[8] A.F. Zakharov, *"Gravitational Lensing and Microlensing,"* Janus-K, Moscow, 1997 (in Russian).

[9] A.F. Zakharov, M.V. Sazhin, *Physics-Uspekhi,* **41**, 945 (1998).

[10] A.F. Zakharov, *Publ. Astron. Obs. Belgrade,* **75**, 27 (2003); astro-ph/0212009.

[11] A.F. Zakharov, 2005, in *"Particle Physics in Laboratory, Space and Universe"* (Proceedings of the Eleven Lomonosov Conference on Elementary Particle Physics), ed. by A.I. Studenikin, World Scientific, p. 106; astro-ph/0403619.

[12] A.F. Zakharov, *Astronomy Reports,* **50**, 79 (2006); astro-ph/0507700.

[13] A.F. Zakharov, A.A. Nucita, F. De Paolis, G. Ingrosso, *New Astronomy,* **10**, 479 (2005).

[14] A.F. Zakharov, A.A. Nucita, F. De Paolis, G. Ingrosso, in *"General Relativity and Gravitational Physics"* (Proceedings of the 16th SIGRAV Conference on General Relativity and Gravitational Physics), ed. by G. Vilasi, G. Esposito, G. Lambiase, G. Marmo, G. Scarpetta, AIP Conference Proceedings, **751,** p. 227 (2005).

[15] A.F. Zakharov, A.A. Nucita, F. De Paolis, G. Ingrosso, in *"Black Holes on Earth and in Space"* (Proceedings of the XXVII Workshop on the Fundamental Problems of High Energy and Field Theory), Institute for High Energy Physics, Protvino, ed. by V.A. Petrov, p. 21, 2005; gr-qc/0507118.

[16] A.F. Zakharov, A.A. Nucita, F. De Paolis, G. Ingrosso, 2005e, in Proceedings of the 22nd Texas Symposium on Relativistic Astrophysics at Stanford University, ed. by P. Chen, E. Bloom, G. Madejski, V. Petrosian, SLAC-R-752, 2005, eConf:C041213, http://www.slac.stanford.edu/econf/C041213, paper 1226.

[17] A.F. Zakharov, F. De Paolis, G. Ingrosso, A.A. Nucita, *A & A,* **442**, 795 (2005).

[18] A.F. Zakharov, L.Č. Popović , P. Jovanović, *A & A,* **420**, 881 (2004).

[19] A.F. Zakharov, L.Č. Popović, P. Jovanović, *"Gravitational Lensing Impact on Cosmology"* (Proceedings of the 225th IAU Symposium), ed. by Y. Mellier and G. Meylan, **225**, Cambridge, UK, Cambridge University Press, p. 363, 2005.

[20] A.F. Zakharov, L.Č. Popović, P. Jovanović, in *"Cosmology: Exploring the Universe"* (Proceedings of the XXXIX Rencontres de Moriond), ed. by Y. Giraud-Heraud and J. Trân Thanh Vân, The GIOI Publishers, p. 41, 2005; astro-ph/0406417.

Search for Gravitational Waves by LIGO Scientific Collaboration

V.P.Mitrofanov (on behalf of LIGO Scientific Collaboration) [a]
Faculty of Physics, Moscow State University, 119992 Moscow, Russia

Abstract

The Laser Interferometer Gravitational Wave Observatory (LIGO) detectors began making dedicated data collecting runs and performing search for gravitational wave. The current situation with LIGO detectors is presented.

1 Introduction

In the last few years a number of long-baseline, laser interferometric gravitational wave detectors have begun operation. These include the Laser Interferometer Gravitational Wave Observatory (LIGO) detectors located in Hanford, Washington and 3000 km away, in Livingston, Louisiana. It is the joint Caltech-MIT project supported by the USA National Science Foundation [1]. LIGO has broadened the participation to include the community of scientists interesting in participating in the LIGO research program by creating the LIGO Scientific Collaboration (LSC). There are some 500 members from 38 institution in both the US and internationally, GEO members from UK and Germany, and from Australia, India, Japan, Russia and Spain.

Gravitational waves predicted in General Relativity are space-time curvature ripples propagate through the Universe at speed of light. Far from the sources gravitational waves will be a small perturbation h on Minkowski metric [2]. The gravitational wave perturbation is transverse to the propagation direction and comes in two polarizations at 45 degrees to each other. The radiation field causes a strain in space itself and leads to tiny quadrupole deformations of test mass structures with which it interacts. The strain $\delta l/l$ is represented by the gravitational wave amplitude h where $h = 2\delta l/l$. The quadrupole formula for gravitational radiation states that the wave amplitude $h_{\mu\nu}$ is proportional to the second time derivative of the quadrupole mass movement of the source $Q_{\mu\nu}$:

$$h_{\mu\nu} = \frac{2G}{rc^4}\frac{d^2}{dt^2}Q_{\mu\nu} \qquad (1)$$

where $Q_{\mu\nu} = \int \rho(x^\mu x^\nu - \frac{1}{3}x^2\delta_{\mu\nu})d^3x$, G is Newton's gravitational constant, c is the speed of light, r is the distance to the source. Note that the lowest-order source of gravitational radiation is quadrupole because mass, unlike electrical charge, comes in only one sign and the momentum of a free system must be conserved. The quadrupole formula leads to the gravitational wave luminosity

[a] e-mail: mitr@hbar.phys.msu.ru

of a source:

$$L = \frac{G}{5c^5}\left[\frac{d^3Q_{\mu\nu}}{dt^3}\right]^2 \tag{2}$$

This formula is scaled by extremely small number G/c^5. Because of this only astrophysical sources can generate detectible gravitational waves. For example, a pair of $1.4M_{solar}$ neutron stars in circular orbit of radius 20 km (imminent coalescence) at orbital frequency 400 Hz generates 800 Hz radiation amplitude $h \approx 10^{-21}$ at the distance of 15 Mpc. The indirect evidence of the existence of gravitational radiation came from discovery and precision measurements, by R. Hulse and J. Taylor, of the binary pulsar system (PSR 1913+16). Monitoring its orbital period they settled that it was gradually speeding up at just the rate predicted for the general-relativistic emission of gravitational waves [3].

The direct detection of gravitational waves from astrophysical sources will open a new window on the Universe and possibility to obtain the information that we have little hope of gaining in any other way. By the time they reach us, the gravitational waves are only very weak perturbations on our local flat space. But they will provide information about the strong field regions where they began, so one can test general relativity in the strong field limit, where it should show us fundamentally new physics. Because gravitational waves are not scattered as they propagate between source and observer, they should provide information about what is happening in the inner and densest regions of the astrophysical source [4].

2 LIGO interferometers

A new generation of detectors based on suspended-mirror Michelson interferometer promises to attain the requisite sensitivity for observing gravitational waves [5]. Each LIGO detector is a power-recycled Michelson interferometer, with 4 km (or 2 km) long Fabry-Perot cavities in each of its orthogonal arms [6]. These interferometers are sensitive to quadrupole oscillations in the space-time metric due to a passing gravitational wave, measuring directly the gravitational wave strain amplitude. Each interferometer is illuminated with stabilized 10 W power Nd:YAG laser, operating at 1.06 microns. To isolate 11 kg test masses mirrors and other optical elements from ground and acoustic vibrations, the detectors implement a combination of passive and active seismic isolation systems, from which the mirrors are suspended as pendulums. The test masses, major optical components, vibration isolation systems, and main optical paths are all enclosed in a high vacuum system. The LIGO interferometers operate in coincidence, so as to filter out local noise. Also one can determine the direction and polarization of a gravitational wave by measuring arrival-time differences between geographically separated detectors.

Four science runs, ranging 2-3 months, have been carried out with LIGO detectors. Some of them have also involved the German/British detector GEO [7] and the Japanese detector TAMA [8].The detectors sensitivity and duty cycle (defined as the fraction of the total run time when the interferometer was locked and operated in its design regime) are being continuously improved. In the first science runs, no evidence of a gravitational-wave events have been found but upper limit results on the signals from a number of potential astrophysical sources such as coalescing compact binary stars, pulsars, as well as on burst events and the level of a stochastic background, have been set. Below we discuss the classes of signal, which LSC is searching.

3 Classes of sources and signals

Chirp signals are emitted by compact binary systems (neutron stars or black holes) as they spiral together and coalesce during the final few seconds . Within the LIGO frequency band, the gravitational waveform produced by binary neutron star system is well described by the restricted second-post-Newtonian approximation. The spins of the neutron stars, tidal coupling and other finite body effects dependent on the equation of state are not expected to significantly affect the waveform. The wave form is therefore parametrized by masses of the companions, the distance to the binary, the inclination of the system to the plane of the sky, and by the initial orbital phase when the waveform enter the LIGO band. Since binary neutron star mass pair would produce a slightly different waveform, a bank of templates with different mass pairs was constructed. Event triggers are generated by filtering the data from each detector with matched filters designed to detect the expected signals. At the design sensitivity of the LIGO detectors, the most probable rate of detectable binary neutron star coalescences lies lower than one per 10 year.

While the detailed noise spectrum of a detector affects different gravitational-wave searches in different way it is reasonable to summarize the sensitivity of a detector for low-mass inspiral signals in terms of the range for archetypal source. Specifically, the range is the distance at which an optimally oriented and located binary system with the mass of each component is equal to $1.4M_{solar}$ would yield an amplitude signal-to-noise ratio of 8 when extracted from the data using optimal filtering. At present the LIGO detectors have the range of about 10 Mpc [9].

Burst signals. Catastrophic events involving solar mass compact objects can produce transient bursts of gravitational radiation in the LIGO frequency band. The collapse of a massive star, after gravitation overwhelms the pressure sustained through the nuclear burning, results in a supernova explosion and the remnant of neutron star or black hole. Type II supernova collapse can generate gravitation radiation if the core collapse is sufficiently asymmetric. Out to distance of 10 Mpc, a supernova occurs about ones per year. Nevertheless the expected rate, strength and wave form morphology for such events is not precisely known. The searches are designed to be general enough to detect other unanticipated sources as well. Gravitational wave bursts associated with gamma ray bursts and other astrophysical events detected by other means are searched using the cross-correlation technique [10].

Periodic signals. A spinning neutron star is expected to emit gravitational waves if it is not symmetric about its rotation axis or due to oscillations of the nuclear fluid on its surface. The search for gravitational waves from rotating neutron stars is of two kinds. The first is a search targeting pulsars whose parameters are known through radio observations. These searches use matched filtering techniques and take into account that the signal is frequency and amplitude modulated by relative motion of the antenna pattern and source. The best strain upper limit of 1.7×10^{-24} was achieved for PSR J1910-5959D. This allowed a limit to be set on the ellipticity of this pulsar: using a value of $10^{45} gcm^2$ for its principal moment of inertia about the rotation axes, it was found to be less than 7×10^{-5}. The second kind of search looks for gravitational waves from rotating neutron stars that are not observed in electromagnetic waves. This search requires long observation time and considerable number of templates. It therefore is computationally limited. Once the detectors operate at design sensitivity for a year, gravitational waves from the Crab pulsar (B0531+21) may be seen, assuming all gravitational wave emission from this pulsar is the cause of the observed spin-down rate [11].

Stochastic signals. A stochastic background of gravitational radiation is analogous to the cosmic microwave background radiation, though its spectrum is unlikely to be thermal. Sources of a stochastic background could be cosmological or astrophysical in origin. Examples of the former are zero-point fluctuations of the spacetime metric amplified during inflation and first- order phase transitions in the early universe. An example of an astrophysical source is the random superposition of many weak signals from binary-star systems. The spectrum of a stochastic background is described by dimensionless quantity $\Omega_{gw}(f)$ which is the gravitational wave energy density per unit logarithmic frequency, divided by critical energy density ρ_c to close the Universe:

$$\Omega_{gw} = \frac{f}{\rho_c} \frac{d\rho_{gw}(f)}{df} \tag{3}$$

A stochastic background is detected by way of cross correlation of the stain data from detectors, taking advantage of the fact that the instrumental noise of one interferometer will, in general, be uncorrelated with that of the other interferometers. The cross correlation is performed in the frequency domain, using a linear filter that optimized the expected signal-to-noise ratio, given the detectors' noise spectra and the targeted stochastic background. At present the upper bound on the energy density stored as gravitational radiation is established. For the flat spectrum it has been found that $\Omega_0 < 8.4 \times 10^{-4}$ in the 69-156 Hz band. It is expected that, with 1 year of data at design sensitivity, the LIGO detectors will be sensitive at a level several times below the nucleosynthesis bound [12].

In late 2005 LSC has begun the fifth science run. It will last for about one year. At present the strain sensitivity achieved by 4 km LIGO detectors is very close to the design value ($h \approx 3 \times 10^{-23} Hz^{-1/2}$ near 120 Hz). There is the large probability to find gravitational wave signals in this run.

4 Plans and prospects

Improvement of the detectors' strain sensitivity will significantly increase this probability. The development of the next generation of the initial LIGO instruments will be pursued after the complete science run of the initial detectors. The proposed design for Advanced LIGO has 25 kg fused silica test masses, suspended by fused silica fibers or ribbons, along with seismic isolation system, increased laser power $\approx 200W$, and signal recycling. The upgrade is expected to accomplish in 2010. The Advanced LIGO detectors will be more than ten times more sensitive over a very broad frequency band. This translates into an increase of event rate by more than 1000 for extragalactic sources, so that several hours of operation will exceed, in physics reach, the integrated observations of the 1-year initial LIGO. The sensitivity of Advanced LIGO detectors will approach to the limitation associated with quantum behavior of the interferometer test masses and the Heisenberg uncertainty principle, which is known as the Standard Quantum Limit (SQL) [13]. A number of systems have been suggested to evade the SQL. Research groups in the field are already looking forward the next generation of sub- quantum-limited interferometric detectors using new quantum non-demolition techniques of measurements [13].

A number of laboratories around the world are developing similar projects for detection of gravitational waves. The Italian-French VIRGO detector of 3 km arm length at Cascina near Pisa is designed to have particularly good low frequency performance and is close to full operation [14]. The Japanese TAMA 300 detector, which has arms of length 300 m, is operating at the Tokyo Astronomical Observatory [8]. The German-British detector, GEO 600, which has arms of length 600 m, is operating near Hannover [7]. The GEO project is a full partner in LIGO, participating at all levels in the effort. GEO detector is operating in coincidence with LIGO detectors. LIGO and GEO interferometers began to form the Global Network. The multiple signal detection will increase a detection confidence and provide better precision on source locations and wave polarizations.

Currently there are efforts underway by both the European SPACE Agency

(ESA) and NASA to design the Laser Interferometer Space Antenna (LISA) for observation of gravitational waves at low frequencies between 10^{-5} - 1 Hz. Three spacecrafts are placed in solar orbit at 1 a.u. The spacecrafts are located at the corners of an equilateral triangle with 5×10^6-km-long sides. Two arms of the triangle comprise a Michelson interferometer. The third arm permits another interferometric observable to be measured, which can determine a second polarization. The interferometers use the same 1 micron light as the terrestrial detectors but need only a single pass in the arms to gain the desirable sensitivity. The end points of the interferometers are referenced to proof masses free floating within the drag-free spacecraft. The spacecraft is incorporated in a feedback loop with precision thrust control to follow the proof masses [15].

Today the Universe is recognized as the largest high-energy laboratory available for scientific study. Besides the traditional astronomy, Cosmic Ray physics and astrophysical neutrino detection, the other new window on the Universe which is soon to open is gravitational wave astronomy. An active collaboration between these branches will be required and will be mutually beneficial for astrophysics and physics.

Acknowledgments

Our collaboration gratefully acknowledges the support of the United States National Science Foundation for the construction and operation of the LIGO Laboratory and the Particle Physics and Astronomy Research Council of the United Kingdom, the Max-Plank-Society and the State of Niedrersachsen, Germany, for support of the construction and operation of the GEO600 detector. We also gratefully acknowledge the support of the research by these agencies and by the Australian Research Council, the Natural Sciences and Engineering Research Council of Canada, the Council of Scientific and Industrial Research of India, the Department of Science and Technology of India, the Spanish Ministerio de Education y Ciencia, the John Simon Guggenheim Foundation, the David and Lucile Packard Foundation, the Research Corporation, and the Alfred P. Sloan Foundation and the Russian Agency of Industry and Science.

References

[1] B.Barish and R.Weiss, *Phys.Today*, **52** (10), 44 (1999).

[2] L.D.Landau and E.M.Lifshitz, *"Theory of Field"*, (Nauka, Moscow) 1995.

[3] R.A.Hulse and J.H.Taylor, *Astrophys.J.* **195**, L51 (1975).

[4] K.S.Thorne, in *"300 Years of Gravitation"*, (Cambridge University Press, Cambridge) 1987.

[5] P.R.Saulson, *"Fundamentals of Interferometric Gravitational Wave Detectors"*, (World Scientific, Singapore) 1994.

[6] B.Abbot et al. (LIGO Scientific Collaboration), *Nucl.Instr.Meth.* A **517** 154 (2004).
[7] B.Willke et al., *Class.Quantum.Grav.* **21** S417 (2004).
[8] R.Takahashi, *Class.Quantum. Grav.* **21** S403 (2004).
[9] B.Abbot et al. (LIGO Scientific Collaboration), *Phys.Rev.* D **72** 062001 (2005)
[10] B.Abbot et al. (LIGO Scientific Collaboration), *Phys.Rev.* D **72** 082001 (2005)
[11] B.Abbot et al. (LIGO Scientific Collaboration), *Phys.Rev.* D **72** 102004 (2005)
[12] B.Abbot et al. (LIGO Scientific Collaboration), *Phys.Rev.Lett.* **95** 221101 (2005)
[13] V.B.Braginsky, *Sov.Phys. Uspehi* **175** (6) 651 (2005).
[14] F.Acernese et al, *Class.Quantum. Grav.* **21** S385 (2004).
[15] R.Weiss, *Rev. Mod. Phys.* **71** (2) S187 (1999).

QUANTUM GRAVITY AS TWISTORIAL UNIFICATION OF QUANTUM AND GRAVITY

A.Burinskii [a]

Gravity Research Group, NSI Russian Acad.Sci., B.Tulskaya 52, 115191 Moscow

*Abstract.*We consider Quantum Gravity as a theory of Unification based on a fundamental role of gravity in the structure of Quantum Theory. A way for such an Unification is hinted by the twistorial structure of the Kerr geometry. We discuss briefly main features of the Kerr spinning particle and new multiparticle Kerr-Schild solutions, which have relation to unification of quantum and gravity and shed some light on a new stringy (twistorial) texture of vacuum space-time.

1 Introduction

We consider Quantum Gravity as a theory of Unification based on a deep fundamental role of gravity in the structure of Quantum Theory. A way for such an Unification is suggested by the structure of the Kerr-Newman solution which gives the most adequate description of the gravitational and electromagnetic fields of a classical sinning and charged particle (electron). It was obtained by Carter (1968) that the Kerr-Newman solution has gyromagnetic ratio $g = 2$ as that of the Dirac electron. If this coincidence is not occasional it points out on a fundamental relation between the one-particle quantum theory of the Dirac electron and the classical Kerr spinning particle which represents a charged and spinning source generating the external Kerr-Newman solution [1–4]. Being adapted to parameters of electron, this solution displays some drastic changes of the space-time structure on the Compton distances. Meanwhile, the one-particle Dirac theory of electron and multi-particle QED theory ignore gravitational field at all, which is at least inconsistent. We discuss briefly main features of the Kerr spinning particle and new multiparticle Kerr-Schild solutions, which have apparently relation to unification of quantum and gravity and shed some light on a new stringy (twistorial) texture of vacuum space-time.

2 Space-time structure of the Kerr solution.

It is widely recognized that gravitational field of elementary particles is very weak and its local action is essential only at the Planck scale. The Kerr geometry is a contr-example to this assertion showing that the local action of gravitational field extends up to the Compton distances due to the stringy structure of the Kerr source. Indeed, the Coulomb and Schwarzschild singular points turn in the Kerr-Newman solution into a singular ring of the radius $a = J/m$. Being adapted to parameters of electron, $J \sim \hbar$, and a turns out to be the Compton radius which exceeds the Schwarzschild one at $\sim 10^{22}$ times. Angular momentum of electron is very high, and we have $a >> m$, which is condition that the related to Kerr-Newman solution black hole horizons disappear and the Kerr singular ring represents a naked singularity. The Kerr singular string is a branch line of the Kerr space, so there appears a strong change in topology on the Compton distances, which affects strongly on the structure of fields in this region. In particular, it may act as a waveguide for electromagnetic field.[b] Since the Kerr singularity is a branch line of space, the Kerr space acquires two-sheeted topology, and the disk spanned on this ring plays the role of gates to a "mirror" anti-world (a "negative" sheet of the Kerr space) where the signs of charges

[a]e-mail: bur@ibrae.ac.ru

[b]There appear stringy excitations generating the spin and mass of the extended particle - "microgeon" model [2]. It was shown, that this ring is indeed a string [3] resembling a heterotic string of superstring theory.

and masses, and the directions of the fields are changed. All the fields interacting with Kerr's particle have to feel these "gates to anti-world" which have the giant Compton radius $a \sim 10^{-11}$ cm. One more remarkable structure of the Kerr geometry is PNC (principal null congruence). It is a vortex of the lightlike rays (twistors) which are falling on the "negative sheet" on the Kerr disk, penetrate it and turn into outgoing "out"-fields on the "positive sheet" of space (see fig.1). The tangent to congruence vector field k^μ determines the Kerr-Schild ansatz for metric

$$g^{\mu\nu} = \eta^{\mu\nu} + 2Hk^\mu k^n \tag{1}$$

(where $\eta^{\mu\nu}$ is the auxiliary Minkowski metric) and the form of vector potential $A_\mu = \mathcal{A}(x)k_\mu$ for electrically charged solution, i.e. it determines polarization of the gravitational and electromagnetic fields around the Kerr source and the directions of radiation for the nonstationary excited solutions. Function H has the form

$$H = (mr - e^2/2)/(r^2 + a^2\cos^2\theta), \tag{2}$$

where r and θ are oblate spheroidal coordinates, so $r > 0$ on the positive sheet of metric and $r < 0$ on the negative sheet. Singular ring corresponds to $r = \cos\theta = 0$.

The Kerr congruence is determined by the Kerr theorem [4,6] which is formulated in terms of twistors (null rays). This theorem allows one to describe all the "geodesic and shear free" congruences which determine the null fields $k^\mu(x)$ describing the metric and electromagnetic field.

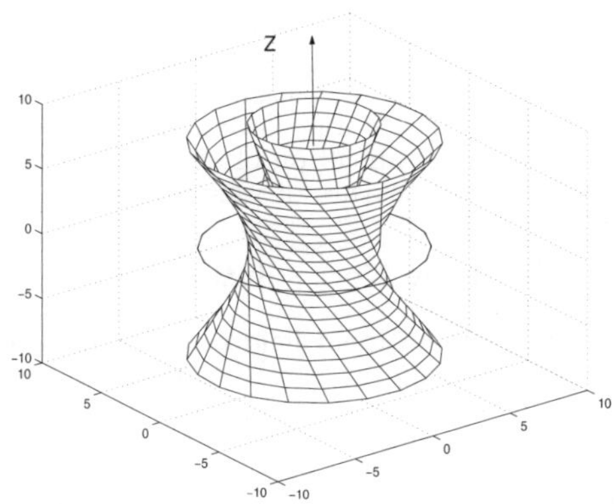

Figure 1: **The Kerr singular ring and the Kerr PNC propagating from "negative" sheet of the Kerr space to "positive " one.**

3 Renormalization by gravity.

For the asymptotically flat particle-like solutions gravity possesses by a remarkable renormalizing effect based on its non-local action. Mass of an isolated source is determined by asymptotic gravitational field only, and therefore, it depends only on the mass parameter m which survives in the asymptotic expansion for metric. On the other hand, the total mass can be calculated as a volume integral which takes into account densities of the electromagnetic energy ρ_{em}, material ("mechanical" mass) sources ρ_m and energy of gravitational field ρ_g. Note, that the last term is not taken into account in QED, but namely this term provides perfect renormalization. For the spherically symmetric systems the expression for mass may be reduced to an integral over radial distance r

$$m = 4\pi \int_0^\infty \rho_{em} dr + 4\pi \int_0^\infty \rho_m dr + 4\pi \int_0^\infty \rho_g dr. \tag{3}$$

Similar is valid in the rotating Kerr-Schild background as a consequence of the Tolman relations, taking into account energy of electromagnetic field ρ_{em} distribution of matter $rho(m)$ and the energy of gravitational field, including the contribution from pressure and rotation [5]. The left side of this relation is the constant which is determined by gravitational asymptotic and does not depend at all on radial distribution of electromagnetic field and matter on the right side. All the exotic matter distributions in a restricted region, including singularities caused by electromagnetic field, will be compensated by gravitational term $4\pi \int_0^\infty \rho_g dr$ to provide the value of constant m on the left side, which depends only on the gravitational asymptotic! It shows that, due to the strong non-local action, gravity may be very essential for elementary particles.

The Kerr-Schild form of metric allows one to consider a broad class of regularized solutions which remove the Kerr singular ring, covering it by a matter source. They retain the Kerr-Schild form of metric and the form of Kerr principal null congruence $k_\mu(x)$, as well as its property to be geodesic and shear-free [5].

Regularized solutions have tree regions:

i) the Kerr-Newman exterior, $r > r_0$, where H is given by (2)

ii) interior $r < r_0 - \delta$, where nominator in H is replaced by new function $f(r) = \alpha r^4$, which suppresses the singularity at $r = 0$, and provide the smoothness of the metric up to the second derivatives.

iii) a narrow intermediate region $r \in [r_0 - \delta, r_0]$ which allows one to get a smooth solution interpolating between regions i) and ii). As it is shown in [4, 5], taking the parameter $\alpha = 8\pi\Lambda/6$, one obtains a regular source of spinning particle. If $a = 0$, the interior solution has the space-time of constant curvature $R = -24\alpha$ which is generated by a source with energy density $\rho = \frac{1}{4\pi}(f'r - f)/\Sigma^2$. It generates a de Sitter interior for $\alpha > 0$ and anti de Sitter interior for $\alpha < 0$. If $\alpha = 0$, we have a flat interior. The resulting sources may be considered as the *bags* filled by a special matter with positive ($\alpha > 0$) or negative ($\alpha < 0$) energy density. The transfer from the external electro-vacuum solution to the internal region (source) may be considered as a phase transition from 'true' to 'false' vacuum [4, 5].

All that is easily generalized to the rotating case $a \neq 0$ [4, 5]. Since parameter a is absent in the nominator of function H, one has only to pass to oblate spheroidal coordinates r and θ.

The source of Kerr-Newman spinning particle with a spin $J = \frac{1}{2}\hbar$, acquires the form of a relativistically rotating disk (the board of the disk has $v \sim c$). For $a >> m$ disk has the form of a highly oblate ellipsoid with the thickness r_0 of the order of classical electromagnetic radius (and the Compton radius of the disk $a = \frac{1}{2}\hbar/m$).

Interior of the disk represents a "false" vacuum having superconducting properties, which is a physical realization of the "Alice" mirror. The charges are concentrated on the surface of this disk, at $r = r_0$. Inside the disk the local gravitational field is negligible.

4 Multiparticle Kerr-Schild solutions and twistorial structure of vacuum.

In the obtained recently multiparticle Kerr-Schild solutions [6] the Kerr twosheetedness is generalized to a multisheeted twistorial structure over the auxiliary Minkowski space-time. It was shown that particles have independent twistorial structures and

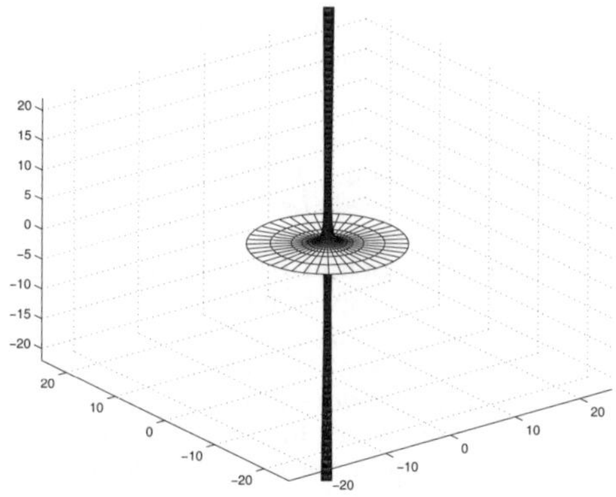

Figure 2: **Rotating disk-like source of the Kerr spinning particle and two singular pp-waves.**

interact via the common twistor lines, forming the relating them null strings - singular pp-wave solutions. In a system containing many particles, such a stringy structure covers all the space, forming a twistorial web, in which singular rings of the spinning particles (loops) play the role of nodes. It is partially reminiscent the Penrose spin-network twistorial ideas [7] and some elements of the loop quantum gravity approach. Also, it supports the point of view that the radiating twistorial structure of the Kerr PNC belongs to the vacuum zero-point field and hints a twistorial texture of vacuum (see details in [4, 6]).

Acknowledgments

This work was supported by RFBR project 04-02-17015-a. Author thanks Organizers for invitation to participate in this Conference.

References

[1] W. Israel, *Phys.Rev.* **D** 2, 641 (1970). C.A. López,*Phys.Rev.* **D** 30, 313 (1984).
[2] A.Ya. Burinskii, *Sov. Phys. JETP*, 39, 193 (1974);
[3] A. Burinskii, *Phys.Rev.* **D** 68, 105004 (2003); *Phys.Rev.* **D** 70, 086006 (2004).
[4] A. Burinskii, The Dirac-Kerr electron, hep-th/0507109.
[5] A. Burinskii, *Grav.& Cosmology.* 8, 261 (2002).
A. Burinskii, E. Elizalde, S. R. Hildebrandt and G. Magli, *Phys.Rev.* **D** 65, 064039 (2002), gr-qc/0109085.
[6] A. Burinskii, The Kerr theorem and Multiparticle Kerr-Schild solutions, hep-th/0510246; *Wonderful consequences of the Kerr theorem*, hep-th/0506006.
[7] R. Penrose, "Quantum Theory and Beyond", ed. E. Bastin, Cambridge Univ. Press, Cambridge 1969. C. Rovelli and L. Smolin, *Phys.Rev.* **D** 52, 5743 (1995), gr-qc/0006061.

QUANTUM COSMOLOGY AND THE GLOBAL ROTATION PROBLEM

M.Fil'chenkov [a],
Institute of Gravitation and Cosmology, Peoples' Friendship University of Russia, 117198 Moscow, Russia

Abstract. The global rotation problem is expected to be solved in the framework of quantum geometrodynamics by quantizing Raychaudhuri's equation for a homogeneous expanding universe with rotation, shear and acceleration. Quantum birth probabilities have been calculated for a multicomponent cosmological model with nonzero angular momentum.

1 Introduction

The problem of the origin of rotation of astronomical objects has not been resolved so far [1,2]. The point is that the existing mechanisms are insufficient for explaining the observable value of angular momenta of spiral galaxies. On the other hand, for the specific angular momentum of the Universe as a whole we have

$$\frac{J}{M} \sim \omega R^2, \quad \omega < H\frac{\Delta T}{T},$$

where

$$H = 65 kms^{-1}Mpc^{-1}, \quad \frac{\Delta T}{T} \sim 10^{-5}, \quad R \sim 10^{28} cm.$$

Hence $\frac{J}{M} < 2 \cdot 10^{32} cm^2 s^{-1}$, which three orders exceeds our Galaxy's specific angular momentum.

2 Classical approach

A homogeneous universe is described classically by Raychaudhuri's equation:

$$\dot{\theta} + \frac{1}{3}\theta^2 - A^i_{;i} + 2\sigma^2 - 2\omega^2 = -\frac{4\pi G}{c^2}(\varepsilon + 3p), \tag{1}$$

where θ is the expansion, A^i is the acceleration, σ is the shear, ω is the angular velocity, ε is the energy density, p is the pressure.
The above quantities are given by the formulae:

$$\theta = \frac{3\dot{a}}{a},$$

$$A^i = c\alpha\theta(u^i - g^{0i}),$$

$$\sigma = \begin{cases} \frac{\Sigma}{a^3} & -\frac{1}{3} < \alpha \leq 1 \\ 0 & -1 \leq \alpha \leq -\frac{1}{3}, \end{cases}$$

[a] e-mail: fmichael@mail.ru

$p = \alpha\varepsilon$ (equation of state), $\frac{1}{c^2}(p+\varepsilon)a^5\omega = J = const$ (angular momentum conservation law), where a is the scale factor, u^i is the four velocity, g^{ik} is the metric tensor. For multicomponent quasi-Friedmann metrics, with

$$|A^i_{;i}| \ll \frac{4\pi G}{c^2}|\varepsilon + 3p|, \tag{2}$$

$$\sigma^2 \ll \omega^2 \ll \frac{2\pi G\varepsilon}{c^2}, \tag{3}$$

$$\dot{\varepsilon}a + 3(\varepsilon + p)\dot{a} = 0 \tag{4}$$

(energy conservation law) being satisfied, Raychaudhuri's equation admits the first integral as follows

$$\dot{a}^2 = -\frac{8\pi G}{3c^2}a^2\sum_{n=0}^{6}\frac{(n-2)\varepsilon_n}{(2-n)\left[1-\left(\frac{n}{3}-1\right)(1-g^{00})\right]-4\left(\frac{n}{3}-1\right)(1-g^{00})}$$
$$+\frac{2}{3}a^2\sum_{n=0}^{6}\frac{\omega_n^2}{(n-4)\left[1-\left(\frac{n}{3}-1\right)(1-g^{00})\right]-2\left(\frac{n}{3}-1\right)(1-g^{00})}$$
$$+\frac{1}{3}a^2\sum_{n=0}^{6}\sigma_n^2 - kc^2\sum_{n=0}^{6}C_n\left(\frac{a}{r_0}\right)^{\frac{4\left(\frac{n}{3}-1\right)(1-g^{00})}{1-\left(\frac{n}{3}-1\right)(1-g^{00})}}, \tag{5}$$

where

$$\varepsilon_n = \varepsilon_0 B_n\left(\frac{r_0}{a}\right)^n, \tag{6}$$

$$\omega_n^2 = \frac{9c^4 J_n^2}{\varepsilon_0^2 r_0^{10} n^2 B_n^2}\left(\frac{r_0}{a}\right)^{2(5-n)}, \tag{7}$$

$$\sigma_n^2 = \begin{cases} 0 & n = 0,1,2 \\ \frac{\Sigma_n^2}{a^6}, & n = 3,4,5,6, \end{cases} \tag{8}$$

$$n = 3(1+\alpha) \tag{9}$$

($n = 0$ de Sitter's vacuum, $n = 1$ domain walls, $n = 2$ strings, $n = 3$ dust, $n = 4$ radiation, $n = 5$ perfect gas, $n = 6$ ultrastiff matter), $k = 0, \pm 1$ is the model parameter, ε_0 is the vacuum energy density related to de Sitter's horizon r_0 by the formula

$$\frac{1}{r_0^2} = \frac{8\pi G\varepsilon_0}{3c^4}, \tag{10}$$

B_n are contributions of the matter components to the total energy density satisfying the condition $\sum B_n = 1$, C_n are coefficients related to matter component accelerations satisfying the condition $\sum C_n = 1$. Since de Sitter's vacuum can be neither rotated nor accelerated, we assume that $J_0 = C_0 = 0$.

Friedmann's limit requires a dependence $1-g^{00} = \beta(n-2)^\gamma$, where $\beta = const$ and $\gamma > 1$.

3 Quantization

For Friedmann's world the standard quantum-geometrodynamical quantization procedure is as follows [3]:

1) $\frac{\partial L}{\partial N} = p_N$ is the primary constraint for the Lagrangian $L(a, \dot{a}, N)$, where N is the lapse function of the metric $ds^2 = N^2c^2t^2 - dl^2$,

2) $\dot{p}_N = 0$ is Dirac's constaint conservation,

3) $\dot{p}_N = -\frac{\partial H}{\partial N}$ is due to one of Hamilton' equations for the Hamiltonian $H = p_a\dot{a} - L$,

4)$H = N\mathcal{H}$, where $\mathcal{H} = \frac{p_a^2}{2} + \frac{ka}{2} - \frac{4\pi G\varepsilon a^3}{c^4}, \quad p_a = \frac{a\dot{a}}{N}$,

5)$\mathcal{H} = 0$ is the Hamiltonian constraint,

6) $\mathcal{H}\psi = 0$ is Wheeler-DeWitt's equation.

For quasi-Friedmann's world in the conformal time η we have $N = a$ and $p_a = \frac{da}{d\eta}$, where $ad\eta = cdt$. Substituting p_a in Wheeler-DeWitt's equation by the operator

$$\hat{p}_a = \frac{1}{i} l_{pl}^2 \frac{d}{da}, \tag{11}$$

where $l_{pl} = \sqrt{\frac{\hbar G}{c^3}}$ is the Planckian length, we obtain a Schrödinger-like equation

$$-\frac{\hbar^2}{2m_{pl}} \frac{d^2\psi}{da^2} + [U(a) - E]\psi = 0. \tag{12}$$

In case Friedmann's world containing only de Sitter's vacuum ($p = -\varepsilon$), strings ($p = -\frac{\varepsilon}{3}$) and radiation ($p = \frac{\varepsilon}{3}$), for a slow rotation of radiation, dust ($p = 0$) and perfect gas ($p = \frac{2}{3}\varepsilon$) we have calculated the Universe's birth probability using Gamow's formula

$$D = \exp\left\{ -\frac{2}{\hbar} \left| \int_{a_1}^{a_2} \sqrt{2m[E - U(a)]}da \right| \right\}, \tag{13}$$

where $m = m_{pl}$ and $U(a_1) = U(a_2) = E$. The energy is equal to

$$E = \frac{m_{pl}c^2}{2} \left(\frac{r_0}{l_{pl}}\right)^2 \left(\frac{3B_4}{4 - g^{00}} - \frac{2J_3^2c^2}{3\varepsilon_0^2 r_0^8 B_3^2}\right). \tag{14}$$

Calculating the integral near the maximum of the potential

$$U(a) = \frac{m_{pl}c^2a^2}{2l_{pl}^2} \left\{ k - B_2 - \left(\frac{a}{r_0}\right)^2 \left[\frac{B_0}{4 - 3g^{00}} + \frac{18J_5^2c^2}{25\varepsilon_0^2 r_0^8 B_5^2(5 - 2g^{00})} \right] \right\} \tag{15}$$

we obtain

$$D = \exp\left\{ -\frac{\pi}{c^2 l_{pl}^2} \frac{\left| \frac{j_3^2}{3B_3^2} - \frac{27 j_5^2 B_4(6 - g^{00})}{25 B_0 B_5^2 (4 - g^{00})(5 - 2g^{00})} \right|}{\sqrt[4]{\frac{3B_0 B_4}{(4 - 3g^{00})(4 - g^{00})}}} \right\}, \tag{16}$$

where $j_n = \frac{c^2 J_n}{\varepsilon_0 r_0^3}$ is the specific angular momentum.

It is easy to see from (5) that in a nonflat world with a nonzero acceleration of radiation we obtain

$$kC_4(1 - g^{00}) = \frac{27 j_4^2}{64 r_0^2 c^2 B_4^2}, \tag{17}$$

i.e. the model geometry is related to the rotation and acceleration of radiation.

4 Conclusion

Quantum cosmology sets initial conditions for quantum field models in curved space-time, which explains generation of initial perturbations that determine a further classical evolution of the Universe and galaxy formation.

The conventional scenario of matter creation does not give a seed momentum at all, since for de Sitter's vacuum $J = \frac{p+\varepsilon}{a^2} a^5 \omega = 0$ due to $p + \varepsilon = 0$. If the matter being born of de Sitter's vacuum had no angular momentum, a cosmological origin of rotation in astronomy is impossible. At the same time the nonvacuum-like matter that had tunnelled through the barrier, in principle, could induce a slow cosmological rotation.

The slow rotation of the multicomponent quasi-Friedmann world results both in an increase and decrease in its quantum birth probability, depending on the equations of state of the matter components. The model geometry is related to the angular momentum and acceleration of radiation.

Acknowledgments

I am grateful to A.I. Studenikin for providing me the possibility of participating the 12th Lomonosov Conference on Elementary Particle Physics.

References

[1] W. Godłowski, M. Szydłowski et al., *Gen. Rel. Grav.* 35 907 (2003).
[2] M.L. Fil'chenkov, R.Kh. Saibatalov, *Grav. & Cosmol.* 11, 116 (2005).
[3] A. Vilenkin, *Phys. Rev.* **D** 50 2581 (1994).

SELF-INTERACTION OF CHARGED PARTICLES OUTSIDE BRANE TOPOLOGICAL DEFECTS

Yu. Grats [a], V. Dmitriev [b]
Faculty of Physics, Moscow State University, 119992 Moscow, Russia
Abstract. We calculate gravitationally induced self-force on a charged test particle held at rest outside conical defects imbedded into the brane of the RS2 model.

1 Introduction

Over the last forty years a number of authors have investigated the influence of a gravitational field on a charged test particle [1–6]. One of the results of these investigations is that the distortion of charge's long range Coulomb field by the background geometry leads to the appearance of a finite self-force.

The purpose of this paper is the calculation of electrostatic self-force on a charged test particle outside conical defects imbedded into the brane of the Randall-Sundrum model with one brane.

In this paper, we use the metric with the signature $(-+++)$ and the system of units $c = 1$.

2 Self-interaction near brane defects

Formal expression for the electromagnetic self-energy of a point test charged particle held at rest at a point x reads

$$U_{em}(\vec{x}) = -\int d^3x\sqrt{-g}T_{em}{}^0{}_0 = 2\pi e^2 G(\vec{x},\vec{x}) \ , \tag{1}$$

where $G(\vec{x},\vec{x})$ stands for the coincidence limit of the Green function of the Poisson equation

$$\partial_i\left(\sqrt{-g}g^{ij}g^{00}\partial_j\right)G(\vec{x},\vec{x}') = \delta^{(3)}(\vec{x}-\vec{x}') \ , \quad i,j,k\ldots = 1,2,3 \ .$$

Proceeding along the same line as in [5], let us write this equation in the form

$$\delta^{ij}\partial_j\partial_i G(\vec{x},\vec{x}') = -\delta^{(3)}(\vec{x}-\vec{x}') - VG(\vec{x},\vec{x}') \ , \tag{2}$$

where

$$V = -\partial_i\left(\sqrt{-g}g^{ij}g^{00}\partial_j\right) - \delta^{ij}\partial_i\partial_j \ .$$

When the use of perturbation theory is justified, the solution of equation (2) can be written as

$$G = G_0 + G_0VG_0 + \ldots \ , \tag{3}$$

[a] e-mail: grats@string.phys.msu.ru
[b] e-mail: cinq@rambler.ru

where G_0 is the Green function in the Euclidean space.

We start with a short review of the Randall-Sundrum model with one brane [7] and the linearized gravity associated with a matter placed on the brane [8].

The metric can be written as

$$ds^2 = \hat{g}_{\mu\nu}dx^\mu dx^\nu + 2N_\mu dx^\mu dy + (1+\phi)dy^2 \ ,$$

$$\hat{g}_{\mu\nu} = e^{-2k|y|}(\eta_{\mu\nu} + h_{\mu\nu}) \ , \quad \mu, \nu, \ldots = 0,1,2,3,$$

where $x^\mu = (t, \vec{x})$ are the coordinates on the brane and the fifth coordinate y is chosen so that the brane is at $y = 0$. $\hat{g}_{\mu\nu}$ gives the induced metric on the hypersurfaces $y = const$. The parameter k^{-1} is related to the bulk cosmological constant $\Lambda = -6k^2$ and gives the characteristic length scale of the bulk corrections to the $4D$ Einstein gravity.

In the linear approximation $h_{\mu\nu}$, N_μ and ϕ are taken to be small, and gauge conditions for the brane to remain at $y = 0$ read

$$N_\mu = -\frac{\operatorname{sgn} y}{8k} h_{,\mu} \ , \quad \phi = -\frac{\operatorname{sgn} y}{4k} h_{,y} \ , \quad \tilde{h}_{\mu\nu}{}^{,\mu} = 0 \ ,$$

where $h = \eta^{\mu\nu}h_{\mu\nu}$ and $\tilde{h}_{\mu\nu} = h_{\mu\nu} - (1/4)\eta_{\mu\nu}h$ is the traceless part of the metric perturbation $h_{\mu\nu}$.

The energy momentum tensor of the model is

$$T^{\mu\nu} = -\frac{3k}{4\pi}\sqrt{1-\phi}\,\hat{g}^{\mu\nu}\delta(y) + t^{\mu\nu}\delta(y) \ , \quad T^{5\mu} = T^{55} = 0 \ .$$

The first term is the background from the brane, and $t^{\mu\nu}$ is the matter perturbation on the brane.

In the lowest order approximation in $h_{\mu\nu}$ and in the case of a static space-time operator V reads

$$V = -\tilde{h}^{ij}\partial_i\partial_j + \delta^{ij}\partial_i h^{00}\partial_j + \frac{1}{4}\delta^{ij}\left(\partial_i h \partial_j + h\partial_i\partial_j\right) + h^{00}\delta^{ij}\partial_i\partial_j \ . \qquad (4)$$

For $k > 0$ and $q^2 > 0$ the solution for $\tilde{h}_{\mu\nu}(x,0)$ in the momentum space reads $\tilde{h}_{\mu\nu}(x,0) = \tilde{h}^0_{\mu\nu}(x,0) + \tilde{h}^1_{\mu\nu}(x,0)$,

$$\tilde{h}^0_{\mu\nu}(x,0) = 16\pi G_4 \int \frac{d^4q}{(2\pi)^4} e^{iqx} \frac{W_{\mu\nu}(q)}{q^2} \ ,$$

$$\tilde{h}^1_{\mu\nu}(x,0) = 8\pi G_5 \int \frac{d^4q}{(2\pi)^4} e^{iqx} \frac{K_0(|q|/k)}{|q|K_1(|q|/k)} W_{\mu\nu}(q) \ , \qquad (5)$$

where

$$W_{\mu\nu}(q) = t_{\mu\nu}(q) - \frac{1}{3}\left(\eta_{\mu\nu} - \frac{q_\mu q_\nu}{q^2}\right) t(q) \ ,$$

G_4 and G_5 are the four-dimensional and five-dimensional gravitational constants ($G_4 = G_5 k$).

Corresponding solution for the trace h has the form

$$h(x,0) = -\frac{32\pi}{3} G_5 k \int \frac{d^4q}{(2\pi)^4} e^{iqx} \frac{t(q)}{q^2} \,.$$

The energy-momentum tensors of a straight string and a global monopole can be approximated by the expressions

$$t^{tt}_{str} = -t^{zz}_{str} = \frac{\mu}{\rho^2}\delta(x^1)\delta(x^2) \,,$$

$$t^{tt}_{mon} = \frac{\eta^2}{r^2} \,, \quad t^{ik}_{mon} = -\eta^2 \frac{x^i x^k}{r^4} \,. \tag{6}$$

ρ and r are the radial coordinates of the cylindrical and spherical systems of coordinates, respectively.

We obtain that in these particular cases

$$W^{str}_{\mu\nu}(q) = \mu\delta(q_3)\,\delta(q_0)\,\frac{(2\pi)^2}{3}\left[3\sum_{i=1}^{2}\delta_{i\mu}\delta_{i\nu} - \eta_{\mu\nu} - 2\frac{q_\mu q_\nu}{q^2}\right] \,,$$

$$W^{mon}_{\mu\nu}(q) = \frac{\eta^2\pi^2\delta(q_0)\,(2\pi)}{q}\,\frac{1}{3}\left[3\sum_{i=1}^{3}\delta_{i\mu}\delta_{i\nu} - 2\eta_{\mu\nu} - \frac{q_\mu q_\nu}{q^2}\right] \,.$$

Substitution of the expressions above into (5) and then into (4) and (3) enables one to calculate the first order correction to the Green function and to consider its coincidence limit. It is well known that in this limit Green's functions diverge. So, some regularization procedure is needed. We proceed along the same line as in [5] and use the method of dimensional regularization (for more details see [5]). We obtain that near brane string

$$G_{str}(\vec{x},\vec{x}) = \frac{G_4\mu}{8\rho} - \frac{5G_4\mu}{48\rho^3k^2}\left(2 - \ln(4k\rho)\right) \quad \text{for} \quad \rho \gg k^{-1} \,,$$

$$G_{str}(\vec{x},\vec{x}) = \frac{G_4\mu}{8\rho} \quad \text{for} \quad \rho \ll k^{-1} \,, \tag{7}$$

while in the case of a monopole

$$G_{mon}(\vec{x},\vec{x}) = \frac{\pi^2 G_4\eta^2}{8r} - \frac{5\pi^2 G_4\eta^2}{48r^3k^2} \quad \text{for} \quad r \gg k^{-1} \,,$$

$$G_{mon}(\vec{x},\vec{x}) = \frac{\pi^2 G_4\eta^2}{8r} - \frac{5\pi G_4\eta^2}{24r^2k} \quad \text{for} \quad r \ll k^{-1} \,. \tag{8}$$

The electrostatic self-force can easily be calculated as the negative gradient of the electrostatic self-energy (1). We obtain that

$$\vec{F}_{str} = \frac{\pi e^2 G_4 \mu}{4\rho^2}\frac{\vec{\rho}}{\rho} - \frac{5\pi e^2 G_4 \mu\,(7 - 3\ln(4k\rho))}{24\rho^4 k^2}\frac{\vec{\rho}}{\rho} \quad \text{for} \quad \rho \gg k^{-1} \,,$$

$$\vec{F}_{str} = \frac{\pi e^2 G_4 \mu}{4\rho^2}\frac{\vec{\rho}}{\rho} \quad \text{for} \quad \rho \ll k^{-1}$$

at small distances from the string.

Analogous results for the case of a monopole read

$$\vec{F}_{mon} = \frac{\pi^3 e^2 G_4 \eta^2}{4r^2}\frac{\vec{r}}{r} - \frac{5\pi^3 e^2 G_4 \eta^2}{8r^4 k^2}\frac{\vec{r}}{r} \quad \text{for} \quad r \gg k^{-1},$$

$$\vec{F}_{mon} = \frac{\pi^3 e^2 G_4 \eta^2}{4r^2}\frac{\vec{r}}{r} - \frac{5\pi^2 e^2 G_4 \eta^2}{6r^3 k}\frac{\vec{r}}{r} \quad \text{for} \quad r \ll k^{-1}.$$

In both cases the first term corresponds to the result of the nonbrane physics, while the second one depends on the parameter k and gives the corrections from the bulk.

3 Conclusion

In our paper we examined linearized gravitational field around conical defects in the Randall-Sundrum brane world scenario. We showed that possibility of gravity to propagate in extra dimension deforms the cones, and this property slightly modifies the effect of gravitational self-interaction. We showed that for a large ($\gg k^{-1}$) distances contributions from the fifth dimension are small, while for small ($\ll k^{-1}$) values of the radial coordinate these corrections become significant.

References

[1] C.M.DeWitt, B.S.DeWitt, *Physics (N.Y.)* 1, 3 (1964).
[2] B.Linet, *J. Phys.* **A** 9, 1081 (1976).
[3] A.G.Smith, C.M.Will, *Phys. Rev.* **D** 22, 1276 (1980).
[4] M.Barriola, A.Vilenkin, *Phys.Rev.Let.* 63, 341 (1989).
[5] D.V.Gal'tsov, Yu.V.Grats, A.B.Lavrent'ev, *Phys. Atom. Nucl.* 58, 516 (1995).
[6] Yu.V.Grats, A.Garsia, *Class. Quantum Gravity* 13, 189 (1996).
[7] L.Randall, R.Sundrum, *Phys. Rev. Let.* 83, 4690 (1999).
[8] I.V.Aref'eva, M.G.Ivanov, *Nucl. Phys.* **B** 590, 273 (2000).

PHYSICAL DEGREES OF FREEDOM IN A STABILIZED RANDALL-SUNDRUM MODEL

E.E.Boos [a], Yu.S.Mikhailov [b], M.N.Smolyakov [c], I.P.Volobuev [d]
Skobeltsyn Institute of Nuclear Physics, Moscow State University, 119992 Moscow, Russia

Abstract. We study linearized gravity in stabilized RS1 model, find the corresponding equations of motion and decouple them in a suitable gauge. It is shown that for the inverse size of the extra dimension 1 TeV the masses of the graviton excitations may be of the order of 1 TeV, the radion mass of the order of 100 GeV. In this case the inverse coupling of the radion to matter on the negative tension brane is of the order of 50 TeV.

1 Introduction to the model

One of the most interesting brane world models is the Randall-Sundrum model with two branes, - the RS1 model [1]. Almost all the papers on the phenomenology of this model consider the unstabilized model, just putting the radion mass by hand. In the present contribution we are going to show that such an approach is inconsistent, because the backreaction of the scalar field on the metric leads to a renormalization of the parameters of the RS1 model. In particular, the coupling of the radion to matter on the negative tension brane turns out to be smaller, than in the unstabilized model.

Let us denote the coordinates in five-dimensional space-time $E = M_4 \times S^1/Z_2$ by $\{x^M\} \equiv \{x^\mu, y\}$, $M = 0,1,2,3,4$, $\mu = 0,1,2,3$, the coordinate $x^4 \equiv y$, $-L \le y \le L$, parameterizing the fifth dimension. It forms the orbifold, which is realized as the circle of the circumference $2L$ with the points y and $-y$ identified. The branes are located at the fixed points of the orbifold, $y = 0$ and $y = L$.

The action of the stabilized brane world model [2] is a sum of the standard five-dimensional actions for gravity and scalar field with bulk and brane potentials. The standard ansatz for the metric and the scalar field, which preserves the Poincaré invariance in any four-dimensional subspace $y = const$, looks like

$$ds^2 = e^{-2A(y)}\eta_{\mu\nu}dx^\mu dx^\nu + dy^2 \equiv \gamma_{MN}(y)dx^M dx^N, \tag{1}$$

$$\phi(x,y) = \phi(y), \tag{2}$$

$\eta_{\mu\nu}$ denoting the flat Minkowski metric. If one substitutes this ansatz into the equations corresponding to this action, one gets a system of nonlinear differential equations for functions $A(y), \phi(y)$. To find an analytic solution to this system we will use the results of papers [2,3].

[a] boos@theory.sinp.msu.ru
[b] mikhailov@theory.sinp.msu.ru
[c] smolyakov@theory.sinp.msu.ru
[d] volobuev@theory.sinp.msu.ru

For a special choice of the potentials the solution of the equations of motion is given by [2]

$$\phi(y) = \phi_1\, e^{-u|y|}, \quad A(y) = k|y| + \frac{\phi_1^2}{48M^3} e^{-2u|y|}, \tag{3}$$

M being the fundamental five-dimensional energy scale. The interbrane distance L is stabilized, i.e. it is defined by the boundary conditions for the field ϕ and is expressed in terms of the parameters of the model.

2 Linearized gravity

Linearized theory is obtained by representing the metric and the scalar field as

$$g_{MN}(x,y) = \gamma_{MN}(y) + \frac{1}{\sqrt{2M^3}} h_{MN}(x,y), \tag{4}$$

$$\phi(x,y) = \phi(y) + \frac{1}{\sqrt{2M^3}} f(x,y), \tag{5}$$

substituting this representation into the action and keeping the terms of the second order in h_{MN} and f. The Lagrangian of this action is called the second variation Lagrangian.

Varying the action built with this Lagrangian and taking into account the background field equations, we arrive at the equations of motion for the fluctuations of metric and scalar field [4]. These equations are invariant under the gauge transformations, which follow in this approximation from the general coordinate transformations. One can show that the substitution

$$h_{\mu\nu} = b_{\mu\nu} - \frac{1}{2}\gamma_{\mu\nu} h_{44}. \tag{6}$$

and the gauge conditions

$$(e^{-2A} h_{44})' - \frac{1}{3M^3} e^{-2A} \phi' f = 0, \tag{7}$$

$$\partial^\nu b_{\mu\nu} = 0, \quad b = \gamma^{\mu\nu} b_{\mu\nu} = 0, \tag{8}$$

decouple the equations of motion (see [5]).

Equation for the $\mu\nu$-component reduces to an equation for a transverse-traceless tensor field $b_{\mu\nu}(x,y)$:

$$\frac{1}{2}\left(e^{2A(y)} \Box b_{\mu\nu} + \frac{\partial^2 b_{\mu\nu}}{\partial y^2}\right) - b_{\mu\nu}\left(2(A')^2 - A''\right) = 0. \tag{9}$$

This equation does not include the background scalar filed $\phi(y)$ and is absolutely analogous to the corresponding equation in the unstabilized RS1 model [6].

Equations of motion for the fields h_{44} and f can be expressed through a single equation in terms of a new function $g = e^{-2A(y)}h_{44}(x, y)$. In order to find the mass spectrum of the scalar particles let us use the following substitution

$$g(x, y) = e^{ipx} g_n(y), \quad p^2 = -\mu_n^2. \tag{10}$$

Equation for $g_n(y)$ can be written as a single equation on the circle S^1 [5]:

$$\frac{d}{dy}\left(\frac{e^{2A}}{(\phi')^2} g_n'\right) - \frac{\hat{\kappa}^2 e^{2A}}{3} g_n = -\mu_n^2 \frac{e^{4A}}{(\phi')^2}\left(1 + \frac{2}{\beta_1^2 + u}\delta(y) + \frac{2}{\beta_2^2 - u}\delta(y - L)\right) g_n. \tag{11}$$

This equation is written in the Sturm-Liouville form with self-adjoint operator in the eigenvalue problem for this equation. The only characteristic property of this equation is that the weight function is a generalized one. It follows from the general theory [7] that if $\beta_2^2 > u$, then all the eigenvalues of the problem under consideration are real and positive. The eigenfunctions of the problem $\{g_n(y)\}$ make up a complete set and can be chosen to be real. The eigenfunctions corresponding to different eigenvalues are orthogonal with the generalized weight defined by equation (11), and can be appropriately normalized.

Thus five-dimensional scalar field $g(x, y)$ can be expanded in this system as

$$g(x, y) = \sum_{n=1}^{\infty} \varphi_n(x) g_n(y), \tag{12}$$

where the four dimensional real scalar fields $\varphi_n(x)$ have masses μ_n^2.

We will use an approximation $uL \ll 1$, which is rather general and physically interesting [4]. Keeping in A only the terms linear in y, we get

$$A(y) = \tilde{k}\left(|y| - L\right), \quad \tilde{k} = k - \frac{\phi_1^2}{24M^3} u \tag{13}$$

in Galilean coordinates on brane 2. Thus, in this approximation equation (9) for the tensor field coincides with the equation of the unstabilized model, where a substitution $k \to \tilde{k}$ was made. This equation can be solved exactly, and the formulas for eigenfunctions and eigenvalues were discussed in detail in [6].

The normalized functions $g_n(y)$ in approximation (13) look like (see [5] for details)

$$g_n(y) = A_n \left(\frac{\mu_n}{\tilde{k}} e^{\tilde{k}y - \tilde{k}L}\right)^{-\left(1 + \frac{u}{\tilde{k}}\right)} J_\alpha\left(\frac{\mu_n}{\tilde{k}} e^{\tilde{k}y - \tilde{k}L}\right). \tag{14}$$

The mass of the lowest scalar excitation for small $u/\tilde{k}$ has the form

$$\mu_1^2 = \frac{\phi_1^2 u^2}{3M^3} \frac{\beta_2^2 - u}{\beta_2^2 + 4\tilde{k}}, \tag{15}$$

which for $\beta_2^2 \to \infty$ coincides with the results of papers [4,8] in our approximation.

In order to get the correct value of the gravitational constant on the brane at $y = L$ on can take the same values of the parameters, which are used in the unstabilized model, i.e. $M \sim \tilde{k} \sim 1TeV$, $\tilde{k}L \sim 35$ [4]. Then the parameter u may be of the order of ten GeV, and the radion mass (15) may be of the order of hundred GeV. In this case the masses of tensor excitations are of the order of $1TeV$ [6]

Formula (14) gives that the coupling constant of the radion is [5]

$$\epsilon_1 \simeq -\frac{\mu_1}{\sqrt{24M^3\tilde{k}}}\sqrt{\frac{\beta_2^2 + 4\tilde{k}^2}{\beta_2^2 - u}},$$

and turns out to be of the order $\epsilon_1^{-1} \sim 50TeV$ for the above given values of the model parameters, which is approximately ten times smaller, than for the same choice of the parameters in the unstabilized model.

Acknowledgments

The work is partly supported by RFBR 04-02-16476, RFBR 04-02-17448, Universities of Russia UR.02.02.503, and Russian Ministry of Education and Science NS.1685.2003.2 grants. The authors are grateful to Yu. Grats, P. Ermolov, Yu. Kubyshin and V. Rubakov for useful discussions.

References

[1] L.Randall, R.Sundrum, *Phys.Rev.Lett.* 83, 3370 (1999).

[2] O.DeWolfe, D.Z.Freedman, S.S.Gubser, A.Karch, *Phys.Rev.* **D** 62, 046008 (2000).

[3] A.Brandhuber, K.Sfetsos, *JHEP* 9910, 013 (1999).

[4] E.E.Boos, Yu.S.Mikhailov, M.N.Smolyakov, I.P.Volobuev, *Nucl.Phys.* **B** 717, 19 (2005).

[5] E.E.Boos, Yu.S.Mikhailov, M.N.Smolyakov, I.P.Volobuev, *arXiv:hep-th/0511185.*

[6] E.E.Boos, I.P.Volobuev, Yu.A.Kubyshin, M.N.Smolyakov, *Theor.Math.Phys.* 131(2), 629 (2002).

[7] V.M.Babich, M.B.Kapilevich, S.G.Mikhlin, *"Linear Equations of Mathematical Physics", Ch. II, §3* (in Russian) (Nauka, Moscow) 1964.

[8] C.Csaki, M.L.Graesser, G.D.Kribs, *Phys.Rev.* **D** 63, 065002 (2001).

GRAVITATIONAL ENERGY-MOMENTUM TENSORS ACCORDING TO BELINFANTE AND ROSENFELD

A.I. Nikishov [a]
I.E. Tamm Department of Theoretical Physics, Lebedev Physical Institute, Moscow, Russia

1 Introduction

We consider gravitational energy-momentum tensors (EMT) in the lowest approximation. EMT obtained according to Belinfante and according to Rosenfeld differ in regions where particles are present. This is due to the fact that the Lagrangian for particles (and their interaction with the gravitational field) is not taken into account for the time being. How to do this is some problem in itself. It turns out that the EMT, obtained according to Belinfante, is not symmetric in terms describing the interaction of particles with gravitational field. Such terms appear because we assume and use the linearized Einstein equation in the right hand side of which stands the energy-momentum tensor of particles. The introduction of an asymmetric interaction tensor makes the total EMT symmetric. Then the conservation laws of the total EMT give the equations of motion for particles in agreement with general relativity.

The gravitational EMT obtained according to Rosenfeld is symmetric by construction. We consider the consequences of assumption that there is no interaction EMT in this case. Then the conservation laws of the total EMT give the equations of motion for particles in disagreement with general relativity. Yet the difference is so subtle that it hardly can be excluded by the present observational data.

Besides the gravitational Lagrangian of general relativity we consider another Lagrangian which differs from the first one by divergence terms. This second Lagrangian leads to gravitational EMT which gives the positive gravitational energy density of spherically symmetric body *in the privileged coordinate system*, see below. It seems that the use of this EMT in 3-graviton vertex gives the correct value for the perihelion precession of Mercury.

2 Lagrangians

We use the notation $\eta_{\mu\nu} = \mathrm{diag}(-1,1,1,1)$,

$$g_{\mu\nu} = \eta_{\mu\nu} + h_{\mu\nu}, \quad \bar{h}_{\mu\nu} = h_{\mu\nu} - \frac{1}{2}\eta_{\mu\nu}h, \quad h = h_\alpha{}^\alpha, \quad h_{,i} = \frac{\partial h}{\partial x^i}, \tag{1}$$

Both Greek and Latin indices run from 0 to 3, if not stated otherwise. The gravitational Lagrangian in the lowest approximation may be taken either in the form [1]

$$\overset{Thir}{L} = \frac{1}{32\pi G}[-\frac{1}{2}\bar{h}_{\alpha\beta,\gamma}\bar{h}^{\alpha\beta,\gamma} + \frac{1}{4}\bar{h}_{,\alpha}\bar{h}^{,\alpha} + \bar{h}_{\alpha\beta,\lambda}\bar{h}^{\lambda\beta,\alpha}], \tag{2}$$

or [2]

$$\overset{MTW}{L} = \frac{1}{32\pi G}[-\frac{1}{2}\bar{h}_{\alpha\beta,\gamma}\bar{h}^{\alpha\beta,\gamma} + \frac{1}{4}\bar{h}_{,\alpha}\bar{h}^{,\alpha} + \bar{h}_{\alpha\beta}{}^{,\alpha}\bar{h}^{\beta\lambda}{}_{,\lambda}]. \tag{3}$$

The Lagrangian (3) is the equation (7.8b) in Misner, Thorne and Wheeler book [2]; earlier it was obtained by Feynman, see (3.6.1) (3.6.5) in [3]. The expressions (2)

[a] e-mail: nikishov@lpi.ru

and (3) differ by divergence terms:

$$\overset{Thir}{L} = \overset{MTW}{L} + \frac{1}{32\pi G}[(\bar{h}_{\alpha\beta}\bar{h}^{\lambda\beta,\alpha})_{,\lambda} - (\bar{h}_{\alpha\beta}\bar{h}^{\lambda\beta}{}_{,\lambda})^{,\alpha}]. \tag{4}$$

Any combination of these Lagrangians leads to the linearized Einstein equation

$$h_{\mu\nu,\lambda}{}^{\lambda} - h^{\lambda}{}_{\mu,\lambda\nu} - h^{\lambda}{}_{\nu,\lambda\mu} + h_{,\mu\nu} + \eta_{\mu\nu}(h_{\sigma\lambda}{}^{,\sigma\lambda} - h_{,\lambda}{}^{\lambda}) = -16\pi G T_{\mu\nu}. \tag{5}$$

As usual, we assume that the r.h.s. of (5) is obtained from matter Lagrangian, see [4]. In what follows we use the particle energy-momentum tensor

$$\overset{M}{T}{}^{jk} = \sum_a m_a \frac{dx^j}{ds}\frac{dx^k}{dt}\delta(\vec{x} - \vec{x}_a(t)), \tag{6}$$

Using this in (5), we may assume that particles in (6) are free (not interacting with gravitational field), but in general

$$ds^2 = -g_{\mu\nu}dx^\mu dx^\nu \approx ds_0^2 - h_{\mu\nu}dx^\mu dx^\nu. \tag{7}$$

and hence (6) contains interaction with $h_{\mu\nu}$. This fact is important in in several ways. First, for one particle interacting with gravitational field the action integral can written as [1]

$$\frac{1}{2}\int \overset{M}{T}{}^{jk} g_{jk} d^4x = -\frac{1}{2}m\int ds. \tag{8}$$

(The first eq. in (7) is used in obtaining the r.h.s. of (8)). Its variation gives the same equations of motion for particles as in general relativity, see §87 in [4]. Second, the interaction, contained in (6) is essential for obtaining the correct Newtonian limit for the (effective) gravitational energy density ($= -\frac{(\nabla\phi)^2}{8\pi G}$), see [1] and [5].

3 Thirring energy-momentum tensor

From the Thirring Lagrangian (2) by Belinfante method we obtain the gravitational energy-momentum tensor, see [1] and [5]. It turns out to be asymmetric in terms describing interaction of particles with gravitational field [5]. Only total energy-momentum tensor must be symmetric. It consists of gravitational, matter and interaction parts. Assuming with Thirring that the latter part has the form

$$\overset{int}{T}{}^{\mu\nu} = \overset{M}{T}{}^{\mu\alpha} h_\alpha{}^\nu, \tag{9}$$

and adding it to the obtained gravitational energy-momentum tensor, we get the symmetric tensor:

$$\overset{Thir}{t}{}^{jk}_{Bel} = \frac{1}{32\pi G}[\overset{1}{T}{}^{jk} - \frac{1}{2}\overset{2}{T}{}^{jk} + \frac{1}{2}\overset{4}{T}{}^{jk} + \overset{5}{T}{}^{jk} + 2\overset{6}{T}{}^{jk} + 2\overset{7}{T}{}^{jk} - \overset{9}{T}{}^{jk} + 2\overset{10}{T}{}^{jk}$$

$$- \overset{11}{T}{}^{jk} - 2\overset{12}{T}{}^{jk} - 4\overset{13}{T}{}^{jk} + \overset{c}{T}{}^{jk} - 2\overset{i}{T}{}^{jk} + \overset{l}{T}{}^{jk} - 2\overset{m}{T}{}^{jk} - 2\overset{n}{T}{}^{jk} + 4\overset{o}{T}{}^{jk}]. \tag{10}$$

Here we use the notation for "building blocks", of which gravitational energy-momentum tensors can be assembled: for terms with first derivatives

$$\overset{1}{T}{}^{jk} = \eta^{jk}h_{\alpha\beta,\gamma}h^{\gamma\beta,\alpha}; \quad \overset{2}{T}{}^{jk} = \eta^{jk}h_{\alpha\beta,\gamma}h^{\alpha\beta,\gamma}; \quad \overset{3}{T}{}^{jk} = \eta^{jk}h_{,\rho}h^{\rho\sigma}{}_{,\sigma};$$

$$\overset{4}{\tau}{}^{jk} = \eta^{jk}h_{,\rho}h^{,\rho}; \quad \overset{5}{\tau}{}^{jk} = h_{\alpha\beta}{}^{,j}h^{\alpha\beta,k}; \quad \overset{6}{\tau}{}^{jk} = h^{j\alpha,\beta}h^{k}{}_{\beta,\alpha}; \quad \overset{7}{\tau}{}^{jk} = h^{j\alpha,\beta}h^{k}{}_{\alpha,\beta};$$

$$\overset{8}{\tau}{}^{jk} = \frac{1}{2}(h^{j\alpha,k} + h^{k\alpha,j})h_{,\alpha}; \quad \overset{9}{\tau}{}^{jk} = h^{jk,\sigma}h_{,\sigma}; \quad \overset{10}{\tau}{}^{jk} = \frac{1}{2}(h^{j\sigma}{}_{,\sigma}h^{,k} + h^{k\sigma}{}_{,\sigma}h^{,j});$$

$$\overset{11}{\tau}{}^{jk} = h^{,j}h^{,k}; \quad \overset{12}{\tau}{}^{jk} = h^{jk,\alpha}h_{\alpha\sigma}{}^{,\sigma}; \quad \overset{13}{\tau}{}^{jk} = \frac{1}{2}(h^{j\alpha,\beta}h_{\alpha\beta}{}^{,k} + h^{k\alpha,\beta}h_{\alpha\beta}{}^{,j});$$

$$\overset{14}{\tau}{}^{jk} = h^{j\sigma}{}_{,\sigma}h^{k\alpha}{}_{,\alpha}; \quad \overset{15}{\tau}{}^{jk} = \frac{1}{2}(h^{j\alpha,k} + h^{k\alpha,j})h_{\alpha\sigma}{}^{,\sigma}; \overset{16}{\tau}{}^{jk} = \eta^{jk}h_{\alpha\beta}{}^{,\beta}h^{\alpha\sigma}{}_{,\sigma}; \quad (11)$$

and for terms with second derivatives

$$\overset{a}{\tau}{}^{jk} = \eta^{jk}h_{,\sigma}{}^{\sigma}h; \quad \overset{b}{\tau}{}^{jk} = \eta^{jk}h_{\alpha\beta}{}^{,\alpha\beta}h; \quad \overset{c}{\tau}{}^{jk} = \eta^{jk}h^{,\alpha\beta}h_{\alpha\beta}; \quad \overset{d}{\tau}{}^{jk} = \eta^{jk}h^{\alpha\beta,\sigma}{}_{\sigma}h_{\alpha\beta};$$

$$\overset{e}{\tau}{}^{jk} = \eta^{jk}h^{\alpha\sigma}{}_{,\sigma}{}^{\beta}h_{\alpha\beta}; \quad \overset{f}{\tau}{}^{jk} = h^{,jk}h; \quad \overset{g}{\tau}{}^{jk} = h^{jk,\sigma}{}_{\sigma}h; \quad \overset{h}{\tau}{}^{jk} = \frac{1}{2}(h^{j\sigma,k}{}_{\sigma} + h^{k\sigma,j}{}_{\sigma})h;$$

$$\overset{i}{\tau}{}^{jk} = h^{jk,\alpha\beta}h_{\alpha\beta}; \quad \overset{j}{\tau}{}^{jk} = \frac{1}{2}(h^{j\alpha,k\beta} + h^{k\alpha,j\beta})h_{\alpha\beta}; \quad \overset{k}{\tau}{}^{jk} = h^{\alpha\beta,jk}h_{\alpha\beta};$$

$$\overset{l}{\tau}{}^{jk} = h^{jk}h_{,\sigma}{}^{\sigma}; \quad \overset{m}{\tau}{}^{jk} = h^{jk}h_{\alpha\beta}{}^{,\alpha\beta}; \quad \overset{n}{\tau}{}^{jk} = \frac{1}{2}(h^{,j\alpha}h_{\alpha}{}^{k} + h^{,k\alpha}h_{\alpha}{}^{j});$$

$$\overset{o}{\tau}{}^{jk} = \frac{1}{2}(h^{j\sigma,\alpha}{}_{\sigma}h_{\alpha}{}^{k} + h^{k\sigma,\alpha}{}_{\sigma}h_{\alpha}{}^{j}); \quad \overset{p}{\tau}{}^{jk} = \frac{1}{2}(h^{j\alpha,\sigma}{}_{\sigma}h_{\alpha}{}^{k} + h^{k\alpha,\sigma}{}_{\sigma}h_{\alpha}{}^{j});$$

$$\overset{q}{\tau}{}^{jk} = \frac{1}{2}(h^{\alpha\sigma,j}{}_{\sigma}h_{\alpha}{}^{k} + h^{\alpha\sigma,k}{}_{\sigma}h_{\alpha}{}^{j}). \quad (12)$$

The equations of motion for particles are contained in the conservation laws

$$\left(\overset{M}{T}{}^{jk} + \overset{Thir}{t}{}^{jk}_{Bel}\right)_{,k} = 0. \quad (13)$$

In this case the equations of motion for particles coincide with those of general relativity, see [1], [5].

For Newtonian centers $\phi = -G\sum_a \frac{m_a}{|\vec{r}-\vec{r}_a|}$ and we obtain from (5) in Hilbert gauge, $\bar{h}^{\mu\nu}{}_{,\nu} \equiv (h^{\mu\nu} - \frac{1}{2}\eta^{\mu\nu}h)_{,\nu} = 0$, (in linear approximation)

$$\bar{h}^{(1)}_{\mu\nu} = -4\phi\delta_{\mu 0}\delta_{\nu 0}, \quad h^{(1)}_{\mu\nu} = -2\phi\delta_{\mu\nu}, \quad h^{(1)} \equiv h^{(1)\sigma}_{\sigma} = -4\phi = -\bar{h}^{(1)}. \quad (14)$$

This coordinate system coincides with harmonic or isotropic one in general relativity in linear approximation and there are reasons to believe that this is the privileged system in the sense that nonphysical degrees of freedom do not contribute to it. Using (14), we get from (10) the following expression for the gravitational energy density

$$\overset{Thir}{t}{}^{00}_{Bel} = -\frac{7}{8\pi G}(\nabla\phi)^2 - 2\overset{M}{T}{}^{00}\phi. \quad (15)$$

Even the sign of the last term on the r.h.s. in (15) is opposite to the sign of the Newtonian interaction term (which is $\frac{1}{2}\overset{M}{T}{}^{00}\phi$). Nevertheless the Newtonian limit for "effective" gravitational energy density is correct, see [1] and [5]. Thirring have shown how to get agreement with general relativity in G^2-approximation in his field theoretical method: instead of $h_{\mu\nu} \equiv g_{\mu\nu} - \eta_{\mu\nu}$ one must use the variable $\tilde{h}^{\mu\nu} \equiv \eta^{\mu\nu} - \sqrt{-g}g^{\mu\nu}$. Then one can take into account the nonlinear effects by adding to $\overset{M}{t}{}^{jk}$ the $\overset{Thir}{t}{}^{jk}_{Bel}$ in gravitational equation for $\tilde{h}^{\mu\nu}$, for details see [6]. Without switching to $\tilde{h}_{\mu\nu}$ it is impossible to obtain agreement with general relativity neither by Thirring method nor by using $\overset{Thir}{t}{}^{jk}_{Bel}$ in 3-graviton vertex.

Instead of Belinfante method one can use Hilbert-Rosenfeld method, see §94 in [4]. The gravitational EMT, obtained in this way, is symmetric and we find

$$\overset{Thir}{t}{}^{jk}_{Ros} - \overset{Thir}{t}{}^{jk}_{Bel} = -\frac{1}{2}h \overset{M}{T}{}^{jk} - \frac{1}{2}\bar{h}^{jk} \overset{M}{T} =$$

$$= \frac{1}{32\pi G}[\overset{f}{T}{}^{jk} + \overset{g}{T}{}^{jk} - 2\overset{h}{T}{}^{jk} - 2\overset{l}{T}{}^{jk} + 2\overset{m}{T}{}^{jk}], \quad \overset{M}{T} = \eta_{jk}\overset{M}{T}{}^{jk}. \tag{16}$$

For the Newtonian centers this difference is zero. We may assume that the equation of motion for particles are given by (13) with substitution $\overset{Thir}{t}{}^{jk}_{Bel} \to \overset{Thir}{t}{}^{jk}_{Ros}$. The drawback of this assumption is that the resulting equation of motion for particles are in disagreement with general relativity and besides it is not seen haw to obtain these equations from variational principle. We shall discuss these equations in the next section, where they reappear in consideration of EMT obtained from the Lagrangian (3).

4 Misner-Thorne-Wheeler energy-momentum tensor

Starting from $\overset{MTW}{L}$ in (3) and again using (9) as in obtaining (10), we find

$$\overset{MTW}{t}{}^{jk}_{Bel} = \frac{1}{32\pi G}[(-\frac{1}{2}\overset{2}{T}{}^{jk} + \frac{1}{2}\overset{4}{T}{}^{jk} + \overset{5}{T}{}^{jk} - \overset{9}{T}{}^{jk} + 2\overset{10}{T}{}^{jk} - \overset{11}{T}{}^{jk} + 2\overset{12}{T}{}^{jk} - 2\overset{13}{T}{}^{jk}$$

$$-2\overset{14}{T}{}^{jk} + 2\overset{15}{T}{}^{jk} - \overset{16}{T}{}^{jk}) + \overset{c}{T}{}^{jk} - 2\overset{e}{T}{}^{jk} + 2\overset{j}{T}{}^{jk} + \overset{l}{T}{}^{jk} - 2\overset{n}{T}{}^{jk} - 2\overset{p}{T}{}^{jk} + 2\overset{q}{T}{}^{jk}]. \tag{17}$$

Instead of (15) in the coordinate system (14) we find in the Newtonian limit

$$\overset{MTW}{T}{}^{00}_{Bel} = \frac{1}{8\pi G}(\nabla\phi)^2 + 2\overset{M}{T}{}^{00}\phi. \tag{18}$$

The total interaction energy density is $\overset{M}{T}{}^{00}\phi$ because the component $\overset{M}{T}{}^{00}$ in (6) contains term $-\overset{M}{T}{}^{00}\phi$ due to (7) [5]. The reason for this is that in the considered approximation

$$\overset{M}{T}{}^{jk} = \overset{M}{T}{}^{jk}_{0} + \overset{int}{T}{}^{jk}_{M}, \quad \overset{int}{T}{}^{jk}_{M} = \frac{1}{2}\sum_a m_a \frac{dx^j}{ds_0}\frac{dx^k}{dt}\delta(\vec{x} - \vec{x}_a)h_{ln}\frac{dx^l}{ds_0}\frac{dx^n}{ds_0}. \tag{19}$$

Now in (18) the gravitational energy density is positive, as desired, and the interaction energy density $(2T^{00}\phi)$ is negative. For slow particles $\overset{int}{T}{}^{jk}_{M}$ in (19) with the help of (5) and (12) can be put in the form similar to (17). Assuming that this tensor together with (17) may be used in 3-graviton vertex, we get the correct $g_{00} = -(1 + 2\phi + 2\phi^2)$. This is in agreement with the perihelion precession [5].

Going back to Rosenfeld method, we find again (16) with substitution $Thir \to MTW$. If we assume that the equation of motion for particles are contained in

$$\left(\overset{M}{T}{}^{jk} + \overset{MTW}{t}{}^{jk}_{Ros}\right)_{,k} = 0, \tag{20}$$

then these equation will be different from those of general relativity. The substitution $MTW \rightarrow Thir$ in (20) does not change them, but the substitution $Ros \rightarrow Bel$ does change.

For t^{jk}_{Bel} and for t^{jk}_{Ros}, respectively, we get

$$t^{jk}_{Bel,k} = -\frac{1}{2}h^{\alpha\beta,j}\overset{M}{T}_{\alpha\beta} + h^{j\alpha,\beta}\overset{M}{T}_{\alpha\beta}, \tag{21}$$

$$t^{jk}_{Ros,k} - t^{jk}_{Bel,k} = -\frac{1}{2}h_{,\alpha}\overset{M}{T}{}^{\alpha j} + \frac{1}{4}h^{,j}\overset{M}{T} - \frac{1}{2}h^{j\sigma}{}_{,\sigma}\overset{M}{T} + \frac{1}{4}h\overset{M}{T}{}^{,j} - \frac{1}{2}h^{\alpha j}\overset{M}{T}_{,\alpha}. \tag{22}$$

In linear approximation we can put $G = 0$ in $\overset{M}{T}{}^{jk}$ in the r.h.s. of (21) and (22), i.e. this tensor is defined by free particles. The equations of motion for particles are obtained by integrating (20) over d^3x. For the movement of one particle in the field of spherically symmetric body only the term $-\frac{1}{2}h_{,\alpha}\overset{M}{T}{}^{\alpha j}$ remains in the r.h.s. of (22) and it does not effect the deflection of ultrarelativistic particle by the Sun gravitational field, cf. Ch 2, §4 in [8] and and Ch 9 §8 in [9]. The time delay of ultrarelativistic particle is insensible to acceleration because the speed remains almost the same. Using $\overset{MTW}{t}{}^{jk}_{Ros}$ in in the same way as $\overset{MTW}{t}{}^{jk}_{Bel}$ in 3-graviton vertex, we find the same $g_{oo} = -(1 + 2\phi + 2\phi^2)$.

Acknowledgments

I am grateful to V.I.Ritus, for valuable discussions. The work was carried out with financial support of Scientific Schools and Russian Fund for Fundamental Research (Grants 1578.2003.2 and 05-02-17217).

References

[1] W.E.Thirring, *Ann.Phys. (N.Y.)* 16, 96 (1961).
[2] C.W.Misner, K.S.Thorne, J.A.Wheeler, *"Gravitation"*, (San Francisco) 1973.
[3] R.P.Feynman, F.B.Moringo, W.G.Wagner, *"Feynman Lectures on Gravitation"*, (Addison-Wesley Company) 1995.
[4] L.Landau, E.Lifshitz, in *"The Classical Theory of Fields"*, (Addison-Wesley, Cambridge, MA) 1971.
[5] A.I.Nikishov,*Particles and Nuclei*, **32**, 5 (2001).
[6] A.I.Nikishov, gr-qc/0408020.
[7] A.I.Nikishov, gr-qc/0410099.
[8] J Schwinger *Particles, Sources, and Fields.* V.1 (Addison-Wesley) 1970.
[9] M.G. Bowler,*Gravitation and relativity*, (Pergamon Press, Oxford) 1976.

STABLE MATTER OF 4TH GENERATION: HIDDEN IN THE UNIVERSE AND CLOSE TO DETECTION?

K.Belotsky [a], M.Khlopov [b], K.Shibaev [c]

Moscow Engineering Physics Institute, 115409 Moscow, Russia

Abstract. Stable neutrino and U quark of 4th generation are excluded neither by experimental data, nor by astrophysical constraints. Moreover, excess of stable $\bar{U}$ quarks in the Universe can lead to an exciting composite nuclear-interacting form of dark matter, which can even dominate in large scale structure formation.

1 Stable pieces of 4th generation matter

The problem of existence of new families of quarks and leptons is among the most important in the modern high energy physics. If these quarks and/or leptons are stable, they should be present around us and the reason for their evanescent nature should be found. Here we concentrate on the recent development of the approach [1](see also [3], [2,4]), which involves minimal number of parameters, but illustrates all the problems, related with stable matter, composed of heavy charged particles.

Precision data on Standard model parameters admit [3] the existence of 4th generation and the results of this analysis determine our choice for masses of N ($m_N = 50$ GeV) and U ($m_U = 350 S_5$ GeV).

4th generation can follow from heterotic string phenomenology and its difference from the three known light generations can be explained by a new gauge charge (y-charge), possessed only by its quarks and leptons [1,2,4]. Similar to electromagnetism this charge is the source of a long range Coulomb-like y-interaction. Strict conservation of y-charge makes the lightest particle of 4th family (4th neutrino N) absolutely stable, while the lifetime of U can exceed the age of the Universe, if $m_U < m_D$ [2,4].

The following conditions fix y-charges (Q_y) of (N, E, U, D). Cancellation of $Z - \gamma - y$ anomaly implies $Q_{yE} + 2 \cdot Q_{yU} + Q_{yD} = 0$; while cancellation of $Z - y - y$ anomaly needs $Q_{yN}^2 - Q_{yE}^2 + 3 \cdot (Q_{yU}^2 - Q_{yD}^2) = 0$. Proper $N - E$ and $U - D$ transitions of weak interaction assume $Q_{yN} = Q_{yE}$ and $Q_{yU} = Q_{yD}$. From these conditions follows $Q_{yE} = Q_{yN} = -3 \cdot Q_{yU} = -3 \cdot Q_{yD}$ so that y-charges of (N, E, U, D) are $(1, 1, -1/3, -1/3)$.

U-quark can form lightest (Uud) baryon and $(U\bar{u})$ meson. The corresponding antiparticles are formed by $\bar{U}$ with light quarks and antiquarks. Owing to large chromo-Coulomb binding energy ($\propto \alpha_c^2 \cdot m_U$, where α_c is the QCD constant) stable double and triple U bound states (UUq), (UUU) and their antiparticles $(\bar{U}\bar{U}\bar{u})$, $(\bar{U}\bar{U}\bar{U})$ can exist [2,4]. Formation of these double and triple states in

[a] e-mail: k-belotsky@yandex.ru

[b] e-mail: Maxim.Khlopov@roma1.infn.it

[c] e-mail: shibaev01@yandex.ru

particle interactions at accelerators and in cosmic rays is strongly suppressed, but they can form in early Universe and strongly influence cosmological evolution of 4th generation hadrons. As shown in [4], anti- U-triple state called anutium or $\Delta^{--}_{3\bar{U}}$ is of special interest.

2 Cosmological evolution of 4th generation matter

In charge symmetric case, which assumes no primordial excessive particles (or antiparticles) of 4th generation, cosmological evolution was studied separately for N and U in [1,2,4].

In the absence of particle (or antiparticle) excess primordial particles of 4th generation with the considered masses can not contribute significantly to total density. However, N and $\bar{N}$ condensation in galaxies can result in significant effect of their annihilation in cosmic fluxes of positrons, antiprotons and in gamma background [1]. They can even provide simultaneous explanation of these data and results of the direct WIMP searches [1], but in this case annihilation in terrestrial matter of N and $\bar{N}$, captured by Earth, should lead to an up-going muon fluxes, exceeding the observed in underground detectors.

The main problem of new stable quark is that it can form stable positively charged hadrons. In charge symmetric case the pregalactic abundance of +2 charge (UUU) and (UUu) is by 10 orders above the terrestrial limits for anomalous helium ($r < 10^{-19}$). For the abundance of +1 charge (Uud) the situation is even more dramatic, since it exceeds the limits on anomalous hydrogen ($r < 10^{-30}$) by 20 orders of magnitude [2].

To satisfy the experimental upper limits, anomalous isotope abundance in Earth should be reduced. The mechanisms for such reduction [2] imply recombination of U- and $\bar{U}$- hadrons in dense objects of ordinary matter and their rapid annihilation in bound $(U\bar{U})$ states. These mechanisms involve y-attraction, which prevents fractionating of U- and $\bar{U}$- hadrons in matter. In the result the primordial abundance of anomalous isotopes in terrestrial matter can be reduced down to $r < 10^{-28}$, so that the problem of anomalous hydrogen can not be resolved in this way. Moreover, these mechanisms, being sufficiently effective in the case of anomalous helium, are accompanied by effects of energy release, which are strongly constrained by the observed γ background and by the data from large volume detectors. All the above listed problems of charge symmetric N and U matter can be avoided in the case of primordial excess of $\bar{U}$ and $\bar{N}$.

The model [2] admits that in the early Universe an antibaryon asymmetry for 4th generation quarks can be generated [4]. Due to y-charge conservation $\bar{U}$ excess should be compensated by $\bar{N}$ excess. $\bar{U}$-antibaryon density can be expressed through the modern dark matter density $\Omega_{\bar{U}} = k \cdot \Omega_{CDM} = 0.224$ ($k \leq 1$), saturating it at $k = 1$.

In the early Universe at temperatures highly above their masses $\bar{U}$ and $\bar{N}$ were in thermodynamical equilibrium with relativistic plasma. It means that at $T > m_U$ ($T > m_N$) the excessive $\bar{U}$ ($\bar{N}$) were accompanied by $U\bar{U}$ ($N\bar{N}$) pairs.

Due to $\bar{U}$ excess frozen out concentration of deficit U-quarks is suppressed at $T < m_U$ for $k > 0.04$. It decreases further exponentially, when the frozen out U quarks begin to bind with antiquarks $\bar{U}$ into charmonium-like state $(\bar{U}U)$ and annihilate. On this line $\bar{U}$ excess binds by chromo-Coulomb forces dominantly into $(\bar{U}\bar{U}\bar{U})$ anutium states with electric charge $Z_\Delta = -2$ and mass $m_o = 1.05 S_5$TeV, while remaining free $\bar{U}$ anti-quarks and anti-diquarks $(\bar{U}\bar{U})$ form after QCD phase transition normal size hadrons $(\bar{U}u)$ and $(\bar{U}\bar{U}\bar{u})$. Then at $T = T_{QCD} \approx 150$MeV additional suppression of remaining U-quark hadrons takes place in their hadronic collisions with $\bar{U}$-hadrons, in which $(\bar{U}U)$ states are formed and U-quarks successively annihilate.

Owing to weaker interaction, effect of $\bar{N}$ excess in the suppression of deficit N is less pronounced but it still takes place at $T < m_N$ for $k > 0.02$. At $T \sim I_{NN} = \alpha_y^2 M_N/4 \sim 15$MeV (for $\alpha_y = 1/30$ and $M_N = 50$GeV) due to y-interaction the frozen out N begin to bind with $\bar{N}$ into charmonium-like states $(\bar{N}N)$ and annihilate. At $T < I_{NU} = \alpha_y^2 M_N/2 \sim 30$MeV y-interaction causes binding of N with $\bar{U}$-hadrons (dominantly with anutium) but only at $T \sim I_{NU}/30 \sim 1$MeV this binding is not prevented by back reaction of y-photo-destruction.

To the period of Standard Big Bang Nucleosynthesis (SBBN) $\bar{U}$ are dominantly bound in anutium $\Delta^{--}_{3\bar{U}}$ with small fraction ($\sim 10^{-6}$) of neutral $(\bar{U}u)$ and doubly charged $(\bar{U}\bar{U}\bar{u})$ hadron states. The dominant fraction of anutium is bound by y-interaction with $\bar{N}$ in $(\bar{N}\Delta^{--}_{3\bar{U}})$ "atomic" state. Owing to early decoupling of y-photons from relativistic plasma presence of y-radiation background does not influence SBBN processes [2, 4].

After ^4He is formed in SBBN, at $T \leq T_{rHe} \sim I_o/\log{(n_\gamma/n_{He})} \approx I_o/27 \approx 60$keV, where $I_o = Z_\Delta^2 Z_{He}^2 \alpha^2 m_{He}/2 \approx 1.6$MeV [4], neutral Anti-Neutrino-O-helium (ANO-helium, $ANOHe$) ($^4He^{++}[\bar{N}\Delta^{--}_{3\bar{U}}]$) "molecule" with mass $m_o \approx 1 S_5$TeV is produced in the reaction $\Delta^{--}_{3\bar{U}} +^4 He \rightarrow \gamma + (^4He^{++}\Delta^{--}_{3\bar{U}})$. The size of this "molecule" is $R_o \sim 1/(Z_\Delta Z_{He} \alpha m_{He}) \approx 2 \cdot 10^{-13}$ cm and it can play the role of a dark matter component and a nontrivial catalyzing role in nuclear transformations.

In nuclear processes ANO-helium looks like an α particle with shielded electric charge. It can closely approach nuclei due to the absence of a Coulomb barrier and opens the way to form heavy nuclei in SBBN. This path of nuclear transformations involves the fraction of baryons not exceeding 10^{-7} [4] and it can not be excluded by observations.

As soon as ANO-helium is formed, it catalyzes annihilation of deficit U-hadrons and N. Charged U-hadrons penetrate neutral ANO-helium, expel

^4He, bind with anutium and annihilate falling down the center of this bound system. The rate of this reaction is $\langle \sigma v \rangle = \pi R_o^2$ and an $\bar{U}$ excess $k = 10^{-3}$ is sufficient to reduce the primordial abundance of (Uud) below the experimental upper limits. N capture rate is determined by the size of $(\bar{N}\Delta)$ "atom" in ANO-helium and its annihilation is less effective.

Interaction of the ^4He component of $(ANOHe)$ with a ^A_ZQ nucleus can lead to formation of $^{A+4}_{Z+2}Q$ nucleus. The final nucleus is formed in the excited $[\alpha, M(A,Z)]$ state, which can rapidly experience α- decay, giving rise to $(ANOHe)$ regeneration and to effective quasi-elastic process of $(ANOHe)$-nucleus scattering. It leads to possible suppression of ANO-helium catalysis of nuclear transformations in matter.

The composite nature of ANO-helium makes it more close to warm dark matter. This dark matter plays dominant role in formation of large scale structure at $k > 1/2$.

The first evident consequence of the proposed scenario is the inevitable presence of ANO-helium in terrestrial matter, which is opaque for $(ANOHe)$ and stores all its in-falling flux. If its interaction with matter is dominantly quasi-elastic, this flux sinks down the center of Earth. If ANO-helium regeneration is not effective and Δ remains bound with heavy nucleus Z, anomalous isotope of $Z-2$ element appears. This is the serious problem for the considered model.

Even at $k = 1$ ANO-helium gives rise to less than 0.1 [4] of expected background events in XQC experiment [5], thus avoiding for all $k \leq 1$ severe constraints on Strongly Interacting Massive particles SIMPs obtained in [6] from the results of this experiment. In underground detectors $(ANOHe)$ "molecules" are slowed down to thermal energies far below the threshold for direct dark matter detection. However, $(ANOHe)$ destruction can result in observable effects. Therefore a special strategy in search for this form of dark matter is needed.

At $10^{-3} < k < 0.02$ U-baryon abundance is strongly suppressed, while the modest suppression of primordial N abundance might not exclude explanation of DAMA, HEAT and EGRET data in the framework of hypothesis of 4th neutrinos [1], making the effect of N annihilation in Earth consistent with the experimental data.

3 Discussion

Owing to excess of $\bar{U}$ anti-quarks primordial abundance of (Uud) is exponentially suppressed and anomalous isotope over-production is avoided. Excessive anti-U-quarks should retain dominantly in the form of anutium, which binds with excessive $\bar{N}$ and then with ^4He in neutral ANO-helium. Galactic cosmic rays destroy ANO-helium, striking off ^4He. It can lead to appearance of a free [anutium-$\bar{N}$] component in cosmic rays, which can be as large as

$[\bar{N}\Delta^{--}_{3\bar{U}}]/{}^4He \sim 10^{-7}$ and accessible to PAMELA and AMS experiments. In the context of composite dark matter like [4] accelerator search for new stable quarks and leptons acquires the meaning of critical test for existence of its charged components.

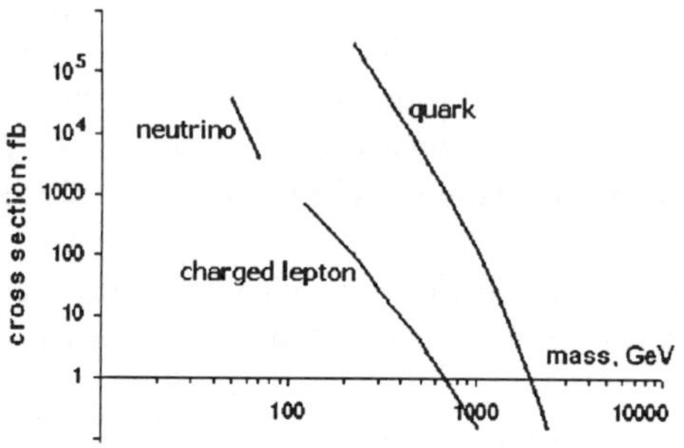

Figure 1. Cross sections of production of 4th generation particles in experiment ATLAS. The level of X-axis corresponds to the lower (conservative) level of LHC sensitivity for 1st year of its operation.

Acknowledgments

The work was supported by Khalatnikov-Starobinsky school. M.Kh. thanks LPSC (Grenoble, France) for hospitality and D. Rouable for help.

References

[1] D.Fargion et al., *JETP Lett.* 69, 434 (1999); astro-ph-9903086; K.M.Belotsky, M.Yu.Khlopov and K.I.Shibaev, *Gravitation and Cosmology* 6 Supplement, 140 (2000); K.M.Belotsky, D.Fargion, M.Yu.Khlopov and R.Konoplich, hep-ph/0411093; K.M.Belotsky, D.Fargion, M.Yu.Khlopov, R.Konoplich, and K.I.Shibaev, *Gravitation and Cosmology* 11, 16 (2005) and references therein.

[2] K.M.Belotsky, D.Fargion, M.Yu.Khlopov, R.Konoplich, M.G.Ryskin and K.I.Shibaev, *Gravitation and Cosmology* 11, 3 (2005).

[3] M.Maltoni et al., *Phys.Lett.* **B**476, 107 (2000); V.A.Ilyin et al., *Phys.Lett.* **B**503, 126 (2001); V.A.Novikov et al., *Phys.Lett.* **B**529, 111 (2002); *JETP Lett.* 76, 119 (2002).

[4] M.Yu.Khlopov, *JETP Lett.* 83, 3 (2006); astro-ph/0511796 and references therein.

[5] D.McCammon et al., *Nucl.Instr.Meth.* **A** 370, 266 (1996); D. McCammon et al., *Astrophys.J.* 576, 188 (2002), astro-ph/0205012.

[6] B.D.Wandelt et al., astro-ph/0006344; P.C.McGuire and P.J.Steinhardt, astro-ph/0105567; G.Zaharijas and G.R.Farrar, *Phys.Rev.* **D** 72, 083502 (2005), astro-ph/0406531.

SMALL-SCALE FLUCTUATIONS OF EXTENSIVE AIR SHOWERS AS THE ORIGIN OF ENERGY ESTIMATION SYSTEMATICS

G.Rubtsov [a]

Faculty of Physics, Moscow State University, 119992 Moscow, Russia
Institute for Nuclear Research of the Russian Academy of Sciences, 60th October Anniversary Prospect 7a, 117312 Moscow, Russia

Abstract. Small-scale fluctuations in extensive air showers were estimated by the Monte-Carlo calculations. It was found that these fluctuations may lead to non-Gaussian error in energy estimation by a ground array.

1 Introduction

Ground array detectors serve today as a precise instrument for detection of extensive air showers (EAS), induced by ultra-high-energy cosmic rays. One of the key procedures for successful analysis of experimental data is estimation of primary particle energy. The area covered by ground detectors usually makes up about 10^{-6} of the total area covered by EAS. Consequently, the reading of each detector represents local particle density and is sensitive to the density fluctuations on the scale of the size of the detector. The latter will hereafter be referred to as small-scale fluctuations.

Typical EAS, induced by a proton with primary energy 10^{20} eV contains about 20 billions of particles at the ground level. Approximate composition of ground level particles is the following: $\sim 90\%$ γ, $\sim 9\%$ electrons, $\sim 1\%$ muons and $\sim 3 \cdot 10^{-4}$ hadrons. Because of the huge number of particles, Monte-Carlo simulations are often performed with some kind of thinning, reducing effective number of particles in calculation and washing out small-scale fluctuations.

2 Simulations

To study fluctuations we simulated two vertical proton-induced extensive air showers with primary energy of 10^{18} eV. Simulations are performed with CORSIKA v6200 [1] *without thinning.* QGSJECT01c [2], GHEISHA [3] and EGS4 [4] models were used for high-energy hadronic, low-energy hadronic and electromagnetic interactions respectively.

The ground detector array was assumed to consist of 100 plastic 5 cm thick scintillators ($1.6\ m \times 1.6\ m$) forming square lattice covering the area of 50 km^2. Each simulated shower has been detected 10000 times with different core position within a ground array and rotation relating to the array.

Our intent is to make the detection procedure close to one used for analysis of AGASA experimental data [5, 6]. Detector responses are estimated using

[a] e-mail: grisha@ms2.inr.ac.ru

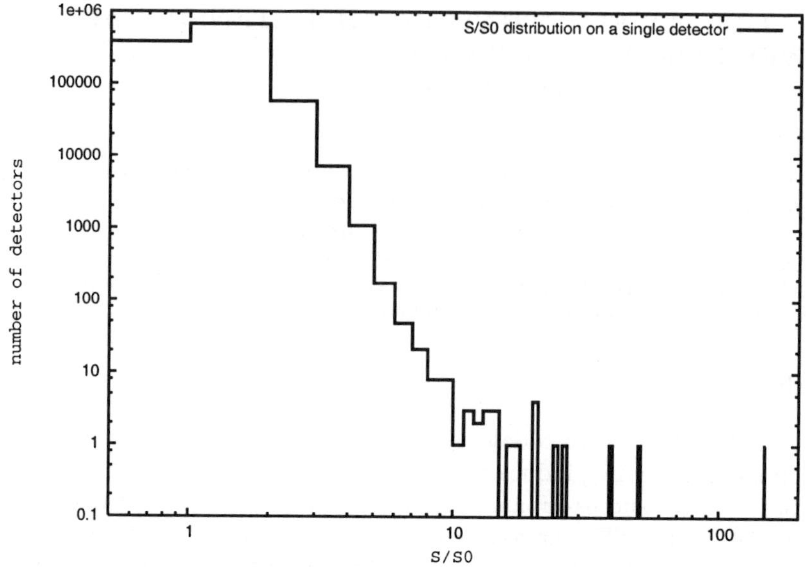

Figure 1: Distribution of the density on a single detector at a core distances from 300 to 1500 meters. S represents a simulated particular detector response and S_0 - an expected average value assuming AGASA experimental lateral distribution.

AGASA average detector response functions [7]. At core distance of 600 meters, an average particles density in vertical 10^{18} eV shower is 63 photons, 4 electrons and 2 muons per square meter. The density distribution on one detector is shown on Fig. 1. It can be seen that there is a non-zero probability that a particular detector response will exceed it's expected average by 10 or 100 times. Nevertheless, such probability is of order of $10^{-5} - 10^{-6}$. Moreover, an average magnitude of density fluctuations on one detector (σ^2) is found to be nearly proportional to the expected average detector response in accordance with experimental results [8].

Fitting responses of detectors at core distances from 300 to 1500 meters with an AGASA experimental LDF [9], we obtain $S(600)$, which is directly related with the reconstructed energy. Following the AGASA procedure to ensure fit quality we exclude worst detector from the fit in case of bad χ^2 ($\chi^2/N >$ 1.5) [10]. Detectors experiencing largest local fluctuations are expected to be excluded by the procedure. The distribution of the number of excluded detectors is presented in Table 1.

The results of our simulations are presented in Fig. 2. It can be seen that while $S(600)$ underestimation obeys Gaussian law, overestimation is substantially non-Gaussian but rather is described by single exponential law.

Number of excluded detectors	% of cases
1	14%
2	2%
≥ 3	0.4%

Table 1: The distribution of a number of excluded detectors in a fit quality assurance procedure

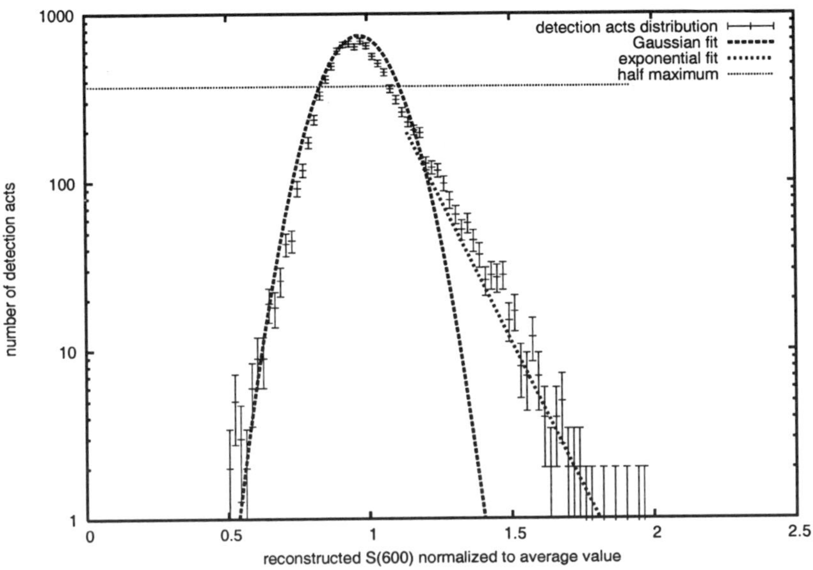

Figure 2: S(600) distribution for a vertical proton-induced EAS ($E_0 = 10^{18}\ eV$), over 10000 detection acts

3 Conclusions

We conclude that small-scale fluctuations of particle density at the ground level may lead to non-Gaussian error in energy estimation. While the error is non-Gaussian it is still not large enough to significantly impact an overall energy spectrum.

Acknowledgments

I am indebted to D. S. Gorbunov, V. A. Kuzmin, V. A. Rubakov, and S. V. Troitsky for inspiring discussions. Work was supported in part by INTAS 03-51-5112, GPRFNS-2184.2003.2 grants and by the fellowship of the "Dynasty" Foundation. Computing facilities of the Department of Theoretical Physics, Institute for Nuclear Research, were used to perform the simulations of air showers.

References

[1] D. Heck, G. Schatz, T. Thouw, J. Knapp and J. N. Capdevielle, *FZKA-6019*
[2] N. N. Kalmykov, S. S. Ostapchenko and A. I. Pavlov, *Nucl. Phys. Proc. Suppl.* **52B**, 17 (1997).
[3] H. Fesefeldt, PITHA 85/02.
[4] W. R. Nelson, H. Hirayama and D. W. O. Rogers, *SLAC-0265*
[5] M. Takeda *et al. Phys. Rev. Lett.* **81**, 1163 (1998)
[6] M. Takeda *et al.*, Astropart. Phys. **19**, 447 (2003).
[7] Sakaki, N. et al., *Proceedings of ICRC 2001*, "Energy estimation of AGASA events", www.hef.kun.nl/nahsa/pdf/ici6938_p.pdf
[8] M. Teshima *et al.*, J. Phys. G **12**, 1097 (1986).
[9] N. Sakaki, PhD thesis, Univ. of Tokyo, 2002.
[10] M. Takeda, PhD thesis, Tokyo University of Technology, 1998.

STATUS OF E391a EXPERIMENT FOR THE RARE DECAY $K_L^0 \to \pi^0 \nu\bar{\nu}$

T.Inagaki [a] (E391a collaboration) [b]
Institute of Particle and Nuclear Studies, High Energy Accelerator Research Organization (KEK), Tsukuba, Ibaraki, 305-0801 Japan

Abstract. The rare K decay $K_L^0 \to \pi^0 \nu\bar{\nu}$ is considered as one of ideal processes in the quark flavor physics for a critical test of the standard model and a search for new physics. A first dedicated experiment for the decay has been performed by the E391a collaboration at the KEK 12-GeV proton synchrotron. Here, we report the status of the E391a experiment together with a preliminary result of a new upper limit of 2.9×10^{-7} using about 10 % sample of the data collected during the first period of run. We also briefly report a status of the J-PARC project.

1 Introduction

The rare decay $K_L^0 \to \pi^0 \nu\bar{\nu}$, which is a direct CP violation process, proceeds through Flavor Changing Neutral Current from strange(s) to down(d) quarks. The branching ratio of the decay is predicted to be $(3 \pm 0.6) \times 10^{-11}$ [1] in the standard model, where the error comes from parametric uncertainties from other experiments and the theoretical ambiguity is only 1-2 %. The decay has been considered as an ideal process to measure precisely ImV_{td} for a critical test of the standard model. The decay is also considered to be valuable for a search of new physics beyond the standard model, because theoretical ambiguity is also small for most of new physics [2].

While measurement of the decay $K_L^0 \to \pi^0 \nu\bar{\nu}$ is a dream of theory, it is a nightmare for experiment. The decay is rare, all concerning particles are neutrals and there is no definite kinematical signature due to three-body decay including two neutrinos. The present experimental limit is only 5.9×10^{-7} [3]. Therefore, we took a redundant step-by-step approach. We plan to improve sensitivity by about three orders of magnitude with E391a at the KEK 12-GeV proton synchrotron, and by another three orders with a next experiment using a high intensity accelerator in order to measure ImV_{td} at the level of theoretical ambiguity. The step of E391a has a role of verifying the experimental method as a pilot experiment for the next experiment as well as uncovering a large unexplored region.

The first data taking (Run-1) of E391a was performed from February to June 2004. The second (Run-2) was in the spring of 2005 and the third (Run-3) is scheduled as in the fall of 2005. In this report, we breifly summarize the characteristic points of the detection method and experimental apparatus, and report a preliminary result obtained from an analysis of about 10 % sample of the data collected in the Run-1. Finally, we add a short introduction of

[a] e-mail: inagaki@post.kek.jp
[b] *http://www-ps.kek.jp/e391*

the J-PARC project, which is now constructing and which is one of promising facilities for the next high-sensitivity experiment.

2 Detection Method

The decay $K_L^0 \to \pi^0 \nu\bar{\nu}$ will be observed as a process of $K_L^0 \to \gamma\gamma + nothing$ in a flight decay of K_L^0 in vacuum. The energies and hit positions of 2γ are measured by a calorimeter, and π^0 vertex, which can be assigned as K_L^0 decay vertex (Z_{VTX}) due to the short life-time of π^0, is obtained by assuming that 2γ come from a $\pi^0 \to \gamma\gamma$ decay on the beam axis. Another kinematical variable is the transverse momentum of π^0 (P_T) with respect to the K_L^0 beam axis. A sizable P_T can be required to match the missing P_T due to escape of two neutrinos. Then, the decay $K_L^0 \to \pi^0 \nu\bar{\nu}$ will be looked for in a signal box of Z_{VTX} and P_T.

One of key elements for this approach is a neutral beam with small size (called pencil beam) in order to achieve good measurements of Z_{VTX} and P_T. A tight collimation having little halo is also important to minimize π^0 productions through interactions of beam halo with detector material. Actually, the beam, which we used, is a cone shaped beam of 2 mradian radius and the averaged momentum of K_L^0 is 2 GeV/c. The resolution of P_T is 5 MeV/c. which is dominantly determined by the beam spread. The P_T window was set from 130 MeV/c and 250 MeV/c. The lower bound removes the background from other multi-body decays of K_L^0, hyperon decays and $K_L^0 \to \gamma\gamma$ decays. The higher P_T bound is determined by the kinematical limit of the decay $K_L^0 \to \pi^0 \nu\bar{\nu}$, which is 231 MeV/$c$. The acceptance for the decay $K_L^0 \to \pi^0 \nu\bar{\nu}$ in the decay fiducial region is expected to be more than 10 %.

Another key element is a tight rejection of additional charged particles and photons (veto) by covering the decay fiducial region with thick calorimeters in the 4π direction to assure the *nothing*. Since all other K_L^0 decays except $K_L^0 \to \gamma\gamma$ accompany at least two charged particles or additionally at least two γ, a tight veto is the most effective tool to reject backgrounds from other K_L^0 decays. For the tight veto it is required to detect particles with high efficiency. We have been aware through various test experiments [4, 5] that high efficiency can be obtained by a detection with low detection threshold. Therefore, one of main concerns in this experiment is how we can understand phenomena of low-energy deposit.

The third key element is ultra-high vacuum in the decay fiducial region. Once a π^0 is produced by an interaction of beam with air, it is hardly discriminated from signal, because the P_T range delivered by a strong interaction is typically 300 Mev/c, which is more than the maximum P_T value of $K_L^0 \to \pi^0 \nu\bar{\nu}$ decays. A high vacuum of 10^{-5} Pa is required for a neglible level of the beam interactions. A technical problem is a compatibility of this high-vacuum requirement

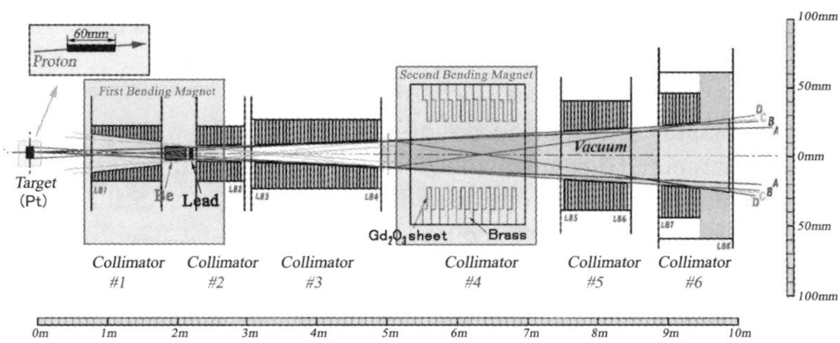

Figure 1: K_L^0 beam line

with the requirement to minimize dead material in front of detectors for high efficient detection.

3 Experimental apparatus

Figure 1 shows the beam line. Neutral beam was extracted in the direction of 4° with respect to the primary proton line. The length of the beam line is 10 m. The averaged momentum of K_L^0 is 2 GeV/c. The neutron-to-K_L^0 ratio in the neutral beam is about 60 and the halo-to-core ratio is about 10^{-5} for both neutron and γ with energy above 10 MeV, as described in the reference [6].

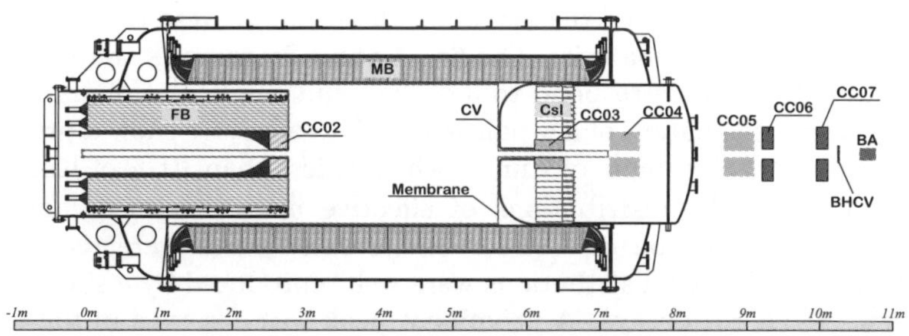

Figure 2: Detection apparatus

The detection apparatus was located at the exit of the beam line. Detectors were cylindrically assembled along the beam axis as shown in Figure 2. The energies and hit positions of 2γ from the decay $K_L^0 \to \pi^0 \nu \bar{\nu}$ were expected to be measured with a calorimeter (CsI). It consists of 576 blocks of un-doped CsI crystal, and the size of the major part of blocks is $7\times7\times30$ cm^2. The detail of the CsI calorimeter is reported elsewhere [7].

A group of plastic scintillation counters (named CV) were placed in front of the CsI calorimeter to remove the decays accompanying charged particles. The decay region is mostly surrounded with two large lead-scintillator sandwich calorimeters, MB and FB. Several collar counters, CC02-CC07, are placed along the beam axis. At the end of detector system beam plug counters, named BA and BHCV, was placed on the beam axis.

We installed most of detectors in a vacuum vessel. The vacuum region was divided into two regions and they were differentially pumped. The pressure of the outer region, where detectors were installed, was less than 1 Pa, and that of the inside region was 1×10^{-5} Pa. Two regions were separated by a sheet of membrane of 20 mg/cm^2 thick.

The number of PMT's in the detector is about one thousand. An analog signal from every PMT was read by a charge-sensitive ADC having a range of 15-bit and a resolution of 50 fC/bit. Every timing information was also read by TDC through a discriminator. The threshold for the discriminator was 0.5 mV, which corresponds to 0.5 MeV deposit for CsI. Trigger was basically formed by the hit multiplicity of the bundles of eight CsI blocks.

4 Data analysis

In an off-line analysis CsI hits were reclustered by drawing a boundary around a crystal having locally maximum energy deposit, and the events were sorted in order of γ cluster count. We raised the detection threshold for each γ cluster to identify γ surely and lowered the detection threshold for veto to achieve higher efficiency in the offline analysis. The 6γ, 4γ and 2γ samples were simultaneously analyzed. They correspond to $3\pi^0$, $2\pi^0$ and single π^0 samples with the assumption that the parent of a γ pair is π^0. Here-after, we show a preliminary result using one week sample of Run-1, which is less than 10 % of Run-1 data.

Figure 3 shows the distributions of effective mass, summed momentum, Z_{VTX} and γ hit position on CsI for $3\pi^0$ events, which were reconstructed from the 6γ sample. All distributions were well reproduced by a simulation for $K_L^0 \to \pi^0\pi^0\pi^0$ ($K\pi3$) decays. A prominent peak was observed in the effective mass distribution. The summed momentum distribution was used for a fine adjustment of the K_L^0 momentum spectrum given into the simulation. Since the Z_{VTX} distribution is quite sensitive to the K_L^0 spectrum and the energy calibration of CsI, it provided a cross-check on these parameters. We defined the outside radius of acceptance for the CsI calorimeter to be 80 cm from the distribution of the γ hit position on CsI.

Figure 4 shows the effective mass distribution of $2\pi^0$ events reconstructed

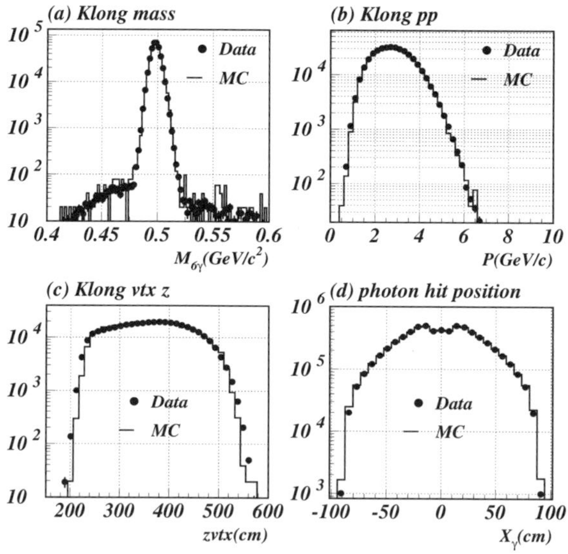

Figure 3: Distributions of (a) effective mass, (b) summed momentum, (c) Z_{VTX} and (d) γ hit position on CsI for $3\pi^0$ events. The dots and histogram show the experimental data and the simulation for the decay $K_L^0 \to \pi^0\pi^0\pi^0$, respectively.

from the 4γ sample. The events around the K_L^0 mass correspond to $K_L^0 \to \pi^0\pi^0$ ($K\pi2$) decays, and their distributions of summed momentum, Z_{VTX} and γ hit position are well reproduced by a simulation for the $K\pi2$ decay as in the case of $3\pi^0$. The bump in low mass region is reproduced by a simulation for the $K\pi3$ decay.

The number of K_L^0, which decayed in the fiducial decay region, was estimated from the $K\pi3$ and $K\pi2$ decays, using the acceptance calculated by the simulation, and the well known values of branching ratio. The number of K_L^0 decays is 1.10×10^9 from the $K\pi3$ decay and 1.17×10^9 from the $K\pi2$ decay. They agree with each other within 6 %.

Figure 5 shows a scatter plot of Z_{VTX} vs P_T for the single π^0 sample obtained from 2γ sample. We set a signal box as follows, The P_T window was set to the region from 130 MeV/c and 250 MeV/c as explained before. The upstream and downstream boundaries were set as to be apart 30 cm from the downstream end of CC02 and 50 cm from the upstream edge of CV.

The event clusters near Z_{VTX}= 270 cm and = 550 cm might come from π^0 's produced by interactions of halo with CC02 and CV, because the downstream end of CC02 and the upstream edge of CV were located at 270 cm and 550 cm,

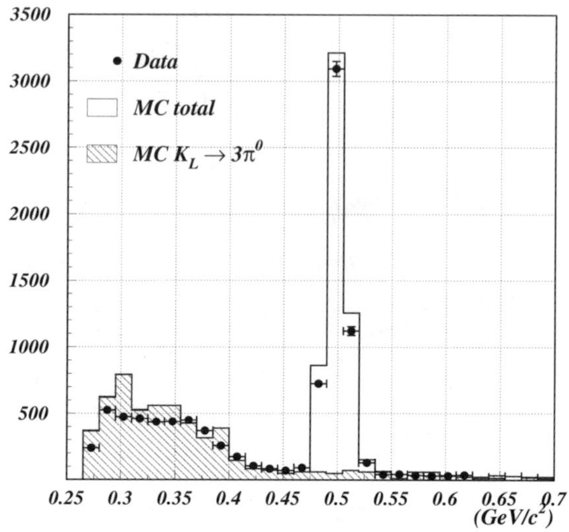

Figure 4: Effective mass distribution of $2\pi^0$ events. The dots show the experimental result and the open and shaded histgrams are those for $K\pi 2$ and $K\pi 3$ simulations.

respectively. We tried to reproduce both event-clusters by a calculation using the neutron flux which had shown the consistency with the results of the beam survey. While the calculation can reproduce the upstream events, it cannot do the downstream ones. The number of data events was much larger than that of calculation by about one order of magnitude, and it suggested an extra material of about 60 mg/cm^2 in the beam near the position of the edge of CV. We confirmed a part of the membrane drooping into the beam core by opening the vacuum vessel after Run-1.

We are trying to understand the background level in the signal box. Tails from the event clusters near Z_{VTX}= 270 cm and = 550 cm were checked to be negligibly small by extrapolating with a Gaussian fitting. The contamination by the $K\pi 2$ decay can be estimated using a simulation, because the simulation of $K\pi 3$ well reproduced the low mass bump in the $2\pi^0$ data, where the contamination process is similar with that of single π^0 events by $K\pi 2$ decays. The expected numbers of events from the $K\pi 2$ decay are 0.02 ± 0.01 in the signal box and 0.05 ± 0.02 in the lower P_T region. These are far below the observed number of events in the lower P_T region. Multi-π^0 produced at the drooping membrane is another background source, because a wrong Z_{VTX} is given, if the 2γ pair from multi-π^0 is an odd combination (γ pair of different π^0). The multi-π^0 productions were calculated using the same neutron flux as

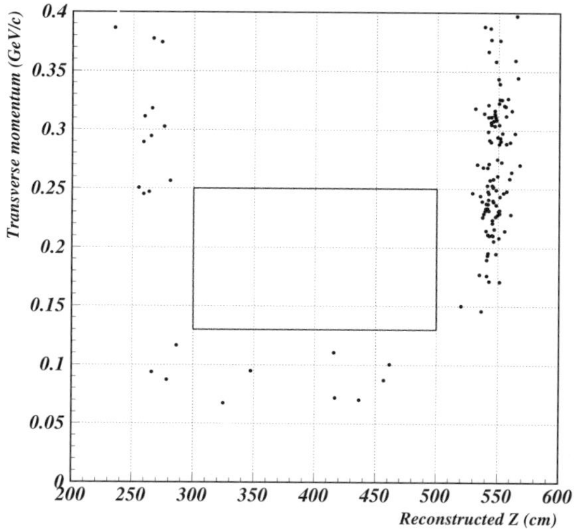

Figure 5: Scatter plot of Z_{VTX} vs P_T for the 2γ events. The box indicates the boundary for signal, which was determined as described in the text.

that used in the case of single π^0 production. The number of events due to multi-π^0 productions at the drooping membrane was estimated to be zero in the signal box and five in the lower P_T region by taking a normalization to the remaining events at Z_{VTX}=550 cm. The observed seven events in the lower P_T region mostly come from this source.

The single event sensitivity for the $K_L^0 \to \pi^0\nu\bar{\nu}$ decay was calculated as the reciprocal of the product of the numbers of K_L^0 decays estimated from $K\pi2$ and $K\pi3$ with the acceptance for the $K_L^0 \to \pi^0\nu\bar{\nu}$ decay. It is 1.17×10^{-7}. Since there is no event in the signal region, we set an upper limit for the $K_L^0 \to \pi^0\nu\bar{\nu}$ decay to be 2.9×10^{-7} at 90 % confidence level by multiplying a factor of 2.3 to the single event sensitivity.

5 Status of J-PARC

A new facility of high intensity proton accelerators is now constructing at the Tokai site in JAERI (Japan Atomic Energy Research Institute) [8] as a joint project of KEK and JAERI. It is a project to construct a facility of the intensity front in both energies at 3 and 50 GeV and to promote the material science using spoliation neutrons and the nuclear and particle researches. The commissioning of 50 GeV PS is scheduled to start in 2007. The first PAC for the day-one experiment will be held in the spring of 2006. The latest status of the J-PARC project is announced in its web site [9]. We submitted a LOI for a $K_L^0 \to \pi^0 \nu\bar{\nu}$experiment, and are now working for a submission of proposal.

References

[1] A. J. Buras, F. Schwab, and S .Uhling, hep-ph/0405132, and references therein.
[2] D. Bryman, A. J. Buras, G. Isidori, and L. Littenberg, hep-ph/0505171.
[3] A. Alavi-Harati *et al.*, *Phys. Rev.*__D61__, 072006 (2000).
[4] S. Ajimura *et al.*, *Nucl. Instr. Meth.*__A 435__, 408 (1999); S. Ajimura *et al.*, *Nucl. Instr. Meth.*__A 552__, 263 (2005).
[5] T. Inagaki *et al.*, *Nucl. Instr. Meth.*__A 359__, 478 (1995).
[6] H. Watanabe *et al.*, *Nucl. Instr. Meth.*__A 545__, 542 (2005).
[7] M. Doroshenko *et al.*, *Nucl. Instr. Meth.*__A 545__, 278 (2005).
[8] *JAERI was re-organized to be JAEA (Japan Atomic Energy Agency).*
[9] *http://jkj.tokai.jaeri.go.jp/*

ON CP EFFECTS GENERATED BY ELECTROWEAK PENGUIN DIAGRAMS IN NON-LEPTONIC K DECAYS

Evgeny Shabalin[a]

State Research Center, Institute for Theoretical and Experimental Physics, 117218, Moscow, Russia

Abstract. It is shown that the contribution of the electroweak penguin (EWP) diagrams increase the CP effects produced by QCD penguin (QCDP) diagram in $K^{\pm} \to \pi^{\pm}\pi^{\pm}\pi^{\mp}$ decay and decrease these effects in $K^{\pm} \to \pi^0\pi^0\pi^{\pm}$ decay. More over, the combination of EWP and QCDP contributions in the last decay turns out to be practically the same as for the parameter ε' in $K \to 2\pi$ decays. It allows to express the value of CP effects in $K^{\pm} \to \pi^0\pi^0\pi^{\pm}$ decay through the known value of ε'.

1 Introduction

28 years ago Shifman, Vainshtein and Zakharov [1] had found that the amplitudes generated by the operators containing the product of left-handed and right-handed quark currents are considerably enhanced comparing with the amplitudes originated by other operators. Such operators were produced by so-called afterwards QCD penguin diagrams. Later on, 16 years ago, it was found [2] that inspite of seeming smallnes of so-called electroweak penguin diagrams (proportional to α_{em} instead of α_s), these diagrams can give a considerable contribution to parameter ε' in $K^0 \to 2\pi$ decays. But the large uncertainties in the theoretical calculations of the ratio ε'/ε giving

$$few \times 10^{-4} \leq \varepsilon'/\varepsilon \leq few \times 10^{-3} \qquad [3]$$

did not allow to fix the relative strength of EWP an QCDP contributions.

It will be shown that this problem may be solved in searches for CP effects in $K^{\pm} \to \pi^{\pm}\pi^{\pm}\pi^{\mp}$ and $K^{\pm} \to \pi^0\pi^0\pi^{\pm}$ decays.

2 Technics of calculations

The amplitudes of $K \to 2\pi$ and $K \to 3\pi$ decays can be expressed in terms of the Wilson coefficients c_i of the effective $\Delta S = 1$ lagrangian [1]

$$L(\Delta S = 1) = \sqrt{2} G_F \sin\theta_C \cos\theta_C \sum c_i O_i \qquad (1)$$

where the operatots O_{1-6} are represented in [1] and [4] and

$$O_7 = \tfrac{3}{2}\bar{s}\gamma_\mu(1+\gamma_5)d(\textstyle\sum_{q=u,d,s} e_q \bar{q}\gamma_\mu(1-\gamma_5)q)$$
$$O_8 = -12\textstyle\sum_{q=u,d,s} e_q \bar{s}_L q_R \cdot \bar{q}_R d_L, \qquad e_q = (\tfrac{2}{3}, -\tfrac{1}{3}, -\tfrac{1}{3})$$

[a] e-mail: shabalin.itep.ru

Using the scheme of bosonization of these operators proposed in [5] and representing the $K \to 2\pi$ matrix elements in the form

$$M(K_1^0 \to \pi^+\pi^-) = A_0 e^{i\delta_0} - A_2 e^{i\delta_2} \tag{2}$$

$$M(K_1^0 \to \pi^0\pi^0) = A_0 e^{i\delta_0} + 2A_2 e^{i\delta_2} \tag{3}$$

we obtain

$$A_0 = \kappa \left[c_1 - c_2 - c_3 + \frac{32}{9}\beta(\Re\tilde{c}_5 + i\Im\tilde{c}_5) \right] \tag{4}$$

$$A_2 = \kappa \left[c_4 + i\frac{2}{3}\beta\Lambda^2\Im\tilde{c}_7(m_K^2 - m_\pi^2)^{-1} \right] \tag{5}$$

where

$$\kappa = G_F F_\pi \sin\theta_C \cos\theta_C (m_K^2 - m_\pi^2)/\sqrt{2}, \qquad \beta = 2m_\pi^4/[\Lambda^2(m_u + m_d)^2],$$
$$\tilde{c}_5 = c_5 + 3c_6/16, \qquad \tilde{c}_7 = c_7 + 3c_8, \qquad \Lambda \approx 1 \text{ GeV}.$$

From data on $K \to 2\pi$ decays

$$c_4 = 0.328, \qquad c_1 - c_2 - c_3 + \frac{32}{9}\beta\Re\tilde{c}_5 = -10.13.$$

Using the data on ε' and the general relation

$$\varepsilon' = ie^{i(\delta_2 - \delta_0)} \left[-\frac{\Im A_0}{\Re A_0} + \frac{\Im A_2}{\Re A_2} \right] \cdot |\frac{A_2}{A_0}|$$

we obtain

$$\beta\Im\tilde{c}_5 \left(1 + \frac{24.36\Im\tilde{c}_7}{(1-\Omega)\Im\tilde{c}_5} \right) = \frac{3.32(1 \pm 015)10^{-4}}{1 - \Omega} \tag{6}$$

The parameter Ω takes into account the effect of isospin violation occuring due $m_d \neq m_u$ and electromagnetic interaction. The recent theoretical calculations [6] give $\Omega = 0.060 \pm 0.077$. Then

$$\beta\Im\tilde{c}_5 \left[1 + (24.36 \div 28.23)\frac{\Im\tilde{c}_7}{\Im\tilde{c}_5} \right] = 3.85(1 \pm 0.18)10^{-4}. \tag{7}$$

This result will allow us to fix a magnitude of CP effect in $K^\pm \to \pi^0\pi^0\pi^\pm$ decay.

3 CP effects in $K^\pm \to 3\pi$ decays

Let's consider the CP-odd asymmetries

$$\Delta g_\tau = \frac{g_\tau^+ - g_\tau^-}{g_\tau^+ + g_\tau^-} \quad \text{and} \quad \Delta g_{\tau'} = \frac{g_{\tau'}^+ - g_{\tau'}^-}{g_{\tau'}^+ + g_{\tau'}^-}, \tag{8}$$

where $g^{\pm}_{\tau,\tau'}$ are determined by the expressions

$$M\left(K^{\pm}(k) \to \pi^{\pm}(p_1)\pi^{\pm}(p_2)\pi^{\mp}(p_3)\right) \sim 1+ia+\frac{1}{2}g^{\pm}_{\tau}Y(1+ib^{\tau}\pm id^{\tau}_{KM})+\ldots \quad (9)$$

$$M\left(K^{\pm}(k) \to \pi^{0}(p_1)\pi^{0}(p_2)\pi^{\pm}(p_3)\right) \sim 1+ia+\frac{1}{2}g^{\pm}_{\tau'}Y(1+ib^{\tau'}\pm id^{\tau'}_{KM})+\ldots \quad (10)$$

where a and $b^{\tau,\tau'}$ are CP-even imaginary parts originated by $\pi\pi$ rescattering and $d^{\tau,\tau'}_{KM}$ are the imaginary parts induced by Kobayashi-Maskawa phase.

$$Y = (s_3 - s_0)/m_\pi^2, \qquad s_3 = (k-p_3)^2, \qquad s_0 = m_K^2/3 + m_\pi^2$$

For $K^{\pm} \to \pi^{\pm}\pi^{\pm}\pi^{\mp}$ (τ-decay): [4]

$$d^{\tau}_{KM} = -2\frac{16c_4}{c_0(c_0+9c_4)}\beta\Im\tilde{c}_5\left(1 - 14.36\frac{\Im\tilde{c}_7}{\Im\tilde{c}_5}\right) \quad (11)$$

$$a = 0.12, \qquad b^{\tau} = 0.71, \qquad g_\tau = -\tfrac{3m_\pi^2}{m_K^2}(1+9c_4/c_0),$$
$$c_0 = c_1 - c_2 - c_3 - c_4 + \tfrac{32}{9}\cdot\Re\tilde{c}_5 = -10.46, \qquad c_4 = 0.328.$$

For $K^{\pm} \to \pi^{0}\pi^{0}\pi^{\pm}$ (τ'-decay): [7]

$$d^{\tau'}_{KM} = \frac{16c_4}{c_0(c_0-9c_4/2)}\beta\Im\tilde{c}_5\left(1 + 27.8\frac{\Im\tilde{c}_7}{\Im\tilde{c}_5})\right) \quad (12)$$

$$b^{\tau'} = 0.49, \qquad g_{\tau'} = \frac{6m_\pi^2}{m_K^2}\left(1 - \frac{9c_4}{2c_0}\right).$$

The CP-odd asymmetry parameters are

$$\Delta g_\tau = \frac{ad^{\tau}_{KM}}{1+ab^{\tau}}, \qquad \Delta g_{\tau'} = \frac{ad^{\tau'}_{KM}}{1+ab^{\tau'}}. \quad (13)$$

The comparison of eq.(12) with eq.(7) shows that the parameter $\Delta g_{\tau'}$ can be evaluated through the experimental value of ε'. From the above equations, it follows

$$\frac{-\Delta g_\tau}{\Delta g_{\tau'}} = 2\frac{c_0 - 9c_4/2}{c_0+9c_4}\cdot\frac{1-14.36\Im\tilde{c}_7/\Im\tilde{c}_5}{1+27.8\Im\tilde{c}_7/\Im\tilde{c}_5}. \quad (14)$$

4 Conclusions

1) At $\Omega = 0.124$, $\Delta g_{\tau'}$ is expressed through ε':

$$\Delta g_{\tau'} = 1.8(1\pm 0.15)10^{-6}.$$

2) If EWP do not play any significant role in direct CP violation, i.e. $\Im\tilde{c}_7/\Im\tilde{c}_5$ is very small, then

$$\frac{-\Delta g_\tau}{\Delta g_{\tau'}} = 3.1, \quad \text{or} \quad -\Delta g_\tau \geq 0.56 \cdot 10^{-5}.$$

3) If EWP cancels the half of QCDP contribution to ε', then

$$-\Delta g_\tau \geq 1.4 \cdot 10^{-5}$$

The experimental measurement of the ratio (14) will allow to fix many parameters of CP violation in K decay.

References

[1] M.A. Shifman, A.I. Vainshtein, V.I. Zakharov, *Zh.Eksp.i Teor.Fiz.(JETP)* **72** 1277 (1977).
[2] J.M. Flinn, L. Randall, *Phys.Lett.B* **224** 221 (1989); G. Buchalla et al., *Nucl.Phys.B* **337** (1990); E.A. Paschos, Y.L. Wu, *Mod.Phys.Lett.A* **6** 93 (1991).
[3] T. Hambye et al., *Nucl.Phys.B* **564** 391 (2000).
[4] E. Shabalin, *Phys.At.Nucl.* **68** 88 (2005).
[5] W.A. Bardeen, A.J. Buras, J.-M. Gerard, *Nucl.Phys.B* **293** 787 (1987).
[6] V. Cirigliano et al., *Eur.Phys.J.C* **33** 369 (2004).
[7] G. Fäldt, E. Shabalin, *Phys.Scripta* **72** 354 (2005).

CP VIOLATION IN $K^{\pm} \to \pi^0\pi^0\pi^{\pm}$

G. Fäldt [a]

Department of nuclear and particle physics, Uppsala University
Box 535, S-75121 Uppsala, Sweden

Abstract. The difference in the decay distributions of $K^+ \to \pi^0\pi^0\pi^+$ and $K^- \to \pi^0\pi^0\pi^-$ is due to CP violation. We argue that this difference can be accurately predicted from a knowledge of the CP-violating parameter ϵ'.

1 Introduction

The slope parameter g in charged kaon decay is defined through

$$|M(K^{\pm}(k) \to \pi^0(p_1)\pi^0(p_2)\pi^{\pm}(p_3))|^2 = \text{const} \cdot [1 + g(K^{\pm})Y + ...] \ , \tag{1}$$

with $Y = (s_3 - s_0)/m_\pi^2$ and $s_i = (k - p_i)^2$. In the presence of CP violation $g(K^+) \neq g(K^+)$. For this reason we introduce the slope-parameter asymmetry

$$\Delta g = \frac{g(K^+) - g(K^-)}{g(K^+) + g(K^-)} \ . \tag{2}$$

The numerical value of Δg is determined by the imaginary parts of the QCD and electroweak(EW) penguin amplitudes [1]. Asymmetry parameters can be measured in two charged kaon decays;

$$\begin{aligned} \tau\ decay: &\quad K^{\pm} \to \pi^{\pm}\pi^{\pm}\pi^{\mp} \\ \tau'\ decay: &\quad K^{\pm} \to \pi^0\pi^0\pi^{\pm} \end{aligned}$$

In reality theoretical predictions for Δg are uncertain due to sizeable cancellations between the contributions from the QCDP and EWP diagrams [2]. However, in τ' decay, we shall argue, that an accurate prediction is possible [3].

2 Effective non-leptonic Lagrangian

We start from the effective non-leptonic Lagrangian of Shifman *et al.* [4]

$$L(\Delta S = 1) = \sqrt{2}G_{\mathrm{F}} \sin\theta_{\mathrm{C}} \cos\theta_{\mathrm{C}} \sum_{i=1}^{i=8} c_i O_i \ . \tag{3}$$

It has two operator contributions from the QCD penguin diagrams

$$O_5 = \bar{s}_L\gamma_\mu\lambda^a d_L(\sum_{q=u,d,s} \bar{q}_R\gamma_\mu\lambda^a q_R) \tag{4}$$

$$O_6 = \bar{s}_L\gamma_\mu d_L(\sum_{q=u,d,s} \bar{q}_R\gamma_\mu q_R) \ , \tag{5}$$

[a] e-mail: faldt@tsl.uu.se

and two from the EW penguin diagrams

$$O_7 = \frac{3}{2}\bar{s}\gamma_\mu(1+\gamma_5)d(\sum_{q=u,d,s} e_q\bar{q}\gamma_\mu(1-\gamma_5)q) \tag{6}$$

$$O_8 = -12\sum_{q=u,d,s} e_q(\bar{s}_L q_R)(\bar{q}_R d_L) \ . \tag{7}$$

It is the imaginary parts of the coefficients c_5-c_8 that cause CP violation. The coefficients c_1-c_4 are all real.

Bosonisation of the quark operators is achieved through the substitutions of Bardeen *et al.* [5],

$$\bar{q}_j(1+\gamma_5)q_k = -\frac{1}{\sqrt{2}}F_\pi r[U - \frac{1}{\Lambda^2}\partial^2 U]_{kj} \tag{8}$$

$$\bar{q}_j\gamma_\mu(1+\gamma_5)q_k = i[(\partial_\mu U)U^\dagger - \frac{rF_\pi}{\sqrt{2}\Lambda^2}m(\partial_\mu U^\dagger) - \text{herm.conj.}]_{kj} \ . \tag{9}$$

Following Cronin [6] the U-matrix is expanded as

$$U = \frac{F_\pi}{\sqrt{2}}\left(1 + \frac{i\sqrt{2}\hat{\pi}}{F_\pi} - \frac{\hat{\pi}^2}{F_\pi^2} - 2\left(\frac{i\hat{\pi}}{\sqrt{2}F_\pi}\right)^4 + ...\right) \tag{10}$$

where $\hat{\pi}$ denotes the 3×3 pseudoscalar nonet-meson-field matrix.

3 The $K^\pm \to \pi^0\pi^0\pi^\pm$ amplitude

The matrix elements for the charged kaon decays are obtained from the effective Lagrangian of eq.(3), as

$$M\left(K^\pm \to \pi^0\pi^0\pi^\pm\right) = \kappa\left[1 \pm ia_{KM} + fY(1 \pm ib_{KM}) + ...\right] \ , \tag{11}$$

with the parameter f as

$$f = \frac{3m_\pi^2}{m_K^2}\left(1 - \frac{9c_4}{2c_0}\right) \ . \tag{12}$$

The terms a_{KM}, b_{KM} are CP odd and generated by the imaginary parts of the penguin-diagram contributions, which in turn are due to the Kobayashi-Maskawa phase. Since CP violation is small, the relevant combination for the slope-parameter asymmetry is the difference

$$d_{KM} = b_{KM} - a_{KM} \tag{13}$$

The overall strength of the decays are regulated by the parameters

$$\kappa = \frac{G_F m_K^2}{6\sqrt{2}} c_0 \sin\theta_C \cos\theta_C \tag{14}$$

$$c_0 = c_1 - c_2 - c_3 - c_4 + \frac{32}{9}\beta \mathrm{Re}\tilde{c}_5 \tag{15}$$

$$\beta = \frac{r^2}{2\Lambda^2} = \frac{2m_\pi^4}{\Lambda^2(m_u + m_d)^2} \tag{16}$$

The penguin-diagram coefficients always appear in the combinations

$$\tilde{c}_5 = c_5 + \frac{3}{16}c_6, \quad \tilde{c}_7 = c_7 + 3c_8 \ .$$

Coefficients f, a_{KM} and b_{KM} are real valued. The imaginary terms in the matrix elements (11) for K^+ and K^- decay only differ in their sings. Thus, after squaring the matrix element no difference remains between K^+ and K^-. To create a visible CP violation we need to introduce the final-state interaction.

4 Final state interactions

Final-state interactions (FSI) among the pions produce a CP-even imaginary part in the amplitude. We calculate the FSI by cutting internal lines of the rescattering graphs. Denoting with a, b the rescattering contributions, the decay amplitudes of eq.(11) are generalised to

$$M\left(K^\pm \to \pi^0\pi^0\pi^\pm\right) = \kappa\left[1 + ia \pm ia_{KM} + fY(1 + ib \pm ib_{KM}) + \ldots\right] . \tag{17}$$

In the decay distribution it is the interference between the CP-even and CP-odd imaginary amplitudes of (17) that gives the slope-parameter asymmetry as

$$\Delta g = \frac{g(K^+) - g(K^-)}{g(K^+) + g(K^-)} = \frac{a(b_{KM} - a_{KM})}{1 + ab} \equiv \frac{a d_{KM}}{1 + ab} . \tag{18}$$

Numerical values for the rescattering contributions are, $a = 0.12$ and $b = 0.49$.

5 Information from $K \to 2\pi$ decay

The isospin $I = 0, 2$ amplitudes in $K \to 2\pi$ decay as derived from the Lagrangian of eq.(3), *i.e.* ignoring strong final-state interactions, are

$$A_0 = \frac{G_F F_\pi}{\sqrt{2}} \sin\theta_C \cos\theta_C (m_K^2 - m_\pi^2)[c_0 + c_4] \ , \tag{19}$$

$$A_2 = \frac{G_F F_\pi}{\sqrt{2}} \sin\theta_C \cos\theta_C (m_K^2 - m_\pi^2)[c_4] \ . \tag{20}$$

Numerical values from data on $K \to 2\pi$ decay yields

$$c_0 = -10.46, \qquad c_4 = 0.328 \ .$$

Important for us is the CP-violating parameter

$$\epsilon' = i \exp^{i(\delta_2 - \delta_0)} \left[-\frac{\mathrm{Im} A_0}{\mathrm{Re} A_0} + \frac{\mathrm{Im} A_2}{\mathrm{Re} A_2} \right] \cdot \left| \frac{A_2}{A_0} \right| \ .$$

From numerical data on ϵ'

$$-\frac{\mathrm{Im}\tilde{c}_5}{\mathrm{Re}\tilde{c}_5} \left(1 - \Omega + 24.4 \frac{\mathrm{Im}\tilde{c}_7}{\mathrm{Im}\tilde{c}_5} \right) = (1.63 \pm 0.50) \cdot 10^{-4} \ . \tag{21}$$

Here, Ω is an isospin-breaking parameter which arises from $\pi - \eta$ mixing. The origin of the numerical parameter is

$$-24.4 = \frac{3(c_0 + c_4)}{16 c_4} \cdot \frac{\Lambda^2}{m_K^2 - m_\pi^2} \ .$$

6 Numerical result for Δg asymmetry

The CP-violating parameter d_{KM} of eq.(13) is calculated from the Lagrangian (3) with the substitutions of eqs (9) and (10). The result is

$$d_{KM} = \frac{16 c_4}{c_0 (c_0 - 9c_4/2)} \times \beta \mathrm{Im}\tilde{c}_5 \left(1 + 27.8 \frac{\mathrm{Im}\tilde{c}_7}{\mathrm{Im}\tilde{c}_5} \right) \tag{22}$$

$$27.8 = \frac{3\Lambda^2}{16 m_K^2 c_4} \left(c_0 - 9c_4 - \frac{c_0 m_\pi^2}{m_K^2 - m_\pi^2} \right) \tag{23}$$

From the numerical values of the coefficients c_0-c_4 we derive

$$\beta \mathrm{Re}\tilde{c}_5 = -2.04 \ .$$

Inserting this number into eq.(21) for ϵ', the following relation emerges

$$\beta \mathrm{Im}\tilde{c}_5 \left(1 + \frac{24.4}{1 - \Omega} \cdot \frac{\mathrm{Im}\tilde{c}_7}{\mathrm{Im}\tilde{c}_5} \right) = \frac{(3.32 \pm 0.50) \cdot 10^{-4}}{1 - \Omega} \ . \tag{24}$$

At the value $\Omega = 0.060 \pm 0.077$, as recently suggested [7], this combination turns into

$$\beta \mathrm{Im}\tilde{c}_5 \left(1 + (24.0 \div 28.2) \frac{\mathrm{Im}\tilde{c}_7}{\mathrm{Im}\tilde{c}_5} \right) = (3.56 \pm 0.61) \cdot 10^{-4} \ . \tag{25}$$

The linear combinations of QCDP and EWP contributions in eqs (22) and (25) are virtually indistinguishable. This leads to a particularly robust prediction for the slope-parameter asymmetry

$$\Delta g = (1.7 \pm 0.3) \cdot 10^{-6} \ .$$

This value is slightly larger than the one suggested in ref. [8] but substantially smaller than the one of ref. [9].

7 Conclusionns

The upcome or our analysis is, that we can with great confidence claim that CP violation in the decay $K^{\pm} \to \pi^0\pi^0\pi^{\pm}$ as manifested in the slope-parameter asymmetry Δg has a value in the range

$$\Delta g = \frac{g(K^+) - g(K^-)}{g(K^+) + g(K^-)} = (1.7 \pm 0.3) \cdot 10^{-6}$$

The basis for our claim is the fact that EWP and QCDP contributions appear in nearly identical linear combinations in ϵ' and Δg.

Acknowledgments

The work described above has been done in collaboration with Evgeny Shabalin. I am deeply indebted to him for years of tutelage and fruitful collaboration.

References

[1] A.A. Bel'kov *et al.*, Phys.Lett. **B300**, 283 (1993).
[2] J.M. Flynn and L. Randall, Phys.Lett. **B224**, 221 (1989); G. Buchalla *et al.*, Nucl.Phys. **B337**, 313 (1990); E.A. Paschos and Y.L. Wu, Mod. Phys.Lett. **A6**, 93 (1991).
[3] G. Fäldt and E. Shabalin, Physica Scripta **72**, 354 (2005).
[4] M.A. Shifman, A.I. Vainshtein and V.I. Zakharov, Zh.Eksp.Teor.Fiz. **72**, 1275 (1977); Sov.Phys.JETP **45**, 670 (1977).
[5] W.A. Bardeen, A.J. Buras and J.-M. Gérard, Nucl.Phys. **B293**, 787 (1987).
[6] J. Cronin, Phys.Rev. **161**, 1483 (1967).
[7] V. Cirigliano *et al.*, Eur.Phys.J. **C33**, 369 (2004).
[8] L. Maiani and N. Paver, In *The Second DAPhNE Physics Handbook*, Eds L. Maiani, G. Pancheri and N. Paver (INFN-LNF, 1995), p.51.
[9] I. Scimemi, E. Gamiz and J. Prades, hep-ph/0405204.

STUDY OF $K^- \to \pi^0 e^- \overline{\nu}_e \gamma$ and $K^- \to \pi^0 \mu^- \overline{\nu}_\mu \gamma$ DECAY WITH ISTRA+ SETUP

V.N. Bolotov[a], E.N. Guschin, V.A. Duk, S.V. Laptev, V.A. Lebedev, A.E. Mazurov, A.Yu. Polyarush, V.E. Postoev.
Institute for Nuclear Research, Moscow, Russia
S.A. Akimenko, G.I. Britvich, K.V. Datsko, A.P. Filin, A.V. Inyakin, V.F.Konstantinov, A.S.Konstantinov, I.Y. Korolkov, V.A.Khmelnikov, V.M. Leontiev, V.P. Novikov, V.F.Obraztsov, V.A.Polyakov, V.I. Romanovsky, V.I. Shelikhov, O.G. Tchikilev, V.A. Uvarov, O.P. Yushchenko
Institute for High Energy Physics, Protvino, Russia

Abstract. This file contains the instructions for the proceedings of the 12th Lomonosov Conference on Elementary Particle Physics. In this place the abstract of the contribution should be placed. Results of study of the $K^- \to \pi^0 e^- \overline{\nu}_e \gamma$ decay at ISTRA+ setup are presented. 3852 events of this decay have been observed. The ratio $Br(K^- \to \pi^0 e^- \overline{\nu}_e \gamma / Br(K^- \to \pi^0 e^- \overline{\nu}_e) = (0.63 \pm 0.02(stat) \pm 0.03(syst)) \cdot 10^{-2}$ for $E^*_\gamma > 30 MeV$, $\theta^*_{e\gamma} > 20°$. $Br(K^- \to \pi^0 e^- \overline{\nu}_e \gamma)$ is found to be $(3.05 \pm 0.02) \cdot 10^{-4}$ (assuming PDG value for K_{e3} branching ratio). Theoretical predictions give $Br = 2.8 \cdot 10^{-4}$ (tree level) and $Br = 3.0 \cdot 10^{-4} (O(p^4)$ level). The obtained value for the asymmetryA_ξ (with the same cuts forE^*_γ and $\theta^*_{e\gamma}$) is $A_\xi = -0.015 \pm 0.021$. At present it is the best estimate of this asymmetry.

1 Introduction

The decay $K^- \to \pi^0 e^- \overline{\nu}_e \gamma$ is good testing ground for the Chiral Pertubation Theory (ChPT) [1,2]. This decay amplitudes are calculated at order ChPT $O(p^4)$ in [1], and branching ratios are evaluated in [3]. Recently next-to-leading $O(p^6)$ corrections were calculated for the corresponding neutral kaon decay [4].

The $K^- \to \pi^0 e^- \overline{\nu}_e \gamma$ decay is one of kaon decays where new physics beyond the Standard Model can be probed. This decay is especially interesting as it is sensitive to T-odd contributions.

According to CPT-theorem observation of T violation is equivalent to observation of CP violating effects. CP violation is a subject of continuing interest in K and B meson decays. In the standard model the source of CP violation is given by the phase in the CKM matrix [5–7]. However it has been argued that this source is not enough to explain the observed baryon asymmetry of the universe and new sources of CP violation have to be introduced [8].

The matrix element for $K^- \to \pi^0 e^- \overline{\nu}_e \gamma$ has general structure

$$T = \frac{G_F}{\sqrt{2}} e V_{us} \varepsilon^\mu(q) \Big\{ (V_{\mu\nu} - A_{\mu\nu}) \overline{u}(p_\nu) \gamma^\nu (1-\gamma_5) v(p_l) + \tag{1}$$

$$+ \frac{F_\nu}{2 p_l q} \overline{u}(p_\nu) \gamma^\nu (1-\gamma_5)(m_l - \not{p}_l - \not{q}) \gamma_\mu v(p_l) \Big\} \equiv \epsilon^\mu A_\mu .$$

First term of the matrix element describes Bremsstrahlung of kaon and direct emission(Fig.1a). The lepton Bremsstrahlung is presented by second part of Eq(1) and (Fig.1b).

The $K^- \to \pi^0 e^- \overline{\nu}_e \gamma$ decay is one of kaon decays where new physics beyond the Standard Model can be probed. This decay is especially interesting as it is sensitive to

[a]E-mail: bolotovvn@mail.ru, bolotov@inr.ru

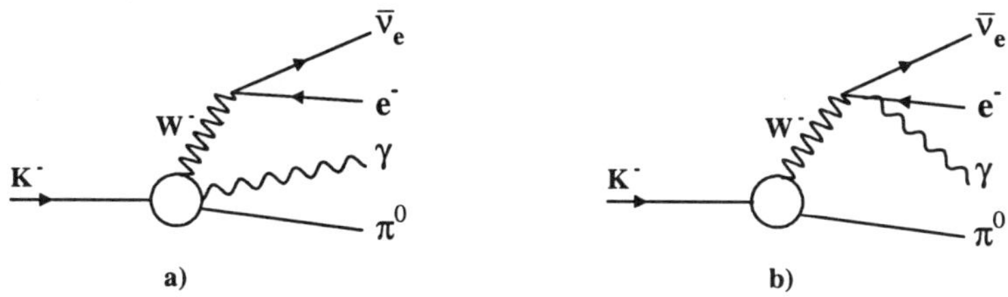

Figure 1: *Diagrammatic repressentation of the $K_{l3\gamma}$ amplitude.*

T-odd contributions. According to CPT-theorem observation of T violation is equivalent to observation of CP-violating effects. CP violation is a subject of continuing interest in K and B meson decays. In the Standard Model the source of CP violation is given by the phase in the CKM matrix [5–7]. However it has been argued that this source is not enough to explain the observed baryon asymmetry of the Universe and new sources of CP violation have to be introduced [8].

Important experimental observable used in CP-violation searches is the T-odd correlation, for $K^- \to \pi^0 e^- \overline{\nu}_e \gamma$ decay it is defined as

$$\xi_{\pi e \gamma} = \frac{1}{M_K^3} p_\gamma \cdot [p_\pi \times p_e] \tag{2}$$

First suggestion to investigate T-odd triple-product correlations was done in [9]

To establish the presence of a nonzero triple-product correlations, one constructs a T-odd asymmetry of the form

$$A_\xi = \frac{N_+ - N_-}{N_+ + N_-} \tag{3}$$

Where N_+ and N_- are number of events with $\xi > 0$ and $\xi < 0$

T-odd correlation vanishes at tree level of SM [10], but the SUSY theory gives rise to CP-odd(T-odd) observables already at tree level [11–13]. T-odd asymmetry value for $SU(2)_L \times SU(2)_R \times U(1)$ model and scalar models was estimated in Ref [14]. In this letter we present first results of the analysis of the $K^- \to \pi^0 e^- \overline{\nu}_e \gamma$ data accumulated by ISTRA+ experiment during the 2001 run.

2 ISTRA+ setup

The experiment was performed using ISTRA+ detector which is modernized ISTRA-M detector [15]. ISTRA+ detector is located in a negative unseparated beam. The measurement of the beam particles, deflected by the beam magnet M1 is performed by four beam proportinal chambers $BPC1 \div BPC4$. The beam momentum is $\sim 25 GeV$ with $\Delta p/p \sim 1.5\%$. Admixture of K^- in the beam is $\sim 3\%$. The beam intensity is $\sim 3 \cdot 10^6$ per 1.9 sec U-70 spill. The kaon identification is performed by $\check{C}1 \div \check{C}4$ threshold Cherenkov counters.

The decay products are deflected by the spectrometer magnet M2 with the field integral of 1Tm. The track measurement is performed by 1-mm-step proportional chambers ($PC1 \div PC3$), 2-cm-cell drift chambers ($DC1 \div DC3$), and by four planes of the 2-cm-diameter drift tubes DT. The photons are measured by lead-glass electromagnetic calorimeter $SP1$ which consists of 576 counters. The counter transverse size is 5.2×5.2 cm and length is about 15 X_0. To veto low energy photons the decay

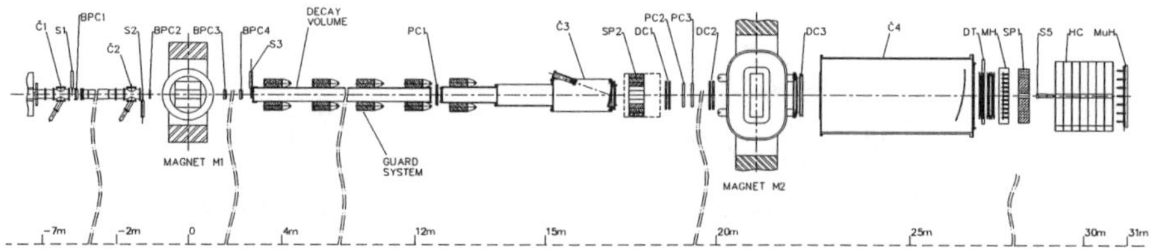

Figure 2: *Plan view of the ISTRA+ detector. S1÷ S5 - scintillation counters; C1 ÷ C4 Cherenkov gas counters; M1, M2 - beam and spectrometer magnets; PC1÷ PC6 - proportional chambers; DV - vacuum decay volume; GS - lead glass guard system;DC-1-16- drift chambers; EC1 and EC2 - lead glass Cherenkov calorimeters; DT - drift tubes; MH - matrix hodoscope; HC - hadron calorimeter*

volume is surrounded by eight lead-glass rings.Lead-glass electromagnetic calorimeter $SP2$ is also used as a part of the veto system.

3 Event selection

During physics run in November-December 2001 350M events were logged on tapes. This information is complemented by 260M events generated with Geant3 [16]. The Monte Carlo simulation includes a realistic description of the experimental setup. The detailed discussion of the simulation and reconstruction procedure is given in our previous publications [17, 18].

Events with one negative track detected in tracking system and four showers detected in electoromagnetic calorimeter SP1 are selected as candidates for $K^- \to \pi^0 e^- \overline{\nu}_e \gamma$ decay. One of this showers must be associated with the charged track. Events with vertex inside interval $400 < z < 1650$ cm, and transverse radius less than 10cm is selected for furser analysis.

The probability of the vertex fit, $CL(\chi^2)$, is required to be more than 10^{-4}. Absence of signals in veto system above noise threshold is required.

The electron identification is done using E/P ratio of the energy of the cluster associated with the track to momentum of this track given by tracking system. This ratio must be inside interval 0.80-1.15(see Fig.3). Another cut used for the suppression of the π^- contamination is that on the distance between the charged track extrapolation to the front plane of the electoromagnetic detector and the nearest shower. This distance must be less than 2,5 cm.

The effective mass $m(\gamma\gamma)$ within ± 30 MeV from π^0 table mass (Fig.4) is required. At the end, the convergence of the 2C $K^- \to \pi^0 e^- \overline{\nu}_e \gamma$ kinematic fit is required.

4 Background suppression

The main background decay channels for the decay $K^- \to \pi^0 e^- \overline{\nu}_e \gamma$ are:

(1) $K^- \to \pi^- \pi^0 \pi^0$ where one of the π^0 photons is not detected and π^- decays to $e^- \nu_e$ or is misidentified as an electron.

(2) $K^- \to \pi^- \pi^0$ with "fake photon" and π^- decayed or misidentified as electron. Fake photon clusters can come from π-hadron interaction in the detector, external bremsstrahlung upstream of the magnet, accidentals. All these sources are included in our MC calculations.

(3) $K^- \to \pi^0 e^- \nu_e$ with extra photon. The main source of extra photons is an electron interactions in the detector.

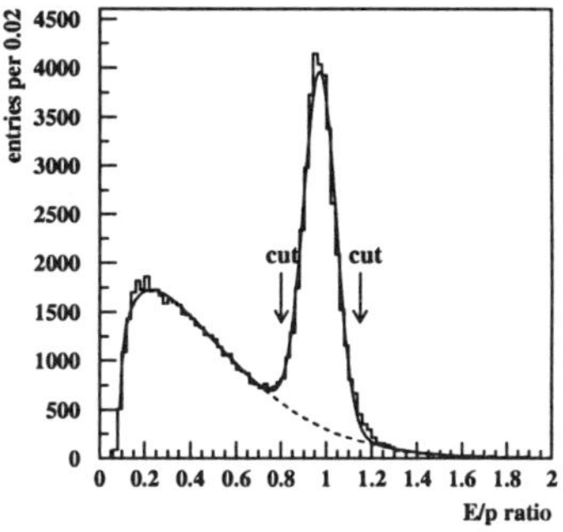

Figure 3: *E/P ratio for the real data. Dotted line is our fit of background.*

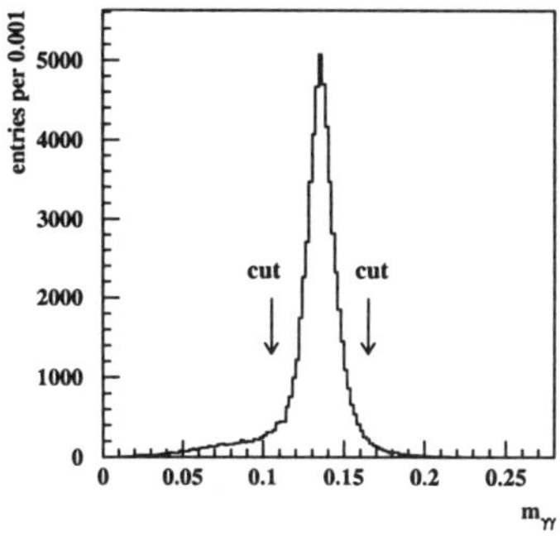

Figure 4: *$\gamma\gamma$ mass for the real data.*

(4) $K^- \to \pi^- \pi^0 \gamma$ when π^- decays or is mis-identified as an electron.

(5) $K^- \to \pi^0 \pi^0 e^- \nu_e$ when one γ is lost

From Fig.3 it is seen that in raw data background contamination from channels with charged pion in final state is about 15%.

Requirement on the missing energy in the decay reduces mainly background (4).

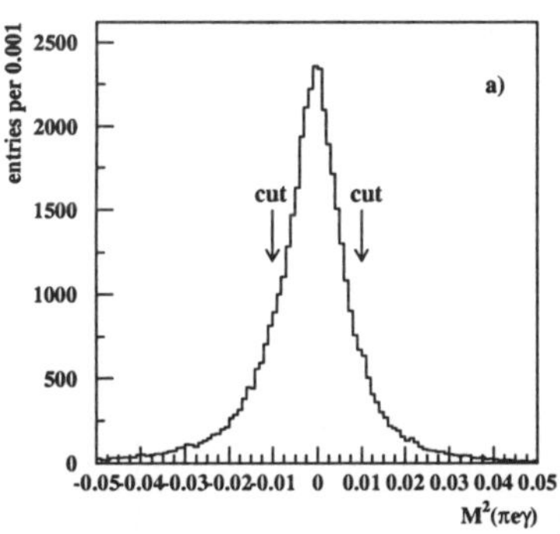

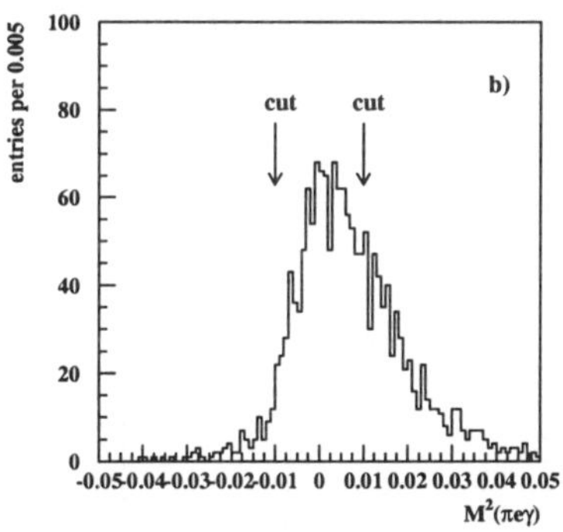

Figure 5: *Missing mass $M^2(\pi^0 e^- \gamma)$ distribution; a) for the real data b) for the background (1).*

Cut1: $E_{miss} > 1.0$ GeV. For the suppression of the backgrounds (1-5) we use a cut on the missing mass squared $M^2(\pi^0 e\gamma) = (P_K - P_{\pi^0} - P_e - P_\gamma)^2$.

For the signal events this variable corresponds to the square of the neutrino mass and must be zero within measurement accuracy (see Fig.5).

Cut	real data	background	signal MC
Number of events selected	40403	24395	9632
$E_{miss} > 1.0GeV$	33436	21837	8646
$-0.01 < M^2(\pi^0 e^- \gamma) < 0.01$	23115	16239	7275
$M^2(\pi^-\pi^0) < 0.009, M^2(\pi^-\pi^0) > 0.024$	20840	14699	6155
$0.002 < \theta_{e\gamma} < 0.030$	5378	1526	3852

Table 1: *Event reduction statistics for the real data, the background MC and signal MC.*

Cut2: $-0.01 < M^2(\pi^0 e^- \gamma) < 0.01$. For the suppression of the background(1) we also use a cut on the missing mass squared $M^2(\pi^-\pi^0) = (P_K - P_{\pi^-} - P_{\pi^0})^2$.

For the background(1) events this variable corresponds to π^0 mass, for the signal events distribution of this variable is rather wide (see Fig.6).

Cut3: The events with $0.009 < M^2(\pi^-\pi^0) < 0.024$ are cuted out.

The dominant background to $K_{e3\gamma}$ arises from K_{e3} with extra photon. The background (3) is suppressed by requirement on the angle between electron and photon in the laboratory frame $\theta_{e\gamma}$ (see Fig.7) The distribution of the K_{e3}-background events has very sharp peak at zero angle. This peak is significantly narrower than that for signal events. This happens, in particular, because the emission of the photons by the electron from K_{e3} decay occures in the setup material downstream the decay vertex, but angle is still calculated as if emission comes from the vertex.

Cut4: $0.002 < \theta_{e\gamma} < 0.030$

Right part of this cut is introduced for suppression of backgrounds (1,2,4,5). After all cuts 5378 event are selected, with a background of 1526 events. Background normalization was done by comparison numbers of events for K_{e3} decay in MC and real data samples.

Event reduction statistics is summarized in Table 1.

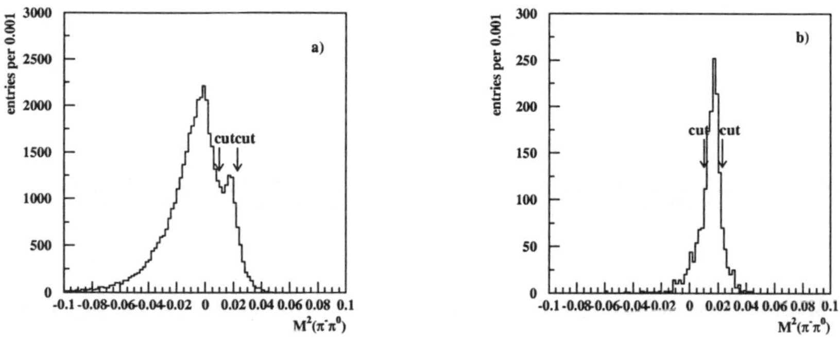

Figure 6: *missing mass distribution $M^2(\pi^-\pi^0)$ a) for real data;b)for the background(1)*

For study of $K^- \to \pi^0 \mu^- \overline{\nu}_\mu \gamma$ decay were used the same selection and background supression procedure as for $K^- \to \pi^0 e^- \overline{\nu}_e \gamma$ decay. Naturely elecrtrons were changed muons by comparision of signals from electromagnetic SP1 and hadron HC calorimeters. Main mass distribution is shown on fig.9

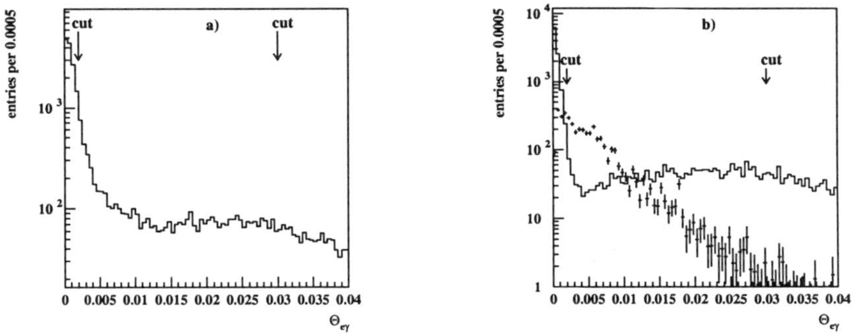

Figure 7: *Distribution over $\theta_{e\gamma}$ - the angle between electron and photon in lab. system. a) real data; b)MC background (histogram) and signal(points with errors)*

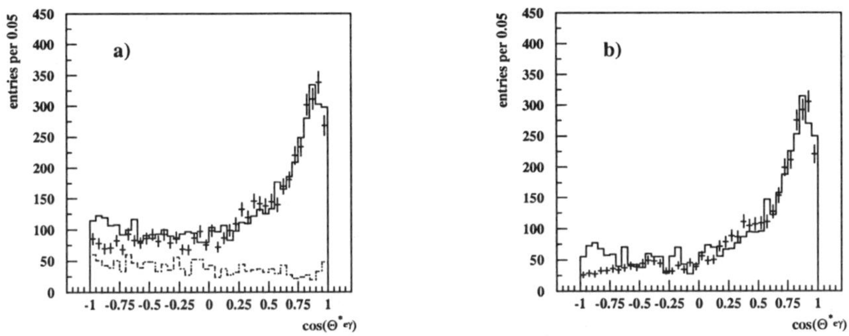

Figure 8: *a) The distribution of the events over $cos\theta^*_{e\gamma}$. Histogram is the real data, points with errors - total MC signal plus background. Dotted line histogram is background. b)the same after background subtraction*

5 Results

The resulting distribution of the selected events over $cos\theta^*{}_{e\gamma}$, $\theta^*{}_{e\gamma}$ being the angle between the electron and the photon in the kaon rest frame, is shown in Fig.8. The distribution over $E^*{}_{e\gamma}$ - the photon energy in the kaon rest frame is shown in Fig. 10. Reasonable agreement of the date with MC is seen. When generating the signal MC, a generator based on $O(p^2)$ [10] is used. A small discrapancy of data and MC in the region of $cos\theta^*{}_{e\gamma} \sim$ -1.0 could be possibly interpreted as a direct emission effect.

To obtain the branching ratio of the $K_{\pi^0 e^- \overline{\nu}_e \gamma}$ relative to the K_{e3} (R), the background and efficiency corrected number of $K_{e3\gamma}$ events is compared to that of 524751 K_{e3} events found with the similar selection criteria. The branching ratio (R) for the cuts

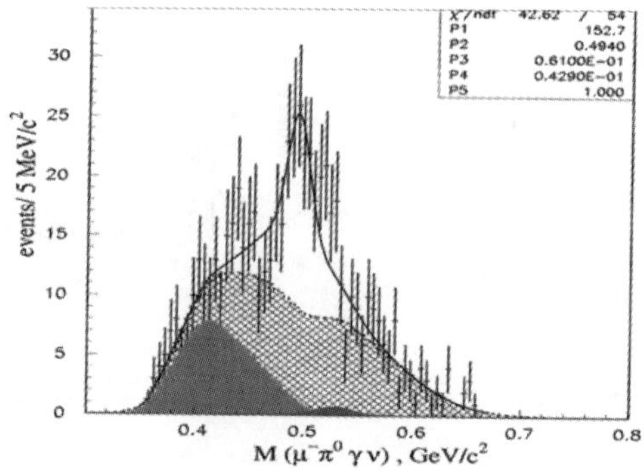

Figure 9: *Effective mass $M(K^- \rightarrow \pi^0\mu^-\overline{\nu}_\mu\gamma)$ spectrum for $30 < E^*_\gamma < 60MeV$. Cross-hatched area shows background. Solid area, left (red) and right (blue) show K_{e3} and $K_{\mu3}$ contributions respectively.*

$E^*_\gamma > 10MeV$ and $\theta^*_{e\gamma} > 10°$ is found to be

$$R = \frac{Br(K^- \rightarrow \pi^0 e^- \overline{\nu}_e \gamma)}{Br(K^- \rightarrow \pi^0 e^- \overline{\nu}_e)} = (1.69 \pm 0.03(stat) \pm 0.07(syst)) \cdot 10^{-2} \qquad (4)$$

Systematic errors are estimated by variation of the cuts of Table 1.

For comparison with previous experiment the branching ratio with cuts $E^*_\gamma >$ 10MeV, $0.6 < cos\theta^*_{e\gamma} < 0.9$ is calculated

$$R = \frac{Br(K^- \rightarrow \pi^0 e^- \overline{\nu}_e \gamma)}{Br(K^- \rightarrow \pi^0 e^- \overline{\nu}_e)} = (0.48 \pm 0.02(stat) \pm 0.03(syst)) \cdot 10^{-2} \qquad (5)$$

The results of previous experiments are given in Table.2
For the cuts $E^*_\gamma > 30MeV$ and $\theta^*_{e\gamma} > 20°$, used in most theoretical papers

$$R = \frac{Br(K^- \rightarrow \pi^0 e^- \overline{\nu}_e \gamma)}{Br(K^- \rightarrow \pi^0 e^- \overline{\nu}_e)} = (0.63 \pm 0.02(stat) \pm 0.03(syst)) \cdot 10^{-2}. \qquad (6)$$

$R_{exp} \times 10^2$	ev numb	experiment
0.48 ± 0.02	1423	this exp.
0.46 ± 0.08	82	XEBC [19]
0.56 ± 0.04	192	ISTRA [20]
0.76 ± 0.28	13	HLBC [21]

Table 2: *$Br(K^- \rightarrow \pi^0 e^- \overline{\nu}_e \gamma)/Br(K^- \rightarrow \pi^0 e^- \overline{\nu}_e)$ for $E^*_\gamma > 10MeV, 0.6 < cos\theta^*_{e\gamma} < 0.9$ in comparison with previous data.*

Using PDG value for K_{e3} decay branchig for $K^- \rightarrow \pi^0 e^- \overline{\nu}_e \gamma$ is calculated $Br = (3.05 \pm 0.02) \cdot 10^{-4}$. It can be compared with theoretical prediction [3] at tree level $Br = 2.8 \cdot 10^{-4}$ and $Br = 3.0 \cdot 10^{-4}$ for $O(p^4)$ level. Theoretical prediction of V.V.Braguta, A.A.Likhoded, A.E.Chalov [10] at tree level is $Br = 3.12 \cdot 10^{-4}$.

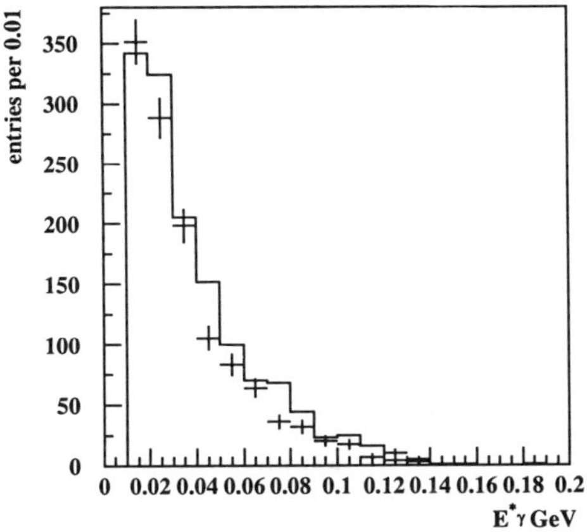

Figure 10: *The distribution of the events over E^*_γ- the energy of the photon in the kaon rest frame. Histogram corresponds to the data, points with errors- total signal plus background MC.*

For the asymmetry A_ξ(for the same cuts as in Table.2) we get

$$A_\xi = -0.015 \pm 0.021 \tag{7}$$

At present it is the best estimate of this asymmetry. It can be compared with an upper limit on the A_ξ value $|A_\xi(K^- \to \pi^0 e^- \overline{\nu}_e \gamma)| < 0.8 \cdot 10^{-4}$ in the $SU(2)_L \times SU(2)_R \times U(1)$ model [14] and $A_\xi = -0.59 \cdot 10^{-4}$ in the Standard Model [10]

6 Results $K^- \to \pi^0 \mu^- \overline{\nu}_\mu \gamma$ decay.

1. First observation of the radiative kaon decay $K^- \to \pi^0 \mu^- \overline{\nu}_\mu \gamma$ is presented.

2. R =Br$(K^- \to \pi^0 \mu^- \overline{\nu}_\mu \gamma)/Br(K^- \to \pi^0 \mu^- \overline{\nu}_\mu) = 0.270 \pm 0.029(stat) \pm 0.026(syst)\%$ for region $K^- \to \pi^0 e^- \overline{\nu}_e \gamma \quad 5 < E^*_\gamma < 30 MeV$.

3. Br$(K^- \to \pi^0 \mu^- \overline{\nu}_\mu \gamma) = (4.48 \pm 0.68(stat) \pm 0.99(syst)) \cdot 10^{-4}$ for region $30 < E^*_\gamma < 60 MeV$ .Theoretical prediction is $4.67 \cdot 10^{-4}$.

4. Asymmetry in the T-odd variable ξ for the region $5 < E^*_\gamma < 30 MeV \quad A(\xi) = -0.03 \pm 0.13$.

Acknowledgments

The authors would like to thank D.S. Gorbunov, V.A. Matveev and V.A. Rubakov, for numerous discussions. V.V.Braguta, A.A.Likhoded, A.E.Chalov for program of matrix element calculation. The work is supported in part by the RFBR grants N03-02-16330 (IHEP group) and N03-02-16135 (INR group).

References

[1] J. Bijnens, G. Echer and J. Gasser, Nucl.Phys. ***B*396**, 81 (1993);
[2] A. Pitch, Rep. Prog. Phys. **58**, 563(1995);
[3] L.Maiani, G.Pancheri and N.Paver, The Second DAFNE Physics Handbook, (INFN-LNF-Divisione Ricerca, SIS-Ufficio Pubblicazioni, Frascati (Roma) Italy, ISBN 88-86409-02-8).
[4] J. Gasser et.al ., arXiv:hep-ph/0412130
[5] Cabibo N. Phys.Rev.Lett **10**, 531(1963).
[6] Kobayashi M., Maskawa T. Progr.Theor. Phys. **49**, 652 (1973).
[7] C. Jarlskog Z.Phys. ***C*29**, 491 (1985).
[8] G.F.Farrar and M.E.Shaposhnikov, Phys.Rev.Lett **70**, 2833(1993) [Erratum ibid 71(1993)210][arXiv:hep-ph/9305274]; P.Het and Sather, Phys.Rev D51, 379(1995) [arXiv:hep-ph/9404302]; M.Carena, M.Quiros and C.E.Wagner, Phys.Lett.***B*380**, 81(1996). [arXiv:hep-ph/9303420]
[9] J.Gevas, J.Iliopolus, J.Kaplan Phys. Lett. **20**, 432 (1966).
[10] V.V.Braguta, A.A.Likhoded, A.E.Chalov, Phys. Rev. ***D*65**, 054038(2002) [arXiv:hep-ph/0106147].
[11] Y.Kuzuruki Phys. Lett., ***B*193**, 339(1987).
[12] A. Bartl, T.Kernreiter and W. Porod Phys Lett., ***B*538**, 59(2002).
[13] N. Oshimo Mod. Phys. Lett. ***A*4**, 145(1989).
[14] V.V.Braguta, A.A.Likhoded, A.E.Chalov Phys.Atom.Nucl.67, 1003-1009(2004), Yad.Fiz. **67**,1025 (2004) [arXiv:hep-ph/0305067]
[15] V.N.Bolotov et.al ., IHEP preprint 8-98, 1998.
[16] R.Brun et al ., Preprint CERN-DD/EE/84-1.
[17] I.V.Ajinenko et al ., Yad. Fiz **67**, 2125 (2002); I.V.Ajinenko et al ., Phys. At. Nucl., **66**, 2064 (2003); I.V.Ajinenko et al ., Phys Lett., ***B*574**, 14 (2003); O.P.Yuschenko et al ., Phys Lett., ***B*589**, 111 (2004).
[18] I.V.Ajinenko et al ., Yad. Fiz **66**, 107 (2003); I.V.Ajinenko et al ., Phys. At. Nucl., **66**, 105 (2003); O.P.Yuschenko et al ., Phys Lett., ***B*581**, 159 (2004).
[19] V.V.Barmin et.al ., SJNP **55**, 547 (1991), **533**, 976 (1991).
[20] V.N.Bolotov et.al ., JETPL **42**, 68 (1986), Yad. Fiz **44**, 108(1986).
[21] F.Romano et.al ., Phys Lett., **36*B***, 525 (1971).

RARE SEMILEPTONIC MESON DECAYS IN $\not R$-PARITY VIOLATING MSSM

A. Ali [a]
Deutsches Electronen-Synchrotron, DESY, 22607 Hamburg, Germany
A.V. Borisov [b], M.V. Sidorova [c]
Faculty of Physics, Moscow State University, 119992 Moscow, Russia

Abstract. We discuss rare meson decays $K^+ \to \pi^- \ell^+ \ell'^+$ and $D^+ \to K^- \ell^+ \ell'^+$ ($\ell, \ell' = e, \mu$) in a supersymmetric extension of the Standard Model with $\not R$-parity violation. Estimates of the branching ratios for these decays are presented.

In the standard model (SM), lepton L and baryon B number conservation take place due to the accidental $U(1)_L \times U(1)_B$ symmetry existing at the level of renormalizable operators. But, for many extensions of the SM, this is not the case. The well known mechanism of lepton number (LN) violation is based on the mixing of massive Majorana neutrinos predicted by various Grand unified theories (GUTs) [1]. The Majorana mass term violates LN by $\Delta L = \pm 2$ [2] and can lead to a large number of LN violating processes. Among them, the most sensitive to the LN violation are processes such as neutrinoless double beta decay $(A, Z) \to (A, Z+2) + e^- + e^-$ (for a recent review, see [3]), rare meson decays (see, e.g., [4,5])

$$M^+ \to M'^- \ell^+ \ell'^+ \,, \tag{1}$$

and like-sign dilepton production in high-energy hadron-hadron and lepton-hadron collisions (see, e.g., the papers and references therein: $pp \to \ell^\pm \ell'^\pm X$ [4, 6], $e^+ p \to \bar{\nu}_e \ell^+ \ell'^+ X$ [7]), which have been extensively studied in the literature.

There exists now convincing evidence for oscillations of solar, atmospheric, reactor, and accelerator neutrinos [8]. The oscillations, i.e., periodic neutrino flavor changes, imply that neutrinos have nonzero masses and they mix among each other: neutrinos ν_ℓ of specific flavors $\ell = e, \mu, \tau$ are the coherent superposition of the neutrino mass eigen-states ν_N of masses m_N,

$$\nu_\ell = \sum\nolimits_N U_{\ell N} \nu_N. \tag{2}$$

Here the coefficients $U_{\ell N}$ are elements of the unitary leptonic mixing matrix – the PMNS matrix [9].

Neutrino flavor changes imply lepton family number L_ℓ nonconservation admissible for neutrinos of both types, Dirac and Majorana, but for the Dirac neutrinos, in contrast to the Majorana ones, the total lepton number $L = \sum_\ell L_\ell$ is conserved. The nature of the neutrino mass is one of the main unsolved problems in particle physics. However, oscillation experiments can not distinguish between the two types of neutrinos.

In Refs. [4,5] we investigated the rare decays (1) of the pseudoscalar mesons $M = K, D, D_s, B$, mediated by light ($m_N \ll m_\ell,\ m_{\ell'}$) and heavy ($m_N \gg m_M$) Majorana neutrinos. It was shown that the present direct experimental bounds on the branching ratios (BRs) are too weak to set reasonable limits on the effective Majorana masses. Taking into account the limits on lepton mixing

[a] e-mail: ahmed.ali@desy.de
[b] e-mail: borisov@phys.msu.ru
[c] e-mail: mvsid@rambler.ru

and neutrino masses obtained from the precision electroweak measurements, neutrino oscillations, cosmological data and searches of the neutrinoless double beta decay, we have derived the indirect upper bounds on the BRs that are greatly more stringent than the direct ones.

In this report, we investigate another mechanism of the $\Delta L = 2$ rare decays (1) based on R-parity violating supersymmetry (SUSY) (for a review see [10]). We recall that R-parity is defined as $R = (-1)^{3(B-L)+2S}$, where S, L, and B are the spin, the lepton and baryon numbers, respectively. In the minimal supersymmetric extension of the SM (MSSM), R-parity conservation is imposed to prevent the L and B violation; it also leads to the production of superpartners in pairs and ensures the stability of the lightest superparticle. However, neither gauge invariance nor supersymmetry require R-parity conservation. There are many generalizations of the MSSM with explicitly or spontaneously broken R-symmetry [10]. We consider a SUSY theory with the minimal particle content of the MSSM and explicit R-parity violation ($\not{R}$MSSM).

The most general form for the R-parity and lepton number violating part of the superpotential is given by [10, 11]

$$W_{\not{R}} = \varepsilon_{\alpha\beta}\left(\frac{1}{2}\lambda_{ijk}L_i^\alpha L_j^\beta \bar{E}_k + \lambda'_{ijk}L_i^\alpha Q_j^\beta \bar{D}_k + \epsilon_i L_i^\alpha H_u^\beta\right). \tag{3}$$

Here $i, j, k = 1, 2, 3$ are generation indices, L and Q are $SU(2)$ doublets of left-handed lepton and quark superfields ($\alpha, \beta = 1, 2$ are isospinor indices), $\bar{E}$ and $\bar{D}$ are singlets of right-handed superfields of leptons and down quarks, respectively; H_u is a doublet Higgs superfield (with hypercharge $Y = 1$); $\lambda_{ijk} = -\lambda_{jik}$, λ'_{ijk} and ϵ_i are constants.

In the superpotential (3) the trilinear ($\propto \lambda,\ \lambda'$) and bilinear ($\propto \epsilon$) terms are present. In this work, we assume that the bilinear terms are absent at tree level ($\epsilon = 0$). They will be generated by quantum corrections [10], but it is expected that the phenomenology will still be dominated by the tree-level trilinear terms.

At first we consider the rare decay $K^+(P) \to \pi^-(P') + \ell^+(p) + \ell'^+(p')$ in the $\not{R}$MSSM (a rough estimate of the width of the decay $\mathrm{B}(K^+ \to \pi^-\mu^+\mu^+)$ in the same theory was obtained in [12]). The leading order amplitude of the process is described by three types of diagrams shown in Fig. 1.

The hadronic parts of the decay amplitude are calculated with the use of a model for the Bethe–Salpeter amplitudes for mesons [13],

$$\chi_P(q) = \gamma^5(1 - \delta_M \not{P})\varphi_P(q), \tag{4}$$

where $\delta_M = (m_1 + m_2)/m_M^2$, m_M is the mass of the meson composed of a quark q_1 and an antiquark $\bar{q}_2$ with the current masses m_1 and m_2, $P = p_1 + p_2$ is the total 4-momentum of the meson, $q = (p_1 - p_2)/2$ is the quark-antiquark relative 4-momentum; $\varphi_P(q)$ is the model-dependent scalar function.

For all mesons in question, $m_M \ll m_{SUSY}$, where $m_{SUSY} \gtrsim 100$ GeV is the common mass scale of superpartners, and for heavy Majorana neutrinos, $m_N \gg m_M$ (the contribution of light neutrinos is strongly suppressed by phenomenology [4, 5]), we can neglect momentum dependence in the propagators (see Fig. 1) and use the effective low-energy current-current interaction. In this approximation the decay amplitude does not depend on the specific form of the functions $\varphi_P(q)$ (see Eq. (4)) and is expressed through the known decay constants of the mesons, f_M, as

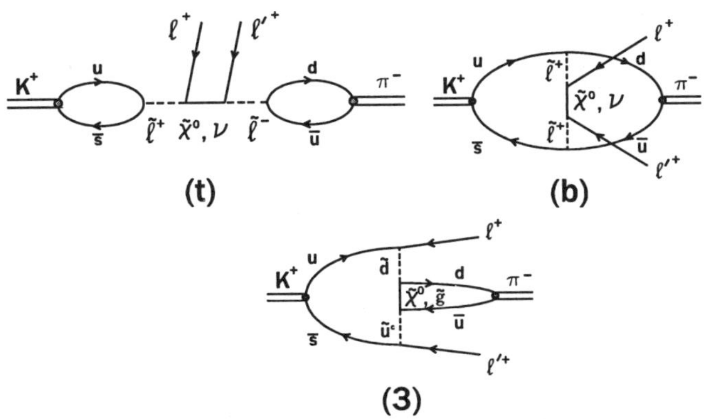

Figure 1: Feynman diagrams for the decay $K^+ \to \pi^- + \ell^+ + \ell'^+$ mediated by Majorana neutrinos ν, neutralinos $\tilde{\chi}^0$, gluinos $\tilde{g}$ with $\tilde{f}$ being the scalar superpartners of the corresponding fermions $f = \ell, u, d$ (leptons and quarks). Bold vertices correspond to Bethe–Salpeter amplitudes for mesons as bound states of a quark and an antiquark. There are also crossed diagrams with interchanged lepton lines.

$$f_M = 4\sqrt{N_c}\,\delta_M\,(2\pi)^{-4}\int d^4q\,\varphi_P(q)\,,$$

where $N_c = 3$ is the number of colors.

For the total width of the decay we obtain

$$\Gamma(K^+ \to \pi^-\ell^+\ell'^+) = \left(1 - \tfrac{1}{2}\delta_{\ell\ell'}\right)\frac{f_K^2 f_\pi^2 m_K^3}{2^{12}\pi^3\delta_K^2\delta_\pi^2}\Phi_{\ell\ell'}$$

$$\times\Bigg|\sum_{i,j,k,k',N}(\lambda^*_{ik\ell}\lambda^*_{jk'\ell'} + \lambda^*_{ik\ell'}\lambda^*_{jk'\ell})\frac{\lambda'_{k12}\lambda'_{k'11}}{m^2_{\tilde{\ell}_{Lk}}\,m^2_{\tilde{\ell}_{Lk'}}}\frac{U_{iN}U_{jN}}{m_N}\left(1 - \frac{1}{2N_c}\right)$$

$$+ (\lambda'_{\ell 11}\lambda'_{\ell' 12} + \lambda'_{\ell' 11}\lambda'_{\ell 12})\Bigg[g_2^2\sum_{\delta=1}^{4}\frac{1}{m_{\tilde{\chi}_\delta}}\Bigg(2\frac{\epsilon^*_{L\delta}(\ell)\epsilon^*_{L\delta}(\ell')}{m^2_{\tilde{\ell}L}\,m^2_{\tilde{\ell}'L}}\left(1 - \frac{1}{2N_c}\right)$$

$$- \frac{1}{N_c}\frac{\epsilon_{R\delta}(d)\epsilon^*_{L\delta}(u)}{m^2_{\tilde{d}_R}\,m^2_{\tilde{u}_L}}\Bigg) + \frac{4g_3^2}{N_c^2}\frac{1}{m^2_{\tilde{d}_R}\,m^2_{\tilde{u}_L}\,m_{\tilde{g}}}\Bigg]\Bigg|^2. \qquad (5)$$

Here $\Phi_{\ell\ell'}$ is the reduced phase space integral ($z = (P - P')^2/m_K^2$):

$$\Phi_{\ell\ell'} = \int_{l_+}^{h_-} dz\left[1 - (l_+ + l_-)(2z)^{-1}\right]\left[(h_+ - z)(h_- - z)(l_+ - z)(l_- - z)\right]^{1/2},$$

and the various parameters are defined as follows:

$$h_\pm = (1 \pm m_\pi/m_K)^2\,,\; l_\pm = [(m_\ell \pm m_{\ell'})/m_K]^2\,;\; \epsilon_{R\delta}(\psi) = Q(\psi)\tan\theta_W N_{\delta 1}\,,$$

$$\epsilon_{L\delta}(\psi) = -T_3(\psi)N_{\delta 2} + \tan\theta_W(T_3(\psi) - Q(\psi))N_{\delta 1}\,,$$

where $Q(\psi)$ and $T_3(\psi)$ are the electric charge and the third component of the weak isospin for the field ψ, respectively, and $N_{\delta\sigma}$ is the 4×4 neutralino mixing matrix. For the numerical estimates of the branching ratios, $B_{\ell\ell'} = \Gamma\left(M^+ \to M'^-\ell^+\ell'^+\right)/\Gamma_{\text{total}}$, we have used the known values for the couplings, decay constants, meson, lepton and current quark masses [5,8], and a typical set of the matrix elements $N_{\delta 1}, N_{\delta 2}$ from Ref. [14]. In addition, we have taken all the masses of superpartners to be equal with a common value m_{SUSY}. Taking into account the present bounds on the effective inverse Majorana masses [5], we find that the main contribution to the decay width comes from the exchange by neutralinos and gluinos (see Fig. 1). The results of the calculations with the use of Eq. (5) for the decays $K^+ \to \pi^- \ell^+ \ell^+$, and an analogous formula for the decays $D^+ \to K^- \ell^+ \ell'^+$, are shown in the fourth column of Table 1 (here $m_{200} = m_{SUSY}/(200\text{GeV})$). In the second and third columns of this table, the present direct experimental upper bounds on the BRs [8] and the indirect bounds for the Majorana mechanism of the rare decays [5] are shown, respectively. Our result for the $\mathrm{B}(K^+ \to \pi^- \mu^+ \mu^+)$ is in agreement with a rough estimate of Ref. [12].

To calculate the upper bounds on the BRs in the $\not{R}$MSSM, we take $m_{200} = 1$ and $|\lambda'_{ijk}\lambda'_{i'j'k'}| \lesssim 10^{-3}$ [15]. It yields

$$\mathrm{B}\left(K^+ \to \pi^- \ell^+ \ell'^+\right) \lesssim 10^{-23}, \ \mathrm{B}\left(D^+ \to K^- \ell^+ \ell'^+\right) \lesssim 10^{-24}.$$

These estimates are much smaller than the corresponding direct experimental bounds but are close (except for the *ee* decay mode) to the indirect bounds based on the Majorana mechanism of the decays (see Table 1).

Table 1: The branching ratios $\mathrm{B}_{\ell\ell'}$ for the rare meson decays $M^+ \to M'^- \ell^+ \ell'^+$.

Rare decay	Exp. upper bound on $\mathrm{B}_{\ell\ell'}$	Ind. bound on $\mathrm{B}_{\ell\ell'}$ (ν_MSM)	$\mathrm{B}_{\ell\ell'} \times m_{200}^{10}$ ($\not{R}$MSSM)
$K^+ \to \pi^- e^+ e^+$	6.4×10^{-10}	5.9×10^{-32}	$5.7 \times 10^{-17} \lvert\lambda'_{111}\lambda'_{112}\rvert^2$
$K^+ \to \pi^- \mu^+ \mu^+$	3.0×10^{-9}	1.1×10^{-24}	$2.0 \times 10^{-17} \lvert\lambda'_{211}\lambda'_{212}\rvert^2$
$K^+ \to \pi^- e^+ \mu^+$	5.0×10^{-10}	5.1×10^{-24}	$1.9 \times 10^{-17} \lvert\lambda'_{111}\lambda'_{212} + \lambda'_{211}\lambda'_{112}\rvert^2$
$D^+ \to K^- e^+ e^+$	1.2×10^{-4}	1.5×10^{-31}	$1.0 \times 10^{-18} \lvert\lambda'_{122}\lambda'_{111} + 0.45\lambda'_{121}\lambda'_{112}\rvert^2$
$D^+ \to K^- \mu^+ \mu^+$	1.3×10^{-5}	8.9×10^{-24}	$9.6 \times 10^{-19} \lvert\lambda'_{222}\lambda'_{211} + 0.45\lambda'_{221}\lambda'_{212}\rvert^2$
$D^+ \to K^- e^+ \mu^+$	1.3×10^{-4}	2.1×10^{-23}	$4.9 \times 10^{-19} \lvert(\lambda'_{122}\lambda'_{211} + \lambda'_{222}\lambda'_{111})$ $+0.45(\lambda'_{121}\lambda'_{212} + \lambda'_{221}\lambda'_{112})\rvert^2$

We thank K. V. Stepanyantz, Kamal Ahmed, Farida Tahir and Werner Porod for helpful discussions.

References

[1] P. Langacker, *Nucl. Phys.* **B** *(Proc. Suppl.)* **100**, 383 (2001).

[2] B. Kayser, F. Gibrat-Debu, F. Perrier, *"The Physics of the Massive Neutrinos"* (World Scientific, Singapore, 1989).

[3] A.S. Barabash, hep-ex/0602037.

[4] A. Ali, A.V. Borisov, N.B. Zamorin, *Eur. Phys. J.* **C21**, 123 (2001).

[5] A. Ali, A.V. Borisov, M.V. Sidorova, *Phys. At. Nucl.* **69**, 475 (2006).

[6] A. Ali, A.V. Borisov, N.B. Zamorin, in *"Frontiers of Particle Physics"* (Proceedings of the 10th Lomonosov Conference on Elementary Particle Physics), ed. by A. Studenikin (World Scientific, Singapore, 2003), p. 74.

[7] A. Ali, A.V. Borisov, D.V. Zhuridov, *Phys. At. Nucl.* **68**, 2061 (2005).

[8] Particle Data Group: S. Eidelman et al., *Phys. Lett.* **B592**, 1 (2004).

[9] B. Pontecorvo, *Sov. Phys. JETP* **6**, 429 (1958); Z. Maki, M. Nakagawa, S. Sakata, *Prog. Theor. Phys.* **28**, 870 (1962).

[10] R. Barbier et al., *Phys. Rep.* **420**, 1 (2005).

[11] D. Aristizabal Sierra, M. Hirsch, W. Porod, *JHEP* **0509**, 033 (2005).

[12] L.S. Littenberg, R. Shrock, *Phys. Lett.* **B491**, 285 (2000).

[13] J.G. Esteve, A. Morales, R. Núñes-Lagos, *J. Phys.* **G9**, 357 (1983).

[14] B.C. Allanach, A. Dedes, H.K. Dreiner, *Phys. Rev.* **D69**, 115002 (2004).

[15] F. Tahir, M. Anwar Mughal, K. Ahmed, *Europhys. Lett.* **54**, 580 (2001).

NEXT TO LEADING ORDER IN SEMI-INCLUSIVE DEEP INELASTIC SCATTERING PROCESSES

A.N. Sissakian, O.Yu. Shevchenko and O.N. Ivanov
Joint Institute for Nuclear Research, 141980 Dubna, Russia

Abstract. It is proposed the modification of the Jacobi polynomial expansion method (MJEM) which is based on the application of the truncated (available to experiment) moments instead of the full ones. The application of MJEM with respect to quark helicity distributions reconstruction from the SIDIS data is considered.

One of very important topics in the modern high energy physics is the investigation of the partonic spin structure of nucleon. In this connection, nowadays, there is a huge growth of interest to semi-inclusive DIS (SIDIS) experiments with longitudinally polarized beam and target such as HERMES [1] and COMPASS [2].

It is argued (see, for example, Ref. [3]) that to obtain the reliable distributions at relatively low average Q^2 available to the modern SIDIS experiments[a], the leading order (LO) analysis is not sufficient and next to leading order analysis (NLO) is necessary. In ref. [4] the procedure allowing the direct extraction from the SIDIS data of the first moments (truncated to the accessible for measurement x_B region) of the quark helicity distributions in NLO QCD was proposed. However, in spite of the special importance of the first moments, it is certainly very desirable to have the procedure of reconstruction in NLO QCD of the polarized densities themselves. At the same time, it is extremely difficult to extract the local in Bjorken x distributions directly, because of the double convolution product entering the NLO QCD expressions for semi-inclusive asymmetries (see [4] and references therein). Fortunately, operating just as in ref. [4], one can directly extract not only the first moments, but the Mellin moments of any required order. In turn, using the truncated moments of parton distribution functions (PDFs) and applying the modified Jacobi polynomial expansion method (MJEM) proposed in Ref. [5] one can reconstruct PDFs themselves in the entire accessible for measurement Bjorken x region.

Let us first apply the proposed procedure [4] to the HERMES data [1] on DIS $A_{p,d}$ and SIDIS $A_{p,d}^{\pi^\pm}$ asymmetries[b]. The procedure in this case allows to directly extract in NLO QCD the first moments $\Delta_1 u$, $\Delta_1 d$ and $\Delta_1 \bar{q} \equiv \Delta_1 \bar{u} =$ ... Certainly, one could choose only three equations (containing any three

[a]For example, HERMES data [1] on semi-inclusive asymmetries is obtained at $Q^2_{\text{average}} = 2.5 GeV^2$

[b]Since the kaon fragmentation functions are stil poorly known, while here we mainly would like to check the validity of the metod itself (irrespectively to this problem), we will consider here the most simple case of the pion production with the assumption $\Delta_1 \bar{u} = \Delta_1 \bar{d} = \Delta_1 s = \Delta_1 \bar{s} \equiv \Delta_1 \bar{q}$. The application of the method to the kaon asymmetries is now in preparation.

Table 1: The results on NLO exctracted $\Delta_1 u$, $\Delta_1 d$ and $\Delta_1 \bar{q}$ from the HERMES data on asymmetries $A_{p,d}$, $A_{p,d}^{\pi^\pm}$ together with the results on $\Delta_1 u$, $\Delta_1 d$ and $\Delta_1 \bar{q}$ obtained from the respective integrals of parametrizations AAC2003 and BB.

Parametrization	$\Delta_1 u$	$\Delta_1 d$	$\Delta_1 \bar{q}$
This work	0.624±0.063	-0.355±0.070	0.016±0.038
AAC2003	0.691	-0.293	-0.034
BB	0.667	-0.274	-0.024

of six considered asymmetries) to obtain the minimal nondegenerate system which can be directly solved with respect to $\Delta_1 u$, $\Delta_1 d$ and $\Delta_1 \bar{q}$. However, to decrease the uncertainties, we, as usual, apply the fitting procedure using all six asymmetries $A_{p,d}$, $A_{p,d}^{\pi^\pm}$ for the χ^2 construction. The respective results are presented in the Table 1. It is seen that the obtained results on NLO QCD extracted first moments $\Delta_1 u$, $\Delta_1 d$ and $\Delta_1 \bar{q}$ are in a good agreement with the results obtained by AAC [6] and BB [7] groups. Thus, the performed analysis argues that the proposed procedure occurs successful for extraction of $\Delta_1 q$ in the next-to-leading QCD order.

The proposed in Ref. [5] modified Jacobi polynomial expansion method is the expansion of the PDF $F(x)$ in the series

$$F(x) \simeq F_{N_{max}}(x) = \left(\frac{x-a}{b-a}\right)^{\beta}\left(1-\frac{x-a}{b-a}\right)^{\alpha}$$

$$\times \sum_{n=0}^{N_{max}} \Theta_n^{(\alpha,\beta)}\left(\frac{x-a}{b-a}\right)\sum_{k=0}^{n} c_{nk}^{(\alpha,\beta)}\frac{1}{(b-a)^{k+1}}\sum_{l=0}^{k}\frac{k!}{l!(k-l)!}M'[l+1](-a)^{k-l} \tag{1}$$

over the Jacobi polynomials $\Theta_n^{(\alpha,\beta)}$ and the truncated to accessible for measurement x_B region $[a,b]$ moments $M'[j] \equiv M'_{[a,b]}[j] \equiv \int_a^b dx x^{j-1} F(x)$. The later is of especial importance since the full Mellin moments $M[j] \equiv \int_0^1 dx x^{j-1} F(x)$ entering the usual Jacobi polynomial expansion method (JEM) [8–11] are unavailable while the substitution in the usual JEM the truncated moments instead of the full ones leads to the strong deviation [5] of the reconstructed $F_{N_{max}}(x)$ from $F(x)$. On the contrary, all numerical tests demonstrate [5] that for the truncated region MJEM yields much better accuracy of the input parametrization reconstruction in comparison with JEM. Thus, dealing with the truncated, available to measurement, x_B region one should apply the proposed MJEM to obtain the reliable results on the local distributions.

Let us now apply MJEM to the HERMES SIDIS data on the pion production. Here we would like just to test the applicability of MJEM to the experimental data, so that, for a moment, we do not like to deal with the such purely known objects such as $D_q^{K^\pm}$ and D_g^h fragmentation functions. From this point of view the most attractive objects are the difference asymmetries (see [4] and references therein), where the fragmentation functions are cancel out in the leading order, while in the next to leading order the difference asymmetry has only weak dependence of the difference of the favored and unfavored fragmentation functions (known with a good precision). At the same time the difference asymmetries are still not constructed[c]. So, let us apply a trick and to express the difference asymmetries

$$A_{p(d)}^{\pi^+-\pi^-} = \frac{1}{P_B P_T f D} \frac{(N_{\uparrow\downarrow}^{\pi^+} - N_{\uparrow\downarrow}^{\pi^-}) - (N_{\uparrow\uparrow}^{\pi^+} - N_{\uparrow\uparrow}^{\pi^-})}{(N_{\uparrow\downarrow}^{\pi^+} - N_{\uparrow\downarrow}^{\pi^-}) + (N_{\uparrow\uparrow}^{\pi^+} - N_{\uparrow\uparrow}^{\pi^-})} \tag{2}$$

via the standard virtual photon SIDIS asymmetries

$$A_{p(d)}^{\pi^\pm} = \frac{1}{P_B P_T f D} \frac{N_{\uparrow\downarrow}^{\pi^\pm} - N_{\uparrow\uparrow}^{\pi^\pm}}{N_{\uparrow\downarrow}^{\pi^\pm} + N_{\uparrow\uparrow}^{\pi^\pm}},$$

which were measured by HERMES [1]. Namely, the difference asymmetries, Eq. (2), can be rewritten as

$$A_{p(d)}^{\pi^+-\pi^-} = \frac{R}{R-1} A_{p(d)}^{\pi^+} - \frac{1}{R-1} A_{p(d)}^{\pi^-}, \tag{3}$$

where the ratio

$$R^{\pi^+/\pi^-} \equiv N^{\pi^+}/N^{\pi^-}, \quad N^{\pi^\pm} \equiv N_{\uparrow\downarrow}^{\pi^\pm} + N_{\uparrow\downarrow}^{\pi^\pm}$$

is taken from the unpolarized SIDIS data. This quantity is well known (much better than the fragmentation functions). We take its value from the LEPTO generator of unpolarized events [13], which reproduces the results on this quantity obtained by EMC collaboration [14]. First, for the sake of testing, we reconstruct the local valence distributions in the leading order. The results are presented in Fig. 1. One can see that reconstructed with MJEM curve is in a good agreement with both HERMES results and with the results of direct LO extraction from the difference asymmetries constructed with the application of Eq. (3).

The NLO extration of quark moments gives the results $\Delta_1 u_V = 0.655 \pm 0.141$, $\Delta_1 d_V = -0.343 \pm 0.196$, which are in very good agreement with the results presented by Table 1. Thus, one can see that the proposed Eq. (3) can be successfully applied, at least for the preliminary analysis. The results of NLO reconstruction with MJEM of Δu_V and Δd_V from the difference asymmetries given by Eq. (3) are presented in Fig. 2 in comparison with the respective LO results. It is seen that the behavior of NLO and LO curves with respect to each other is in agreement with the predictions of existing parametrizations (see, for example, [12]).

[c]At present the such analysis is performed by HERMES collaboration

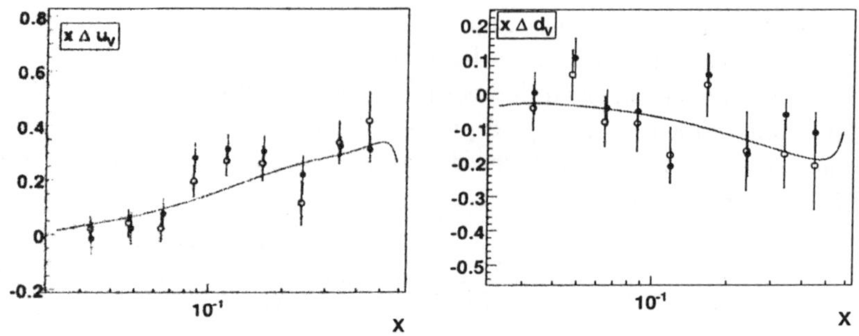

Figure 1. The results of LO reconstruction of $\Delta u_V(x)$ and $\Delta d_V(x)$. The closed circles correspond to the HERMES LO analysis [1]. The open circles correspond to direct LO extraction from the difference asymmetries constructed applying Eq. (3). The solid curves correspond to reconstruction with MJEM ($N_{max} = 4$).

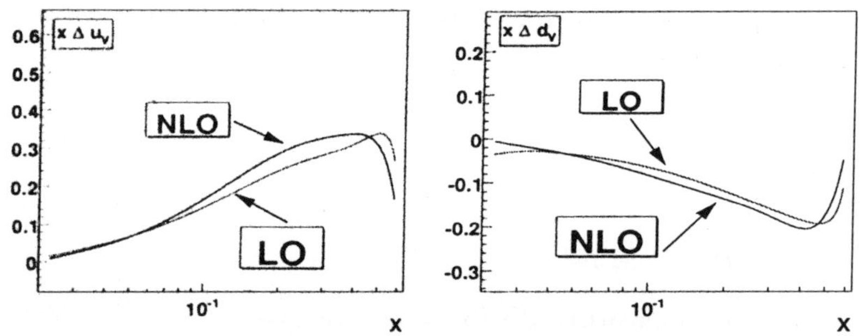

Figure 2. The results of Δu_V and Δd_V NLO reconstruction with MJEM from the HERMES data in comparison with the respective LO results.

Thus, all tests confirm that the proposed modification of the Jacobi polynomial expansion method, MJEM, allows to reconstruct with a high precision the quark helicity distributions in the accessible for measurement x_B region. The performed preliminary analysis of the HERMES data on the pion production gives the reliable results on the valence quark helicity distributions, which are in accordance with the leading order HERMES results as well as with the existing NLO parametrizations on these quantities. Thus, MJEM occurs successful with respect to NLO extraction of the valence PDFs. In future we also plan to apply MJEM to NLO QCD extraction of the sea and strange quark helicity distributions from the published HERMES data and expected COMPASS data on the SIDIS asymmetries with the pion and kaon production.

[1] HERMES collaboration (A. Airapetyan et al.), Phys. Rev. **D71** (2005) 012003

[2] COMPASS collaboration (G. Baum et al.), "COMPASS: A proposal for

a common muon and proton apparatus for structure and spectroscopy", CERN-SPSLC-96-14 (1996).

[3] A.N. Sissakian, O.Yu. Shevchenko and O.N. Ivanov, Phys. Rev. **D68** (2003) 031502

[4] A.N. Sissakian, O.Yu. Shevchenko and O.N. Ivanov, Phys. Rev. **D70** (2004) 074032

[5] A.N. Sissakian, O.Yu. Shevchenko and O.N. Ivanov, JETP Lett. **82** (2005) 53

[6] Asymmetry Analysis Collaboration (M. Hirai et al.), Phys. Rev. **D69** (2004) 054021

[7] J. Blumlein, H. Bottcher, Nucl.Phys. **B636** (2002) 225

[8] G. Parisi, N. Sourlas, Nucl. Phys. **B151** (1979) 421

[9] I.S. Barker, C.S. Langensiepen, G. Shaw, Nucl. Phys. **B186** (1981) 61

[10] V.G. Krivokhizhin et al, Z. Phys. **C36** (1987) 51; JINR-E2-86-564

[11] E. Leader, A.V. Sidorov, D.B. Stamenov, Int. J. Mod. Phys. **A13** (1998) 5573

[12] M. Gluck, E. Reya, M. Stratmann, W. Vogelsang, Phys. Rev. **D63** (2001) 094005

[13] G. Ingelman, A. Edin, J. Rathsman Comput. Phys. Commun. **101** (1997) 108.

[14] EMC collaboration (J, Ashman et al.), Z. Phys. **C 52** (1991) 361 , CERN-PPE-91-53.

$B_d(\bar{B}_d) \to \rho^\pm\pi^\mp, \rho^+\rho^-, \pi^+\pi^-$: HUNTING FOR ALPHA

M.I. Vysotsky
ITEP, Moscow, Russia
talk at 12 Lomonosov Conference

In paper [1] from the data on CP asymmetries in $B_d(\bar{B}_d) \to \rho^\pm\pi^\mp$, $\rho^+\rho^-$ decays and BABAR data on CP asymmetries in $B_d(\bar{B}_d) \to \pi^+\pi^-$ decays we determine the value of angle α of the unitarity triangle:

$$\alpha = 96^o \pm 3^o \ , \tag{1}$$

where only a tree quark decay amplitude $\bar{b} \to u\bar{u}\bar{d}(b \to u\bar{u}d)$ was taken into account. The numerical values of angle α obtained from the considered decays are consistent with each other and with the value which follows from the global CKM fit. This observation testifies to the validity of a proposed approach.

As the next step in the paper [2] we studied what changes in the values of α are induced by QCD penguins. Our aim was twofold. First, in this way we get an estimate of the theoretical uncertainty of the value of α determined in paper [1]. Secondly, we get the formulas for CP violating parameters describing these decays which vanish when penguins are neglected ($C_{\rho\pi}$, $A^{\rho\pi}_{CP}$, $C_{\rho\rho}$, $C_{\pi\pi}$). My talk is based on the paper [2].

The angle shifts $\Delta\alpha$ we are interested in were estimated in paper [3]; however, in that paper FSI phases were neglected, that is why $C_{\rho\pi} = A^{\rho\pi}_{CP} = C_{\rho\rho} = C_{\pi\pi} = 0$ follows from [3] (the nonzero penguin amplitudes are a necessary, but not a sufficient condition for $C_{\rho\pi}$... to be nonzero). We will take these phases into account. The asymmetries depend on the differences of FSI phases in the processes described by the tree and penguin diagrams. One source of these differences is an imaginary part of the quark penguin diagram, the so-called BSS mechanism of the strong phases generation [4] (see also paper [5]). The phase of the penguin diagram depends on the gluon q^2 which is transferred to $u\bar{u}$ pair, each quark of which goes to different $\pi^\pm$- or $\rho^\pm$-mesons. In this way the value of q^2 depends on the light meson wave functions and we can estimate it only roughly. Another source of FSI phases is hadron rescattering and even less is known about the values of the phase shifts between the penguin and tree diagrams generated in this way. In view of this we will determine FSI phases from the experimental data on CPV asymmetries, and investigate to what values of α it will lead.

Let us start from $\bar{B}_d(B_d) \to \rho^\mp\pi^\pm$ decays.

The time dependence of the decay probabilities is given by [6]:

$$\begin{aligned}\frac{dN(B_d(\bar{B}_d) \to \rho^\pm\pi^\mp)}{dt} &= (1 \pm A^{\rho\pi}_{CP})e^{-t/\tau}[1 - q(C_{\rho\pi} \pm \Delta C_{\rho\pi}) \times \\ &\times \cos(\Delta mt) + q(S_{\rho\pi} \pm \Delta S_{\rho\pi})\sin(\Delta mt)] \ , \end{aligned} \tag{2}$$

where $q = -1$ describes the case when at $t = 0$ B_d was produced, while $q = 1$ corresponds to $\bar{B}_d$ production at $t = 0$. In the case of $\Upsilon(4S) \to B_d\bar{B}_d$ decay the flavor of the beauty meson which will decay to $\rho\pi$ is tagged by the charge of a lepton in the other beauty meson semileptonic decay. A partner decay starts clocks as well. τ is $B_d(\bar{B}_d)$ life time, while Δm is the difference of masses of $(B_d, \bar{B}_d)$ system eigenstates (it equals the frequency of $B_d - \bar{B}_d$ oscillations).

The following expression for decay amplitude can be obtained (for more details see paper [2]:

$$\bar{M}^{-+} = AV_{ub}V_{ud}^*[1 - 0.07e^{i(\delta-\alpha)}] = AV_{ub}V_{ud}^*[1 - 0.07\sin\delta + i0.07\cos\delta] \ , \quad (3)$$

$$M^{-+} = BV_{ub}^*V_{ud} \ , \quad (4)$$

$$\lambda^{-+} \equiv \frac{q}{p}\frac{\bar{M}^{-+}}{M^{-+}} = e^{2i\alpha}\frac{A}{B}[1 - 0.07\sin\delta + i0.07\cos\delta] \ , \quad (5)$$

where parameters q and p enter the expressions for $(B_d, \bar{B}_d)$ eigenstates and we have substituted $\alpha = \pi/2$ in the (small) second term in square brackets in Eq. (3). Analogously we get:

$$\bar{M}^{+-} = BV_{ub}V_{ud}^* \ , \quad (6)$$

$$M^{+-} = AV_{ub}^*V_{ud}[1 - 0.07e^{i(\alpha+\delta)}] = AV_{ub}^*V_{ud}[1 + 0.07\sin\delta - 0.07i\cos\delta] \ , \quad (7)$$

$$\lambda^{+-} \equiv \frac{q}{p}\frac{\bar{M}^{+-}}{M^{+-}} = e^{2i\alpha}\frac{B}{A}[1 + 0.07\sin\delta - i0.07\cos\delta]^{-1} \ . \quad (8)$$

From the expressions for the quantities $C_{\rho\pi}$ and $\Delta C_{\rho\pi}$ [6]:

$$C_{\rho\pi} \pm \Delta C_{\rho\pi} = \frac{1 - |\lambda^{\pm\mp}|^2}{1 + |\lambda^{\pm\mp}|^2} \quad (9)$$

we obtain:

$$\begin{aligned} \Delta C_{\rho\pi} &= \frac{a^2 - b^2}{a^2 + b^2} \ , \ \frac{a^2}{b^2} = \frac{1 + \Delta C_{\rho\pi}}{1 - \Delta C_{\rho\pi}} \ , \\ C_{\rho\pi} &= 0.28\sin\delta\frac{(a/b)^2}{(1 + a^2/b^2)^2} \ , \end{aligned} \quad (10)$$

where $a \equiv |A|$, $b \equiv |B|$.

The Belle and BABAR averaged result for $\Delta C_{\rho\pi}$ is [7]:

$$\Delta C_{\rho\pi} = 0.22 \pm 0.10 \ , \quad (11)$$

which leads to:

$$\left(\frac{a}{b}\right)^2 = 1.56 \pm 0.33 \ . \quad (12)$$

From the averaged experimental result [7]

$$C_{\rho\pi} = 0.31 \pm 0.10 \quad (13)$$

and Eq. (10) we get:

$$\sin\delta = 4.6 \pm 1.5 \ . \qquad (14)$$

We see that poor accuracy in the measurement of $C_{\rho\pi}$ does not allow to get any definite information on the value of phase δ.

The next observable we wish to discuss is CP asymmetry $A_{CP}^{\rho\pi}$:

$$\begin{aligned} A_{CP}^{\rho\pi} &= \frac{|M^{+-}|^2 - |\bar{M}^{-+}|^2 + |\bar{M}^{+-}|^2 - |M^{-+}|^2}{|M^{+-}|^2 + |\bar{M}^{-+}|^2 + |\bar{M}^{+-}|^2 + |M^{-+}|^2} = \\ &= 0.14 \sin\delta \frac{(a/b)^2}{1+(a/b)^2} \ , \qquad (15) \end{aligned}$$

which should be compared with the experimental result [7]:

$$A_{CP}^{\rho\pi} = -0.102 \pm 0.045 \ . \qquad (16)$$

From (15) and (16) we get:

$$\sin\delta = -1.2 \pm 0.5 \ , \qquad (17)$$

and it differs from given in Eq.(14) by 3.5 standard deviations. This is the largest discrepancy we encounter. Averaging these two numbers we obtain:

$$\sin\delta = -0.62 \pm 0.47 \ . \qquad (18)$$

Finally we come to the discussion of the observables which are sensitive to the angle α:

$$S_{\rho\pi} \pm \Delta S_{\rho\pi} = \frac{2\mathrm{Im}\lambda^{\pm\mp}}{1+|\lambda^{\pm\mp}|^2} \ , \qquad (19)$$

$$S_{\rho\pi} = \frac{2a/b}{1+a^2/b^2}[\sin 2\alpha \cos\tilde{\delta} - 0.07\cos\delta\cos\tilde{\delta} - 0.07\frac{a^2/b^2-1}{a^2/b^2+1}\sin\delta\sin\tilde{\delta}] \ , \qquad (20)$$

$$\begin{aligned} \Delta S_{\rho\pi} &= \frac{2a/b}{1+a^2/b^2}[-\cos 2\alpha \sin\tilde{\delta} + 0.07\cos\delta\sin\tilde{\delta}\sin 2\alpha - \\ &- 0.07\frac{a^2/b^2-1}{a^2/b^2+1}\sin\delta\cos\tilde{\delta}\sin 2\alpha] \ , \qquad (21) \end{aligned}$$

where the definition of the phase $\tilde{\delta}$ is $A/B = (a/b)e^{i\tilde{\delta}}$ and in the small terms proportional to 0.07 in the expression for $S_{\rho\pi}$ we have substituted $\cos 2\alpha = -1$.

Let us start the analysis of the experimental data from $\Delta S_{\rho\pi}$. According to [7]:

$$\Delta S_{\rho\pi} = 0.09 \pm 0.13 \ , \qquad (22)$$

which is much less than one. According to Eq. (12) the factor which multiplies square brackets in Eq. (20) (and in Eq. (21)) is very close to one, that is why it is the expression in square brackets which should be much less than one. The second and the third terms of this expression are really very small and we can neglect them. What concerns the first term, it is small when $\tilde{\delta}$ is close to zero or π:

$$\sin\tilde{\delta} = \Delta S_{\rho\pi} = 0.09 \pm 0.13 \ , \tag{23}$$

where a small deviation of $\cos 2\alpha$ from -1 is neglected.

Now everything is ready and from Eq. (20) we get:

$$\sin 2\alpha = S_{\rho\pi} / \cos\tilde{\delta} + 0.07 \cos\delta \ , \tag{24}$$

where we omit the last term in square brackets since it is negligibly small. In order to find angle α from the experimental data [7]:

$$S_{\rho\pi} = -0.13 \pm 0.13 \ , \tag{25}$$

we should determine the values of $\cos\delta$ and $\cos\tilde{\delta}$.

To propagate errors from $\sin\delta$ to $\cos\delta$, we consider gaussian distribution for $\sin\delta$ truncated to physical region $|\sin\delta| < 1$, transform it into (non-gaussian) distribution for $\cos\delta$ and take an interval containing 68% of probability.

In this way from Eq. (18) we obtain:

$$|\cos\delta| = 0.88 \pm 0.12 \ . \tag{26}$$

The average value of $|\cos\delta|$ appears to be close to one due to the so-called Jacobian pick.

Concerning $\tilde{\delta}$ it follows from Eq. (23) that $|\cos\tilde{\delta}| = 1$ with very good accuracy. Depending on the values of phases $\tilde{\delta}$ and δ we get the following four domains for the angle α:

$$\tilde{\delta} \approx 0 \ , \quad \delta \approx 0 : \qquad \begin{array}{l} \sin 2\alpha = -0.13 \pm 0.13 + 0.06 \\ \alpha = 92^o \pm 4^o \end{array} \tag{27}$$

$$\tilde{\delta} \approx 0 \ , \quad \delta \approx \pi : \qquad \begin{array}{l} \sin 2\alpha = -0.13 \pm 0.13 - 0.06 \\ \alpha = 96^o \pm 4^o \end{array} \tag{28}$$

$$\tilde{\delta} \approx \pi \ , \quad \delta \approx 0 : \qquad \begin{array}{l} \sin 2\alpha = 0.13 \pm 0.13 + 0.06 \\ \alpha = 84^o \pm 4^o \end{array} \tag{29}$$

$$\tilde{\delta} \approx \pi \ , \quad \delta \approx \pi : \qquad \begin{array}{l} \sin 2\alpha = 0.13 \pm 0.13 - 0.06 \\ \alpha = 88^o \pm 4^o \ . \end{array} \tag{30}$$

Thus QCD penguins split values of α obtained without taking them into account: $94^o \to 92^o, 96^o$; $86^o \to 84^o, 88^o$. If BSS mechanism is valid, then only the domains given by Eqs.(27) and (29) remain (see also [8]).

Next to consider are $\bar{B}_d(B_d) \to \rho^+\rho^-$ decays.

The time dependence of CP violating asymmetry is described by the following formula:

$$a_{CP}(t) \equiv \frac{\frac{dN(\bar{B}_d \to \rho_L^+ \rho_L^-)}{dt} - \frac{dN(B_d \to \rho_L^+ \rho_L^-)}{dt}}{\frac{dN(\bar{B}_d \to \rho_L^+ \rho_L^-)}{dt} + \frac{dN(B_d \to \rho_L^+ \rho_L^-)}{dt}} = -C_{\rho\rho} \cos(\Delta m t) + S_{\rho\rho} \sin(\Delta m t) \ . \tag{31}$$

Let us remind that the longitudinal polarization fraction $f_L = 0.98 \pm 0.01 \pm 0.02$ [9]; $f_L = 0.95 \pm 0.03 \pm 0.03$ [10] and its closeness to one greatly simplify the extraction of CPV parameters from $B_d(\bar{B}_d) \to \rho^+ \rho^-$ decay data. These parameters are given by the following expressions:

$$C_{\rho\rho} = \frac{1 - |\lambda_{\rho\rho}|^2}{1 + |\lambda_{\rho\rho}|^2} \ , \ \ S_{\rho\rho} = \frac{2\mathrm{Im}\lambda_{\rho\rho}}{1 + |\lambda_{\rho\rho}|^2} \ , \tag{32}$$

$$\lambda_{\rho\rho} \equiv \frac{q}{p} \frac{\bar{M}_{\rho\rho}}{M_{\rho\rho}} = e^{2i\alpha} \frac{1 - 0.07 e^{i(\delta - \alpha)}}{1 - 0.07 e^{i(\delta + \alpha)}} = e^{2i\alpha}(1 + 0.14 i \sin\alpha e^{i\delta}) \ , \tag{33}$$

where we use the same letter δ for FSI phase difference of the amplitudes generated by penguin and tree diagrams as in the case of $B \to \rho\pi$ decays. These differences would be really the same if BSS mechanism dominates.

Comparing the averaged experimental result for $C_{\rho\rho}$

$$C_{\rho\rho} = -0.03 \pm 0.18 \pm 0.09 \quad [9], \tag{34}$$

$$C_{\rho\rho} = 0.0 \pm 0.30 \pm 0.10 \quad [10]; \tag{35}$$

$$C_{\rho\rho}^{exp} = -0.02 \pm 0.17 \tag{36}$$

with the theoretical expression which follows from Eqs. (32), (33)

$$C_{\rho\rho} = 0.14 \sin\alpha \sin\delta \approx 0.14 \sin\delta \tag{37}$$

we get:

$$\sin\delta = -0.15 \pm 1.2 \ ; |\cos\delta| = 0.88 \pm 0.12 \ , \tag{38}$$

where the same procedure of error propagation as in the case of $B \to \rho\pi$ was used and we sum statistical and systematic errors of $\sin\delta$ as independent.

From Eq. (32) we obtain:

$$S_{\rho\rho} = \sin 2\alpha + 0.14 \sin\alpha \cos 2\alpha \cos\delta = \sin 2\alpha - 0.14 \cos\delta \ . \tag{39}$$

According to the recent measurements:

$$S_{\rho\rho} = -0.33 \pm 0.26 \quad [9], \tag{40}$$

$$S_{\rho\rho} = 0.09 \pm 0.43 \quad [10], \tag{41}$$

$$S_{\rho\rho}^{exp} = -0.21 \pm 0.22 \ , \tag{42}$$

and we get two domains for α:

$$\delta \approx 0 : \ \alpha = 92^o \pm 7^o \tag{43}$$

$$\delta \approx \pi : \ \alpha = 100^o \pm 7^o \ . \tag{44}$$

Just as in the case of $B \to \rho\pi$ decays only the first domain remains if $|\delta| < \pi/2$ [8].

Let us note that using the isospin analysis (which allows to prove the smallness of the penguin contribution) it was obtained:

$$\alpha = 100^o \pm 13^o \quad [9], \tag{45}$$

$$\alpha = 87^o \pm 17^o \quad [10], \tag{46}$$

where the error is mainly due to the uncertainty of the penguin contribution. Extracting this uncertainty and averaging last two numbers we get:

$$\alpha = 96^o \pm 7^o(exp) \pm 11^o(penguin) \ . \tag{47}$$

Finally we come to the $\bar{B}_d(B_d) \to \pi^+\pi^-$ decays.

The time dependence of CP violating asymmetry is given by the formula analogous to Eq. (31) with the evident substitution of π instead of ρ. Eq. (32) with the same substitution is valid as well, while for the quantity $\lambda_{\pi\pi}$ we obtain:

$$\lambda_{\pi\pi} \equiv \frac{q}{p}\frac{\bar{M}_{\pi^+\pi^-}}{M_{\pi^+\pi^-}} = e^{2i\alpha}(1 + 0.28 i \sin\alpha e^{i\delta}) \ , \tag{48}$$

and what concerns letter δ we should repeat the comment made after Eq. (33).

The experimental data of BABAR and Belle for CPV parameters $S_{\pi\pi}$ and $C_{\pi\pi}$ were controversial though at present (with the latest Belle results) the divergence diminishes. In view of this we will perform a two step analysis, taking at the beginning only BABAR results and then the averaged results of two collaborations.

Comparing the theoretical expression

$$C_{\pi\pi} = 0.28 \sin\alpha \sin\delta \approx 0.28 \sin\delta \tag{49}$$

with BABAR result [7]

$$C_{\pi\pi}^{\text{BABAR}} = -0.09 \pm 0.15 \tag{50}$$

we get:

$$\sin\delta = -0.32 \pm 0.54 \ , \tag{51}$$

while for $S_{\pi\pi}$ we have:

$$S_{\pi\pi} = \sin 2\alpha - 0.28 \cos\delta \ , \ S_{\pi\pi}^{\text{BABAR}} = -0.30 \pm 0.17 \ . \tag{52}$$

From Eq. (51) we get:

$$|\cos\delta| = 0.9 \pm 0.1 \tag{53}$$

and two domains of α corresponding to two signs of $\cos\delta$[a]:

$$\delta \approx 0: \;\; \alpha = 91^o \pm 5^o(\text{exp}) \pm 1^o(\text{theor}) \tag{54}$$

$$\delta \approx \pi: \;\; \alpha = 107^o \pm 5^o(\text{exp}) \pm 1^o(\text{theor}) \;\; . \tag{55}$$

Belle result:

$$C_{\pi\pi}^{\text{Belle}} = -0.56 \pm 0.13 \tag{56}$$

deviates by 2.5σ from BABAR and (if correct) would require considerably larger P/T ratio than we use in our paper.

Finally, averaging (50) and (56) one gets [7]:

$$C_{\pi\pi}^{\text{ex}} = -0.37 \pm 0.10 \;\; , \tag{57}$$

and comparing with the theoretical expression (49) we obtain:

$$\sin\delta = -1.32 \pm 0.35 \;\; , \tag{58}$$

which leads to:

$$|\cos\delta| = 0.55 \pm 0.25 \;\; . \tag{59}$$

From the averaged experimental result [7]:

$$S_{\pi\pi}^{\text{ex}} = -0.50 \pm 0.12(\text{exp}) \tag{60}$$

we get two domains[b]:

$$\delta \approx 0: \;\; \alpha = 100^o \pm 4^o(\text{exp}) \pm 2^o(\text{theor}) \tag{61}$$

$$\delta \approx \pi: \;\; \alpha = 110^o \pm 4^o(\text{exp}) \pm 2^o(\text{theor}) \;\; . \tag{62}$$

In conclusion, we have analyzed CPV asymmetries in $B_d(\bar{B}_d) \to \rho^\pm\pi^\mp$, $\rho_L^+\rho_L^-$ and $\pi^+\pi^-$ decays induced by the charmless strangeless b-quark decay $b \to u\bar{u}d$. This decay can proceed through a tree or penguin diagram. As it was noted in [1] when the penguin diagram is neglected, one obtains the values of the unitarity triangle angle α from CPV asymmetries in these decays which are consistent with each other as well as with the value of α which follows from the global CKM fit. However, in order to determine the theoretical accuracy of α extracted from the decays under study one should take the penguin amplitude into account.

In order to determine numerical value of α one should average the values which follow from the considered decays. Since the phase shifts δ can be different in $B \to \rho\rho, \pi\pi$ and $\rho\pi$ decays, we get too many possibilities. That is why let us limit ourselves to the theoretically motivated case $|\delta| < \pi/2$.

---

[a] Assuming $|\delta| < \pi/2$ [8] we would get only the first domain.

[b] Assuming $|\delta| < \pi/2$ [8] which follows from BSS mechanism of δ generation, we would get the first domain only.

Averaging Eqs. (43) and (61) we obtain:

$$\delta \approx 0: \ \alpha_{\rho\rho,\pi\pi} = 98^o \pm 4^o \ . \tag{63}$$

Averaging it with (27) and (29) we get two possibilities:

$$\begin{aligned} \alpha_{b\to u\bar{u}d} &= 95^o \pm 3^o \ , \ \text{or} \\ \alpha_{b\to u\bar{u}d} &= 91^o \pm 3^o \ , \end{aligned} \tag{64}$$

and the last one corresponds to the smallest possible value of α. In the case $|\delta_i| > \pi/2$, $\tilde{\delta} = 0$ averaging Eqs.(28), (44), (62) we get the largest possible value: $\alpha_{b\to u\bar{u}d} = 102^o \pm 3^o$.

The global fit results for α are:

$$\alpha_{\rm UTfit}^{[11]} = 94^o \pm 8^o \ , \ \alpha_{\rm CKMfitter}^{[12]} = 94 \pm 10^o \ . \tag{65}$$

Thus the accuracy of the present day knowledge of α can be close to that of β:

$$\beta = 23^o \pm 2^o \ . \tag{66}$$

I am grateful to Alexander Studenikin for the invitation to participate in the Lomonosov Conference which he wonderfully organizes.

[1] G. G. Ovanesyan and M. I. Vysotsky, Pis'ma v ZhETF, **81**, 449 (2005).
[2] M.I. Vysotsky, hep-ph/0506322 (2005).
[3] R. Aleksan, F. Buccella, A. Le. Yaouanc, L. Oliver, O. Pene and J. - C. Raynal, Phys. Lett. B **356**, 95 (1995).
[4] M. Bander, D. Silverman and A. Soni, Phys. Rev. Lett. **43**, 242 (1979).
[5] G. M. Gérard, W. - S. Hou, Phys. Rev. D **43**, 2909 (1991).
[6] M. Gronau, Phys. Lett. B **233**, 479 (1989).
[7] Heavy Flavor Averaging Group (HFAG), hep-ex/0505100 (2005).
[8] M. Gronau, E. Lunghi and D. Wyler, Phys. Lett. B **606**, 95 (2005).
[9] BABAR Collaboration, B. Aubert et al., Phys. Rev. Lett. **93**, 231801 (2005).
[10] Belle Collaboration, K. Abe et al., hep-ex/0507039 (2005).
[11] M. Bona et al. (UT fit Collaboration), hep-ph/0501199 (2005).
[12] J. Charles et al. (The CKMfitter Group), Eur. Phys. J. C **41**, 1 (2005).
[13] A. J. Buras et al., Nucl. Phys. B **370**, 69 (1992).
[14] M. Beneke, G. Buchalla, M. Neubert and C. T. Sachrajda, Phys. Rev. Lett. **83**, 1914 (1999).

DEEP INELASTIC SPIN STRUCTURE FUNCTIONS AT SMALL x

B.I. Ermolaev
Ioffe Physico-Technical Institute, 194021 St.Petersburg, Russia
M. Greco
Department of Physics and INFN, University Rome III, Rome, Italy
S.I. Troyan
St.Petersburg Institute of Nuclear Physics, 188300 Gatchina, Russia

Abstract. Explicit expressions for the non-singlet and singlet spin-dependent structure function g_1 in the small-x region are obtained. They include the total resummation of the double- and single- logarithms of x and account for the running QCD coupling α_s effects. Both the non-singlet and singlet structure functions are Regge behaved asymptotically, with the intercepts predicted in agreement with experiments. A detailed comparison with the DGLAP evolution equations for different values of x and Q^2 is performed. Finally, the role played by singular terms in DGLAP fits for the initial quark densities is discussed and explicitly shown to mimic the resummation of leading logarithms at small-x.

PACS numbers: 12.38.Cy

1 Introduction

In the standard theoretical approach for investigating the DIS structure function $g_1(x, Q^2)$, namely DGLAP [1], g_1^{DGLAP} is a convolution of the coefficient functions C_{DGLAP} and the evolved parton distributions. The latter are also expressed as a convolution of the splitting functions P_{DGLAP} and initial parton densities, which are fitted from experimental data at large x, $x \sim 1$ and $Q^2 \sim 1$ GeV^2. However there is an obvious asymmetry in treating the Q^2- and x- logarithmic contributions in DGLAP. Indeed, the leading Q^2- contributions, $\ln(Q^2)$, are accounted to all orders in α_s whereas $C_{DGLAP}(x)$ and $P_{DGLAP}(x)$ are known in first two orders of the perturbative QCD. On the other hand, in the small-x region the situation looks opposite: logarithms of x, namely double logarithmis(DL), i.e. the terms $(\alpha_s \ln^2(1/x))^k$, and single logarithms (SL), the terms $(\alpha_s \ln(1/x))^k$, with $k = 1, 2, ..$, are becoming quite sizable and should be accounted to all ordersi in α_s. The total resummation of DL terms was first done [2] in the fixed α_s approximation, and led to a new expression g_1^{DL}, for g_1, that in the small-x asymptotics was of the Regge (power-like) form and much greater than the well-known small-x asymptotics of g_1^{DGLAP}.

Strictly speaking, the results of Refs. [2] could not be compared in a straightforward way with DGLAP because instead of the running α_s, with the parameterization

$$\alpha_s^{DGLAP} = \alpha_s(Q^2), \tag{1}$$

Refs. [2] had used α_s fixed at an unknown scale. A closer investigation of this matter [5] led us to conclude that the DGLAP- parametrization of Eq. (1) can be a good approximation at x not far from 1 only. Instead, a new parametrization was suggested, where the argument of α_s in each of the ladder rungs of the involved Feynman diagram is the virtuality of the horizontal gluon (see Ref. [5] for detail). Indeed this parametrization works well both for small and large x. It converges to the DGLAP- parametrization at large x but differs from it at small x, and it allowed us to obtain in Refs. [7] the expressions for g_1 accounting for all-order resuummations of DL and SL terms, including the running α_s effects [a]. This led us to predict the

[a] The parametrization of see Ref. [5] was used later in Refs. [6] for studying the small-x contribution to the Bjorken sum rule.

numerical values of the intercepts of the singlet and non-singlet g_1. These results were then confirmed [8] by several independent groups who have analyzed the HERMES data and extrapolated them at small x.

On the other hand, it is well known that, despite missing the total resummation of $\ln x$, DGLAP works quite successfully at $x \ll 1$. This might suggest that the total resummation of DL contributions performed in Refs. [7] should not be relevant at available values of x and might be of some importance at extremely small x reachable in the future. In Ref. [9] we made a detailed numerical analysis and explained why DGLAP fits can be successful at small x. Indeed in order to describe the available experimental data, singular expressions (see for example Refs. [10,11]) are introduced for the initial parton densities. These singular factors (i.e. the factors which $\to \infty$ when $x \to 0$) introduced in the fits mimic the total resummaton of Refs. [7]. Then using the results of Ref [7] for incorporating the total resummation of $\ln x$ it allows to simplify the rather sophisticated structure of the standard DGLAP fits down to a normalization constant at small x.

2 Difference between DGLAP and our approach

In DGLAP, g_1 is expressed through convolutions of the coefficient functions and evolved parton distributions. As convolutions look simpler in terms of integral transforms, it is convenient to represent g_1 in the form of the Mellin integral. For example, the non-singlet component of g_1 can be represented as follows:

$$g_1^{NS}{}_{DGLAP}(x,Q^2) = \tag{2}$$
$$(e_q^2/2)\int_{-\imath\infty}^{\imath\infty}\frac{d\omega}{2\imath\pi}(1/x)^{\omega}C_{DGLAP}(\omega)\delta q(\omega)\exp\Big[\int_{\mu^2}^{Q^2}\frac{dk_\perp^2}{k_\perp^2}\gamma_{DGLAP}(\omega,\alpha_s(k_\perp^2))\Big]$$

with $C_{DGLAP}(\omega)$ being the non-singlet coefficient functions, $\gamma_{DGLAP}(\omega,\alpha_s)$ the non-singlet anomalous dimensions and $\delta q(\omega)$ the initial non-singlet quark densities in the Mellin (momentum) space. The expression for the singlet g_1 is similar, though more involved. Both γ_{DGLAP} and C_{DGLAP} are known in first two orders of the perturbative QCD. Technically, it is simpler to calculate them at integer values of $\omega = n$. In this case, the integrand of Eq. (2) is called the n-th momentum of g_1^{NS}. Once the moments for different n are known, g^{NS} at arbitrary values of ω is obtained by interpolation. The expressions of the initial quark densities are obtained from phenomenological consideration, by fitting the experimental data at $x \sim 1$. Eq. (2) shows that γ_{DGLAP} govern the Q^2- evolution whereas C_{DGLAP} evolve $\delta q(\omega)$ in the x-space from $x \sim 1$ into the small x region. When in the x-space the initial parton distributions $\delta q(x)$ are regular in x, i.e. do not $\to \infty$ when $x \to 0$, the small-x asymptotics of $g_{1\ DGLAP}$ is given by the well-known expression:

$$g_1^{NS}{}_{DGLAP},\ g_1^{S}{}_{DGLAP} \sim \exp\Big[\sqrt{\ln(1/x)\ln\Big(\ln(Q^2/\mu^2)/\ln(\mu^2/\Lambda^2_{QCD})\Big)}\ \Big]. \tag{3}$$

On the contrary, when the total resummation of the double-logarithms (DL) and single- logarithms of x is done [5], the Mellin representation for g_1^{NS} is

$$g_1^{NS}(x,Q^2) = (e_q^2/2)\int_{-\imath\infty}^{\imath\infty}\frac{d\omega}{2\pi\imath}(1/x)^{\omega}C_{NS}(\omega)\delta q(\omega)\exp\left(H_{NS}(\omega)\ln(Q^2/\mu^2)\right), \tag{4}$$

with new coefficient functions C_{NS},

$$C_{NS}(\omega) = \frac{\omega}{\omega - H_{NS}^{(\pm)}(\omega)} \tag{5}$$

and anomalous dimensions H_{NS},

$$H_{NS} = (1/2)\Big[\omega - \sqrt{\omega^2 - B(\omega)}\Big] \tag{6}$$

where

$$B(\omega) = (4\pi C_F(1 + \omega/2)A(\omega) + D(\omega))/(2\pi^2)\ . \tag{7}$$

$D(\omega)$ and $A(\omega)$ in Eq. (7) are expressed in terms of $\rho = \ln(1/x)$, $\eta = \ln(\mu^2/\Lambda^2_{QCD})$, $b = (33 - 2n_f)/12\pi$ and the color factors $C_F = 4/3$, $N = 3$:

$$D(\omega) = \frac{2C_F}{b^2 N}\int_0^\infty d\rho e^{-\omega\rho}\ln\big(\frac{\rho+\eta}{\eta}\big)\Big[\frac{\rho+\eta}{(\rho+\eta)^2+\pi^2} \mp \frac{1}{\eta}\Big]\ , \tag{8}$$

$$A(\omega) = \frac{1}{b}\Big[\frac{\eta}{\eta^2+\pi^2} - \int_0^\infty \frac{d\rho e^{-\omega\rho}}{(\rho+\eta)^2+\pi^2}\Big]. \tag{9}$$

H_S and C_{NS} account for DL and SL contributions to all orders in α_s.
When $x \to 0$,

$$g_1^{NS} \sim \left(x^2/Q^2\right)^{\Delta_{NS}/2},\ g_1^S \sim \left(x^2/Q^2\right)^{\Delta_S/2} \tag{10}$$

where the non-singlet and singlet intercepts are $\Delta_{NS} = 0.42$, $\Delta_S = 0.86$. The x-behaviour of Eq. (10) is much steeper than the one of Eq. (3). Obviously, the total resummation of logarithms of x leads to a faster growth of g_1 when x is decreasing, compared to what is predicted by DGLAP, provided the input initial parton density δq in Eq. (2) is a regular function of ω at $\omega \to 0$.

3 Role of the initial parton densities

There are various forms in the literature for $\delta q(x)$, but all available fits include both a regular and a singular factor when $x \to 0$ (see e.g. Refs. [10, 11] for detail). For example, one of the fits from Ref. [10] is given by

$$\delta q(x) = N\eta x^{-\alpha}\Big[(1-x)^\beta(1+\gamma x^\delta)\Big], \tag{11}$$

with N, η being normalization factors, $\alpha = 0.576$, $\beta = 2.67$, $\gamma = 34.36$ and $\delta = 0.75$. In the ω -space Eq. (11) is a sum of pole contributions:

$$\delta q(\omega) = N\eta\Big[(\omega-\alpha)^{-1} + \sum m_k(\omega+\lambda_k)^{-1}\Big], \tag{12}$$

with $\lambda_k > 0$, and the first term in Eq. (12) corresponds to the singular factor $x^{-\alpha}$ of Eq. (11). When Eq. (11) is substituted in Eq. (2), the singular factor $x^{-\alpha}$ affects the small -x behavior of g_1 and changes its asymptotics Eq. (3) for g_1 for the Regge asymptotics. Indeed, the small- x asymptotics is governed by the leading singularity $\omega = \alpha$, so

$$g_{1\ DGLAP} \sim C(\alpha)(1/x)^\alpha\Big((\ln(Q^2/\Lambda^2))/(\ln(\mu^2/\Lambda^2))\Big)^{\gamma(\alpha)}. \tag{13}$$

Obviously, the actual DGLAP asympotics of Eq. (13) is of the Regge type, and differs a lot from the conventional DGLAP asympotics of Eq. (3). Indeed it looks similar to our asymptotics given by Eq. (10), namely by incorporating the singular factors into DGLAP initial parton densities it leads to the steep rise of g_1^{DGLAP} and therefore to a successful description of DGLAP at small x. In Ref. [9] it is shown that without the singular factor $x^{-\alpha}$ in the fit of Eq. (11), DGLAP would not be able to work successfully at $x \leq 0.05$. In other words, the singular factors in DGLAP fits mimic the total resummation of logarithms of x of Eqs. (4,10). To be more specific, although both (13) and (10) predict the Regge asymptotics for g_1, there is a numerical difference in the intercepts: Eq. (13) predicts that the intercept of g_1^{NS} should be $\alpha = 0.57$, a value which is greater than our predicted non-singlet intercept $\Delta_{NS} = 0.42$. Therefore the non-singlet g_1^{DGLAP} grows, when $x \to 0$, faster than our predictions. Such a rise however is too steep and contradicts the results obtained in Refs. [7] and confirmed in Refs. [8].

4 Conclusions

We have explicitly shown, by direct comparison of Eqs. (3) and (13), that the singular factor $x^{-\alpha}$ in the Eq. (11) for the initial quark density converts the exponential DGLAP-asympotics into the Regge one. On the other hand, comparison of Eqs. (10) and (13) also shows that this singular factor in the DGLAP fits mimics the total resummation of logarithms of x. This type of factors can be dropped when the total resummation of logarithms of x performed in Ref. [7] is taken into account. The remaining terms, which are regular in x in the DGLAP fits (the terms in squared brackets in Eq. (11)), can obviously be simplified or even dropped at small x and replaced by constants. A more detailed analysis as well as a suggestion to combine the leading logarithms resummation at small x with DGLAP can be found in Ref. [9]. The above results lead to an interesting conclusion: the expressions for the initial parton densities δq used in DGLAP analyses have been commonly believed to be related to non-perturbative QCD effects; indeed they actually mimic the contributions of the perturbative QCD, so the whole impact of the non-perturbative QCD effects on g_1 at small x is not large and can be approximated by a normalization constant.

[1] G. Altarelli and G. Parisi, Nucl. Phys.B126 (1977) 297; V.N. Gribov and L.N. Lipatov, Sov. J. Nucl. Phys. 15 (1972) 438; L.N.Lipatov, Sov. J. Nucl. Phys. 20 (1972) 95; Yu.L. Dokshitzer, Sov. Phys. JETP 46 (1977) 641.

[2] B.I. Ermolaev, S.I. Manaenkov and M.G. Ryskin. Z.Phys.C69(1996)259; J. Bartels, B.I. Ermolaev and M.G. Ryskin. Z.Phys.C 70(1996)273; J. Bartels, B.I. Ermolaev and M.G. Ryskin. Z.Phys.C 72(1996)627.

[3] J. Blumlein, A. Vogt. Acta Phys. Polon. B27 (1996) 1309. J. Blumlein, S. Riemersma, A. Vogt. Nucl.Phys.Proc.Suppl.51C (1996) 30; Acta Phys. Poloon. B28 (1997) 577.

[4] J. Kwiecinski and B. Ziaja. Phys. Rev. D60(1999)054004; J. Kwiecinski, M. Maul. Phys.Rev.D67(2003)03401; B. Ziaja. Phys.Rev.D66(2002)114017; A. Kotlorz, D. Kotlorz. Acta Phys.Polon. B34(2003)2943; Acta Phys.Polon. B35(2004)705.

[5] B.I. Ermolaev, M. Greco and S.I. Troyan. Phys.Lett.B 522(2001)57.

[6] D. Kotlorz A. Kotlorz. Acta Phys. Polon. B35(2004)2503; hep-ph/0407040.

[7] B.I. Ermolaev, M. Greco and S.I. Troyan. Nucl.Phys.B 594 (2001)71; ibid 571(2000)137; Phys.Lett.B579(321),2004.

[8] J. Soffer and O.V. Teryaev. Phys. Rev.56(1997)1549; A.L. Kataev, G. Parente, A.V. Sidorov. Phys.Part.Nucl 34(2003)20; Nucl.Phys.A 666(2000)184; A.V. Kotikov, A.V. Lipatov, G. Parente, N.P. Zotov. Eur.Phys.J.C26(2002)51; V.G. Krivohijine, A.V. Kotikov, hep-ph/0108224; A.V. Kotikov, D.V. Peshekhonov hep-ph/0110229; N.I. Kochelev, K. Lipka, W.D. Novak, A.V. Vinnnikov. Phys. Rev. D67(2003) 074014.

[9] B.I. Ermolaev, M. Greco and S.I. Troyan. Phys. Lett. B622 (2005) 93 (hep-ph/0503019).

[10] G. Altarelli, R.D. Ball, S. Forte and G. Ridolfi, Nucl. Phys. B496 (1997) 337; Acta Phys. Polon. B29(1998)1145;

[11] A. Vogt. hep-ph/0408244.

$B^0 - \bar{B}^0$ MIXING[a]

A.A.Pivovarov [b]

Institute for Nuclear Research of the Russian Academy of Sciences, 117312 Moscow, Russia

Abstract. The results of calculations of three-loop α_s corrections relevant for extraction of the matrix element of $B^0 - \bar{B}^0$ mixing with next-to-leading order accuracy are reported [1].

Presently the pattern of CP violation in the Standard Model is under thorough experimental study in dedicated experiments by BABAR collaboration at SLAC (e.g. [2]) and BELLE collaboration at KEK (e.g. [3]). The systems of $B_d(\bar{d}b)$ and $B_s(\bar{s}b)$ mesons are the most promising laboratory for performing a precision analysis of CP violation and mixing both experimentally and theoretically [4].

Mixing in any system of neutral pseudoscalar mesons is described by a 2x2 effective Hamiltonian or mass operator $(M - i\Gamma/2)_{ij}$, $\{i,j\} = \{1,2\}$ where M is related to the mass spectrum of the system and Γ describes the widths of the mesons. In the presence of flavor violating interactions ($\Delta B = 2$ in our particular case) the effective Hamiltonian acquires non-diagonal terms. The difference between the values of the mass eigenstates of B mesons $\Delta m = M_{heavy} - M_{light} \approx 2\,|M_{12}|$ is precisely measured $\Delta m = 0.489 \pm 0.005(stat) \pm 0.007(syst)\ ps^{-1}$ [5]. With an accurate theoretical description of the mixing, it can be used to extract the top quark CKM parameters.

The expression for the effective Hamiltonian describing $\Delta B = 2$ transitions in the Standard Model reads [6]

$$H_{\rm eff}^{\Delta B=2} = \frac{G_F^2 M_W^2}{4\pi^2} \left(V_{tb}{}^* V_{td}\right)^2 \eta_B S_0(x_t) \times \left[\alpha_s^{(5)}(\mu)\right]^{-6/23} \mathcal{O}(\mu)$$

where G_F is a Fermi constant, M_W is the W-boson mass, $\eta_B = 0.55 \pm 0.1$ [7], $S_0(x_t)$ is the Inami-Lim function [8], and $\mathcal{O}(\mu) = (\bar{b}_L\gamma_\sigma d_L)(\bar{b}_L\gamma_\sigma d_L)(\mu)$ is a local four-quark operator at the normalization point μ. Mass splitting of heavy and light mass eigenstates is

$$\Delta m = 2|\langle \bar{B}^0| H_{\rm eff}^{\Delta B=2} |B^0\rangle| \qquad (1)$$

The largest uncertainty in calculation of the mass splitting is introduced by the hadronic matrix element $\mathcal{A} = \langle \bar{B}^0|\mathcal{O}(\mu)|B^0\rangle$ that is poorly known. The evaluation of this matrix element is a genuine non-perturbative task, which should be approached with some non-direct techniques. The simplest approach ("factorization") [9] reduces the matrix element $\mathcal{A}$ to the product of simpler

[a] supported in part by the Russian Fund for Basic Research under contract 05-01-00992
[b] e-mail: aapiv@ms2.inr.ac.ru

matrix elements measured in leptonic B decays

$$\mathcal{A}^f = \frac{8}{3}\langle \bar{B}^0|\bar{b}_L\gamma_\sigma d_L|0\rangle\langle 0|\bar{b}_L\gamma^\sigma d_L|B^0\rangle = \frac{2}{3}f_B^2 m_B^2 \tag{2}$$

where the decay constant f_B is defined by $\langle 0|\bar{b}_L\gamma_\mu d_L|B^0(\mathbf{p})\rangle = ip_\mu f_B/2$ and m_B is the B^0 meson mass. A deviation from the factorization ansatz is usually described by the parameter B_B defined as $\mathcal{A} = B_B\mathcal{A}^f$; in factorization $B_B = 1$. The evaluation of this parameter has long history [10–16].

In my talk I report on the results of the calculation of the B_B parameter using QCD sum rule techniques for three-point functions [1,11–13,17,18]. This approach is very close in spirit to lattice computations [15].

To start with let me introduce the three-point correlation function

$$\Pi(p_1, p_2) = \int dxdy\langle 0|TJ_{\bar{B}}(x)\mathcal{O}(0)\bar{J}_B(y)|0\rangle e^{ip_2x-ip_1y} \tag{3}$$

of the relevant $\Delta B = 2$ operator $\mathcal{O}(\mu)$ and interpolating currents for the B^0-meson $J_B = (m_b + m_d)\bar{d}i\gamma_5 b$. Here m_b is the b quark mass. The main relevant property of this current is $\langle 0|J_B(0)|B^0(p)\rangle = f_B m_B^2$ where m_B is the B-meson mass. A dispersive representation of the correlator reads

$$\Pi(p_1, p_2) \equiv \Pi(p_1^2, p_2^2, q^2) = \int \frac{\rho(s_1, s_2, q^2)ds_1 ds_2}{(s_1 - p_1^2)(s_2 - p_2^2)} \tag{4}$$

where $q = p_2 - p_1$.

The matrix element $\langle \bar{B}^0|\mathcal{O}(\mu)|B^0\rangle$ determines the contribution of the B-mesons in the form of a double pole

$$\Pi(p_1^2, p_2^2, q^2) = \frac{\langle J_{\bar{B}}|\bar{B}^0\rangle}{m_B^2 - p_1^2}\langle \bar{B}^0|\mathcal{O}(\mu)|B^0\rangle\frac{\langle B^0|\bar{J}_B\rangle}{m_B^2 - p_2^2} + \dots \tag{5}$$

Because of technical difficulties of calculation, a practical way of extracting the $B^0 - \bar{B}^0$ matrix element is to analyze the moments of the correlation function at $p_1^2 = p_2^2 = 0$ at the point $q^2 = 0$

$$M(i,j) \equiv \frac{\partial^{i+j}\Pi(p_1^2, p_2^2, 0)}{i!j!\partial p_1^{2i}\partial p_2^{2j}} = \int \frac{\rho(s_1, s_2, 0)ds_1 ds_2}{s_1^{i+1}s_2^{j+1}}\,.$$

A theoretical computation of these moments reduces to an evaluation of single scale vacuum diagrams and can be done analytically with available tools for the automatic computation of multi-loop diagrams.

The expression for the "theoretical" moments reads

$$M_{th}(i,j) = \frac{m^6 a_{ij}}{m^{2(i+j)}}\left(1 + \frac{\alpha_s}{4\pi}\left(b_{ij}^f + b_{ij}^{nf}\right)\right) \tag{6}$$

where the quantities a_{ij}, b^f_{ij} and b^{nf}_{ij} represent LO, NLO factorizable and NLO nonfactorizable contributions. The analytical result for the lowest finite moment $M_{th}(2,2)$ reads

$$a_{22} = \frac{1}{(16\pi^2)^2}\left(\frac{8}{3}\right), \quad b^f_{22} = \frac{40}{3} + \frac{16\pi^2}{9}, \tag{7}$$

$$b^{nf}_{22} = S_2\frac{8366187}{17500} - \zeta_3\frac{84608}{875} - \pi^2\frac{33197}{52500} - \frac{426319}{315000}.$$

Here $S_2 = \frac{4}{9\sqrt{3}}\mathrm{Cl}_2\left(\frac{\pi}{3}\right) = 0.2604\ldots$, $\zeta_3 = \zeta(3)$, and $\mu^2 = m^2$. Higher moments contain the same transcendental entries S_2, ζ_3, π^2 with different numerical coefficients. The numerical values for the moments are b^{nf}_{ij}: $b^{nf}_{2(2345)} = \{0.68, 1.22, 1.44, 1.56\}$ and $b^{nf}_{3(34)} = \{1.96, 2.25\}$. The above theoretical results are used to extract the non-perturbative parameter B_B from the sum rules analysis. We extract ΔB by a combined fit of several "theoretical" and "phenomenological" moments. The final formula for the determination of ΔB reads

$$\frac{\alpha_s}{4\pi}b^{nf}_{ij} = \Delta B + \Delta R(z^{j-2} + z^{i-2}) + \Delta C z^{i+j-4} \tag{8}$$

where ΔR and ΔC are free parameters of the fit, $z = m_B^2/(m_B^2 + \Delta)$ is a parameter that describes the suppression of higher state contributions. Δ is a gap between the squared masses of the B-meson and higher states. ΔR, ΔC are parameters of the model for higher state contributions within the sum rule approach. We take $\Delta = 0.4m_B^2$ for the B meson two-point correlator. This corresponds to the duality interval of 1 GeV in energy scale for the analysis based on finite energy sum rules [19]. Estimating all uncertainties we finally find the NLO non-factorizable QCD corrections to ΔB due to perturbative contributions to the sum rules to be

$$\Delta B = (6 \pm 1)\frac{\alpha_s(m)}{4\pi}$$

For $m_b = 4.8$ GeV and $\alpha_s(m_b) = 0.2$ [20] one finds $\Delta B = 0.1$.

In conclusion, the $B^0 - \bar{B}^0$ mixing matrix element has been evaluated in the framework of QCD sum rules for three-point functions at NLO in perturbative QCD. The effect of radiative corrections on B_B is under complete control within pQCD and amounts to approximately +10% of the factorized value.

References

[1] J. G. Körner, A. I. Onishchenko, A. A. Petrov and A. A. Pivovarov, *Phys. Rev. Lett.* **91**, 192002 (2003).

[2] B. Aubert *et al.* [BABAR Collaboration], *Phys.Rev.* D **66**, 032003 (2002).
[3] K. Abe *et al.* [Belle Collaboration], *Phys.Rev.Lett.* **86**, 3228 (2001).
[4] A. Ali and D. London, arXiv:hep-ph/0002167.
[5] K. Hagiwara *et al.*, *Phys.Rev.* D **66**, 010001 (2002).
[6] F. J. Gilman and M. B. Wise, *Phys.Rev.* D **27**, 1128 (1983).
[7] A. J. Buras, M. Jamin and P. H. Weisz, *Nucl.Phys.* B **347**, 491 (1990); J. Urban, F. Krauss, U. Jentschura and G. Soff, *Nucl.Phys.* B **523**, 40 (1998).
[8] T. Inami and C. S. Lim, *Prog. Theor. Phys.* **65**, 297 (1981) [Erratum-ibid. **65**, 1772 (1981)].
[9] M. Gaillard and B. Lee, *Phys. Rev.* D **10**, 897 (1974).
[10] J. Bijnens, H. Sonoda and M. B. Wise, *Phys. Rev. Lett.* **53**, 2367 (1984); W. A. Bardeen, A. J. Buras and J. M. Gerard, *Phys. Lett.* B **211**, 343 (1988).
[11] K. G. Chetyrkin, A. L. Kataev, A. B. Krasulin and A. A. Pivovarov, *Phys.Lett.* B **174**, 104 (1986); L. J. Reinders and S. Yazaki, *Nucl. Phys.* B **288**, 789 (1987); K. G. Chetyrkin and A. A. Pivovarov, *Nuovo Cim.* A **100**, 899 (1988); A. A. Pivovarov, *Nucl. Phys. Proc. Suppl.* B **23**, 419 (1991).
[12] A. A. Ovchinnikov and A. A. Pivovarov, *Sov. J. Nucl. Phys.* **48**, 120 (1988), *Phys. Lett.* B **207**, 333 (1988).
[13] L. J. Reinders and S. Yazaki, *Phys. Lett.* B **212**, 245 (1988).
[14] A. Pich, *Phys. Lett.* B **206**, 322 (1988); S. Narison and A. A. Pivovarov, *Phys. Lett.* B **327**, 341 (1994).
[15] D. Becirevic et al., *Phys. Lett.* B **487**, 74 (2000); D. Becirevic et al., *Nucl. Phys.* B **618**, 241 (2001); M. Lusignoli, G. Martinelli and A. Morelli, *Phys. Lett.* B **231**, 147 (1989); M. Lusignoli, L. Maiani, G. Martinelli and L. Reina, *Nucl.Phys.* B **369**, 139 (1992); C. T. Sachrajda, *Nucl. Instrum. Meth.* A **462**, 23 (2001); J. M. Flynn and C. T. Sachrajda, *Adv. Ser. Direct. High Energy Phys.* **15**, 402 (1998).
[16] A. Hiorth and J. O. Eeg, *Eur.Phys.J.* C **30**, 006 (2003).
[17] M. A. Shifman, A. I. Vainshtein and V. I. Zakharov, *Nucl.Phys.* B **147**, 448 (1979).
[18] N. V. Krasnikov, A. A. Pivovarov and A. N. Tavkhelidze, *Pisma Zh. Eksp. Teor. Fiz.* **36**, 272 (1982); *Z. Phys.* C **19**, 301 (1983).
[19] N. V. Krasnikov and A. A. Pivovarov, *Phys. Lett.* B **112**, 397 (1982), *Yad. Fiz.* **35**, 1270 (1982); A. A. Pivovarov, *Yad. Fiz.* **62**, 2077 (1999).
[20] J. H. Kühn, A. A. Penin and A. A. Pivovarov, *Nucl. Phys.* B **534**, 356 (1998); A. A. Penin and A. A. Pivovarov, *Phys. Lett.* B **435**, 413 (1998), *Nucl. Phys.* B **549**, 217 (1999).

MASS SPECTRA OF RADIALLY AND ORBITALLY EXCITED STATES OF MESONS

V.Khruschov [a], V.Savrin [b], S.Semenov [c]

[a,c] *RRC "Kurchatov Institute", Kurchatov Sq., 123182 Moscow, Russia*

[b] *Skobeltsyn Institute of Nuclear Physics, Moscow State University*

Abstract. Meson mass spectra, evaluated in the framework of the relativistic model of quasi-independent quarks, are presented. Mass values are obtained with the help of numerical calculations based on the Dirac equation and by phenomenological mass formulae. The Dirac equation involves the potential, which is sum of the vector quasi-Coulombic potential and the scalar linear rising confinement potential. The phenomenological mass formulae are applied to excited meson states consisting of $u-$, $d-$quarks and antiquarks with isotopical spin $I = 1$. A comparison of the evaluated mass spectra with existing data is performed. Problems of identification of some meson states in vector and scalar channels are discussed.

It is known that the calculation of hadron mass spectra on the level of experimental data precision [1] still remains among the unsolved QCD problems due to some technical and conceptual difficulties, which are related mainly with the nonperturbative effects, such as confinement and spontaneous chiral symmetry breaking. Nowadays hadron characteristics calculations are frequently carried out with the help of phenomenological models. Among phenomenological models the potential quark models are the simplest ones, they make it possible to represent the different stages of calculations in terms of commonly used physical quantities. Although these models are suitable rather well for the heavy quark hadrons treating of hadrons containing the light quarks is more complicated task, which demands relativistic and nonpotential effects to be taken into account [2–4]. The relativistic model of quasi-independent quarks is applicable to the description of properties of light and heavy hadrons [5–8]. Below we present the results of calculations in the framework of this model [8–10] of mass spectra of quarkonium type mesons, which are orbitally and radially excited states. We compare results obtained with data, make predictions for mass values of unobserved yet meson states and discuss possible structures of some mesons in scalar and vector channels.

According to the main statement of the independent-quark model a hadron is considered as a system composed of a few non-interacting with each other directly valency constituents (quarks, diquarks and constituent gluons) with the coordinates $\mathbf{r_i}$, $i = 1, ..., N$, which moving in some mean colour singlet confining field. It is assumed that the field is spherically symmetric and its motion in space is determined by motion of its center with the coordinate $\mathbf{r_0}$. We treat the mean field as a quasi-classical object possessing some energy ϵ_0. Meantime each of N constituents interacting with the mean field gets the state

[a] e-mail: khru@imp.kiae.ru

[b] e-mail: savrin@theory.sinp.msu.ru

[c] e-mail: semenov@imp.kiae.ru

with a definite value of its energy ϵ_i, which can be determinated after solving of the model equation. We choose the Dirac equation as the model equation for quarks (the Klein-Gordon equation for diquarks and constituent gluons):

$$\sqrt{\lambda_i + m_i^2}\psi_i(\mathbf{r_i}) = [(\boldsymbol{\alpha_i}\mathbf{p_i}) + \beta_{\mathbf{i}}(m_i + V_O) + V_I]\,\psi_{\mathbf{i}}(\mathbf{r_i}), \qquad (1)$$

with $E_i(n_i^r, j_i) = \sqrt{\lambda_i + m_i^2}$, $i = 1, 2$, $V_0(r) = \sigma r/2$ and $V_1(r) = -2\alpha_s/3r$, where the model parameters σ and α_s have meanings of the string tension and the strong coupling constant at small distances, correspondingly. In order to exclude superfluous meson states, the following selection rules for $n^{2S+1}L_J$ -states with quark masses m_1, m_2 and quantum numbers n_1^r, j_1, n_2^r, j_2 are used

$$\begin{array}{l} n_1^r = n_2^r = n-1, \quad j_1 = j_2 = J + 1/2, \quad if \quad J = L + S, \\ j_1 = j_2 + 1 = J + 3/2, \quad if \quad J \neq L + S, \quad m_1 \leq m_2, \end{array} \qquad (2)$$

After the elimination the angular dependence the radial equation obtained can be solved only by numerical methods. For calculating its eigenvalues the code algorithm, which was based on the Numerov three point recurrent relation [11], has been used jointly with the regularization procedure for the singular potential at $r = 0$. In order to estimate the spin-spin interaction between a quark and an antiquark we use the expression for this interaction from the relativized quark model [3]: $V_{S_1S_2} = \frac{32\pi\alpha_s \boldsymbol{s_1 s_2}}{9E_1E_2}\delta^3(\mathbf{r})$, where $\boldsymbol{s_1}$ and $\boldsymbol{s_2}$ are the spin operators and E_1, E_2 are the energies of quark and antiquark, respectively. Evaluation of the expectation value of this operator in the first order of perturbation theory shows that a contribution of the spin-spin interaction is a slowly varying function versus E_1 and E_2 and depends mainly on the total spin of quark-antiquark system and the α_s. For masses of S-wave mesons [9]:

$$M_m = E_{0m} + E_1(n_1^r, 1/2) + E_2(n_2^r, 1/2) + 4 < \boldsymbol{S_1 S_2} >_{q_1\bar{q}_2} V_{SS}, \qquad (3)$$

where V_{SS} is approximately proportional to α_s with constant v_{0S}. As calculations show, the E_{0m} are negligible for the majority of 0^{-+} and 1^{--} mesons, excluding only a few cases, which will be discussed below. It should be noted, that for ground states of $\overline{q}q'-$ mesons the form of relation (3) is analogous (after the substitution $E_i \to m_i^c$, $i = 1, 2$) to the mass formula, which depends on constituent quark masses m_i^c [12]. Although for vector heavy mesons the contributions of spin-spin interaction are on the level of the model absolute uncertainty $\sim 40MeV$, nevertheless, for pseudoscalar mesons these contributions should be taken into account. The magnitude of the spin-spin interaction constant v_{0S} is approximately equal to $100MeV$. In the case of isoscalar mesons, which are neutral with respect to flavour quantum numbers, the terms due to the annihilation interaction for quark and antiquark or instanton interaction appear [13, 14]. But in the framework of our model the contributions of such type can be included into E_{0m} terms.

We have found that a suitable fit within the model accuracy for masses of ground state mesons, which are composed of quark and antiquark with $u-$, $d-$, $s-$, $c-$ or $b-$ flavours [1], is provided by the following values of E_{0m} parameter: $E_0(\eta_c) = E_(J/\Psi) = -150MeV$, $E_0(K) = -200MeV$, $E_0(\pi) = E_0(\eta_b) = E_0(\Upsilon) = -450MeV$. For the α_s and the quark masses we obtain the following values: $\bar{m} = (0.007 \pm 0.005)$ GeV, $\alpha_s^n = 0.60 \pm 0.15$, $m_s = (0.14 \pm 0.03)$ GeV, $\alpha_s^s = 0.47 \pm 0.10$, $m_c = (1.28 \pm 0.05)$ GeV, $\alpha_s^c = 0.33 \pm 0.03$, $m_b = (4.60 \pm 0.10)$ GeV, $\alpha_s^b = 0.27 \pm 0.02$, $\sigma = (0.20 \pm 0.01) GeV^2$. Within the model accuracy data for the well known experimental mass values for 1^{--} and 0^{-+} mesons with u-, d-, s-, c- or b quark and antiquark are consistent with the presented above model parameters. Thus in the framework of the model the flavor independence of the confinement potential [15–17] is confirmed as well as the concordance with the asymptotic freedom behaviour of the α_s.

In spite of difficulties which exist in for evaluations of mass spectra of light hadrons a number of relations among their masses have been obtained using phenomenological considerations, such as the mass relations for Regge trajectories, daughter Regge trajectories [18–21] and the mass formulae, obtained on the base of finite energy QCD sum rules [22]. In the framework of relativistic quasi-independent quarks model the quark and antiquark energies are determined from the solutions of the Dirac equation. However, for the superlight $\overline{q}q'-$mesons with $u-$ and $d-$ quarks and antiquarks only, the phenomenological energy spectral functions $E_i(n_i, j_i, c, \kappa)$ are suitable within the relative uncertainty of the order of a few percents [7]:

$$M(n^{2S+1}L_J) = E_1(n_1^r, j_1, c, \kappa) + E_2(n_2^r, j_2, c, \kappa), \tag{4}$$

where

$$E_i(n_i, j_i, c, \kappa) = \begin{cases} c + \kappa\sqrt{2n^r + L + j - 1/2}, & L + j - 1/2 = 2k, \quad k = 0, 1, \ldots \\ \kappa\sqrt{2n_r + L + j - 1/2}, & L + j - 1/2 = 2k + 1, \quad k = 0, 1, \ldots \end{cases}$$

Note that nonpotential corrections, which are connected mainly with c value, are the most important for mass spectra of light mesons [4, 13]. When evaluating masses of unobserved yet excited meson states we use the formulae written above together with the values of two parameters c and κ, which have been obtained by fitting the mass values of experimentally detected meson states. The c and κ values are $c = 69\ MeV$, $\kappa = 382 \pm 4\ MeV$ [10].

The mass formulae presented above permit to explain the degeneracy of mass values for different J^{PC} mesons. For example, we get $M(\rho'') = M(\rho_3)$. If one substitute the data, then $1720 \pm 20MeV = 1688.8 \pm 2.1MeV$. In the high mass value region the degeneracy of such kind will be increased and it is seen in the $\sim 2315MeV$ region. The mass value calculated for orbital excitation of the 1^{--} meson is equal to $1506MeV$. So mixing between the radial and orbital excitations for the 1^{--} mesons should take place, which leads to the experimentally visible vector meson states V'_R and V'_O. Taking into account the experimental data the following cases are most preferable. In the first case the observed bump at $1465 MeV$ consists of one resonance V'_R, that is mainly the radial excitation of $\rho-$meson. In the second case the observed bump at $1465 MeV$ consist of two resonances V'_R and V'_O (see also [23]). Another importain problem is the interpretation of mesons discovered in scalar channel. The results of our evaluations confirm the existence of $P-$wave $\overline{q}q'-$mesons with masses $\sim 980\ MeV$. Another fact in favour of $P-$wave $\overline{q}q'-$mesons with masses $\sim 980\ MeV$ is the coincidence of the evaluated mass for the first radial excitation of $a_0(980)$ meson with the mass value of the $a_0(1450)$ meson [1]. This meson has been predicted for the first time in the framework of finite energy QCD sum rules [24].

Acknowledgments

We are grateful to B.A. Arbuzov, Yu.V. Gaponov, V.N. Melnikov, M.I. Kalinin, E.E. Boos, A.E. Dorokhov, A.V. Leonidov, A.L. Kataev, A.M. Snigirev and V.E. Troizky for valued discussions.

References

[1] S.Eidelman et al. (Particle Data Group), *Phys.Lett.* **B** 592, 1 (2004).
[2] A.A.Bykov, I.M.Dremin, A.V.Leonidov, *Usp.Fis.Nauk* 143, 3 (1984).
[3] S.Godfrey and N.Isgur, *Phys.Rev.* **D** 32, 189 (1985).
[4] B.Diekmann, *Phys.Rep.* 159, 99 (1988).
[5] P.N.Bogoliubov, *Ann.Inst. H.Poincare* 8, 2 (1968).
[6] P.N.Bogoliubov, A.E.Dorokhov, *Part.& Nucl.* 8, 917 (1987).
[7] V.V.Khruschev, *Yad.Fiz.* 55, 773 (1992).
[8] V.V.Khruschev, V.I.Savrin, S.V.Semenov, *Phys.Lett.* **B** 525, 283(2002).
[9] V.V.Khruschev, S.V.Semenov, *Part.Nucl.Lett.* 5, 5 (2002).
[10] V.V.Khruschov, *prep. IAE-6359/2*, Moscow, 2005; [*hep-ph/*0504077].
[11] B.V.Numerov, *Proc.Centr.Astrophys.Observ.* 2, 188 (1933).
[12] Ya.B.Zeldovich, A.D.Sakharov, *Yad.Fiz.* 4, 395 (1966).
[13] A.T.Filippov, *Yad. Fiz.* 29, 1035 (1979).
[14] A.E.Dorokhov, Ya.A.Zubov, N.I.Kochelev, Part.&Nucl. 23 (1992) 1192.
[15] A.E.Dorokhov, *prep. JINR* R2-121159, Dubna, 1979.
[16] A.Martin, *prep. CERN* TH-2741, Geneva, 1979.
[17] C.Quigg, J.L.Rosner, H.B.Thacker, *FERMILAB* 79-52(THY), 1979.
[18] M.Toller in Proc. 8 Nobel Symp. on Elem. Part. Theory, N.Y. 1968.
[19] G.Veneziano in Proc. Coral Gables Conf. on Fund. Inter., N.Y. 1969.
[20] Y.Nambu in Lect. at the 1970 Copenhagen Summer Symp., 1970.
[21] A.Anisovich, V.Anisovich, A.Sarantsev, *Phys.Rev.* **D** 62, 051502(2000).
[22] A.L.Kataev, *arxiv: hep-ph/*0406305.
[23] N.N.Achasov, A.A.Kozhevnikov, *Yad.Fiz.* 65, 158 (2002).
[24] S.G.Gorishny, A.L.Kataev, S.A.Larin, *Phys.Lett.* **B** 135, 457 (1984).

INCLUSIVE Θ^+ AND $\Lambda(1520)$ PRODUCTION IN HADRON COLLISIONS AT HIGH ENERGY

I.M.Narodetskii [a], M.A.Trusov [b]
ITEP, 117218 Moscow, Russia

Abstract. We estimate the $\Theta^+(1540)$ and $\Lambda(1520)$ production cross sections in inclusive pp collisions at high energy using the K exchange diagrams. Our conclusion is that inclusive Θ^+ production should be at the level of 1 μb at energies $\sqrt{s}$ > 10 GeV and is about 1% relative to inclusive $\Lambda(1520)$ production. The ratio of the $\Lambda(1520)$ yields in pp and Σp collisions at high energies is ~ 4.

1 Introduction

The possible existence of the Θ^+ pentaquark remains one of the puzzling mysteries of recent years. To date there are more than 20 experiments with evidence for this state, but criticism for the Θ^+ claim arises because similar number of high energy experiments did not find any evidence for the Θ^+, even though the other "conventional" three-quark hyperons such as $\Lambda(1520)$ hyperon resonance are seen clearly [c].

Most of negative high energy experiments are high statistic hadron beam experiments. *E.g.* HERA-B, a fixed target experiment at the 920 GeV proton storage ring of DESY [2] finds no evidence for narrow signals in the $\bar{K}^0 p$ channel and only sets modest upper limits for Θ^+ production of less than 16 μb/N and less than about 12% relative to $\Lambda(1520)$ in mid-rapidity region. This negative result would present serious rebuttal evidence to worry about. However, without obvious production mechanism of the Θ^+ (if it exists) or even $\Lambda(1520)$ the rebuttal is not very convincing.

In scattering hadronic probes at high energy from nuclear target the only positive signal for the Θ^+ decaying to $K^0_S p$ was reported by the SVD Collaboration, using 70 GeV proton in a fixed target arrangement $pA \to \Theta^+ X$ at a center-of-mass energy of about 11.5 GeV. Their initial report [3] recently was supported by a more detailed analysis [4] which increased their pentaquark signal by a factor of about 8.

In this paper we estimate the high-energy behavior of the Θ^+ and $\Lambda(1520)$ production cross sections in inclusive pp processes using the K exchange mechanism, which we believe is at work for the Θ^+ and $\Lambda(1520)$ inclusive production in the fragmentation region. At high energy limit the production cross sections do not depend on energy. We show that the cross section of the Θ^+-production is suppressed compared to the production of $\Lambda(1520)$. This suppression is mainly due to smallness of the coupling constant $G^2_{\Theta KN}$ compared to $G^2_{\Lambda KN}$ that in turn is related to the small width of the Θ^+. As a byproduct we also estimate the inclusive $\Lambda(1520)$ production in Σp collisions.

---

[a] e-mail: naro@itep.ru
[b] e-mail: trusov@itep.ru

[c] The situation is getting more intriguing, as recently CLAS collaboration reported negative results on Θ^+ photoproduction off proton and deuteron with high statistics. Meanwhile LEPS collaboration reported the new evidence of the Θ^+ in $\gamma d \to \Theta^+ \Lambda^*(1520)$. Also, STAR collaboration observed the doubly charged exotic baryon in the pK^+ decay channel. For a recent reviews see [1]. Therefore pentaquark existence is still under the question and new experiments are needed to confirm or refute it.

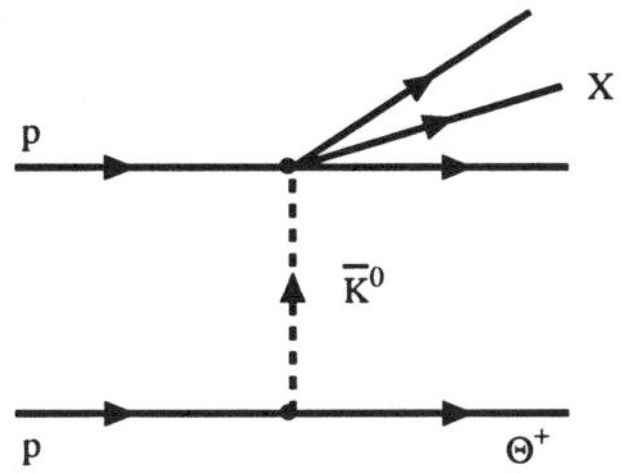

Figure 1: Θ^+ production in inclusive pp collisions

2 Inclusive cross sections

Consider Θ^+ production in processes of the type

$$p + p \to \Theta^+ + X, \tag{1}$$

where X is unspecified inclusive final state carrying the strangeness -1. The $\bar{K}^0$-exchange diagram for $pp \to \Theta^+ X$ is shown in fig. 1. The standard expression for the single inclusive Θ^+ hadroproduction cross section in pp collisions in terms of the 4-momentum transfer squared $t = q^2$ and the invariant mass $W = \sqrt{s_1}$ of the $\bar{K}^0 p$ system is well known (see *e.g.* [5]). At high energy, it is more convenient to convert this expression to an integral over the Feynman variable x_F, the fraction of the incident proton momentum carried by the Θ^+ in the initial direction of the proton (in the center-of-mass system), and $k_\perp$, the transverse momentum of Θ^+ relative to the initial proton direction. Then the contribution of the K meson exchange to the double differential cross section for the Θ^+ inclusive production reads

$$\frac{d\sigma}{dx_F dk_\perp^2} = \frac{1}{4\pi} \frac{G^2_{\Theta KN}}{4\pi} J \frac{p}{E_\Theta} \Phi_\Theta(t) F^4(t)\, \sigma^{tot}_{K^0 p}(s_1), \tag{2}$$

where

$$s_1 = s(p) + m_\Theta^2 - 2E_\Theta \sqrt{s(p)}, \quad E_\Theta = \sqrt{x_F^2 p^2 + k_\perp^2 + m_\Theta^2}, \tag{3}$$

$\sqrt{s(p)} = 2\sqrt{p^2 + m_p^2}$ being the center-of-mass energy, and

$$t = m_p^2 + m_t^2 - 2E_\Theta \sqrt{p^2 + m_p^2} + 2x_F p^2. \tag{4}$$

The factor J in (2) is the ratio of flux factors in the pp and $\bar{K}^0 p$ reactions

$$J = \sqrt{\frac{\lambda(s_1, q^2, m_p^2)}{\lambda(s, m_p^2, m_p^2)}}, \tag{5}$$

where

$$\lambda(s, m_1^2, m_2^2) = s^2 - 2s(m_1^2 + m_2^2) + (m_1^2 - m_2^2)^2. \tag{6}$$

The function $\Phi_\Theta(t)$ is the squared product of the off-mass-shell $p \to \Theta^+ K^0$ vertex and the kaon propagator

$$\Phi_\Theta(t) = \frac{(m_N - m_\Theta)^2 - t}{(t - m_K^2)^2}. \tag{7}$$

To evaluate the cross sections away from the pole position $t = M_K^2$ we include the phenomenological form factor $F_K(t)$.

In the high energy limit with accuracy $\mathcal{O}(1/p^2)$

$$J\,\frac{p}{E_\Theta} \approx \frac{1-x_F}{x_F}, \quad s_1 \approx (1-x_F)s, \quad q^2 \approx m_\Theta^2 + m_p^2(1-x_F) - \frac{m_\Theta^2 + k_\perp^2}{x_F}, \tag{8}$$

and Eq. (9) written in terms x_F and $k_\perp^2$ reads

$$\frac{d\sigma}{dx_F dk_\perp^2} = \frac{1}{4\pi}\,\frac{G_{\Theta KN}^2}{4\pi}\,\frac{1-x_F}{x_F}\,\Phi_\Theta(x_F, k_\perp^2)\,F^4(t)\,\sigma_{\bar{K}^0 N}(s_1). \tag{9}$$

Because of (8) all the energy dependence of the right hand side of Eq. (9) is due to the factor $\sigma_{\bar{K}^0 N}(s_1)$. Since $\sigma_{\bar{K}^0 N}(s_1)$ is a constant (up to logarithms) at large enough s, Eq. (9) asymptotically does not depend on energy.

The cross section for the inclusive $\Lambda(1520)$ production has the similar form with the substitutions

$$\frac{G_{\Theta KN}^2}{4\pi} \to \frac{G_{\Lambda KN}^2}{4\pi}, \quad \Phi_\Theta \to \Phi_\Lambda = \frac{(m_p + m_\Lambda)^2 - t}{6 m_\Lambda^2 m_K^2} \cdot \frac{((m_p - m_\Lambda)^2 - t)^2}{(t - m_K^2)^2}. \tag{10}$$

The $KN\Theta^+$ vertex for $J^P(\Theta^+) = \frac{1}{2}^+$ is

$$L_{KN\Theta} = iG_{KN\Theta}(K^\dagger \bar{\Theta}\gamma_5 N + \bar{N}\gamma_5 \Theta K). \tag{11}$$

The Lagrangian of Eq. (11) corresponds with the Θ^+ being a p-wave resonance in the $K^0 p$ system. The partial decay width $\Gamma_{\Theta\to K^0 p}$ is

$$\Gamma_{\Theta\to K^0 p} = \frac{G_{KN\Theta}^2}{4\pi} \cdot \frac{2p_K^3}{(m_\Theta + m_p)^2 - m_K^2}, \tag{12}$$

where $p_K = 260$ MeV/c is the kaon momentum in the rest frame of Θ^+. To extract a value for $G_{KN\Theta}$, we need the experimental information of the width $\Gamma_{\Theta\to KN}$, which is not known precisely but whose measurement is the subject of several planned dedicated experiments, see *e.g.* [6]. To provide numerical estimates, we will use the value $\Gamma_{\Theta\to K^0 p} = 1$ MeV, which is consistent with the upper limit for the width derived from elastic KN scattering [7]. Evaluating Eq. (12) with $\Gamma_{\Theta\to K^0 p} = 1$ MeV, we extract the value $G_{KN\Theta}$

$$\frac{G_{\Theta KN}^2}{4\pi} = 0.167 \cdot \frac{\Gamma_{\Theta\to K^0 p}}{1\ \mathrm{MeV}}, \tag{13}$$

which will be used in the following estimates for the inclusive cross section.

The $KN\Lambda(1520)$ vertex is

$$L_{KN\Lambda} = \frac{G_{KN\Lambda}}{m_K}\left(\bar{\Lambda}^\mu \gamma_5 N \partial_\mu K + \bar{N}\gamma_5 \Lambda^\mu \partial_\mu K^\dagger\right), \tag{14}$$

where Λ^μ is the vector spinor for the spin 3/2 particle. The Lagrangian of Eq. (14) corresponds with the $\Lambda(1520)$ being a d-wave resonance in the $K^- p$ system.

We treat $\Lambda(1520)$ relativistically using the Rarita-Schwinger vector-spinor formalism with the density matrix

$$L^{\mu\nu} = \frac{1}{4}(\hat{p}_\Lambda + m_\Lambda)\left[-g^{\mu\nu} + \frac{1}{3}\gamma^\mu\gamma^\nu + \frac{1}{3m_\Lambda}(\gamma^\mu p_\Lambda^\nu - \gamma^\nu p_\Lambda^\mu) + \frac{2}{3m_\Lambda} p_\Lambda^\mu p_\Lambda^\nu\right]. \tag{15}$$

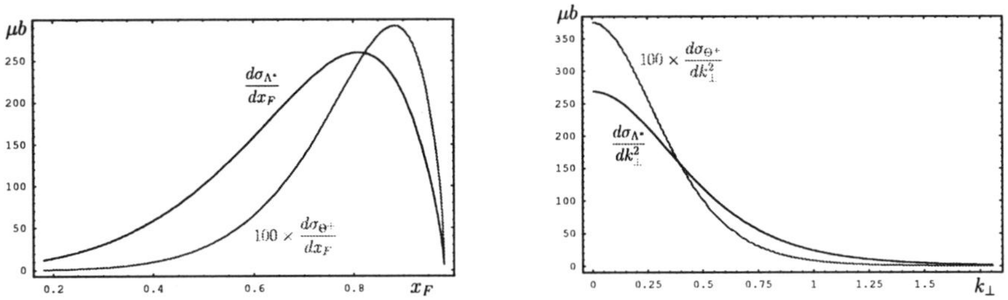

Figure 2: x_F (left figure) and $k_\perp^2$ (right figure) dependence of inclusive cross sections

The $\Lambda(1520) \to pK^-$ width is

$$\Gamma_{\Lambda^* \to K^- p} = \frac{G_{\Lambda NK}^2}{4\pi} \cdot \frac{2p_K^5}{3m_K^2} \cdot \frac{1}{(m_\Lambda + m_p)^2 - m_K^2}, \tag{16}$$

where $p_K = 246$ MeV/c is the kaon momentum in the rest frame of $\Lambda(1520)$. Using the PDG values of $\Gamma_{tot}(\Lambda(1520)) = 15.6$ MeV and $BR(\Lambda(1520) \to N\bar{K}) = 45\%$ we obtain

$$\Gamma_{\Lambda^* \to K^- p} = \frac{1}{2} \cdot BR(\Lambda^* \to N\bar{K}) \cdot \Gamma_{tot} = \frac{1}{2} \cdot 0.45 \cdot 15.6 \text{ MeV} = 3.51 \text{ MeV}, \tag{17}$$

and

$$\frac{G_{\Lambda NK}^2}{4\pi} \approx 8.14. \tag{18}$$

3 Results

The total cross sections can be obtained by integrating (9) over $k_\perp^2$ and x_F. The kinematical limits of integration in (9) are given by equation

$$x_F^2 + \frac{k_\perp^2}{p^2} < \frac{(s(p) + m_\Theta^2 - (m_p + m_K)^2)^2 - 4s(p)m_\Theta^2}{4p^2 s(p)}. \tag{19}$$

For estimation we take the total cross sections $\sigma_{tot}(\bar{K}^0 p)$ and $\sigma_{tot}(K^- p)$ to be a constant ($\sigma_{tot}(\bar{K}^0 p) \sim \sigma_{tot}(K^- p) \sim 20$ mb at $\sqrt{s_1} > 10$ GeV). We employ two representative examples for the form factor $F(t)$ in (9):

$$\text{A}: \quad F(t) = \frac{\Lambda^2 - m_K^2}{\Lambda^2 - t}, \qquad \text{and} \qquad \text{B}: \quad F(t) = \frac{\Lambda^4}{\Lambda^4 + (t - m_K^2)^2}, \tag{20}$$

with the cut-off parameter Λ being a typical hadronic scale $\Lambda = 1$ GeV. Then we obtain for the production cross sections [d]

$$\sigma(pp \to \Theta^+ X) = 0.8\,(1.6) \times \frac{\Gamma_{\Theta KN}}{1 \text{ MeV}}\ \mu\text{b}, \quad \sigma(pp \to \Lambda(1520)^+ X) = 106\,(126)\ \mu\text{b}, \tag{21}$$

where the first values refer to the form factor (A) and the second ones to the form factor (B). The x_F distributions and transverse momentum distributions for the

[d] These values correspond for the IHEP energy $\sqrt{s} = 12$ GeV and practically coincide with the asymptotic values of the integrals provided $\sigma_{tot}(KN)$ is a constant.

$\Theta^+(1540)$ and $\Lambda(1520)$ are shown in fig. 2. For the average transfer momenta squared of Θ^+ and $\Lambda(1520)$ we get $\langle k_\perp^2 \rangle = 0.25$ GeV^2 and $\langle k_\perp^2 \rangle = 0.25$ GeV^2, respectively.

The result for $\sigma(pp \to \Theta^+ X)$ matches well that of Ref. [8] for the inclusive $pp \to \Theta^+ X$ production at $\sqrt{s} < 10$ GeV [e]. One can expect that the K^* exchange yields the cross sections of the similar order of magnitude. The prediction for $\sigma(pp \to \Lambda(1520)^+ X)$ agrees with the preliminary result of the SVD-2 collaboration [4], but our $\sigma(pp \to \Theta^+ X)$ is lower than the preliminary cross section estimation (for $x_F > 0$) of Ref. [4]: $\sigma \cdot BR(\Theta^+ \to pK^0) \sim 6$ μb. The ratio of $\Theta^+(1540)$ to $\Lambda(1520)$ production cross-sections is $\sim 1\%$. Probably today our calculations can be useful to explain why the Θ^+ production is suppressed in some high energy experiments.

In the same way it is possible to make quantitative predictions for other type of colliding particles. As an example, we estimate the cross section for the inclusive $\Lambda(1520)$ production in $\Sigma p \to \Lambda(1520)$ collisions at 600 GeV/c studied in the the fixed target Fermilab experiment E781. This reaction can proceed via the π-meson exchange. Using the experimental value for the branching ratio $BR(\Lambda(1520) \to \Sigma\pi) = 55\%$ and $\sigma_{\rm tot}(\pi N) = 60$ mb we obtain for the ratio

$$\frac{\sigma(pp \to \Lambda(1520)X)}{\sigma(\Sigma p \to \Lambda(1520)X} \approx 4, \tag{22}$$

that qualitatively agrees with the preliminary experimental result of the SELEX collaboration [9].

4 Conclusions

Let us recall that our estimations are no more than a guess, as they may somehow depend on our specific assumptions regarding for instance the K-meson exchange dominance at forward direction, not to say the intrinsic uncertainty coming from the choice of the form factor. As an outlook, it would be interesting to go beyond the present calculation and to perform a systematic study of K, K^* and π (reggeized) exchanges into inclusive production of (anti)strange baryons in pp collisions. We plan to come back to these issues in a next publication.

Acknowledgments

We are grateful to K.G.Boreskov, A.G.Dolgolenko, B.L.Ioffe and A.B.Kaidalov for the discussions. This work was supported by RFBR grants 03-02-17345, 04-02-17263, 05-02-17869 and by the grant for leading scientific schools 1774.2003.2.

References

[1] A.Sandorfi, Experimental Review of Pentaquarks, P. Stoler, Experimental summary, Pentaquark 2005 Workshop, Jlab, Newport News, VA USA, October 20-22, 2005: http://conferences.jlab.org/pentaquark/
[2] I.Abt *et al.* [HERA-B Collaboration], *Phys.Rev.Lett.* **93**, 212003 (2004).
[3] A.Aleev *et al.*, hep-ex/0401024
[4] A.Aleev *et al.*, hep-ex/0509033
[5] T.Yao, *Phys.Rev.* **125**, 1048 (1962)
[6] JLab Hall A experiment E-04-012
[7] R.N.Cahn and G.H.Trilling, *Phys.Rev.* **D** 69, 011501 (2004) [arXiv:hep-ph/0311245].
[8] V.Yu.Grishina, *et al.*, *Eur.Phys.J.* **A** 25, 141 (2005)
[9] A.G.Dolgolenko *et al.*, SELEX Collaboration, Private communication

[e] The authors of Ref. [8] used the standard expression for the inclusive cross section [5] and $\sigma_{\rm tot}(\bar{K}^0 p)$ taken from the parameterization of experimental data at low energies.

TESTING THE k_T-FACTORISATION APPROACH AT THE LHC IN QUARKONIUM PRODUCTION PROCESSES

S.P.Baranov [a]

P.N.Lebedev Institute of Physics, 119991 Moscow, Russia

*Abstract.*We consider the production of Υ and χ_b mesons at the LHC conditions and compare the predictions of the k_T-factorization approach and collinear parton model.

1 Introduction

Over the years, the k_T-factorization approach [1, 2] has demonstrated its capability in describing a large body of experimental data (see [3] and references therein). In many cases, however, the data can also be described within the usual (collinear) parton model, if the relevant next-to-leading order QCD corrections are taken into account or the so called colour-octet mechanism is included. In this context, it would be interesting and important to search for examples, where the difference between the collinear and noncollinear approaches manifests in a clear and unumbiguous way.

Here, we focus attention on two processes, which can be investigated in the experiments planned at the LHC and where the ideological difference between the "collinear" and "k_T-factorised" interpretations of parton model leads to potentially well recognizable effects. The guideline is shown by the theoretical predictions on the J/ψ spin alignment at HERA [4, 5] and the Tevatron [6] where the collinear and k_T-factorization approaches demonstrate qualitatively different behaviour. The production of P-wave charmonium and bottomonium states considered in Ref. [7] gives another useful hint. In the present paper we continue the line of the above works and analyse the production of Υ and χ_b mesons at the LHC energies.

2 Theoretical framework

In the k_T-factorization scheme, the production of Υ and χ_b mesons is dominated by the gluon-gluon fusion partonic subprocesses $g+g \to \Upsilon + g$ and $g+g \to \chi_b$. In the latter case, the nonzero transverse momentum of the final state originates from the initial momenta of the colliding gluons. It has been shown already [6] that the Tevatron data can be successfully described within the colour-singlet model [8, 9] alone, and so, there is no need in the colour-octet contributions. Thus, in the k_T-factorization approach, the production of a heavy meson is described at the purely perturbative level as the production of a colour-singlet $b\bar{b}$ pair with quantum numbers of the quarkonium state under consideration.

[a]e-mail: baranov@sci.lebedev.ru

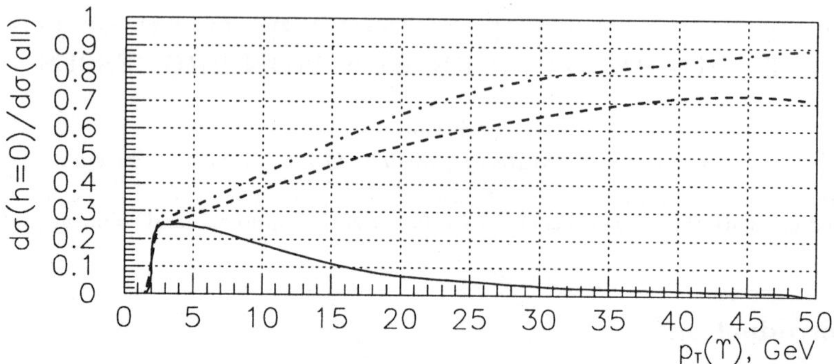

Figure 1: Theoretical predictions on the polarisation of Υ mesons at the LHC conditions. Solid line – collinear parton model (singlet + octet); dashed and dash-dotted lines – k_T-factorization approach with gluon densities from [1, 12] and [11], respectively (singlet only).

In the case of collinear factorisation scheme, we have to invoke the colour-octet model [10], which implies that the heavy $b\bar{b}$ pair is perturbatively created as an octet colour state and then evolves into a physical quarkonium state via emitting soft (nonperturbative) gluons. The dominant contribution to the Υ production at high p_t comes from the 'gluon fragmentation' subprocess via the intermediate ${}^3S_1^{(8)}$ octet state: $g+g \to g+g$, $g \to b\bar{b}[{}^3S_1^{(8)}] \to \Upsilon + ng$. A similar production mechanism is assumed for χ_b states: $g+g \to g+b\bar{b}[{}^3S_1^{(8)}]$, $b\bar{b}[{}^3S_1^{(8)}] \to \chi_b + ng$. Note that, in contrast with the k_T-factorization approach, the production of χ_b mesons refers to a $2 \to 2$ subprocess: otherwise, the final state meson would have no transverse momentum and could not be detected.

The evaluation of the partonic matrix elements is straightforward and is described in every detail in Refs. [6, 7]. The ideological difference between the collinear and k_T-factorization approaches is connected with the initial gluon off-shellness, which leads to the presence of longitudinal components in the gluon polarisation vectors. We take the initial gluon spin density matrix in the form $\overline{\epsilon^\mu \epsilon^{*\nu}} = k_T^\mu k_T^\nu / |k_T|^2$, similar to that of virtual photons in electrodynamics. The partonic matrix elements squared have to be convoluted with the so called un-integrated gluon distribution functions $\mathcal{F}(x, k_T^2, \mu_F^2)$ showing the probability to find a gluon carrying the longitudinal momentum fraction x and transverse momentum k_T at a given probing scale μ_F^2. In the present calculations we use two different parametrisations of gluon densities proposed in Ref. [11] and in Refs. [1, 12]. The 'collinear' calculations are based on the GRV parametrisation [13]. The latter set is also used as the boundary condition for the k_T-dependent gluon distributions. The setting of the other model parameters is as in our previous publications [6, 7]: the quark mass $m_b = m_\Upsilon/2 = 4.5$ GeV, the factorisation and renormalisation scales $\mu_F^2 = \mu_R^2 = m_\Upsilon^2 + p_{T,\,\Upsilon}^2$.

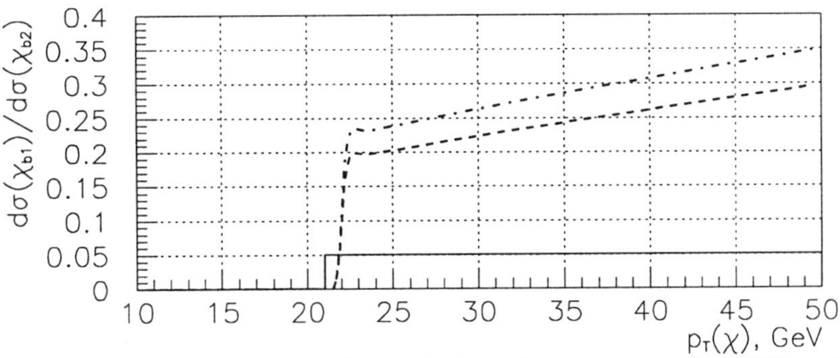

Figure 2: Theoretical predictions on the production of χ_b mesons at the LHC conditions. Solid line – collinear parton model (singlet + octet); dashed and dash-dotted lines – k_T-factorization approach with gluon densities from [1, 12] and [11], respectively (singlet only).

3 Numerical results and discussion

Fig. 1 displays our predictions on the Υ polarisation at the LHC conditions. The calculations are performed for the typical ATLAS trigger cuts requiring one muon with $p_t > 6$ GeV and another muon with $p_t > 3$ GeV. One can see the dramatic disagreement between the k_T-factorization approach and collinear parton model. The key point lies in the gluon spin density matrix.

Typically, the k_T values of the two colliding gluons are very much different, as the parton evolution is equivalent to the random walk in the $\ln |k_T|$ plane, not in k_T plane. Roughly speaking, the k_T of one of the gluons can be neglected in comparison with that of the other. Thus, the initial state is represented by one nearly on-shell (transversely polarised) gluon and one off-shell (longitudinally polarised) gluon. After the interaction, they convert into one on-shell gluon and one vector meson. Simple helicity conservation arguments show that the polarisation of vector meson must be longitudinal, in contrast with the ordinary parton model, where the initial gluons are both on-shell.

Important effects are also seen in the production of P-wave bottomonium states with different spins χ_{b1} and χ_{b2}, see Fig. 2. We have found that the shapes of the transverse momentum distributions and, also, the ratio of the production rates $\sigma(\chi_1)/\sigma(\chi_2)$ are qualitatively different in the k_T-factorization and the ordinary parton model. One major difference is seen the behaviour of the differential cross section $d\sigma(\chi_2)/dp_T$ at low transverse momentum. This quantity remains finite in the k_T-factorization approach, while it diverges in the collinear parton model when p_T goes to zero. The latter prediction seems to be not supported by the available experimental data on the bottomonium production at the Tevatron.

Another well pronounced difference is seen in the ratio of the production rates $d\sigma(\chi_1)/d\sigma(\chi_2)$. The underlying physics is, again, connected with gluon off-shellness. In the collinear parton model, the relative suppression of χ_1 states becomes stronger with increasing p_T because of the increasing role of the colour-octet contribution: in this approach, the leading-order fragmentation of an on-shell transversely polarised gluon into a vector meson is forbidden. In contrast with that, in the k_T-factorization approach, the increase in the final state p_T is mainly due to the increasing transverse momenta (resp., virtualities) of the initial gluons, and, consequently, the suppression motivated by Landau-Yang theorem becomes weaker at large p_T.

4 Conclusion

We have analysed the effect of the initial gluon virtuality on the production properties of Υ an χ_b mesons at the LHC conditions. We have found that the polarisation of Υ mesons and, also, the ratio of the production rates of χ_{b1} and χ_{b2} mesons can serve as good indicators of the parton dynamics. An experimental measurement of these characteristics can help in discriminating between the k_T-factorization approach and collinear parton model.

References

[1] L.V.Gribov, E.M.Levin, M.G.Ryskin, *Phys.Rep.* 100, 1 (1983);

[2] E.M.Levin, M.G.Ryskin, *Phys.Rep.* 189, 267 (1990); S.Catani, M.Ciafaloni, F.Hautmann, *Phys.Lett.* **B** 242, 97 (1990); *Nucl.Phys.* **B** 366, 135 (1991); J.C.Collins, R.K.Ellis, *Nucl.Phys.* **B** 360, 3 (1991).

[3] B.Andersson et al. (Small-x collab.), *Eur.Phys.J.* **C** 25, 77 (2002).

[4] S.P.Baranov, *Phys.Lett.* **B** 428, 377 (1998).

[5] A.V.Lipatov, N.P.Zotov, *Eur.Phys.J.* **C** 27, 87 (2003).

[6] S.P.Baranov, *Phys.Rev.* **D** 66, 014003 (2002).

[7] S.P.Baranov, *Phys.Lett.* **B** 594, 277 (2004).

[8] C.-H.Chang, *Nucl.Phys.* **B** 172, 425 (1980); R.Baier, R.Rückl, *Phys.Lett.* **B** 51, 153 (1981); E.L.Berger, D.Jones, *Phys.Rev.* **D** 23, 1521 (1981).

[9] H.Krasemann, *Z.Phys.* **C** 1, 189 (1979); G.Guberina, J.Kühn, R.Peccei, R.Rückl, *Nucl.Phys.* **B** 174, 317 (1980).

[10] E.Braaten, S.Fleming, *Phys.Rev.Lett.* 74, 3327 (1995); M.Cacciari, M.Greco, M.L.Mangano, A.Petrelli, *Phys.Lett.* **B** 356, 560 (1995); G.T.Bodwin, E.Braaten, G.P.Lepage, *Phys.Rev.* **D** 51, 1125 (1995); P.Cho, A.K.Leibovich, *Phys.Rev.* **D** 53, 150 (1996), *Phys.Rev.* **D** 53, 6203 (1996).

[11] J.Blümlein, DESY preprint DESY 95-121 (1995).

[12] N.Nikolaev, B.Zakharov, *Nucl.Phys.* **B** 333, 250 (1994).

[13] M.Glück, E.Reya, A.Vogt, *Z.Phys.* **C** 67, 433 (1995).

ON THE RELATION BETWEEN x-DEPENDENCE OF THE HIGHER TWIST CONTRIBUTION TO THE STRUCTURE FUNCTIONS F_3 AND $g_1^p - g_1^n$

A.V. Sidorov[a]

Bogoliubov Theoretical Laboratory

Joint Institute for Nuclear Research, 141980 Dubna, Russia

Abstract. We compare the higher twist (HT) contribution to the unpolarized structure function F_3 with that one to the nonsinglet combination $g_1^p - g_1^n$ of the polarized proton and neutron structure functions using the assumption that the HT contributions to the Gross-Llewellyn Smith and the Bjorken sum rules are similar. We have found, that the relation $\frac{1}{3x}h^{xF_3}(x) \approx \frac{6}{g_A}h^{g_1^p-g_1^n}(x)$ is valid for $x \geq 0.1$ and for $x \geq 0.2$ in the case of LO and NLO QCD approximations, respectively.

The structure functions in deep inelastic lepton nucleon scattering are presently a subject of intensive experimental and theoretical investigations. While the leading twist (LT) part of the structure functions related with the parton distributions and their Q^2-evolution is studied in detail in pQCD, the higher twist corrections ($\sim 1/Q^2$) are of a big interest and an intensive study in the last years. The higher twist effects are especially important in the case of polarized structure functions because the most of the precise data (JLAB, HERMES, SLAC) are in the region of $Q^2 \sim 1$ GeV^2.

In this note we consider the relation between the HT contribution to the unpolarized structure function F_3 and $g_1^p - g_1^n$ which are pure non-singlets. As it was shown in the paper [1] the Q^2-evolutions of the F_3 and the nonsinglet part of the g_1 structure functions are identical up to NLO order. Moreover the x-shapes of the F_3 and nonsinglet part of g_1 are also similar [b]. By analogy, one could suppose that the HT contributions to F_3 and $g_1^p - g_1^n$ are similar too. Such an assumption was recently used for the first moments of the HT corrections in the Gross-Llewellyn Smith and Bjorken sum rules in the infrared renormalons approach [3]:

$$GLS(Q^2) = \int_0^1 dx F_3(x,Q^2) \quad = \quad 3(GLS_{pQCD} - \frac{\langle\langle O_1 \rangle\rangle}{Q^2}) \quad (1)$$

$$Bjp(Q^2) = \int_0^1 dx[g_1^p(x,Q^2) - g_1^n(x,Q^2)] \quad = \quad \frac{g_A}{6}(Bjp_{pQCD} - \frac{\langle\langle O_2 \rangle\rangle}{Q^2}) \quad (2)$$

where

$$\langle\langle O_1 \rangle\rangle \quad \approx \quad \langle\langle O_2 \rangle\rangle \quad (3)$$

Here GLS_{pQCD} and Bjp_{pQCD} are the leading twist contribution to correspond-

---

[a]e-mail: sidorov@theor.jinr.ru

[b]This property is intensively used in the phenomenological applications [2].

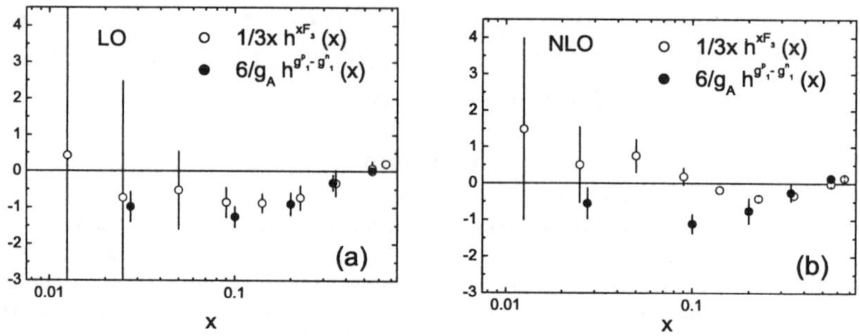

Figure 1: Comparison of the LO and NLO($\overline{\rm MS}$) results for $\frac{1}{3x}h^{xF_3}(x)$ based on the analysis of the CCFR data [6, 7] - (open cycles), and for $\frac{6}{g_A}h^{g_1^p-g_1^n}(x)$ based on the results of [12] - black cycles.

ing sum rules:

$$GLS_{LO} = Bjp_{LO} = 1 \tag{4}$$

$$GLS_{NLO} = Bjp_{NLO} = 1 - \alpha_S(Q^2)/\pi \tag{5}$$

In this note we are going to verify if the relation (3) between the lowest moments of the HT contribution can be generalized for the higher twists themselves, namely:

$$\frac{1}{3x}h^{xF_3}(x) \approx \frac{6}{g_A}h^{g_1^p-g_1^n}(x) \tag{6}$$

To test this relation we will use the values of HT obtained in the QCD analyses of the corresponding structure functions in model independent way. In the QCD analyses of DIS data when the higher twist corrections are taken into account, the structure functions are given by:

$$xF_3(x,Q^2) = xF_3(x,Q^2)_{\rm LT} + h^{xF_3}(x)/Q^2 \tag{7}$$

$$g_1^p(x,Q^2) = g_1^p(x,Q^2)_{\rm LT} + h^{g_1^p}(x)/Q^2 \tag{8}$$

$$g_1^n(x,Q^2) = g_1^n(x,Q^2)_{\rm LT} + h^{g_1^n}(x)/Q^2 \ . \tag{9}$$

In (9) $h^{xF_3}(x)$, $h^{g_1^p}(x)$ and $h^{g_1^n}(x)$ are the *dynamical* higher twists corrections to xF_3, g_1^p and g_1^n, which are related to multi-parton correlations in the nucleon. They are non-perturbative effects and can not be calculated without using models. The target mass corrections, which are also corrections of inverse powers of Q^2, are calculable [4, 5] and effectively belong to the leading twist term. A model independent determination of $h^{xF_3}(x)$ was done in [6][c] on the

[c]See Table 12 in [6]

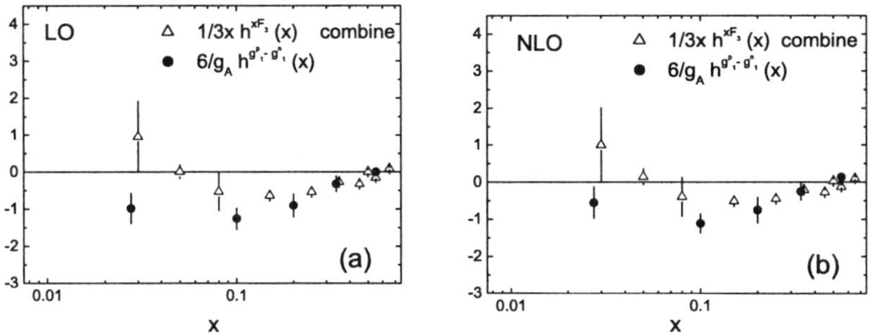

Figure 2: Comparison of the LO and NLO($\overline{\text{MS}}$) results for $\frac{1}{3x}h^{xF_3}(x)$ based on the combined data analysis [?,8] - (open triangles), and for $\frac{6}{g_A}h^{g_1^p-g_1^n}(x)$ based on the results of [12] - black cycles.

basis of the analysis of CCFR-NuTev (anti-)neutrino deep–inelastic scattering data [7] at $Q^2 \geq 5\ GeV^2$ and in [8] using the combine set of data different from that of CCFR at $Q^2 \geq 0.5\ GeV^2$. [d] The values of $h^{g_1^p}(x)$ and $h^{g_1^n}(x)$ in LO [e] and NLO($\overline{\text{MS}}$) are given in [12], where the results of the analysis of the world data on polarized structure function g_1 [13] at $Q^2 \geq 1\ GeV^2$, including the precise JLab g_1^n [14] data, are presented. Using these results and taking into account the coefficients in (1) and (2) one could construct the l.h.s. and r.h.s. of Eq. (6).

In Fig. 1 and 2 we compare the results on HT in l.h.s. and r.h.s. of Eq. (6). One can see (Fig. 1), that while in the polarized case the values of HT change slightly from LO to NLO approximation, in the unpolarized one the shape of $h^{xF_3}(x)$ depends on the order of pQCD used, especially for $x \leq 0.1$. As seen from Fig. 1 and 2 the equality (6) is approximately valid for $x \geq 0.1$ and $x \geq 0.2$ for the LO and NLO approximations, respectively. It means that the higher Mellin moments of the both parts of equation (6) should approximately coincide:

$$\int_0^1 dx\ x^N \frac{1}{3x} h^{xF_3}(x) \quad \approx \quad \int_0^1 dx\ x^N \frac{6}{g_A} h^{g_1^p - g_1^n}(x),\ N - large. \qquad (10)$$

We would like to mention, that equality (3) is suggested in the framework of the infrared renormalon approach, so the violation of equality (4), which is

[d] See [9] for comparison of the results of [10] where the infrared renormalon model approach for HT contribution was applied in analysis of combine set of IHEP-JINR [11] and CCFR-NuTev data.

[e] It should be stressed, that LO approach for QCD analysis of polarised structure function g_1 is not reliable enough. See discussion in [15]

shown in the Fig. 1b, Fig. 2a and 2b at $x < 0.1$, could be due to the contribution of the dynamical twists connected with the non-perturbative structure of the nucleon in this x region.

Finally, it should be noted, that there is additional sources of uncertainties which should be taken into account in a more detailed test of Eq. (6): the contribution of $\mathcal{O}(1/Q^4)$); the separation of the twist-3 contribution in the polarized case, which is effectively included in $h^{g_1}(x)$; the Q^2 dependence of the functions $h(x)$, etc.

This research was supported by the RFBR grants N 05-02-17748 and N 05-01-00992. I am grateful to D.B. Stamenov and C. Weiss for discussions.

[1] J. Kodaira, et al., Phys. Rev. D **20**, 627 (1979); Nucl. Phys. B **159**, 99 (1979).
[2] A.V. Kotikov, D.V. Peshekhonov, Eur. Phys. J. **C 9** (1999) 55.
[3] A.L. Kataev, hep-ph/0505108, to be published in JETP Lett., v. 81(12).
[4] S. Wandzura and F. Wilczek, Phys. Lett. **B 72** (1977) 195.
[5] S. Matsuda and T. Uematsu, Nucl. Phys. **B 168** (1980) 181; H. Kawamura and T. Uematsu, Phys. Lett. **B 343** (1995) 346; A. Piccione and G. Ridolfi, Nucl. Phys. **B 513** (1998) 301; J. Blumlein and A. Tkabladze, Nucl. Phys. Proc. Suppl. **79** (1999) 541; S. Simula, M. Osipenko, G. Ricco and M. Taiuti, Phys. Rev. **D 65** (2002) 034017.
[6] A. L. Kataev, G. Parente and A. V. Sidorov, Phys. Part. Nucl. **34** (2003) 20 [Fiz. Elem. Chast. Atom. Yadra **34** (2003) 43].
[7] W. G. Seligman et al., Phys. Rev. Lett. 79 (1997) 1213.
[8] A.V. Sidorov, JINR Rapid Commun., 80 (1996) 11, hep-ph/9609345.
[9] A. V. Sidorov, arXiv:hep-ph/0505246.
[10] S.I. Alekhin *et al.*, Phys. Lett. **B 512** (2001) 25.
[11] A.V. Sidorov *et al.*, Eur.Phys. J. **C10** (1999) 405.
[12] E. Leader, A.V. Sidorov and D.B. Stamenov, JHEP **06** (2005) 033.
[13] SLAC E142 Coll., P.L. Anthony *et al.*, Phys. Rev. **D 54** (1996) 6620; SLAC/E154 Coll., K. Abe *et al.*, Phys. Rev. Lett. **79** (1997) 26; SMC, B. Adeva *et al.*, Phys. Rev. **D 58** (1998) 112001; HERMES, K. Ackerstaff *et al.*, Phys. Lett. **B 404** (1997) 383; *ibid* **B 442** (1998) 484; SLAC E143 Coll., K. Abe *et al.*, Phys. Rev. **D 58** (1998) 112003; SLAC/E155 Coll., P.L. Anthony *et al.*, Phys. Lett. **B 463** (1999) 339, *ibid* **B 493** (2000) 19.
[14] JLab/Hall A Coll., X. Zheng *et al.*, Phys. Rev. Lett. **92** (2004) 012004.
[15] E. Leader, A.V. Sidorov and D.B. Stamenov, Eur. Phys. J. **C 23** (2002) 479; Phys. Rev. **D 67** (2003) 074017.

PRODUCTION OF HEAVY BARYONS

S.P.Baranov [a], V.L.Slad [b]

[a] *P.N.Lebedev Institute of Physics, Leninskii prosp. 53, Moscow 119991, Russia,*
[b] *Institute of Nuclear Physics, Moscow State University, Moscow 119992, Russia*

Abstract. The total and differential cross sections for the production of Ω_{scb} and Ω_{ccc} baryons in electron-positron and photon-photon collisions are calculated.

1 Introduction

Baryons with two or three heavy quarks are interesting because of many reasons. Investigation of the production mechanisms is of interest from the theoretical point of view as this provides the possibility of further verifying QCD testing both its perturbative aspect (which is used to describe the simultaneous production of several quark pairs) and QCD-inspired nonperturbative models of bound states. On the other hand, derivation of theoretical cross-section estimates is important for planning experimental searches for such particles and investigation of their properties. The presence of two or more heavy quarks in a hadron substantially affects the properties of its weak decays. At the same time, the reliability of theoretical predictions is higher for such hadrons, and this makes it possible to test model concepts more thoroughly. Baryons involving two or three heavy quarks have not yet been observed experimentally.

Calculations presented in the literature devoted to the production cross sections of baryons containing two heavy quarks either rely on the fragmentation approach [1] - [2], or treat the production of unbound four-quark states in a bounded phase space [3], or consider (as is done in the majority of the most recent studies on this subject) the production of the relevant heavy diquark [4] - [6]. No attention has so far been given to the production of triply heavy baryons.

An overview of theoretical investigations regarding the mass spectra of hadrons containing two or more heavy quarks, the cross sections of their production in various processes, and their lifetimes and decay modes is given in [7].

2 Calculations and numerical results

In the present study we consider some features of the production of Ω_{scb} and Ω_{ccc} baryons in elecrton-positron and photon-photon collisions:

$$e^+e^- \to \Omega_{scb}; \quad e^+e^- \to \Omega_{ccc}; \quad \gamma\gamma \to \Omega_{scb}. \tag{1}$$

[a]e-mail: baranov@lebedev.sci.ru
[b]e-mail: vslad@sinp.msu.ru

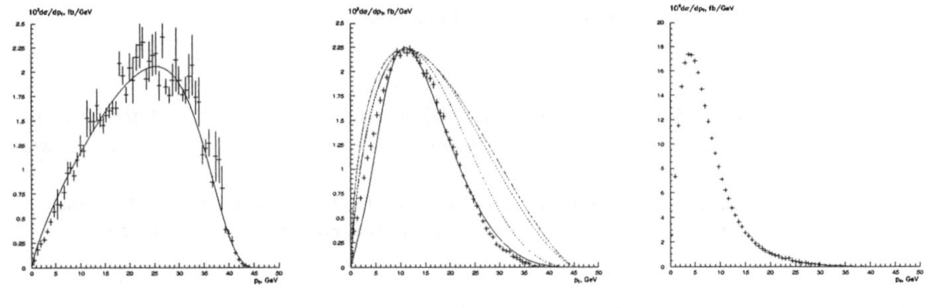

Figure 1: Transverse momentum distributions of Ω_{scb}-baryons produced in e^+e^- annihilation (left panel), Ω_{ccc}-baryons produced in e^+e^- annihilation (middle panel), and Ω_{scb}-baryons produced in $\gamma\gamma$ collision (right panel) at $\sqrt{s} = 91.2$ GeV.

The total numder of Feynman diagrams describing these processes is 84, 504, and 552, respectively. The choice of the purely leptonic or photonic initial state was motivated, on one hand, by the fact that the relevant calculations are simpler than for the production in hadronic processes, and, on the other hand, by the fact that they offer a number of advantages for experimental observation, which include favorable background conditions and the precisely known initial energy.

At the quark level, the considered subprocesses are of the order $\mathcal{O}(\alpha^2\alpha_s^4)$ (for example: $e^+(k_1) + e^-(k_2) \to c(p_1) + s(p_2) + b(p_3) + \bar{s}(p_4) + \bar{c}(p_5) + \bar{b}(p_6)$). In evaluating the square of the relevant matrix element, we used the method proposed in [8], which is referred to as the method of orthogonal amplitudes. Previously, this method was employed, for example, in [5]. In this method, the squared modulus of the relevant matrix element, summation over the polarizations of final state particles and averaging over the polarizations of colliding particles is performed by the formula

$$\overline{\mathcal{M}^2} = \sum_N |c_N|^2 (w_N, w_N), \quad \text{where} \quad c_N = \sum_{polar} \{\mathcal{M} \cdot w_N^*\}/(w_N, w_N). \tag{2}$$

Here the orthogonal amplitues are taken in the form $w_1 = \overline{u}(p_1)u(p_2)$, $w_2 = \overline{u}(p_1)\hat{R}u(p_2)$, $w_3 = \overline{u}(p_1)\hat{Q}u(p_2)$, $w_4 = \overline{u}(p_1)\hat{R}\hat{Q}u(p_2)$. The 4-vectors R and Q must satisfy the orthogonality conditions $R^\mu Q_\mu = 0$, $R^\mu p_{1\mu} = 0$, $R^\mu p_{2\mu} = 0$, $Q^\mu p_{1\mu} = 0$, $Q^\mu p_{2\mu} = 0$ and are arbitrary in other respects.

In order to describe the bound state of heavy quarks, we make use of the nonrelativistic approximation [9] - [11], according to which the relative velocities of the quarks in a heavy hadron are assumed to be low. In the case of S-wave states, these velocities can be set to zero. Accordingly, the velocities of all three quarks in the final state of the considered processes are taken to be identical, so that the momenta of the quarks are proportional to their masses,

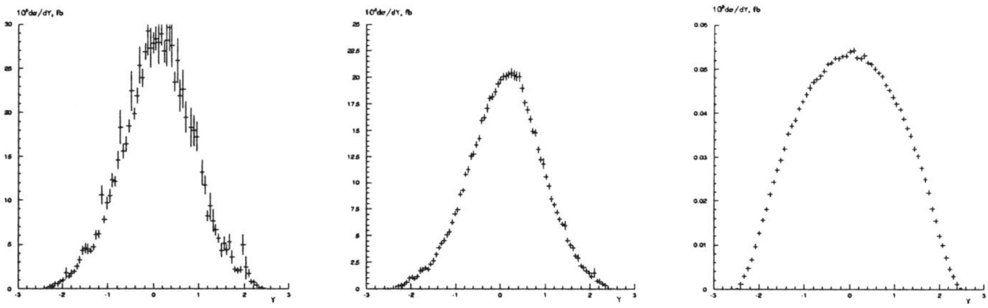

Figure 2: Rapidity distributions of Ω_{scb}-baryons produced in e^+e^- annihilation (left panel), Ω_{ccc}-baryons produced in e^+e^- annihilation (middle panel), and Ω_{scb}-baryons produced in $\gamma\gamma$ collision (right panel) at $\sqrt{s} = 91.2$ GeV.

that is,

$$p_1{=}(m_1/M)\,p, \quad p_2{=}(m_2/M)\,p, \quad p_3{=}(m_3/M)\,p, \quad M{=}m_1{+}m_2{+}m_3. \qquad (3)$$

Thus, the six-particle phase space of the final state reduces to the four-particle phase space, as for example

$$e^+(k_1) + e^-(k_2) \to \Omega_{scb}(p) + \bar{s}(p_4) + \bar{c}(p_5) + \bar{b}(p_6). \qquad (4)$$

The probability of bound-state formation is controlled by the value of the baryon wave function at the origin of coordinates, the only model parameter in this approach. Eventually, the differential cross section takes the form

$$\begin{aligned} d\sigma &= \frac{(2\pi)^4\overline{|\mathcal{M}|^2}}{2s} \cdot \frac{|\Psi(0)|^2}{M^2}\delta^4(k_1 + k_2 - p_4 - p_5 - p_6 - p) \times \qquad (5) \\ &\times \frac{d^3p_4}{(2\pi)^3 2E_4} \cdot \frac{d^3p_5}{(2\pi)^3 2E_5} \cdot \frac{d^3p_6}{(2\pi)^3 2E_6} \cdot \frac{d^3p}{(2\pi)^3 2E}. \end{aligned}$$

To carry out the multidimensional integration over the phase space in (5) we employed the codes from the CompHEP package [12].

In numerical calculations, we used the following parameter setting: baryon masses $m_{\Omega(scb)} = 6.5$ GeV, $m_{\Omega(ccc)} = 4.5$ GeV; quark masses: $m_s = 0.3$ GeV, $m_c = 1.5$ GeV, $m_c = 4.8$ GeV; baryon wave functions [13]: $|\Psi(0)|^2_{\Omega(scb)} = 0.90 \cdot 10^{-3}$ GeV^6, $|\Psi(0)|^2_{\Omega(ccc)} = 0.36 \cdot 10^{-3}$ GeV^6; and coupling constants $\alpha = 0.128$, $\alpha_s = 0.134$.

Our numerical results are displayed in Figs. 1 and 2 and in Table 1. A more detailed discussion of the theoretical approach and numerical results can be found in papers [14].

Table 1. Total production cross sections of heavy baryons

Process	$\sqrt{s}$, GeV	Cross section, fb
$e^+e^- \rightarrow \Omega_{scb}$	91.2	0.0534 ± 0.0014
$e^+e^- \rightarrow \Omega_{ccc}$	91.2	0.0404 ± 0.0004
$\gamma\gamma \rightarrow \Omega_{scb}$	200	$(0.0410 \pm 0.0002) \cdot 10^{-3}$
$\gamma\gamma \rightarrow \Omega_{scb}$	91.2	$(0.1636 \pm 0.0005) \cdot 10^{-3}$

Acknowledgments

We are grateful to A.E. Pukhov for offering the CompHEP package and assistance in adapting it to the problem considered in this article.

References

[1] A. Falk, M. Luke, M. Savage, and M. Wise, *Phys. Rev.* **D 49**, 555 (1994)
[2] M.A. Gomshi Nobary and R. Sepahvand, hep-ph/0406148
[3] V. V. Kiselev, A. K. Likhoded, and M. V. Shevlyagin, *Phys. Lett.* **B 332**, 411 (1994)
[4] A. V. Berezhnoy, V. V. Kiselev, and A. K. Likhoded, *Yad. Fiz.* **59**, 909 (1996) Phys. At. Nucl. 59, 870 (1996); A. V. Berezhnoy, V. V. Kiselev, A. K. Likhoded, and A. I. Onichshenko, *Yad. Fiz.* **60**, 2048 (1997) Phys.At. Nucl. 60, 1875 (1997); A. V. Berezhnoy, V. V. Kiselev, A. K. Likhoded, and A. I. Onishchenko, *Phys. Rev.* **D 57**, 4385 (1998)
[5] S. P. Baranov, *Phys. Rev.* **D 54**, 3228 (1996); S. P. Baranov, *Phys. Rev.* **D 56**, 3046 (1997)
[6] V. V. Braguta and A. E. Chalov, hep-ph/0005149
[7] V. V. Kiselev and A. K. Likhoded, *Usp. Fiz. Nauk* **172**, 497 (2002).
[8] R. E. Prange, *Phys. Rev.* **110**, 240 (1958)
[9] C.-H. Chang, *Nucl. Phys. B* **172**, 425 (1980)
[10] R. Baier and R. Ruckl, *Phys. Lett.* **B 102B**, 364 (1981)
[11] E. L. Berger and D. Jones, *Phys. Rev.* **D 23**, 1521 (1981)
[12] A. Pukhov et al., hep-ph/9908288
[13] E. Bagan, H. G. Dosch, P. Godzinsky, et al., *Z. Phys.* **C 64**, 57 (1994).
[14] S.P. Baranov and V.L. Slad, *Phys. At. Nuclei* **66** , 1730 (2003); S.P. Baranov and V.L. Slad, *Phys. At. Nuclei* **67** , 808 (2004); S.P. Baranov and V.L. Slad, *Phys. At. Nuclei* **68** , 1183 (2005); S.P. Baranov and V.L. Slad, *Phys. At. Nuclei* **69** , 59 (2006).

SELF-ENERGY OF KAONS IN PION MATTER

M. I. Krivoruchenko [a b]
Institute for Theoretical and Experimental Physics, B. Cheremushkinskaya 25
117259 Moscow, Russia

Abstract. The modifications of kaon properties in pion matter are investigated to one loop of ChPT at small temperatures and phenomenologically at temperatures up to $T \sim 200$ MeV using the experimentally measured phase shifts of the πK scattering within the optical potential method framework. At $T \sim 170$ MeV, we obtain the kaon mass shift $\delta M_K \sim -33$ MeV and the mean-field potential $V_K \sim 21$ MeV. Their sum is negative, so the kaons are slightly bound in pion matter similar to nucleons in nuclear matter. The kaon in-medium broadening results in an acceleration of the $\phi \to K\bar{K}$ decay. The increased apparent dilepton branching of the ϕ-mesons, observed by NA50, NA49, and the PHENIX collaborations at RHIC, is interpreted in terms of rescattering of secondary kaons inside of the pion matter

In heavy-ion reactions at RHIC energies, the medium is baryon dilute but meson rich. E.g., at $\sqrt{s} = 130$ GeV/A, the ratio $\pi^-/p \sim 8$ experimentally and is well understood assuming the thermal equilibrium of the produced hadrons [2]. Since by far the most abundant particles are pions, one can speak about pion matter created in heavy-ion collisions at RHIC energies.

The ϕ-meson is a promising mesonic probe for the fireball formation at RHIC conditions [3,4]. Possible modifications of the ϕ-meson dilepton branching in a pion gas have been discussed one decade ago by Lissauer and Shuryak [5] and by Blaizot and Galain [6]. The $\mu^+\mu^-$ yield from ϕ-meson decays was recently measured in central Pb+Pb collisions at CERN/SPS energies by the NA50 Collaboration [7] and the $K\bar{K}$ yield was measured by the NA49 Collaboration [8]. The number of ϕ-mesons detected through the dilepton channel was found to be a factor of two to four greater than that of ϕ-mesons detected trough the kaon channel. This difference might be attributed to in-medium ϕ and K mass shifts [6] and/or rescattering of the secondary kaons in hadronic matter [9,10]. Recently, preliminary ϕ-meson production data in Au + Au collisions at RHIC energies were reported by the PHENIX Collaboration [11]. In this experiment, the ϕ yield was simultaneously measured in the $\phi \to e^+e^-$ and $K\bar{K}$ channels. The result is consistent with an increased dilepton branching.

We consider modifications of the kaon self-energy operator $\Sigma(p^2, E)$ and of the $\phi \to K\bar{K}$ decay mode inside hot pion matter. ChPT, proposed for description of interactions of pseudoscalar mesons at low energies, is an adequate tool for studying the problem at low temperatures and useful to control the low temperature limit of phenomenological models.

The in-medium mass operator of kaons, $\Sigma(p^2, E)$, can be expressed in terms of the πK forward scattering amplitudes for on-shell pions and off-shell kaons.

[a]e-mail: mikhail.krivoruchenko@itep.ru

[b]This talk is based on the work made in collaboration with Boris Martemyanov, Amand Faessler and Christian Fuchs [1].

The on-shell πK amplitudes have been calculated in ChPT to the order p^4 by several authors (see e.g. [12] and references therein). To lowest order p^2, the off-shell πK amplitudes are given in Ref. [13]. Near the threshold, the isospin-even and odd πK scattering amplitudes can be written as

$$A^{\pm}(s,t,p^2) \quad = \quad 8\pi\sqrt{s}\left(a_0^{\pm} + p^{*2}(b_0^{\pm} + 3a_1^{\pm}) + \frac{3}{2}ta_1^{\pm}\right) + c^{\pm}(p^2 - M_K^2) \quad (1)$$

where $a_\ell^{\pm}$ and $b_\ell^{\pm}$ are the πK scattering lengths and effective ranges, $p^* = p^*(\sqrt{s}, M_\pi, M_K)$ is the c.m. momentum of the πK system, $s = (p + p_\pi)^2$, $t = (p'_\pi - p_\pi)^2$, $p = (E, \mathbf{p})$ is the kaon momentum and $p^2 \neq M_K^2$ in general.

The number densities of pions are given by Bose distributions $dn_{v\pi} = (2\pi)^{-3} d^3p_\pi(e^{E_\pi/T} - 1)^{-1}$. At RHIC energies the charged pion number ratio n_{π^+}/n_{π^-} is very close to unity [2], so the pion chemical potentials are equal to zero. The scalar pion density is defined by $dn_{s\pi} = dn_{v\pi}/(2E_\pi)$.

The sum of the forward π^+K, π^0K, and π^-K scattering amplitudes can be integrated over the pion momenta in the rest frame of the medium to obtain the K-meson self-energy operator:

$$-\Sigma(p^2, E) = 3\int A^{+}(s, 0, p^2)dn_{s\pi}. \quad (2)$$

Near threshold the amplitude $A^{+}(s,t,p^2)$ can be expanded up to $\sim O(s - (M_\pi + M_K)^2) + O(p^2 - M_K)^2)$, in which case the self-energy operator can be written as follows

$$-\Sigma(p^2, E) = (p^2 - M_K^2)(Z_K^{-1} - 1) - \delta M_K^2 - 2EV_K. \quad (3)$$

This representation allows to identify δM_K as a mass shift and V_K as a mean-field potential. The poles of the propagator appearing at $p^2 - M_K^2 - \Sigma(p^2, E) = 0$ determine the in-medium kaon dispersion law

$$E_K^{(\pm)}(\mathbf{p}) = \pm\sqrt{\mathbf{p}^2 + M_K^2 + \delta M_K^2} + V_K \quad . \quad (4)$$

Using current algebra predictions for the threshold πK-scattering parameters [13], one gets $Z_{K^\pm}^{[0]-1} = 1 + 3n_{s\pi}/(2F^2)$, $\delta M_{K^\pm}^{[0]} = V_{K^\pm}^{[0]} = 0$ where $F = 92$ MeV is the pion decay constant. Ambiguities in the off-shell amplitudes due to different parameterizations of the pion field can be fixed at the tree level using the Adler self-consistency condition. The results for K^0 and $\bar{K}^0$ can be obtained from isospin symmetry.

It is worthwhile to notice that the kaon dispersion law in nuclear matter has a similar structure with a negative mass shift $\delta M_K^2 < 0$ and a potential $V_{K^+} = -V_{K^-} > 0$ [14–16].

To lowest order ChPT isospin symmetric pion matter does not change the kaon dispersion law. The leading order effect appears at the one loop level.

The effective mass of a quasiparticle has then to be determined from the relation $1/m^{\text{eff}} = \partial^2\varepsilon(\mathbf{p})/\partial|\mathbf{p}|^2$ at $\mathbf{p}=0$ where $\varepsilon(\mathbf{p})$ is the single particle energy. The vacuum ChPT corrections to the kaon self-energy are absorbed by M^2 and the vacuum renormalization constant of the kaon propagator and renormalize the scattering amplitudes (1). To lowest order in $\Sigma(p^2,E) = O(n_v)$ one obtains $\varepsilon(\mathbf{p}) = E_{\mathbf{p}} + \Sigma(M_K^2, E_{\mathbf{p}})/(2E_{\mathbf{p}})$ with $E_{\mathbf{p}} = +\sqrt{\mathbf{p}^2 + M_K^2}$. The mean-field potential

$$V_K = \frac{1}{2}\frac{\partial\Sigma(M_K^2,E)}{\partial E}\bigg|_{E=M_K} \tag{5}$$

entering the dispersion law (4) determines the first-order shift of the self-energy operator from the kaon mass shell and leads automatically to the correct dispersion law to order $O(\mathbf{p}^2 n_v)$. Given that the potential V_K is known, the mass shift can be found from equation

$$\delta M_K + V_K = \frac{\Sigma(M_K^2, M_K)}{2M_K}. \tag{6}$$

The values $\Sigma(M_K^2, M_K)$, δM_K, and V_K can be expressed in terms of the s- and p-wave scattering lengths and the s-wave effective ranges. Using Eqs. (1) and (2), we obtain

$$\Sigma(M_K^2, M_K) = -4\pi n_v \frac{M_\pi + M_K}{M_\pi} a_0^+ \tag{7}$$

and

$$V_K = -\frac{2\pi n_v}{M_\pi + M_K}\left(a_0^+ + 2M_\pi M_K (b_0^+ + 3a_1^+)\right) \quad . \tag{8}$$

The corresponding mass shift can be found from Eq. (6). The self-energy operator (2) for kaons in isotopically symmetric pion matter has the same form as for antikaons due to C-parity. The values δM_K and V_K are also isoscalars, and so $\delta M_{K^+} = \delta M_{K^0} = \delta M_{K^-} = \delta M_{\bar{K}^0}$ and $V_{K^+} = V_{K^0} = V_{K^-} = V_{\bar{K}^0}$, as distinct from the nuclear matter case [14–17].

Current algebra predicts [13] $a_0^+ = b_0^+ + 3a_1^+ = 0$. The parameters a_0^+, b_0^+, and a_1^+ entering into Eqs. (7) - (8) receive in ChPT corrections to the order p^4. The calculations of Ref. [18] give $a_0^+ = (0.023 \pm 0.012)/M_\pi$ and $b_0^+ + 3a_1^+ = (0.054 \pm 0.008)/M_\pi^3$.

According to ChPT the pseudoscalar meson masses increase with temperature as expected for collective modes. A similar effect exists in the NJL model [6]. With increasing temperature, the constituent quark masses decrease, whereas the occupation numbers increase, pushing the K mass up and the ϕ mass down. A decrease of the ϕ-meson mass was also predicted by Asakawa and Ko [19].

Equations (7) - (8) are valid at $T << M_\pi$. The chemical freeze-out temperature at RHIC $T \sim 170$ MeV [2] is high, and so we should use a more phenomenological approach. We use the experimental πK scattering phases in the lowest partial waves to saturate the amplitudes $A^I(s,t)$.

At $T = 170$ MeV, we obtain $\delta M_K = -33$ MeV and $V_K = 21$ MeV. The positive mass shift at low temperatures where ChPT is applicable becomes negative with increasing temperature. The value of $\Re\Sigma(M_K^2, M_K)/(2M_K)$ remains relatively small up to $T = 200$ MeV. The analogy with the Walecka model for nucleons is remarkable: The kaon mass shift at high temperatures is negative, the mean-field potential is positive, their sum is relatively small and negative. The kaons are therefore slightly bound in pion matter. The mean-field potential is, however, C-even, as distinct from the case of nucleons and kaons in nuclear matter [14–16].

The ϕ-meson collision broadening has been discussed in Refs. [19–21]. The $\phi \to K\bar{K}$ decay might be suppressed due to a increasing kaon mass and/or a decreasing ϕ-meson mass. Collisions of the ϕ-meson and the kaons keep, however, the $\phi \to K\bar{K}$ decay channel open even at $M_\phi < 2M_K$ and result in an increase of the total ϕ width, both at $M_\phi < 2M_K$ and $M_\phi > 2M_K$.

The kaon collision width is expressed in terms of the on-shell partial wave projections of the πK amplitudes with total isospin $I = 1/2$ and $3/2$. The K^* increases the width by about 40 MeV. As a result, we get $\Gamma_K^* = 81$ MeV at $T = 170$ MeV.

The ϕ-meson in-medium width can be written as $\Gamma^{\rm med}_{\phi\to K\bar{K}} = \eta \Gamma^{\rm vac}_{\phi\to K\bar{K}}$ where

$$\begin{aligned}\eta &= \frac{1}{\pi^2}\int \frac{p^{*3}(M_\phi^*, m_1^*, m_2^*)}{p^{*3}(M_\phi, M_K, M_K)} \frac{M_K^*\Gamma_K^* dm_1^{*2}}{(m_1^{*2} - M_K^{*2})^2 + (M_K^*\Gamma_K^*)^2} \\ &\times \frac{M_K^*\Gamma_K^* dm_2^{*2}}{(m_2^{*2} - M_K^{*2})^2 + (M_K^*\Gamma_K^*)^2}.\end{aligned}$$

We set $M_\phi^* = M_\phi - 2V_K$ and $M_K^* = M_K + \delta M_K$. Any modification of the real part of the kaon self-energy does not significantly alter the in-medium ϕ-meson decay rate as long as $\Re\Sigma(M_K^2, M_K) << \Im\Sigma(M_K^2, M_K) = M_K\Gamma_K^*$. This is the present case. At $T = 170$ MeV, we get $\eta \sim 3$ and $\Gamma_\phi^{\rm med} \sim 12$ MeV. The in-medium ϕ-meson width increases, since the Breit-Wigner distribution allows an effective reduction of the kaon masses and, as a consequence, an effective increase of the available phase space.

Dileptons from ϕ decays leave the pion matter essentially undistorted by final-state interactions, whereas the secondary kaons rescatter and contribute to the experimental background. This can result in an increase of the apparent dilepton branching. The apparent dilepton branching can be estimated as [1]

$$B^{\rm app} = B\frac{1 + (e^{\Gamma_\phi^*\tau} - 1)\Gamma_\phi/\Gamma_\phi^*}{1 + \frac{\Gamma_\phi}{2\Gamma_K^* - \Gamma_\phi^*}(1 - e^{-(2\Gamma_K^* - \Gamma_\phi^*)\tau})}\ . \tag{9}$$

According to transport calculations $e^{-\tau\Gamma_\phi} \sim 1/2$ at RHIC energies [22]. In such a case, varying the temperature $T = 120 \div 170$ between the commonly accepted values for thermal and chemical freeze-out, we get $B^{\rm app}/B \sim 2 \div 3$ which is in qualitative agreement with the observations from NA50, NA49, and PHENIX collaborations who report an increased apparent dilepton branching. Refined studies incorporating the fireball dynamics might help to bring more insight in the ϕ- and K-meson propagation in a dense and hot medium.

References

[1] B.V. Martemyanov, Amand Faessler, C. Fuchs, M.I. Krivoruchenko, Phys. Rev. Lett. **93**, 052301 (2004).

[2] P. Braun-Munzinger, D. Magestro and K. Redlich, Phys. Lett. **B518**, 41 (2001).

[3] P. Koch, B. Müller, and J. Rafelski, Phys. Rep. **142**, 1 (1986).

[4] A. Shor, Phys. Rev. Lett. **54**, 1122 (1985).

[5] D. Lissauer and E.V. Shuryak, Phys. Lett. **B253**, 15 (1991).

[6] J.-P. Blaizot and R. M. Galain, Phys. Lett. **B271**, 32 (1991).

[7] N. Willis [for NA50 Collaboration], Nucl. Phys. **A661**, 534c (1999).

[8] C. Hoehne [for NA49 Collaboration], Nucl. Phys. **A661**, 485c (1999).

[9] S.C. Johnson, B.V. Jacak, and A. Drees, Eur. Phys. J. **C18**, 645 (2001).

[10] S. Soff *et al.*, J. Phys. **G27**, 449 (2001);
S. Pal, C.M. Ko, and Z.-W. Lin, Nucl. Phys. **A707**, 525 (2002).

[11] J.L. Nagle *et al.*, PHENIX Collaboration, nucl-ex/0209015.

[12] A. G. Nicola and J. R. Peláez, Phys. Rev. **D65**, 054009 (2002).

[13] J.A. Cronin, Phys. Rev. **161**, 1483 (1967);
R.W. Griffith, **176**, 1705 (1968).

[14] D.B. Kaplan and A.E. Nelson, Phys. Lett. **B175**, 57 (1986).

[15] G.E. Brown and M. Rho, Phys. Reports **269**, 333 (1996).

[16] G.Q. Li *et al.*, Phys. Rev. Lett. **74**, 235 (1995).

[17] T. Waas, N. Kaiser, and W. Weise, Phys. Lett. **B365**, 12 (1996);
M. Lutz, Phys. Lett. **B426**, 12 (1998).

[18] V. Bernard, N. Kaiser and Ulf-G. Meeissner, Phys. Rev. **D43**, R2757 (1991); Nucl. Phys. **B357**, 129 (1991).

[19] M. Asakawa and C.M. Ko, Phys. Lett. **B322**, 33 (1994); Phys. Rev. **C50**, 3064 (1994).

[20] D. Seibert and C. Gale, Phys. Rev. **C52**, 490 (1995).

[21] L. Alvarez-Ruso and V. Koch, Phys. Rev. **C65**, 054901 (2002).

[22] L. Bravina *et al.*, Nucl. Phys. **A715**, 665c (2003).

STRUCTURE FUNCTIONS RESULT IN CHORUS EXPERIMENT

Meltem Serin [a]
On behalf of the CHORUS Collaboration,
Department of Physics, Middle East Technical University,
06531, Ankara, Turkey

Abstract. In the 1998 run CHORUS lead-scintillator calorimeter was used for the measurements of (anti-)neutrino-nucleon differential cross sections in the range of $0.01 < x < 0.7$, $0.05 < y < 0.95$, $10 < E_\nu < 200\,\text{GeV}$ and thus, for the extraction of the neutrino-nucleon structure functions, $F_2(x,Q^2)$, $F_3(x,Q^2)$ and $R(x,Q^2)$.

1 Introduction

For 1998 run the CHORUS detector was used for a measurement of muon (anti-)neutrino-lead nucleon structure functions through charged-current interactions. The full kinematics can be reconstructed from the energy of the hadronic shower E_{had}, muon momentum p_μ and angle θ_μ:

$$E_\nu = E_\mu + E_{\text{had}} \quad , \quad x = \frac{Q^2}{2M_{\text{N}}E_{\text{had}}} \quad , \quad y = \frac{E_{\text{had}}}{E_\nu}$$

$$Q^2 = 2E_\nu(E_\mu - p_\mu \cos(\theta_\mu) - m_\mu^2) \approx 4E_\nu - p_\mu \sin^2(\frac{\theta_\mu}{2})$$

where E_ν is the neutrino energy, M_{N} is the nucleon mass, $Q^2 = -q \cdot q$ is the 4-momentum transfer squared and the x and y are the Björken variables. In the single W exchange approximation, the differential cross-section can be described in terms of structure functions, $F_1(x,Q^2)$, $F_2(x,Q^2)$ and $xF_3(x,Q^2)$,

$$\begin{aligned}\frac{d^2\sigma}{dxdy} &= \frac{G_{\text{F}}^2 M_{\text{N}} E_\nu}{\pi(1+Q^2/M_{\text{W}}^2)^2}[\frac{y^2}{2}2xF_1 + (1 - y - \frac{M_{\text{N}}xy}{2E_\nu})F_2 \\ &\quad \pm(y - \frac{y^2}{2})xF_3]\end{aligned} \tag{1}$$

where G_{F} is the Fermi constant and M_{W} is the W mass. In the Parton Model the structure functions can be represented in terms of quark distributions. For an isoscalar target and a symmetric sea ($s = \bar{s}$, $c = \bar{c}$), we can write,

$$2xF_1^{\nu N} = 2xF_1^{\bar{\nu} N} = F_2^{\nu} = F_2^{\bar{\nu}} = x(d^p + \bar{d}^p + u^p + \bar{u}^p + 2(s^p + 2c^p)) \tag{2}$$

$$xF_3^{\nu N} = xF_3^{\nu N} + xF_3^{\bar{\nu} N} = 2x(d^p + \bar{d}^p + x(u^p - \bar{u}^p) \tag{3}$$

$$\Delta xF_3^{\nu N} = xF_3^{\nu N} - xF_3^{\bar{\nu} N} = 4(s - c) \tag{4}$$

where $u(\bar{u})$, $d(\bar{d})$, and $s(\bar{s})$ represent up, down, charm, strange quark (anti-quark) distributions respectively. Since top and bottom quarks are too heavy to be formed up in this experiment they have been left out of the equations.

[a] e-mail: serinm@metu.edu.tr

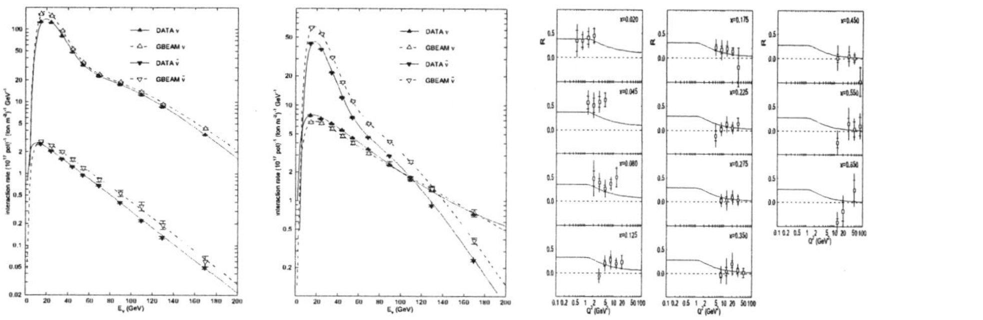

Figure 1: a-b) The beam flux fits for neutrino and anti-neutrino beam modes respectively. c) $R(x, Q^2)$ calculated from 3-parameter fit where the points are from data and the curves from SLACR90 parametrization. Numerical values of these measurements can be found from [5].

2 Experiment

In 1998 run the CHORUS detector [1], primarily designed to search for neutrino oscillations, was well suited to perform measurements of neutrino-nucleon differential cross-sections [2] with a high-resolution calorimeter mainly made of lead and plastic scintillators as target and a large acceptance muon spectrometer to measure charge, momentum and direction of the muon.

3 Analysis

In this analysis a sample of charge current interactions has been used. The detailed description of the full procedure can be found in the references [3, 4] and a brief explanation is given as follows.

3.1 Beam Flux Study

The neutrino flux and its energy spectrum is measured from the data and compared with the prediction from the beam simulation. The fit results for both beam modes are shown in Fig.1-a and Fig.1-b.

3.2 Cross-section Measurement

For the determination of differential cross-sections, the neutrino and anti-neutrino data are counted in bins of x, y, E_ν. Since the beam flux is not known with enough accuracy the differential cross-section measurement is normalized to the total neutrino-nucleon cross-section and is thus, not affected by the discrepancies between the beam simulation and the beam flux measurement. The muon reconstruction efficiency and the momentum resolution smearing are calculated from Monte-Carlo for the applied cuts. Then, the differential cross-section is determined by

$$\frac{1}{E_\nu}\frac{d^2\sigma}{dxdy} = \frac{\sigma_{\text{tot}}}{E_\nu}\frac{N(E_i, x_j, y_k)}{N(E_i)\Delta x_j \Delta y_k}, \tag{5}$$

where $N(E_i, x_j, y_k)$ is the number of events in bin E_i, x_j, y_k, $N(E_i)$ is the total number of events in energy bin E_i and Δx_j, Δy_k are the sizes of the bins in x and y, and σ_{tot} is the total ν-nucleon cross-section [6], corrected for the neutron excess in the target for each bin. The numerical results of the differential cross-section measurement for both beam modes can be found from [5]

3.3 Structure Function Extraction

The y dependence of both neutrino and anti-neutrino data is then used to extract the structure functions. A linear 3-parameter fit is done by averaging neutrino and anti-neutrino structure functions, $2xF_1$, F_2 and assigning the value of ΔxF_3 only to the strange quark distribution. This fit result was used to calculate $R(x, Q^2)$, the expression for the violation of Callan-Gross relation,

$$R = (1 + \frac{4M_N^2 x^2}{Q^2})\frac{F_2}{2xF_1} \tag{6}$$

The comparison of calculated R value to its world data parametrization is shown in Fig.1-c. Then, 3-parameters are reduced to 2 parameters by using the world data parametrization of R and leaving F_2 and xF_3 as free parameters. The comparison of the results to the other experiments is also shown in Fig.2. The sources of systematic uncertainties can be found in Ref. [5].

4 Conclusion

From the data of 1998 CHORUS run we have performed the first high-statistics measurement of the differential (anti-)neutrino cross-sections on a target predominantly made of lead, as a function of the neutrino energy and the Björken variables x and y. We have extracted the structure functions $F_2(x, Q^2)$, $xF_3(x, Q^2)$ and calculated $R(x, Q^2)$. Our measured R value is in agreement with SLAC90 [7] and our measurement of xF_3 agrees with both CCFR [8] and CDHSW [9] experiments while measurement of F_2 favors the CCFR experiment over CDHSW.

Acknowledgments

I would like to thank Dr. Jaap Panman who has given me the chance to talk in this conference on behalf of the CHORUS Collaboration, the organizers for

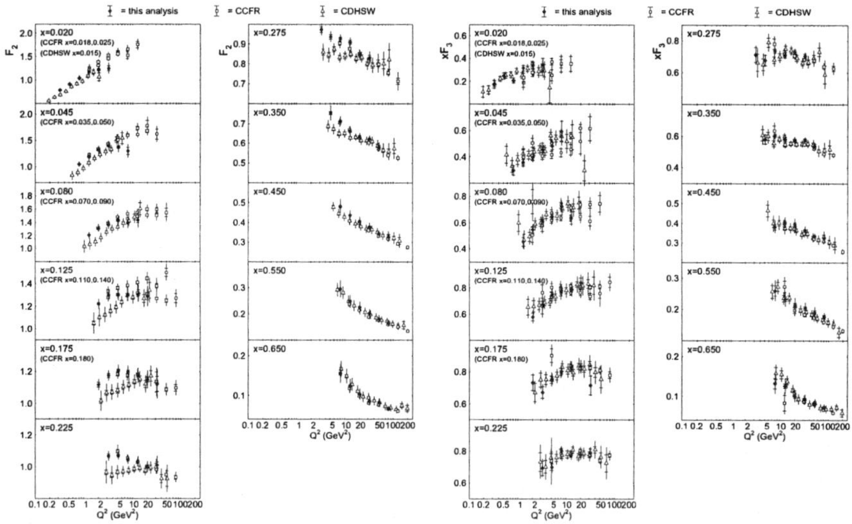

Figure 2: Comparison of our $F_2(x, Q^2)$ and $xF_3(x, Q^2)$ results with measurements from CCFR and CDHSW.

giving me this opportunity and The Turkish Atomic Energy Authority (TAEK) and Middle East Technical University (METU) for their financial supports which made my attendance to the conference possible.

References

[1] Chorus Collab. *Nucl. Instrum. Meth. A* **401** *(1997) 7.*

[2] A. Kayis-Topaksu *et al.*,CHORUS Collaboration,Eur.Phys.J.C(2003)159.

[3] R.Oldeman, *"Measurement of differential cross-sections and structure functions in neutrino and antineutrino scattering on lead"*, (PhD Thesis, Univ. of Amsterdam), June 2000.

[4] S.Kama, *"Measurement of (Anti-)Neutrino-Nucleon Structure Functions in CHORUS Experiment"*, (MSc Thesis, METU), August 2005.

[5] G.Önengüt *et al.*, CHORUS Collaboration, Submitted to Phys. Lett.B.

[6] J. M. Conrad, M. H. Shaevitz, T. Bolton, Rev. Mod. Phys. **70**(1998)1341.

[7] L.W.Whitlow *et al.* Phys.Lett.B **250**(1990)193.

[8] W.G.Seligman *et al.*, CCFR Collaboration, Phys.Rev.Lett.**79**(1997)1213.

[9] J.P.Berge *et al.*,CDHSW Collaboration, Z. Phys. C **49** (1991) 187.

NEW RELATIONS BETWEEN THE QCD SUM RULES FOR MESON – BARYON COUPLINGS

T. Aliev [a], A.Özpineci [b],
Physics Dept., Middle East Technical Univ., 06531, Ankara, Turkey
S. Yakovlev [c], V. Zamiralov [d]
D.V. Skobeltsyn Institute of Nuclear Physics, Moscow State University, 119992, Moscow, Russia

Abstract. New relations between QCD Borel sum rules for strong coupling constants of mesons to baryons are derived. It is shown that starting from the sum rule for the coupling constants $g_{\pi\Sigma\Sigma}$ and $g_{\pi\Sigma\Lambda}$ it is straightforward to obtain corresponding sum rules for the g_{KYN}, $g_{KY\Xi}$ couplings, $Y = \Sigma, \Lambda$.

Since the advent of the $SU(3)$ symmetry all the meson-baryon coupling constants were usually expressed in terms of F and D constants which gave possibility to construct a reliable phenomenological aproach.

As soon as in [1] QCD sum rules (SR's) were proposed, they were used not only for baryon masses and magnetic moments starting from the works [2] but also for baryon-meson coupling constants. Naturally, a pion-nucleon coupling attracted the most attention (see, for example, [3], [4], [5]). Coupling constants of π^0- and η- mesons to baryons were studied recently in various QCD SR approaches [6], [7]. Also QCD sum rule for the η coupling to the Λ hyperon was written [8] which was usually absent in these approaches.

As for K-mesons they also were studied in the framework of the QCD sum rules (see, e.g., [9]– [13]). Usually these sum rules are not related straightforwardly to those treating π and η couplings to baryons.

We would like to propose here QCD sum rules for meson – baryon couplings through some universal $\mathcal{F}$ and $\mathcal{D}$ functions written in a unified manner.

In the unitary model all the pion-baryon coupling constants can be expressed in terms of F and D coupling constants. Couplings of the Σ-like baryons $B(qq, q'), q, q' = u, d, s$ to π^0 meson related in the quark model to the current $j^{\pi^0} = \frac{1}{\sqrt{2}}[\bar{u}\gamma_5 u - \bar{d}\gamma_5 d]$ can be put in the form $g(\pi^0 BB) = g_{\pi qq} 2F + g_{\pi q'q'}(F - D)$, where $g_{\pi uu} = +\sqrt{\frac{1}{2}}$, $g_{\pi dd} = -\sqrt{\frac{1}{2}}$ and $g_{\pi ss} = 0$ are read off the quark current. The coupling $\pi^0\Sigma^0\Lambda$ can be obtained through the relation (see [14]):

$$g(\pi^0\tilde{\Sigma}^{0,ds}\tilde{\Sigma}^{0,ds}) - g(\pi^0\tilde{\Sigma}^{0,us}\tilde{\Sigma}^{0,us}) = \sqrt{3}g(\pi^0\Sigma^0\Lambda), \tag{1}$$

where subscripts 'ds' and 'us' mean that we change $(d \leftrightarrow s)$ and $(u \leftrightarrow s)$ and

$$\begin{pmatrix} -\tilde{\Sigma}^0_{ds,us} \\ \tilde{\Lambda}_{ds,us} \end{pmatrix} = \begin{pmatrix} 1/2 & \pm\sqrt{3}/2 \\ \mp\sqrt{3}/2 & 1/2 \end{pmatrix} \begin{pmatrix} \Sigma^0 \\ \Lambda \end{pmatrix}. \tag{2}$$

[a] e-mail: taliev@metu.edu.tr
[b] e-mail: ozpineci@newton.physics.metu.edu.tr
[c] yakovlev@depni.sinp.msu.ru
[d] zamir@depni.sinp.msu.ru

Kaon and charged pion couplings to baryons can be related by using (2):

$$2[g(\pi^-\Sigma^+\bar{\Lambda}_{ds})] = -\sqrt{3}g(\pi^-\Sigma^+\bar{\Sigma}^0) + g(\pi^-\Sigma^+\bar{\Lambda}) = \sqrt{\frac{2}{3}}(3F+D). \quad (3)$$

With $d \leftrightarrow s$ exchange Λ_{ds} returns to Λ while π^- goes to K^- and Σ^+ to $-p$, so

$$2[g(\pi^-\Sigma^+\Lambda_{ds})]_{ds} = -2[g(K^-p\Lambda)] = \sqrt{\frac{2}{3}}(3F+D). \quad (4)$$

This is the unitary symmetry result. Similarly one can start from $[g(\pi^-\Sigma^+\bar{\Lambda}_{us})]$, $[g(\pi^-\Sigma^+\bar{\Sigma}^0_{ds,us})]$ to obtain $g(K^0\Xi^0\bar{\Lambda})$, $g(K^-p\bar{\Sigma}^0)$, $g(K^0\Xi^0\bar{\Sigma}^0)$, respectively. The way to proceed in the formalism of QCD sum rules is now clear.

We use as the example QCD sum rules based on the formalism developed in [6] where unitary symmetry is broken but formules are rather transparent. The sum rule for the $\mathcal{M}\Sigma^0\Sigma^0$ coupling reads:

$$\frac{1}{\sqrt{2}}m^2_{\mathcal{M}}\lambda^2_\Sigma g(\mathcal{M}\Sigma^0\Sigma^0)e^{-(m^2_\Sigma/M^2)}[1+A_\Sigma M^2] =$$

$$g_{\mathcal{M}ss}\frac{m^2_{\mathcal{M}}}{f_{\mathcal{M}}}[M^4\frac{E_0(x)}{12\pi^2} - \frac{M^2}{m^2_{\mathcal{M}}}(m_d\langle\bar{u}u\rangle + m_u\langle\bar{d}d\rangle) - \frac{1}{72}\langle\frac{\alpha_s}{\pi}\mathcal{G}^2\rangle]\langle\bar{s}s\rangle$$

$$+\frac{m_0^2}{6f_{\mathcal{M}}}[\langle\bar{s}s\rangle(m_d g_{\mathcal{M}uu}\langle\bar{u}u\rangle + m_u g_{\mathcal{M}dd}\langle\bar{d}d\rangle) + m_s(g_{\mathcal{M}uu}+g_{\mathcal{M}dd})\langle\bar{u}u\rangle\langle\bar{d}d\rangle]. \quad (5)$$

where m_q, $q = u, d, s$ are current quark masses, $f_{\mathcal{M}}$ is a $\mathcal{M}$-meson decay constant, $\mathcal{M} = \pi^0, \eta$, quark condensates are $\langle\bar{u}u\rangle = \langle\bar{d}d\rangle = -(0.23)^3$ GeV^3, $\langle\bar{s}s\rangle/\langle\bar{d}d\rangle = 0.8$, while $m_0^2 = 0.8$ GeV^2. In the factor $E_0(x) = (1-e^{-x})$ $x = W^2/M^2$ [2] we take $W^2 = 2.0$ GeV^2. The overlap amplitude is taken as $\lambda_B^2 = C \cdot M_B^6$ GeV^6 [6], with $C = 5.48 \times 10^{-4}$. We neglect in calculations $f_{3\mathcal{M}}$. Parameter A_B accounts for high-resonance contributions.

With isotopic invariance we construct Borel sum rule for the $\pi^-\Sigma^+\bar{\Sigma}^0$:

$$-m^2_\pi\lambda^2_\Sigma g(\pi^-\Sigma^+\Sigma^0)e^{-(m^2_\Sigma/M^2)}[1+A_\Sigma M^2] =$$

$$\frac{m_0^2}{6f_\pi}[(m_u\langle\bar{s}s\rangle + m_s\langle\bar{u}u\rangle)(\langle\bar{u}u\rangle + \langle\bar{d}d\rangle)] \equiv \sqrt{2}\mathcal{F}^{(-)}(\pi^-;M^2;u,d;s) \quad (6)$$

and a similar sum rule for $\pi^+\Sigma^-\Sigma^0$ coupling (upon $u \leftrightarrow d$).

Using analogue of the Eq.(1)

$$\sqrt{3}RHS(\pi^0\Sigma^0\Lambda) = RHS(\pi^0\Sigma^0_{ds}\Sigma^0_{ds}) - RHS(\pi^0\Sigma^0_{us}\Sigma^0_{us})$$

we construct QCD Borel sum rule for $\pi^0\Sigma\Lambda$ coupling [8]

$$\sqrt{3}m^2_\pi\lambda_\Lambda\lambda_\Sigma g(\pi^0\Sigma^0\Lambda)\frac{M^2}{M^2_\Sigma - M^2_\Lambda}(e^{-M^2_\Lambda/M^2} - e^{-M^2_\Sigma/M^2})[1+A_{\Sigma\Lambda}M^2] =$$

$$\frac{m_\pi^2}{12\pi^2 f_\pi}[-M^4E_0(x)+\frac{\pi^2}{6}\langle\frac{\alpha_s}{\pi}\mathcal{G}^2\rangle](\langle\bar{d}d\rangle+\langle\bar{u}u\rangle)$$
$$+\frac{(6M^2+m_0^2)}{6f_\pi}[(m_s\langle\bar{u}u\rangle+m_u\langle\bar{s}s\rangle)\langle\bar{d}d\rangle+(m_d\langle\bar{s}s\rangle+m_s\langle\bar{d}d\rangle)\langle\bar{u}u\rangle] \qquad (7)$$

Isotopic invariance allows to deduce the expression for the $\pi^-\Sigma^+\bar{\Lambda}$ coupling:

$$\sqrt{3}m_\pi^2\lambda_\Lambda\lambda_\Sigma g(\pi^-\Sigma^+\bar{\Lambda})\frac{M^2}{M_\Sigma^2-M_\Lambda^2}(e^{-M_\Lambda^2/M^2}-e^{-M_\Sigma^2/M^2})[1+A_{\Sigma\Lambda}M^2]=$$
$$\frac{m_\pi^2}{12\pi^2 f_\pi}[-M^4E_0(x)+\frac{\pi^2}{6}\langle\frac{\alpha_s}{\pi}\mathcal{G}^2\rangle](\langle\bar{u}u\rangle+\langle\bar{d}d\rangle)+$$
$$\frac{(m_0^2+6M^2)}{6f_\pi}[(m_u\langle\bar{s}s\rangle+m_s\langle\bar{u}u\rangle)(\langle\bar{u}u\rangle+\langle\bar{d}d\rangle)]$$
$$\equiv\sqrt{2}\mathcal{D}^{(-)}(\pi^-;M^2;s,d;u)\rightarrow|_{exact\quad SU(3)}\ \sqrt{2}D. \qquad (8)$$

Now we derive Borel sum rules for K-meson couplings to baryons from those for $\pi\Sigma\Lambda$ and $\pi\Sigma\Sigma$ couplings. We form auxiliary couplings upon using quantities $\Lambda_{ds},\Sigma^0_{ds}$ and $\Lambda_{us},\Sigma^0_{us}$ given by the Eq. (2), and then return to those usual ones with $d\leftrightarrow s$ and $u\leftrightarrow s$. First we construct a formal sum rule with Λ changed to Λ_{ds} by using Eq.(2) and retain for a moment only RHS of this sum rule:

$$RHS(\pi^-\Sigma^+\bar{\Lambda}_{ds})_{ds}=[-\frac{\sqrt{3}}{2}RHS(\pi^-\Sigma^+\bar{\Sigma}^0)+\frac{1}{2}RHS(\pi^-\Sigma^+\bar{\Lambda})]_{ds}=$$
$$\sqrt{\frac{1}{6}}(3\mathcal{F}^{(-)}(\pi^-;M^2;u,s;d)+\mathcal{D}^{(-)}(\pi^-;M^2;d,s;u)). \qquad (9)$$

or in full notation

$$m_K^2 g_{K^-p\bar{\Lambda}}\frac{\lambda_\Lambda\lambda_N M^2}{(M_\Lambda^2-M_N^2)}(e^{-M_N^2/M^2}-e^{-M_\Lambda^2/M^2})(1+A_{\Lambda N}M^2)$$
$$=-\frac{1}{2\sqrt{3}}[m_K^2(-M^4E_0(x)+\frac{\pi^2}{6}\langle\frac{\alpha_s}{\pi}\mathcal{G}^2\rangle)\frac{(\langle\bar{u}u\rangle+\langle\bar{s}s\rangle)}{12\pi^2 f_K}+$$
$$\frac{(2m_0^2+3M^2)}{3f_K}[(m_u\langle\bar{d}d\rangle+m_d\langle\bar{u}u\rangle)(\langle\bar{u}u\rangle+\langle\bar{s}s\rangle)]. \qquad (10)$$

In a similar way constructing a formal sum rule with Λ_{us} we obtain:

$$g_{\bar{K}^0\Xi^0\bar{\Lambda}}\sim\sqrt{\frac{1}{6}}(3\mathcal{F}^{(-)}(K^0;M^2;s,d;u)-\mathcal{D}^{(-)}(K^0;M^2;u,d;s)), \qquad (11)$$

$$g_{K^-p\bar{\Sigma}^0}\sim-\sqrt{\frac{1}{2}}\mathcal{F}^{(-)}(K^-;M^2;u,s;d)+\sqrt{\frac{1}{2}}\mathcal{D}^{(-)}(K^-;M^2;d,s;u), \qquad (12)$$

$$g_{\bar{K}^0\Xi^0\bar{\Sigma}^0} \sim -\sqrt{\frac{1}{2}}\mathcal{F}^{(-)}(K^0;M^2;s,d;u) - \sqrt{\frac{2}{2}}\mathcal{D}^{(-)}(K^0;M^2;u,d;s). \qquad (13)$$

Sum rules for other $g_{\bar{K}N}$ and $g_{\bar{K}\Xi Y}$ couplings are obtained with change $(u \leftrightarrow d)$.

Thus we have constructed QCD sum rules for the Lorenz structure $i\gamma_5$ for coupling constants g_{KNY} and $g_{K\Xi Y}$, $Y = \Sigma, \Lambda$ starting from those for $g_{\pi\Sigma\Sigma}$ and $g_{\pi\Sigma\Lambda}$ ones arriving at $|g_{KN\Lambda}| = 0.80$,$|g_{KN\Sigma}| = 1.20$, $|g_{K\Xi\Lambda}| = 1.35$ and $|g_{K\Xi\Sigma}| = 3.10$ with $g_{\pi NN} = 13.4$. The sum rules in terms of the generalized functions $\mathcal{F}$ and $\mathcal{D}$ reveal a simple $SU(3)_f$ pattern. The relations obtained here indicate in what way one can change and use the concept of the unitary symmetry in the framework of QCD sum rules.

Acknowledgments

We are grateful to F.Hussain, B.L.Ioffe, G.Thompson for useful discussions. One of us (V.S.Z.) is grateful to to Abdus Salam ICTP (Trieste, Italy) for financial support.

References

[1] M. A. Shifman, V. I. Vainshtein, V. I. Zakharov, *Nucl. Phys.* **B** 147,385 (1979).
[2] V.M.Belyaev, B.L.Ioffe, *JETP* 56, 493 (1982); B.L.Ioffe, A.Smilga, *Nucl.Phys.* **B** 232, 109 (1984); I. I. Balitsky and A. V. Yung, *Phys. Lett.* **B** 129, 328 (1983).
[3] H.Sciomi and T.Hatsuda, *Nucl.Phys.* **A** 594, 294 (1995).
[4] M.C.Birse and B.Krippa, *Phys.Lett.* **B** 373, 9 (1996); *Phys.Rev.* **C** 54,3240 (1996).
[5] T.Doi, H.Kim and M.Oka, *Phys.Rev.* **C** 62, 055202 (2000).
[6] H.Kim, T.Doi, M.Oka, S.H.Lee, *Nucl.Phys.* **A** 662, 371 (2000); ibid. **A** 678, 295 (2000).
[7] T.M.Aliev, A.Ozpineci, and M.Savci, *Phys.Rev.* **D** 64, 034001 (2001).
[8] A.Özpineci, S.B.Yakovlev, V.S.Zamiralov, Vestnik MGU, 2006 (in press); NPI MSU Preprint 2004-19/758.
[9] S.Choe, M.K.Cheoun, and Su H.Lee, *Phys.Rev.* **C** 53, 1363 (1996).
[10] S.Choe ,*Phys.Rev.* **C** 57, 2061 (1998).
[11] B.Krippa, *Phys.Lett.* **B** 420, 13 (1998).
[12] M.E.Bracco, F.S.Navarra and M.Nielsen, *Phys.Lett.* **B** 454, 346 (1999).
[13] T.Aliev, *Phys.Rev.* **C** 61, 045201 (2000).
[14] A.Özpineci, S.B.Yakovlev, V.S.Zamiralov, *Mod.Phys.Lett.* **A** 20,1 (2005); *Phys.Atom.Nucl.* 68, 279 (2005).

ANALYTICAL APPROACH TO CONSTRUCTING EFFECTIVE HADRON-HADRON INTERACTION OPERATORS AND ITS APPLICATION TO NUCLEON-NUCLEON SCATTERING AT LOW AND INTERMEDIATE ENERGIES

A.N.Safronov [a]

Institute of Nuclear Physics, Moscow State University, 119992 Moscow, Russia

*Abstract.*The new relativistic approach to constructing effective hadron-hadron interaction operators in an states with arbitrary angular momenta, based on analytical theory of S-matrix and methods of inverse scattering problem, is applied to NN-scattering in S- and P- partial-wave states at low and intermediate energies.

1 Introduction

The path-integral method in quantum chromodynamics together with idea of spontaneous violation of chiral invariance leads at low energies to effective field theory, which formulates the problem in terms of mesons and baryons degrees of freedom [1]. For a long time the meson conception of strong interactions was used in nuclear physics [2,3]. Now it is not quite clear, up to what distances the meson picture of nuclear force is valid and what degrees of freedom determine the basic properties of the short-range components of hadron-hadron interactions. In our recent work [4] the new relativistic approach to constructing the effective hadron-hadron interaction operators, based on analytical theory of S-martix and methods of inverse scattering problem, was suggested. In the present work on the basis of this approach the new method to constructing interaction operators in an states with arbitrary angular momenta, taking into account the effects of inelasticity, is developed.

2 Analytical approach to constructing effective local hadron-hadron interaction operators

As follows from principle of analyticity, partal-wave S-matrix in complex $s = k^2$-plane (k is the center-of-mass momentum of colliding particles) has i) the poles corresponding to bound states, ii) the unitary (right-hand) cut and iii) the dynamic (left-hand) cuts caused by exchange processes in t- and u-channels of scattering. For description of dynamics of the hadron-hadron interaction we shall use the relativistic three-dimensional equation of quasi-potential type which can be obtained by reconstruction of initial Bethe-Salpeter equation [5]. Using the special form of the three-dimensional reduction of Bethe-Salpeter-equation we can obtain the equation of a Lippmann-Shwinger type. The basic idea of the suggested approach is to define effective potential as such local

[a]e-mail: safron@srd.sinp.msu.ru

operator in Lippmann-Shwinger partial-wave equation, which generates S - matrix, having required discontinuities on dynamic cuts.

Partial-wave S-matrix in a state with the angular momentum l can be represented in the form $S_l = \tilde{S}_l(k)\exp[2ik^{2l+1}J_l(k^2)/\pi]$, where $\tilde{S}_l(k)$ is reduced S-matrix satisfying "elastic" unitarity condition. The function $J_l(s)$ is determined by dispersion integral $J_l(s) = \int_{s_a}^{\infty} \ln \eta_l(s')(s^2 - s'^2)^{-1} s'^{1/2-l} ds'$, where s_a is a threshold of inelastic channels in s-plane and $\eta_l(s') = |S_l|$ is inelasticity parameter. It is analytic function in a complex s-plane with a cut above a threshold of inelastic channels ($s_a \leq s < \infty$). It also has left-hand cut ($-\infty < s \leq -s_a$), generated by connection with inelastic channels. The on-mass-shell reduced partial-wave scattering amplitude is defined as $\tilde{S}_l(k) = 1 - 2ikT_l(k^2)$. Corresponding scattering operator $\hat{T}_l(s)$ satisfies the Lippmann-Shwinger - type equation $\hat{T}_l(s) = \hat{W}_l + \hat{W}_l\hat{G}_{0l}(s)\hat{T}_l(s)$, where $\hat{G}_{0l}(s) = (2/\pi)\int_0^{\infty} |ql\rangle\langle ql|(s - q^2)^{-1}q^2 dq$ is free Green operator. The matrix element of local interaction operator $\hat{W}_l$ in configuration representation has the form $\langle r'|\hat{W}_l|r\rangle = \delta(r'-r)V_l(r)$. Connection between dynamic singularities of the partial-wave S-matrix and effective local potential $V_l(r)$ can be established by methods of inverse scattering problem. The kernel of Marchenko equation [6] can be expressed [4] through the discontinuity of the partial-wave S-matrix along the dynamic cut $S_l(i\xi+\varepsilon)-S_l(i\xi-\varepsilon) = 2iD_l(\xi)\theta(2\xi-\mu)$, where $\xi = -ik$. The parameter μ determines the position of the point of a dynamic cut nearest to physical region. The required potential can be represented in the form

$$V_l(r) = \int_{\mu}^{\infty} C_l(\alpha)\exp(-\alpha r)d\alpha. \tag{1}$$

We formulate the equations to calculate the weight functions $C_l(\alpha)$ for scattering in states with arbitrary angular momenta. As an example we shall write down the formulas for a case of P-wave scattering:

$$C_1(\alpha) = z(\alpha)+\alpha\int_{\mu}^{\alpha} \frac{d\gamma z(\gamma)}{\gamma^2}\left[2+\gamma p(\gamma,\alpha-\gamma+\mu)+2\int_{\alpha}^{\gamma} p(\gamma,\beta-\gamma+\mu)d\beta\right],$$

where $z(\alpha) = \alpha D_1(\alpha/2)/\pi$ and function $p(\alpha,\beta)$ is solution of equation

$$p(\alpha,\beta) = \frac{2}{\alpha} - \theta(\beta-2\mu)\left\{\frac{z(\beta-\mu)}{(\beta-\mu)(\alpha+\beta-\mu)}+\right.$$

$$\left.+\frac{2}{\alpha}\int_{\mu}^{\beta-\mu}\frac{d\gamma z(\gamma)}{\gamma^2}\left[1+\frac{\alpha\gamma p(\gamma,\beta-\gamma)}{2(\alpha+\gamma)}+\int_{\mu}^{\beta-\gamma} d\xi p(\gamma,\xi)\right]\right\}. \tag{2}$$

It follows from obtained equations that i) at the fixed value of argument α function $C_l(\alpha)$ is determined by discontinuity $D_l(\xi)$ only at $\xi \leq \alpha/2$ and ii) to find $C_l(\alpha)$ finite number of iteration is needed.

3 Application of analytical approach to NN-scattering

The developed approach has been applied to constructing effective nucleon-nucleon potentials in different partial-wave states. The one-boson-exchange model was used to calculate the discontinuities of the partial-wave scattering amplitudes taking into account $\pi, \rho, \sigma, \omega, \eta, a_0$-meson-exchange mechanisms. The NN-potentials in our approach (as against one-boson-exchange model in usual sense) contain nonlinear contributions relative to dynamic discontinuities of partial-wave scattering amplitudes, which play essential role at small distances. In the formula (1) we have replaced the upper infinite limit of integration by final one $\mu\tau$. The cut-off parameter τ was chosen equal 15. That meets to the account of nuclear forces with radius more than 0.1 fm. The inelasticity parameters were chosen on the basis of available phase shift analyses in the region of kinetic energies of incident nucleon T up to 6 GeV. In Tab.1 the predictions of developed approach for phase shifts in S- and P- partial-wave channels and the available data of phase shift analyses are represented. The used coupling constants (defined as in [3, 4]) and meson masses are listed in Tab.2. Note that a σ-mesons with various masses and coupling constants in different scattering channels are considered by us as a method of taking into account of the two-π-meson exchange cuts. In the further investigations we believe to improve the contribution of two-π-meson exchange mechanism. In particular, we have intension to use for these purposes effective field theory.

Table 1. S- and P-wave nucleon-nucleon phase shifts (in degrees). The first (upper) value in each cell is prediction of present work, the second and third values are solutions of energy-dependent and single energy partial-wave analyses [7]. T is the kinetic energy (in MeV) of incident nucleon in laboratory frame.

T	100	200	300	400	600	800	1000	1100	1200
	26.9	8.5	-4.6	-15.1	-31.9	-45.3	-56.1	-60.7	-64.7
$^1S_0$	28.5	8.1	-5.3	-14.9	-31.6	-45.3	-54.8	-58.1	-60.6
	27.9	8.7	-4.8	-15.3	-31.1	-44.3	-57.7	-57.5	-
	41.8	21.5	8.8	-0.8	-15.8	-27.7	-38.0	-42.9	-47.6
$^3S_1$	42.2	20.6	7.3	-3.2	-20.8	-34.5	-43.8	-46.8	-48.8
	43.7	21.2	6.7	-3.0	-21.9	-35.3	-45.7	-50.4	-
	-15.9	-21.9	-27.3	-32.0	-38.6	-41.1	-40.5	-38.4	-33.8
$^1P_1$	-14.6	-21.7	-26.6	-30.3	-36.5	-41.6	-42.9	-39.9	-33.4
	-11.3	-21.0	-27.5	-30.6	-35.8	-43.2	-45.6	-37.1	-
	7.9	-2.7	-11.9	-20.2	-33.4	-45.1	-57.4	-62.9	-67.8
$^3P_0$	8.8	-1.6	-11.7	-20.9	-35.7	-46.1	-54.7	-57.8	-60.1
	9.0	-2.1	-12.1	-20.4	-35.4	-50.0	-58.8	-	-
	-14.5	-22.7	-29.4	-35.2	-44.6	-52.0	-58.2	-61.1	-63.8
$^3P_1$	-13.5	-21.7	-28.1	-34.0	-44.9	-52.3	-56.4	-57.5	-58.4
	-13.7	-21.7	-28.0	-34.1	-44.4	-52.5	-58.9	-59.0	-

Table 2. The meson masses and meson-nucleon coupling constants.

λ	m_λ, MeV	$g^2_{\lambda NN}/4\pi$	σ	m_σ, MeV	$g^2_{\sigma NN}/4\pi$
π	138.03	14.0	$^1S_0$	550	8
ρ	769	0.84 ($\kappa = 5$)	$^3S_1$	550	11
ω	782.6	7.56	$^3P_1$	350	1.3
η	548.8	1.73	$^3P_0$	750	33
a_0	983	2.5	$^1P_1$	350, 550	1.4, 7

4 Conclusion

The new relativistic approach to constructing effective hadron-hadron interaction operators in an states with arbitrary angular momenta is applied to NN-scattering in S- and P-partial-wave states. We define effective potential as such local operator in the partial-wave Lippman-Shwinger equation, which generates relativistic (Feynman) on-mass-shell scattering amplitude having required discontinuities on dynamic cuts. The theoretical predictions for nucleon-nucleon phase shift parameters are in good agreement with available experimental data at kinetic energies of incident nucleon up to 1.2 GeV. It is shown that the nonlinear contributions (relative to dynamic discontinuities of partial-wave scattering amplitudes) generate in our approach the additional repulsion in NN-interaction at small distances, not related to ω -meson exchange mechanism. Note that in Bonn potential model [3] the short-range repulsion is due to ω-meson exchange contribution. It is required in this theory non-realistically large value ($\simeq 20$) of a coupling constant $g^2_{\omega NN}/4\pi$. In our approach the value of this coupling constant is consistent with available experimental data and also with theoretical quark-model calculations.

Acknowledgments

This work was supported by the Russian Foundation for Basic Research under the project No 04-02-16967.

References

[1] S.Weinberg *Nucl. Phys.* **B** 363, 3 (1991).
[2] M.Lacomb, B.Loiseau, J.M.Richard, et al. *Phys. Rev.* **C** 21, 861 (1980).
[3] R.Machleidt *Phys. Rev.* **C** 63, 024001 (2001).
[4] A.N.Safronov, A.A.Safronov *Izv. RAN, ser. fiz.* 68, 1195 (2004).
[5] R.Blankenbecler, R.Sugar *Phys. Rev.* 142, 1051 (1966).
[6] V.A.Marchenko *Rep. Sov. Acad. Sci.* 104, 695 (1955).
[7] R.A.Arndt, et al. *Phys. Rev.* **C** 62, 034005 (2000); http://gwdac.phys.gwu.edu.

TOP QUARK PHYSICS

E.Boos[a]
Skobeltsyn Institute of Nuclear Physics, Moscow State University, 119992 Moscow, Russia

*Abstract.*A short summary of selected topics on top quark physics at hadron and lepton colliders is presented. More details can be found in a number of reviews (see, [1–9] and references therein).

1 Introduction. Discovery and Puzzles

In the Standard Model the top quark is the spin-$\frac{1}{2}$ fermion with the electric charge $Q^t_{em} = \frac{2}{3} \mid e \mid$, the weak isospin partner of the b quark, and a color triplet.

Top quark has been found by the Fermilab CDF and D0 collaborations. The main Tevatron RUN1 results are summarized as follows:

- Top mass: $M_t = 174.3 \pm 3.2(stat) \pm 4.0(syst)$
- Top pair production cross section: $\sigma_{t\bar{t}}(CDF\ M_t = 175GeV) = 6.5^{+1.7}_{-1.4}pb$ $\sigma_{t\bar{t}}(D0\ M_t = 172GeV) = 5.9 \pm 1.7pb$
- Top quark Yukawa coupling: $\lambda_t(M_t) = 1.00 \pm 0.03$
- The CKM parameter: $|V_{tb}| > 0.78$ (90% CL)
- The 95% Confidence Level Limit on single top production cross section: 13.5 pb by CDF, 39 pb (17 pb Neural Network) (s-channel) and 58 pb (22pb Neural Network) (W-gluon fusion) by D0. (SM prediction: $\sigma_{SM} = 2.43 \pm 0.32$ pb)

Top quark is the heaviest elementary particle found so far with a mass slightly less than the mass of the gold nucleus. In various respects the top quark is a very unique object.

- Since the top quark is so heavy and its decay width is proportional to the third power of its mass, the top quark decays ($\tau_t \sim 5 \times 10^{-25}$ sec) much faster than a typical time-scale for a formation of the strong bound states ($\tau_{QCD} \sim 3 \times 10^{-24}$ sec). So, top provides, in principle, a very clean source for a fundamental information.
- Top is so heavy and point like at the same time. So, one might expect a possible deviations from the SM predictions more likely in the top sector.
- Top Yukawa coupling $\lambda_t = 2^{3/4} G_F^{1/2} m_t$ is very close to unit. Studies of top may shed a light on an origin of the mechanism of the EW symmetry breaking.

Top quark physics will be a very important part of research programs for all future hadron and lepton colliders including studies of top quark properties, various new physics via the top quark, and kinematical characteristics of top quark events as significant backgrounds to a number of other processes.

2 Basic production processes at colliders

At hadron and lepton colliders top quarks may be produced either in pairs or singly.

The representative diagrams for the top pair in quark-antiquark and glue-glue collisions (right) and single top (left) production at the Tevatron and LHC are shown in the Fig. 1.

[a]e-mail: boos@theory.sinp.msu.ru

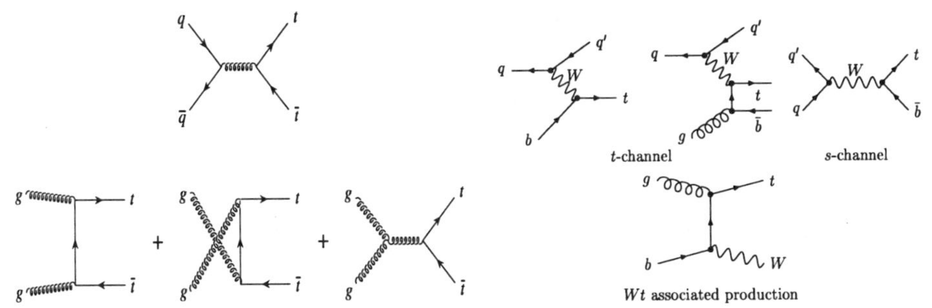

Figure 1:

	$\sigma_{\rm NLO}$ (pb)	$q\bar{q} \to tt$	$gg \to tt$
Tevatron ($\sqrt{s} = 1.8$ TeV $p\bar{p}$)	$4.87 \pm 10\%$	90%	10%
Tevatron ($\sqrt{s} = 2.0$ TeV $p\bar{p}$)	$6.70 \pm 10\%$	85%	15%
LHC ($\sqrt{s} = 14$ TeV pp)	$833 \pm 15\%$	10%	90%

Table 1:

Three mechanisms of the single top production are distinguished by the virtuality Q_W^2 of the W-boson involved: t-channel ($Q_W^2 < 0$), s-channel ($Q_W^2 > 0$), associated tW ($Q_W^2 = M_W^2$).

The NLO cross sections for basic top pair and single top production processes are collected in the Tables 1,2.

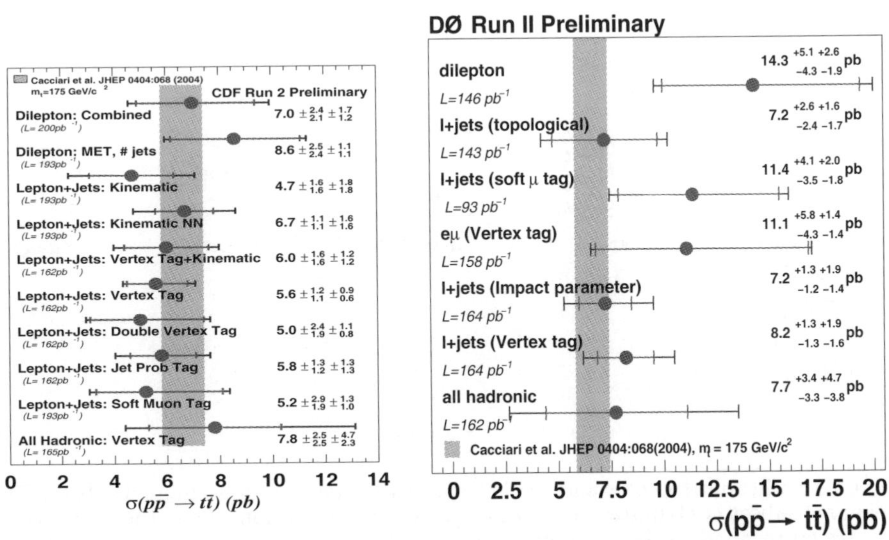

Figure 2:

The Figures 2 show recent RUN2 results for the top pair production cross section

	s channel	t channel	Wt
Tevatron ($\sqrt{s} = 2.0$ TeV $p\bar{p}$)	$0.90 \pm 5\%$	$2.1 \pm 5\%$	$0.1 \pm 10\%$
LHC ($\sqrt{s} = 14$ TeV pp)	$10.6 \pm 5\%$	$250 \pm 5\%$	$75 \pm 10\%$

Table 2:

measurement in various signatures being in an agreement with the NLO computations within the error intervals.

The best 95% confidence level upper limits on single top production cross sections in RUN2 by D0 collaboration are 6.4 pb in the s-channel and 5.0 pb in the t-channel.

The first Single Top observation is expected at the Tevatron RUN2 rather soon when accumulated integrated luminosity will be about 1-1.5 fb^{-1}. The main problems hare are large backgrounds ($W + jets, Wb\bar{b}, t\bar{t}$ etc.) and complicated analysis to extract the signal.

The contributing diagrams to both pair and single top quark production in e^+e^- collisions are shown in the Fig. 3,4 for $e\nu_e bt$ final state. These diagrams form so-called CC20 set of diagrams which splits into two gauge invariant subsets of 10 diagrams, s-channel and t-channel.

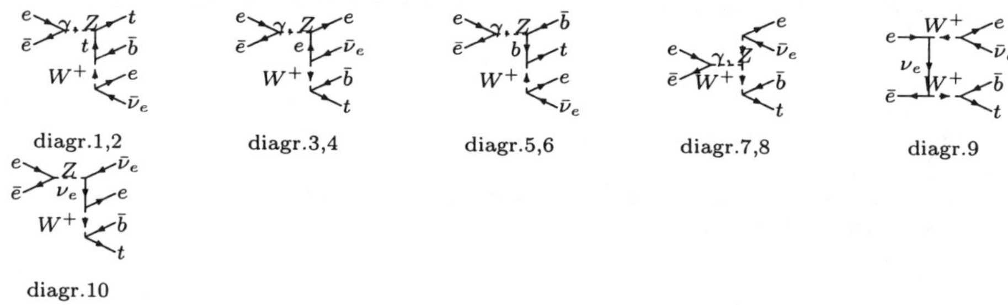

Figure 3:

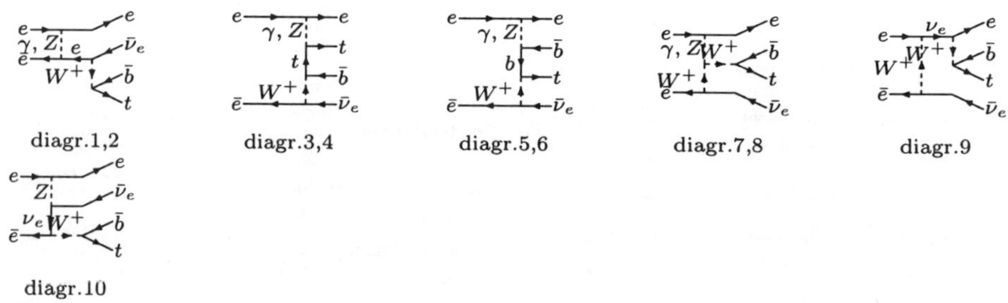

Figure 4:

One could split top pair and single top contributions in the s-channel subset. The t-channel subset contributes only to the single top production. The other possible final states correspond to the top pair production and follow from possible decay modes of W-boson $e^+e^- \to t\bar{t} \to WWb\bar{b}$, $\quad W \to f\bar{f}'$,where e.g. for W^+ $f = u, c, \nu_e, \nu_\mu, \nu_\tau \nu_\mu$; $f' = d, s, e, \mu, \tau$.

In case of $\gamma\gamma$ collisions there are no nontrivial gauge invariant subsets. A situation is similar to single top at the LHC in Wt mode.

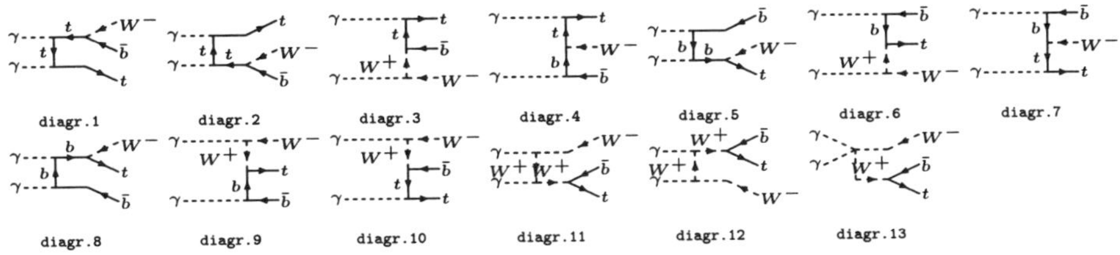

The top pair and single top contributions could be separated in a gauge invariant way as above.

In γe collisions the top quarks could be produced only singly, the corresponding diagrams are shown in the following figure. This is one of so called "gold plated" processes in γe collision mode of ILC.

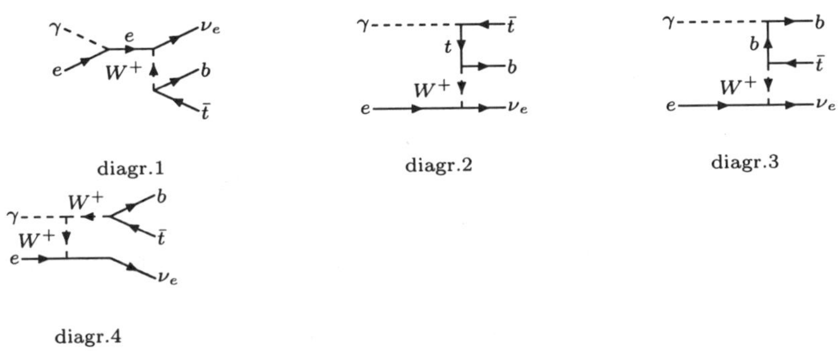

All main production cross sections of the top quarks production in various linear collider collision modes are shown in the Figure 5 Note that at LEP2 energies the rate of kinematically allowed single top production is too small, about 10^{-5} pb. In the region of 0.5-1.5 TeV discussed for the ILC the top pair production is main production mehanism while at very high energies of 3-5 TeV expected for the CLIC project the single top production dominates.

3 Decays and spin correlations

In SM top decays to W-boson and b-quark practically with 100% probability. Due to (V-A) structure of the SM charged currents the top quark decay matrix element has a very simle form

$d\Gamma \sim |\mathcal{M}|^2 \sim (t + ms)\cdot \ell b \cdot \nu$, where, in the top-quark rest frame, the spin four-vector is $s = (0, \hat{s})$, and $\hat{s}$ is a unit vector that defines the spin quantization axis of the top quark. In the top quark rest frame: $\frac{1}{\Gamma}\frac{d\Gamma}{d\cos\theta_\ell} = \frac{1}{2}(1 + \cos\theta_\ell)$. Therefore, the charged lepton tends to point along the direction of top spin.

Since the top quark decays before hadronization its spin properties are not spoiled. Therefore the spin correlation in top production and decays is an interesting issue of top quark physics. In particular, the SM predicts probabilities for three different helicity states ($h_W = \pm 1, 0$) of W-boson in top decays. The average polarization of a top quark produced in QCD subprocesses equals zero. However, in top pair

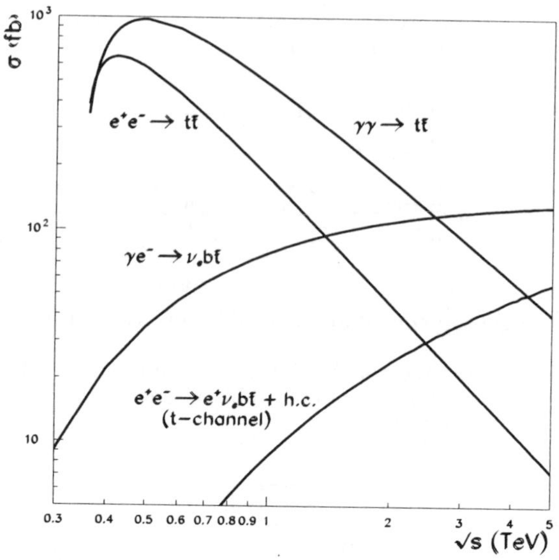

Figure 5:

production one can investigate the $t\bar{t}$ spin correlations. In the single top quark t-channel and s-channel production processes the top quark is produced in SM through a left-handed interaction and spin correlations survive in the final decay products. Hence, single-top-quark production offers an opportunity to observe the polarization of the top-quark in production. The top-quark spin in each event should follow the direction of the down-type quark momentum in the top-quark rest frame. This is the direction of the initial $\bar{d}$-quark for the s-channel, and the direction of the final d-quark for the t-channel single-top-quark production processes. This follows from the above result on polarized top decays, because two single-top production modes could be considered as top-quark decays "backward in time". Therefore the best variable to observe maximal top-quark spin correlations in single-top-quark s-channel or t-channel production and the subsequent decay of the top-quark is the angle between this down-type quark direction in the production processes and the charged lepton (or d, s-quark) direction from the top-quark decay in the top-quark rest frame. Spin properties in the tW production process are more involved. Here one can find a kinematic region in which top quarks are produced with the polarization vector preferentially close to the direction of the charged lepton or the d, s-quark momentum from the associated W decay. In this kinematic region the direction of the produced charged lepton or the d, s-quark should be as close as possible to the direction of the initial gluon beam in the top-quark rest frame.

4 Top mass, V_{tb}, Top Yukawa coupling

In SM the W and Z boson masses get corrections via the top quark and H boson loop contributions to W and Z propagators as shown in the Fig.6.

Figure 6:

$M_W^2 = \frac{\frac{\pi\alpha}{\sqrt{2}G_F}}{s_W^2(1-\Delta r)}$ where Δr contains the one-loop corrections. $(\Delta r)_{\rm top} \approx -\frac{3G_F m_t^2}{8\sqrt{2}\pi^2}\frac{1}{t_W^2}$ where $t_W^2 \equiv \tan^2\theta_W$. This one-loop correction depends quadratically on the top-quark mass. The experimental accuracy of the top and EW gauge boson mass measurements allow to get indirect bounds on the Higgs boson mass, see the Fig.7 At

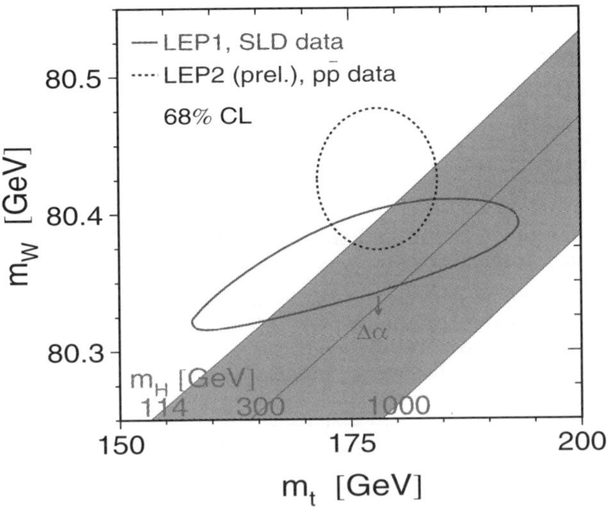

Figure 7:

the Tevatron Run II with $2fb^{-1}$ one expects: $M_W \sim 27$ MeV, $M_t \sim 3$ GeV yielding a prediction for the Higgs mass with an uncertainty of $\frac{M_h}{M_h \sim 40}\%$ At the LHC with 10 fb^{-1} one expects $M_t \sim 0.7$ GeV and at the ILC with 500 fb^{-1} from the top pair threshold scan one can get $M_t \sim 0.1$ GeV . This would lead to much better bound to the Higgs mass, however by the time of this measurements one expects a light Higgs boson discovery, otherwise all the above limits would by irrelevant.

At LHC and Tevatron Run2 via single top V_{tb}^2 could be measured with an accuracy of 10% dominated by systematics. At ILC (1 TeV, 500 fb^{-1}) in $e\gamma$ collisions - 2-3 % accuracy dominated by statistics.

Top Yukawa coupling $t\bar{t}H$ measurements could be done at the LHC with an accu-

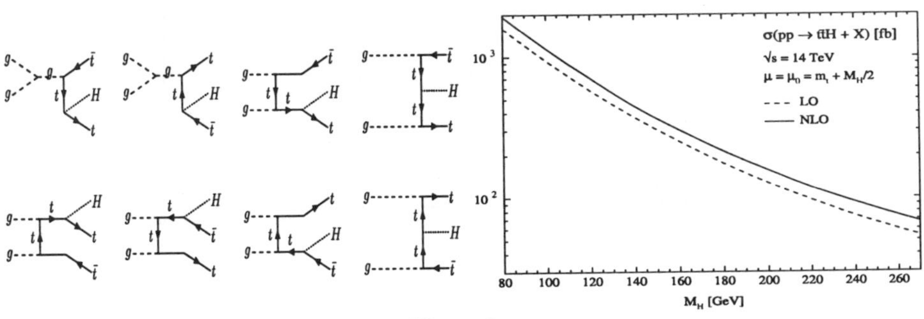

Figure 8:

racy from 16% at low luminosity to 11% at high luminosity regime. An important ingredient here is the NLO computation (see the Fig.8).

5 "New Physics" via top quark

Various expected aspects of New Physics could be tested via top quark involved processes:

- W_{tb} anomalous couplings
- FCNC couplings
- Various SUSY effects without and with R-parity violation
- Charged Higgs in top decays
- New strong dynamics (W', Z', π_T, ρ_T, topgluon, $W_L W_L \to t\bar{t}$ ...)
- Kaluza-Klein graviton excitations and radion in ADD and RS scenarios
- Kaluza-Klein gauge boson and third generation quark excitations in universal extra dimension scenarios etc

As an example we show here in the Fig.9 only results for generic searches for new resonances decaying to the top quark pair. Measured $t\bar{t}$ invariant mass distribution for reconstruction of a narrow resonance of mass 1600 GeV decaying to $t\bar{t}$ and value of $\sigma\times$ BR required for a 5σ discovery potential.

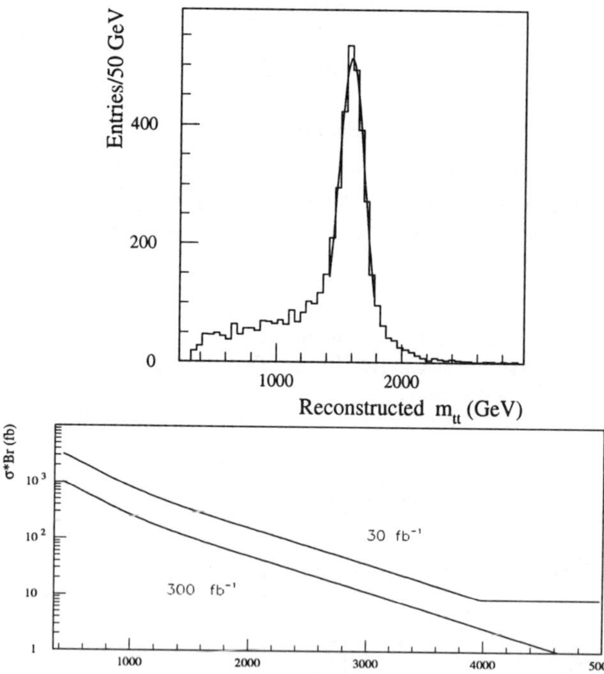

Figure 9:

6 Conclusions

Discovery of the top quark has opened up many new avenues to interesting physics

- Precision measurements of top quark characteristics such as mass, production cross sections, decay width and branching fractions, spin correlations are needed to test the SM
- Tests and understanding all possible deviations from the SM expectations to check if top is exotic in some way
- Precise calculations and simulations, and measurements of the top event kinematical characteristics to understand backgrounds to many other possible New physics processes
- Possible discovery and study of various New physics effects via top production and/or decay

The top quark physics is a very wide subject. In a very short review presented here we did not discuss many aspects and results, such as details of the NLO and NNLO computations, LO and NLO top quark event generators, details of spin correlations, very interesting threshold effects etc. We where not able to give the references on the original publications which could be found in the mentioned reviews.

Acknowledgments

The work is partly supported by RFBR 04-02-17448 and University of Russia UR.02.03.028 grants, and Russian Ministry of Education and Science NS.1685.2003.2. The author thanks the Organizing Committee of 12th Lomonosov Conference on Elementary Particle Physics and, especially, Prof. Alexander Studenikin for a pleasant scientific atmosphere of the conference.

References

[1] M. Beneke *et al.*, "Top quark physics," arXiv:hep-ph/0003033.
[2] E. E. Boos, "Top quarks at photon colliders," Nucl. Instrum. Meth. A **472** (2001) 22 [arXiv:hep-ph/0009100].
[3] J. A. Aguilar-Saavedra *et al.* [ECFA/DESY LC Physics Working Group], "TESLA Technical Design Report Part III: Physics at an e+e- Linear Collider," arXiv:hep-ph/0106315.
[4] S. Willenbrock, "The standard model and the top quark. ((U))," arXiv:hep-ph/0211067.
[5] D. Chakraborty, J. Konigsberg and D. L. Rainwater, "Review of top quark physics," Ann. Rev. Nucl. Part. Sci. **53** (2003) 301 [arXiv:hep-ph/0303092].
[6] S. Dawson, "The top quark, QCD, and new physics," arXiv:hep-ph/0303191.
[7] G. Weiglein *et al.* [LHC/LC Study Group], "Physics interplay of the LHC and the ILC," arXiv:hep-ph/0410364.
[8] E. W. N. Glover *et al.*, "Top quark physics," Acta Phys. Polon. B **35** (2004) 2671 [arXiv:hep-ph/0410110].
[9] W. Wagner, "Top quark physics in hadron collisions," Rept. Prog. Phys. **68** (2005) 2409 [arXiv:hep-ph/0507207].

PRECISION MEASUREMENT OF THE TOP QUARK MASS FROM $M_{b\ell}$ DISTRIBUTION IN $t \to b\,\ell\nu$ DECAYS

M.L.Nekrasov [a]

Institute for High Energy Physics, 142284 Protvino, Russia

Abstract. The method of the top quark mass measurement from $M_{b\ell}$ distribution in $t \to b\,\ell\nu$ decays can be considerably improved if applying the technique of moments and proceeding to moments of high degree. At the LHC this allows one to reduce the systematic and statistical errors of the top mass by more than a factor of two.

The precision measurement of the top quark mass is necessary for testing of the SM and/or selecting the probable scenario of its extension. The current accuracy of the top mass, attained at Tevatron, is $\Delta M_t = 2.9$ GeV, and a further progress is expected mainly at LHC and a future International linear collider. At the LHC the accuracy of measurement of the top mass is anticipated at the level of 1–2 GeV and in view of large statistics the main difficulties are expected from systematic errors. An analysis of [1] shows that the most promising method from the point of view of optimization of the errors is based on an investigation of the invariant-mass distribution of $J/\psi + \ell$ system produced in decays $t \to bW \to b\ell\nu$ with b-jet formation and the J/ψ being a member of the b-jet. The evaluation made by Monte-Carlo modelling gives approximately 0.7 GeV for systematic error of the top mass measured by this method and of about 1 GeV for statistical error for 4 years of LHC operation [2]. In this paper we discuss an opportunity to improve this result by means of optimization of the data handling within the same experiment.

So, following [2] we consider the processes

$$q\bar{q},\, gg \to t\bar{t} \to bW\, bW \to b\ell\nu\; bq_1q_2 \Rightarrow \{b\text{-jet} + \ell\} + \{3 \text{ jets}\}\,, \qquad (1)$$

with the b-jet and the isolated lepton $\ell = \{e, \mu\}$ coming from one t quark, and the remaining three jets coming from another t quark. In the experiment the above states are registered and the invariant-mass distribution of $J/\psi + \ell$ system is measured. The $M_{J/\psi\,\ell}$ distributions can then be compared with a template of shapes parametrized by the top mass and thus M_t can be fitted [2]. One could note, however, that the invariant-mass distribution of the b+ℓ system inevitably emerges at a certain stage of this analysis. Therefore on the equal rights one can carry out the fitting in terms of the data converted to the form of $M_{b\ell}$ distribution. In what follows we adhere to the latter option and hence consider that (as if) a distribution $F(q) = \sigma^{-1}\,\mathrm{d}\sigma/\mathrm{d}q$ is measured, where σ is the cross-section of the process (1) and $q = M_{b\ell}$ is the reconstructed invariant mass of the $b + \ell$ system.

By proceeding in this manner, we simulate the data under the following suppositions. First we suppose that there is a satisfactory method for extracting

[a] e-mail: Maksim.Nekrasov@ihep.ru

signal from the data. (Actually this means the existence of a satisfactory model for background processes that survive after setting of kinematic cuts [2].) Secondly we describe the signal in the Born approximation, identifying the b-jet with the b quark. Finally, on the basis of the results of [2], we disregard the effects of the finite width of the top quark. The latter assumption means that $\sigma^{-1}\,\mathrm{d}\sigma/\mathrm{d}q$ is equal to $\Gamma_{b\ell\nu}^{-1}\,\mathrm{d}\Gamma_{b\ell\nu}/\mathrm{d}q$, where $\Gamma_{b\ell\nu}$ is a partial width of the decay $t \to b\ell\nu$. (Thus the distribution F becomes process-independent.)

At Fig. 1(a) the distribution $F(q)$ is represented for various values of the top mass. We see from the figure that in the absolute value the distribution is most sensitive to M_t in a region located between the maximum of the distribution and a large-q tail where $F(q)$ is almost vanishing. So in this region of q-variable one could expect the highest accuracy for the M_t determination. A practical way to enhance the role of this region is to proceed to high moments over the distribution,

$$\langle q^n \rangle = \int_0^M \mathrm{d}q\, q^n F(q)\,. \tag{2}$$

Here M is a technical parameter fixed close to M_t, and the normalization of $F(q)$ is adjusted so as to satisfy the equality $\langle 1 \rangle = 1$. From Fig. 1(b) it is obvious that with increasing n the distribution $q^n F(q)$ is concentrating more and more in the above intermediate region, increasing thus the sensitivity of $\langle q^n \rangle$ relative to M_t.

So, by basing on the above observation, we define the experimentally measured top-quark mass as a solution to the equation

$$\langle q^n \rangle_{\mathrm{exp}} = \langle q^n \rangle\,. \tag{3}$$

Here the moment in the l.h.s. is determined on the basis of the experimental data, and that in the r.h.s. on the basis of the theoretical distribution, which depends on the parameter M_t. Let, for a given n, a solution to (3) be $M_t = M_{t(n)}$. Then the error of the solution is

$$\Delta M_{t(n)} = \Delta \langle q^n \rangle_{\mathrm{exp}} \bigg/ \frac{\mathrm{d}\langle q^n \rangle}{\mathrm{d}M_t}\bigg|_{M_t = M_{t(n)}}\,. \tag{4}$$

Our aim is to estimate $\Delta M_{t(n)}$ and find an optimal value of n which would minimize $\Delta M_{t(n)}$. Since in view of (2) the derivative $\mathrm{d}\langle q^n \rangle/\mathrm{d}M_t$ is known, the problem is reduced to the estimation of the error $\Delta \langle q^n \rangle_{\mathrm{exp}}$ or its components, the statistical and systematic experimental errors.

The statistical error can be estimated, in almost model-independent way, by means of the formula [3]

$$\Delta^{\mathrm{stat}} \langle q^n \rangle_{\mathrm{exp}} = \sqrt{\frac{1}{N} \langle q^{2n} \rangle}\,. \tag{5}$$

Here N is the representative sample of events, and $\langle q^{2n} \rangle$ is a theoretically calculated moment. For the consistent determination of the systematic error the application of a proper MC event generator is needed. The analysis [2] made with the aid of the PYTHIA and HERVIG events generators showed that the major uncertainties in the $M_{J/\psi\,\ell}$ distribution were caused by (*i*) the uncertainties in the b-quark fragmentation and (*ii*) the background processes. It is obvious that at solving the inverse problem, the reconstructing the $M_{b\ell}$ distribution from the $M_{J/\psi\,\ell}$ one, the errors should be of the same origin.

With this in mind we consider first the error resulting from the uncertainty in the b-quark fragmentation. For brevity we call this the type I error. At the level of the $M_{b\ell}$ distribution it appears as an uncertainty Δq in the q variable. Let us suppose that Δq is sufficiently small, and let us neglect nonlinear effects. Then we have

$$\Delta^{\mathrm{sys\ I}} \langle q^n \rangle_{\mathrm{exp}} = \int_0^M \mathrm{d}q \ \ [q^n F(q)]' \ \Delta q \, . \tag{6}$$

Here the prime means the derivative with respect to q. The Δq is determined by the relation $\Delta q = rq$ with r is a coefficient, $r \simeq \frac{1}{2}\Delta E_b/E_b$, where E_b is the energy of the b quark in the laboratory frame and ΔE_b is its error. In fact the mentioned relation for Δq follows from the observation that q^2 actually is a doubled scalar product of 4-momenta of the b quark and of the lepton ℓ whose momentum is considered precisely determined.

The systematic error arising after subtraction of the background processes—we call it the type II error—appears in the absolute value of the distribution function. Denote the corresponding error by δF, then we have

$$\Delta^{\mathrm{sys\ II}} \langle q^n \rangle_{\mathrm{exp}} = \int_0^M \mathrm{d}q \ q^n \, \delta F(q) \, . \tag{7}$$

The function $\delta F(q)$ should be vanishing at the boundaries of the phase space. We also assume that δF only once changes the sign when q runs the values. The simplest function satisfying to these requirements is a polynomial, $\delta F = h\, q(q-M/2)(q-M)$ with parameter h describing the amplitude of the error.

Now we turn to numerical results. We assign the following values for the parameters with global meaning: $M_W = 80.4$ GeV, $\Gamma_W = 2.1$ GeV, $M_t = M = 175$ GeV. The parameters N, r, h take the values depending on the conditions of observation. Since in [2] the $M_{J/\psi}$ distribution was determined at $N = 4000$ (with kinematic cuts and for 4 years of LHC operation), we set this value for N, as well. The parameters r and h may be fixed by basing on the properties of the $M_{J/\psi\,\ell}$ distribution. We omit here the respective discussion and show only the result: $r \simeq 0.005$, $h \simeq 1.7 \times 10^{-10}$ GeV^{-4} [3]. Note that the same estimation for r follows from the above mentioned formula $r \simeq \frac{1}{2}\Delta E_b/E_b$ when taking into consideration the 1%-precision of the determination of the energy of the b jets expected at LHC [1].

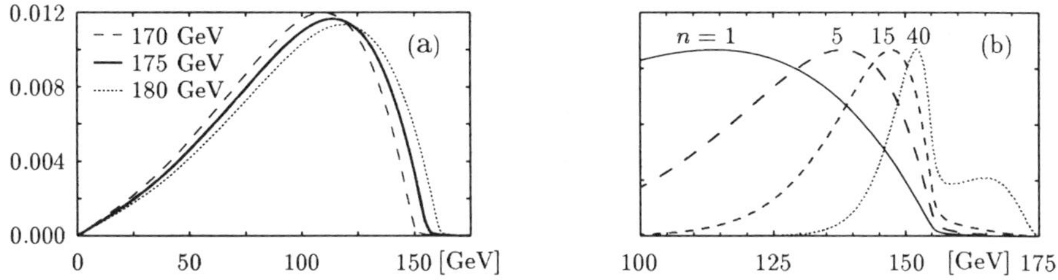

Figure 1: The distribution $F(q)=\Gamma^{-1}\,\mathrm{d}\Gamma/\mathrm{d}q$, $q \equiv M_{b\ell}$, at $M_t = 170$, 175, 180 GeV (a), and the shape of $q^n F(q)$ at $n = 1$, 5, 15, 40 ($M_t = 175$ GeV), arbitrary normalization (b).

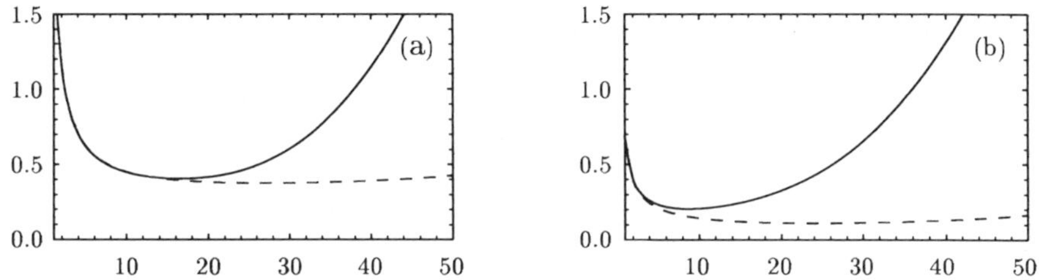

Figure 2: The $\Delta^{\rm stat} M_{t(n)}$ (a) and $\Delta^{\rm sys} M_{t(n)}$ (b) in GeV depending on n.

Now, as we know the parameters, we can calculate the errors of the top mass. At Fig. 2(a) we show the behavior of $\Delta^{\rm stat} M_{t(n)}$ versus degree n of the moments. Fig. 2(b) shows the behavior of the summed-quadrature systematic error $\Delta^{\rm sys} M_{t(n)}$. The dashed lines in both figures represent the errors obtained with usage of effective moments determined by way of introducing a cut-off Λ_n instead of M in formulas (2), (6) and (7); the aim of introducing the cut-off is to isolate the tail of $q^n F(q)$ with an unphysical peak arising at large n, see Fig. 1(b). Having the errors $\Delta^{\rm stat} M_{t(n)}$ and $\Delta^{\rm sys} M_{t(n)}$, it is easily shown that near $n = 15$ the total error reaches the minimum $\Delta M_{t(15)} = 0.48(0.42)$ GeV, with $\Delta^{\rm stat} M_{t(15)} = 0.41(0.40)$ GeV and $\Delta^{\rm sys} M_{t(15)} = 0.24(0.08)$ GeV. Here we put into the brackets the errors calculated by the effective-moment method. It can be shown by the naive counting method [3] that the theoretical errors will not spoil the above accuracy provided that the $M_{b\ell}$ distribution will be calculated within the one-loop electroweak and two-loop QCD precision.

In summary, we discussed the potentialities inherent in the technique of moments and demonstrated that at the LHC the proceeding to moments of high degree allows one to reduce the total error of M_t to the level of about 500 MeV.

References

[1] M. Beneke et al. (conveners), A.Ahmadov et al., *"Top quark physics"*, CERN 2000-004 [hep-ph/0003033].

[2] A.Kharchilava, *Phys.Lett* **B** 476, 73 (2000).

[3] M.L.Nekrasov, *Eur.Phys.J.* **C** 44, 233 (2005).

REVIEW OF CKM RESULTS FROM BABAR

E.I.Rosenberg [a]
Department of Physics and Astronomy, Iowa State University, Ames, Iowa 50011-3160, U.S.A.

Abstract. We present the latest results, as of August 2005, on CKM matrix elements from the BaBar experiment. Results projected on the $\bar{\rho}-\bar{\eta}$ plane from the CKM fitter group [1] are presented to show how the measurements compare with the Standard Model fits.

1 Introduction

The unitary CKM matrix [2] describes the coupling of the quark flavors and is given by:

$$V_{\rm CKM} = \begin{pmatrix} V_{ud} & V_{us} & V_{ub} \\ V_{cd} & V_{cs} & V_{cb} \\ V_{td} & V_{ts} & V_{tb} \end{pmatrix}.$$

The unitarity of this matrix can be expressed in six equations of the type $V_{ud}V_{ub}^* + V_{cd}V_{cb}^* + V_{td}V_{tb}^*$ (the equation most useful in b-physics). Each of these six equations can be represented in geometric form as a triangle in the complex plane.
It is usual to parameterize this matrix in terms of four real parameters: A, λ, ρ, η as proposed by Wolfenstein [3].

$$V_{\rm CKM} = \begin{pmatrix} 1-\frac{1}{2}\lambda^2 & \lambda & A\lambda^3(\rho - i\eta) \\ -\lambda & 1-\frac{1}{2}\lambda^2 & A\lambda^2 \\ A\lambda^3\left[1-(\rho+i\eta)\right] & -A\lambda^2 & 1 \end{pmatrix} + O(\lambda^4).$$

In this parameterization the elements $V_{ub} = A\lambda^3(\rho - i\eta) = |V_{ub}|e^{-i\gamma}$ and $V_{td} = A\lambda^3(1-(\rho+i\eta)) = |V_{td}|e^{-i\beta}$ are complex and we can represent the unitarity triangle in the $\rho-\eta$ plane, or normalizing the base of this triangle to 1, in the $\bar{\rho}-\bar{\eta}$ plane as shown in Figure 1 where ($\bar{\rho} = (1-\lambda^2/2)\rho$ and $\bar{\eta} = (1-\lambda^2/2)\eta$. In the Wolfenstein approximation $V_{tb} = 1$ and $V_{ts} = -V_{cb}$. The angles of the triangle, $\beta = -\arg(V_{td}); \gamma = \arg(V_{ub}^*); \alpha = \pi - \gamma - \beta$, are determined by CP asymmetries. The sides and angles of this triangle parameterize the CKM matrix and in this report we present determinations of the sides and angles from the BABAR experiment. A separate report on measurements of the side labelled R_b will be given by V. Azzolini at this conference.

2 Experimental details

The data described in this report were taken with the BABAR detector at the SLAC PEP-II Asymmetric B-factory by the BABAR collaboration. The collaboration consists of over 600 physicists from 81 institutions in 11 countries. The detector is characterized by precision tracking, high-resolution calorimetry and excellent particle-type identification and has been described in detail elsewhere [4]. The PEP-II Asymmetric

[a] e-mail: redmount@iastate.edu

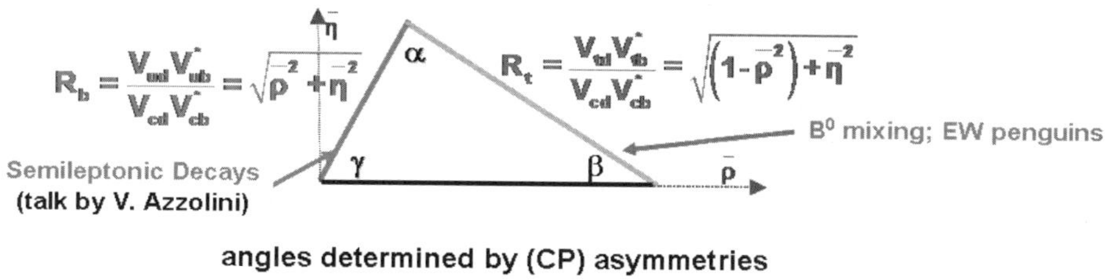

Figure 1: The CKM triangle

B-factory collides e^- of $\sim$ 9 GeV with e^+ of $\sim$ 3.1 GeV so that the center of mass energy corresponds to the mass of $\Upsilon(4S)$ resonance. At the $\Upsilon(4S)$, $\sigma_{b\bar{b}}/\sigma_{\mathrm{hadrons}} \approx 1/4$. The $B\bar{B}$ events of interest can be distinguished from the hadronic background due to the difference of the topological and kinematic characteristics of the events. Analysis specific experimental details can be found in the publications cited and the references therein. The bulk of the results present here are based on a sample of over 200 million $B\bar{B}$ pairs.

3 Determination of R_T [5,6]

Radiative penguin dominated decays can be used to extract the ratio of CKM matrix elements $|V_{td}/V_{ts}|$. While not yet competitive with the constraint on this side of the unitarity triangle from the ratio $\Delta m_d/\Delta m_s$, recent BABAR results are reported here. BABAR has an upper limit on the branching ratios for the decays $B \to \rho(770)\gamma$ and $B^0 \to \omega(782)\gamma$ [5]. The SU(3) weighted branching ratio

$$BR(B \to (\rho/\omega)\gamma) = \frac{1}{2}\{BR(B^+ \to \rho^+\gamma) + \frac{\tau_{B^+}}{\tau_{B^0}}[BR(B^0 \to \rho^0\gamma) + BR(B^0 \to \omega\gamma)]\}$$

is then combined with the previously reported BABAR measurement of $BR(B \to K^*\gamma)$ [5] to calculate a limit of $BR(B \to (\rho/\omega)\gamma)/BR(B \to K^*\gamma) < 0.029$ at 90% C.L. This limit constrains the ratio of CKM elements $|V_{td}/V_{ts}|$ by means of the equation [7,8]:

$$\frac{BR(B \to (\rho/\omega)\gamma)}{B \to K^*\gamma} = \left|\frac{V_{td}}{V_{ts}}\right|^2 \left(\frac{1 - m_\rho^2/M_B^2}{1 - m_{K^*}^2/M_B^2}\right)^3 \zeta^2[1 + \Delta R],$$

Both ζ, which describes the flavor-SU(3) breaking between ρ/ω and K^*, and ΔR, which accounts for annihilation diagrams, must be taken from theory. [7–9] The resulting limit on R_T in the $\bar{\rho} - \bar{\eta}$ plane is shown in figure 2

4 Determination of $\beta = \arg(V_{td}^*)$ [10–16]

We now turn to the determination of the angles of the unitarity triangle. Since the angles correspond to complex phases in the elements of the CKM-matrix, each of

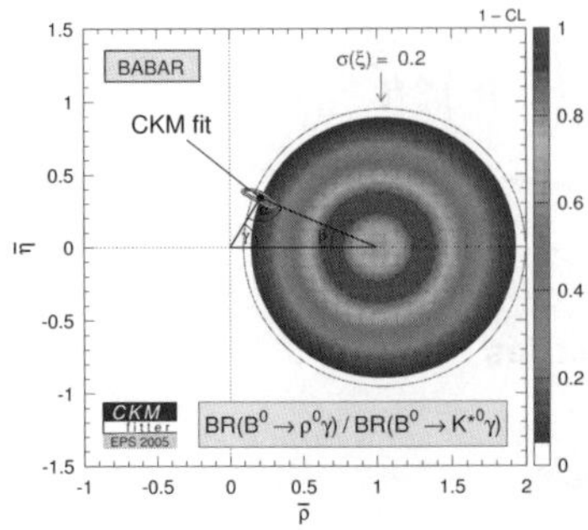

Figure 2: CKM fitter group constraint in the $\bar{\rho} - \bar{\eta}$ for R_t based on the ratio of the penguin branching ratios as described in the text.

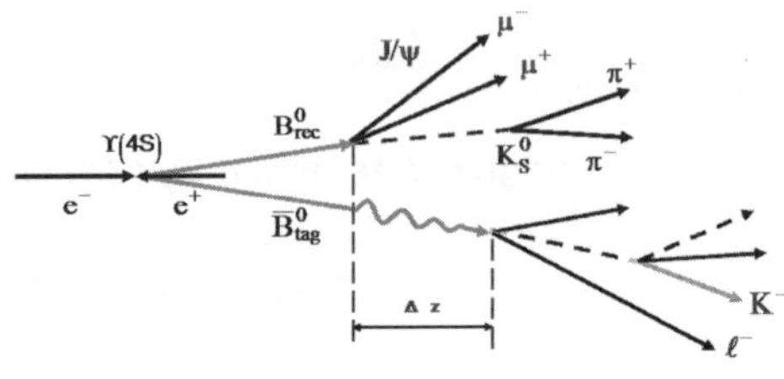

Figure 3: Time dependencies are measured by using the fact that the at the $\Upsilon(4S)$ the $B\bar{B}$ is produced coherently. By tagging the flavor of one B-meson (using the lepton charge, kaon charge, or a neural network) one determines the flavor of the other B-meson, B_{rec}, at the same time (set to t=0).

these measurements require interfering amplitudes. Because their mass- and weak-eigenstates differ, B^0 and $\bar{B}^0$ mesons can mix via the second order process (i.e. box diagrams). The loops in these diagrams are dominated by t-quark exchanges so that amplitude is sensitive to V_{td} and thus to the angle β. The mixing creates a linear combination of the two states $pB^0 + q\bar{B}^0$, where $(q/p) \approx e^{2i\beta}$. Since absolute phases cannot be measured for a single amplitude, we must search for processes in which there is a competing amplitude which will allow sensitivity to this phase via the interference term. For example, the B and $\bar{B}$ may decay to a common final state, f, so that B may reach that state either by direct decay or by first becoming a $\bar{B}$ and then decaying. In such a situation we can have interference between the direct decay and the decay via the mixing. We introduce the parameter

$$\lambda_f = (\frac{q}{p})\frac{\bar{A}_f}{A_f},$$

where the first term comes from the mixing and the second from the decay. The time-dependent CP asymmetry is then given by

$$A_{CP}(t) = \frac{N(\overline{B^0}(t) \to f_{CP}) - N(B^0(t) \to f_{CP})}{N(\overline{B^0}(t) \to f_{CP}) + N(B^0(t) \to f_{CP})} = S_f rmsin(\Delta mt) - C_f \cos(\Delta mt),$$

where $S_f = 2\text{Im}(\lambda_f))/(1 + |\lambda_f|^2)$ and $C_f = (1 - |\lambda_f|^2)/(1 + |\lambda_f|^2)$ can then be measured. Note that for a single weak phase $|\lambda| = 1$ and $S_f = Im(\lambda_f); C_f = 0$. For the general case a factor η_f represents the CP value of the state f, and we can compare the values of $-\eta_f S$ and C_f obtained in different final states.

These time dependent measurements are made possible by the boost of cm frame at the PEP-II Asymmetric b-factory. The principle of the time dependent measurements is shown in Figure 3. The interference between the $B^0\bar{B}^0$ mixing and the tree diagram for the transitions $b \to c\bar{c}s$ can be studied in the B decays modes $J/\psi K_S, J/\psi K_L, \psi(2s)K_S, \chi_{c1}K_S, \eta_c K_S, J/\psi K^{*0}(K_S\pi^0)$. Here the amplitudes $A_f \sim V_{cb}V_{cs}$ so that $\lambda \sim e^{-2i\beta}$. The time dependent asymmetry is then given by $A_{CP} \approx \sin(2\beta)\sin(\Delta mt)$. A BABAR analysis of these charmonium modes yields $\sin(2\beta) = 0.722 \pm 0.040(\text{stat.}) \pm 0.034(\text{syst.})$ [10]. The results for the individual modes are shown in the upper part of Figure 4.

BABAR also has studied β in the transitions $b \to c\bar{c}d$ by looking for charmed mesons in the final states: $D^{(*)+}D^{(*)-}$, where the notation $D^{(*)}$ indicates either a D or a D^* meson. Unlike the charmonium case where there is only a single tree-level amplitude, in this case there is a competing penguin process. In the Standard Model the penguin contribution is predicted to be small, in this limit $|C| \to 0$. The data are consistent with this limit. The results for the parameter S in a given final state f, with the CP of the final state accounted for, η_f, are shown in Figure 4. The band indicates the value from the charmonium modes only.

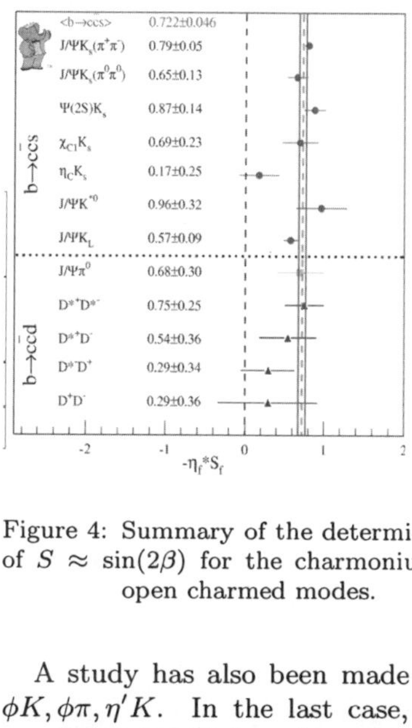

Figure 4: Summary of the determinations of $S \approx \sin(2\beta)$ for the charmonium and open charmed modes.

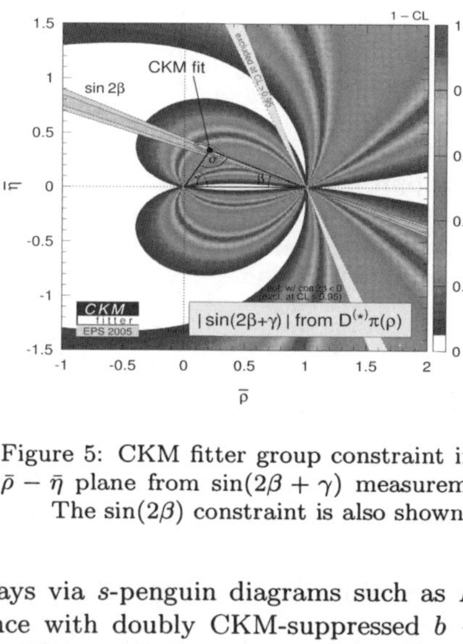

Figure 5: CKM fitter group constraint in the $\bar{\rho} - \bar{\eta}$ plane from $\sin(2\beta + \gamma)$ measurements. The $\sin(2\beta)$ constraint is also shown.

A study has also been made of of decays via s-penguin diagrams such as $B \to \phi K, \phi\pi, \eta' K$. In the last case, interference with doubly CKM-suppressed $b \to u$ amplitudes which introduce the angle γ can influence the extraction of $\sin(2\beta)$. Recall that for $\lambda \approx 0, -\eta S = \sin(2\beta)$ and $C = 0$ for each mode. The results for ηS and C are given in Figure 6. While C is consistent with 0, the naive average for S is $\sim 2.7\sigma$ below the value obtained from the charmonium measurements. Additional data should clarify the cause of this discrepancy.

The four-fold ambiguity in extracting β from $\sin(2\beta)$ can be reduced to a two-fold ambiguity by utilizing the interference of the CP-even and CP-odd amplitudes in the decays of the B^0 to a vector-vector final state such as $J/\psi K^{*0}$ to determine $\cos(2\beta)$. The ambiguity due to the strong phases are removed by observing the S- and P-wave interference in the the $K\pi$ final state. BABAR finds $\cos(2\beta) > 0$ at 85% CL.

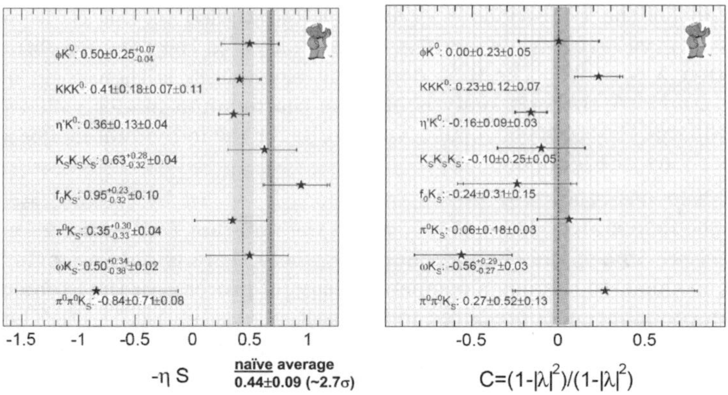

Figure 6: The values of η_s and C obtained from s-penguin decays. The dark bands indicate the results obtained from the charmonium measurements.

5 $\sin(2\beta + \gamma)$ analysis [20]

Constraints on the combination of CKM angles, $\sin(2\beta + \gamma)$ can be obtained from the interference of the tree-level CKM-favored $b \to c$ and $B\bar{B}$ mixing followed by the CKM suppressed $b \to u$ transitions by studying the time evolution of decays of the type $B \to D^{(*)} + \pi(\rho)$. The parameter λ characterizing the mixing is then given by $\lambda \approx re^{-2i\beta}e^{-i\gamma}$ where r is given by,

$$r = \left| \frac{A(B^0 \to D^{(*)+}h^-)}{A(\overline{B}^0 \to D^{(*)+}h^-)} \right|.$$

Since the $b \to u$ transitions are doubly CKM suppressed, r^2 is expected to be small ($\approx .02$), in which case $S \approx 2r\sin(2\beta + \gamma + \delta)$ and $C \approx 1$. BABAR has presented results based on a maximum likelihood fit to the time dependent decay distributions. The resulting constraints in the $\bar{\rho} - \bar{\eta}$ plane are shown in Figure 5.

6 Determination of $\alpha = \pi - \beta - \gamma$ [17–19]

The angle α is accessible via the combination of mixing and a $b \to u$ transition in decays such as $B^0 \to \pi\pi, \rho\rho$ and $\rho\pi$. In this case the amplitude $A_f \sim V_{ub}V_{ud} \sim e^{-i\gamma}$ so that $\lambda \sim e^{-2i\gamma}e^{-2i\beta} \sim e^{-2i\alpha}$. In the absence of gluonic penguins, $S = \sin(2\alpha)$; $C = 0$ and $A_{CP}(t) \approx \sin(2\alpha)\sin(\Delta mt)$. The presence of the penguin contribution means that

$$\lambda = e^{2i\alpha}\frac{T + P\,e^{+i\gamma}e^{i\delta}}{T + P\,e^{-i\gamma}e^{i\delta}},$$

where T and P are the tree and penguin contributions to A_f and δ is a relative strong interaction phase. In this case $C \neq 0$ and $S = \sqrt{1 - C^2}\sin(2\alpha_{eff}) \neq \sin(2\alpha)$. In order to extract α from α_{eff}, we need to unravel the effect of the strong phase. This can be done using an isospin analysis. The limits on α from these analyses are shown in Figure 7. An additional constraint on α comes from a time-dependent Dalitz plot

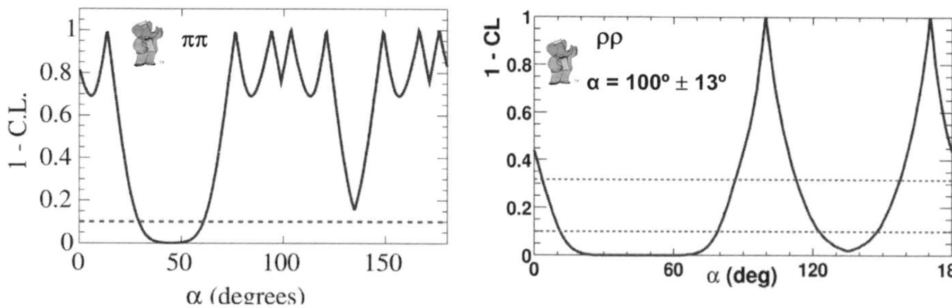

Figure 7: Confidence level as a function of alpha for the $\pi^+\pi^-$ and $\rho^+\rho^-$ analyses.

analysis of $B^0 \to \pi^+\pi^-\pi^0$ decays. Interference at equal two-body invariant masses on the Dalitz plot gives information on the strong phases and yields an unambiguous fit of α and tree and penguin amplitudes. The combined results are shown in Figure 8.

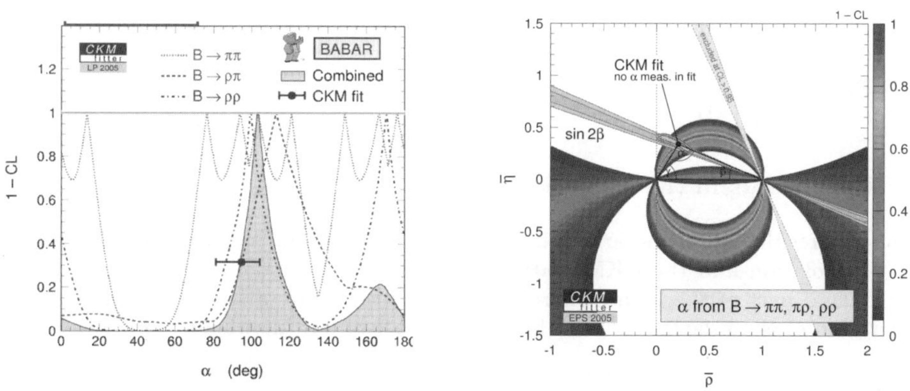

Figure 8: Combined results for α: (left) the CL plot for the BABAR data which yield $\alpha = 103^o + 11^o - 9^o$ (the horizontal bar is the result of the CKM fitter group); (right) the result in the $\bar{\rho} - \bar{\eta}$ plane.

7 Determination of $\gamma = \arg(V_{ub})$ [21–23]

The angle γ can be extracted using the interference that occurs when the D^0 and $\bar{D}^0$ decay to the same final state. The three methods exploiting these processes are: (1) D^0 goes to a CP eigenstate (GLW analysis [24]) ; (2) D^0 goes to a CKM suppressed $K\pi$ final state (ADS analysis [25]); and D^0 goes to a three-body final state and one uses the full three-body Dalitz plot (Dalitz analysis [26]). In these analyses one measures asymmetries, $A = (\Gamma(B^- \to D^{(*)0}K^-) - \Gamma(B^+ \to D^{(*)0}K^+))/(\Gamma(B^- \to D^{(*)0}K^-) - \Gamma(B^+ \to D^{(*)0}K^+))$ and the ratios of branching ratios.

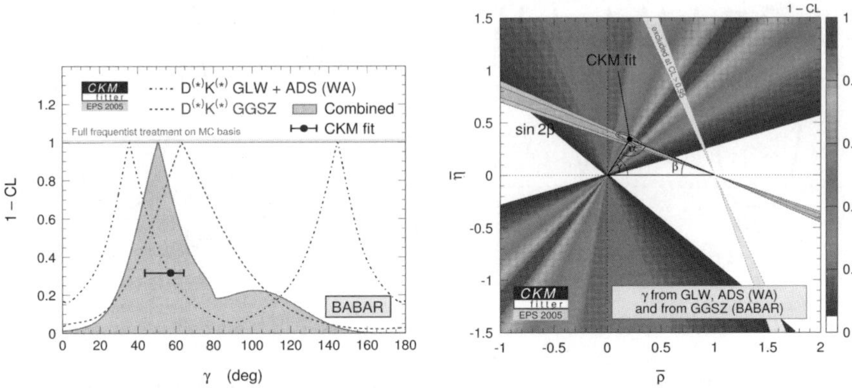

Figure 9: (left) CL for the values of γ from the world average for the GLW and ADS methods; the Dalitz method from BABAR and the combined result compared with the CKM fit ;(right) The constraint from γ in the $\bar{\rho} - \bar{\eta}$ plane.

To date the first two methods have given limited information about γ. The BABAR Dalitz plot analysis alone yields $\gamma = 67^\circ \pm 28^\circ(\text{stat.}) \pm 13^\circ(\text{expt.syst.}) \pm 11^\circ(\text{Dalitz method})$. The combined results for γ are shown in Figure 9 and yield a value of $\gamma = (51^{+23}_{-18})^\circ$.

8 Constraints in the $\bar{\rho} - \bar{\eta}$ plane

The net results are summarized in Figure 10. The progress in our understanding of CP-violation and the CKM-matrix in the Standard Model are illustrated in the right-hand part of the figure where we compare the state of knowledge prior to the operation of the b-factories with the current state of our knowledge.

Acknowledgments

The over 600 hard working physicists in the BABAR collaboration have all been instrumental in collecting and analyzing the data whose results are presented. The CKM fitter group [1] has been instrumental in putting these results in the context of the Standard Model.

References

[1] CKMfitter Group – J. Charles (*et al.*), *Eur. Jour. Phys.* **C41**, 1-131 (2005) [hep-ph/0406184], updated results and plots available at: http://ckmfitter.in2p3.fr

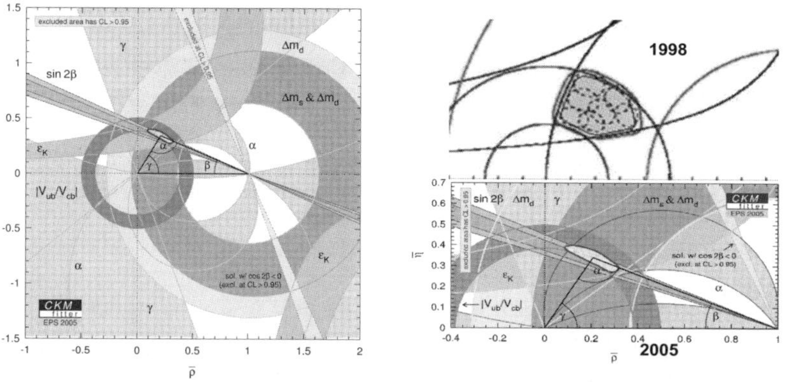

Figure 10: All measurements on the $\bar{\rho} - \bar{\eta}$ plane as of summer 2005. (left) The full plane; (right) the region of the CKM triangle with a comparison to the state of knowledge 1998.

[2] N. Cabbibo, *Phys. Rev. Lett.* **10**, 531 (1963); M. Kobayashi and T. Maskawa, *Prog. Theoret. Phys.* **49**, 652 (1973).

[3] L. Wolfenstein, *Phys. Rev. Lett.* **51**, 1945 (1983).

[4] B. Aubert *et al.* [BABAR Collaboration], *Nucl. Instr. Meth.* **A479**, 1 (2002). 179 (1936).

[5] B. Aubert *et al.* [BABAR Collaboration],*Phys. Rev. Lett.* **94**, 011801 (2005).

[6] B. Aubert *et al.* [BABAR Collaboration], *Phys. Rev.* **D70**, 112006 (2004).

[7] A. Ali, E. Lunghi, and A. Parkhomenko, *Phys. Lett. B* **595**, 323 (2004).

[8] A. Ali and A. Parkhomenko, *Eur. Jour. Phys.* **C23**, 89 (2002).

[9] B. Grinstein and D. Pirjol, *Phys. Rev.* **D62**, 093002 (2000).

[10] B. Aubert *et al.* [BABAR Collaboration], *Phys. Rev. Lett.* **94**, 161803 (2005).

[11] B. Aubert *et al.* [BABAR Collaboration], *Phys. Rev. Lett.* **95** 151804 (2005).

[12] B. Aubert *et al.* [BABAR Collaboration], *Phys. Rev. Lett.* **94** 191802 (2005).

[13] B. Aubert *et al.* [BABAR Collaboration], *Phys. Rev. Lett.* **95** 011801 (2005).

[14] B. Aubert *et al.* [BABAR Collaboration], hep-ex/0507021.

[15] B. Aubert *et al.* [BABAR Collaboration], hep-ex/0508017.

[16] B. Aubert *et al.* [BABAR Collaboration],*Phys. Rev.* **D71**, 032005 (2005).

[17] B. Aubert *et al.* [BABAR Collaboration],*Phys. Rev. Lett.* **95**, 151803 (2005).

[18] B. Aubert *et al.* [BABAR Collaboration],*Phys. Rev. Lett.* **94**, 131801 (2005).

[19] B. Aubert *et al.* [BABAR Collaboration],*Phys. Rev. Lett.* **95**, 041805 (2005).

[20] B. Aubert *et al.* [BABAR Collaboration], *Phys. Rev.* **D71**, 112003 (2005).

[21] B. Aubert *et al.* [BABAR Collaboration], *Phys. Rev.* **D72**, 071103 (2005).

[22] B. Aubert *et al.* [BABAR Collaboration], *Phys. Rev.* **D72**, 032004 (2005).

[23] B. Aubert *et al.* [BABAR Collaboration], *Phys. Rev. Lett.* **95**, 121802 (2005).

[24] M. Gronau and D. London, *Phys. Lett.* **B253**,483(1991); M. Gronau and D. Wyler, *Phys. Lett.* **B265**172 (1991).

[25] D. Atwood, I. Dunietz, and A. Soni, *Phys. Rev. Lett.* **78**, 3257 (1997); *Phys. Rev.* **D63**, 036005 (2001).

[26] A. Giri, Yu. Gros

LEPTON FLAVOR VIOLATION IN τ DECAYS AT *BABAR* [a]

Swagato Banerjee [b]
University of Victoria, British Columbia, V8W 3P6 Canada.
E-mail: swaban@slac.stanford.edu

Abstract. Searches for lepton flavor violating $\tau \to \ell\gamma$, $\tau \to \ell\ell\ell$ and $\tau \to \ell hh$ decays at the *BABAR* experiment are presented. Upper limits on the branching ratios are obtained at the level of $O(10^{-7})$ at 90% confidence level.

1 Introduction

Lepton flavor violating (LFV) processes such as the neutrinoless decay of the τ lepton have long been identified as unambiguous signatures of new physics, because no known fundamental local gauge symmetry forbids such a decay. While forbidden in the Standard Model (SM) because of vanishing neutrino mass in the three lepton generations, extensions to include current knowledge of neutrino mass and mixing imply $\mathcal{B}(\tau \to \mu\gamma) \sim \mathcal{O}(10^{-54})$ [1], which is many orders of magnitude below the experimental sensitivity. However, many new theories, as tabulated below, allow for LFV decays: $\tau \to \ell\gamma$, $\tau \to \ell\ell\ell$, $\tau \to \ell hh$ (where $\ell = e, \mu; h = \pi, K$) up to their existing experimental bounds $\sim \mathcal{O}(10^{-7})$:

	$\mathcal{B}(\tau \to \ell\gamma)$	$\mathcal{B}(\tau \to \ell\ell\ell)$
mSUGRA + seesaw [2,3]	10^{-7}	10^{-9}
SUSY + SO(10) [4,5]	10^{-8}	10^{-10}
SM + seesaw [6]	10^{-9}	10^{-10}
Non-Universal Z′ [7]	10^{-9}	10^{-8}
SUSY + Higgs [8,9]	10^{-10}	10^{-7}

Feynman diagrams for $\tau \to \mu\gamma$ and $\tau \to \mu\mu\mu$ decays via s-neutrino mixing in minimal supergravity model with heavy ν_R (seesaw mechanism) and via neutral Higgs exchange in supersymmetric model are shown in Figure 1, respectively.

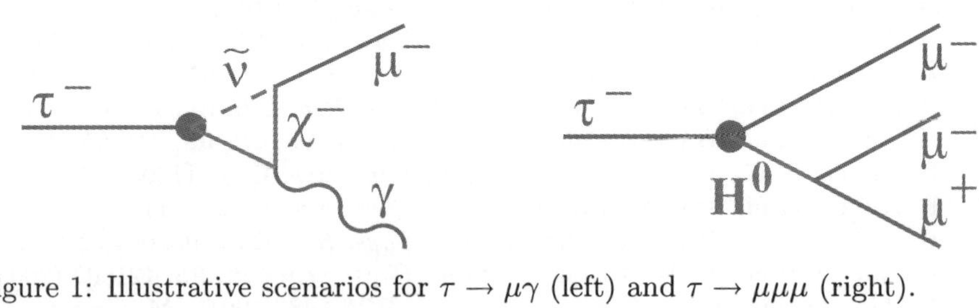

Figure 1: Illustrative scenarios for $\tau \to \mu\gamma$ (left) and $\tau \to \mu\mu\mu$ (right).

[a] To appear in the proceedings of XII Lomonosov Conference on Elementary Particle Physics, Moscow, Russia (25 - 31 August 2005).
[b] On behalf of the *BABAR* Collaboration.

2 Signal Identification

Searches for LFV decay in modes: $\tau \to \ell\gamma$ [1,10], $\tau \to \ell\ell\ell$ [11] and $\tau \to \ell hh$ [12] have been performed with 232.2 fb^{-1}, 91.6 fb^{-1} and 221.4 fb^{-1} of data collected by the *BABAR* experiment at $\sqrt{s} \approx 10.58$ GeV, respectively. The characteristic feature of these decays is that both the energy and the mass of the τ-daughters are known in such an e^+e^- annihilation environment. In terms of the two independent variables: beam energy constrained mass (m_{EC}) and the energy variable $\Delta E = E_\tau - \sqrt{s}/2$, where E_τ is the energy of the τ-daughters in center-of-mass system, the signal is clustered around $(m_\tau, 0)$ in the $(m_{EC}, \Delta E)$ plane.

The identification of daughters from signal τ decays are optimized for searches for each decay mode separately. The electrons are identified from the energy deposited in the electromagnetic calorimeter and momentum of the track measured in the drift chamber with an efficiency of 91% for $\tau \to e\gamma$ and $\tau \to \ell\ell\ell$ searches. The muons are identified by its minimal ionizing particle signature in the calorimeter and hits in the instrumented flux return with an efficiency of 82%, 63% and 44% in $\tau \to \mu\gamma$, $\tau \to \ell\ell\ell$ and $\tau \to \ell hh$ searches respectively. The kaons are identified using the measured rate of ionization loss in the drift chamber and the measured Cherenkov angle in a ring-imaging detector with an efficiency of 81%. The mis-identification rates for a pion track to be identified as an electron, a muon or a kaon are 0.1%, 1.0 − 4.8%, 1.4% respectively.

3 Background estimation

The primary backgrounds are from Bhabha or di-muon events, which are restricted to a narrow band at small values of $|\Delta E|$, or the $e^+e^- \to \tau^+\tau^-$ events, which are restricted to negative values of ΔE, because the signal topology reconstruction does not account for the missing neutrino's. The remaining backgrounds from $e^+e^- \to q\bar{q}$ are uniformly distributed.

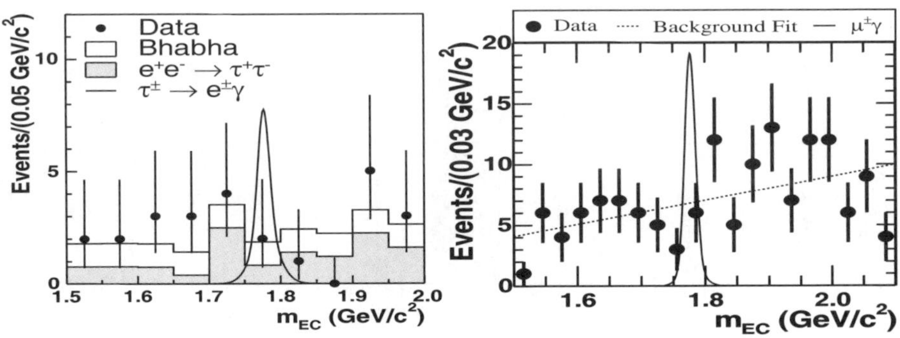

Figure 2: m_{EC} distribution inside a $\pm 2\sigma$ band in ΔE for $\tau \to \ell\gamma$ searches.

For $\tau \to \ell\gamma$ searches, the signal probability density function (PDF) is described by a double Gaussian shape in m_{EC}, and the background is well described by a constant PDF or with a small slope in m_{EC} inside a $\pm 2\sigma$ band in ΔE, as shown in Figure 2 for $\tau \to e\gamma$ (left) and $\tau \to \mu\gamma$ (right) decays.

For $\tau \to \ell\ell\ell$ (ℓhh) searches, the background PDF's are analytically parameterized as function of ΔE and $\Delta M = m_{dau} - m_{\tau}$, where m_{dau} is the reconstructed mass of the τ daughters. The background rates are determined by un-binned maximum likelihood fits to the data, shown along with the selected signal MC events in Figure 3 and 4 for $\tau \to \ell\ell\ell$ and $\tau \to \ell hh$ decay modes respectively.

All these searches are performed in a blinded manner, where the background predictions from sideband data are compared to the data inside the signal region, only after the optimization and systematic studies have been completed.

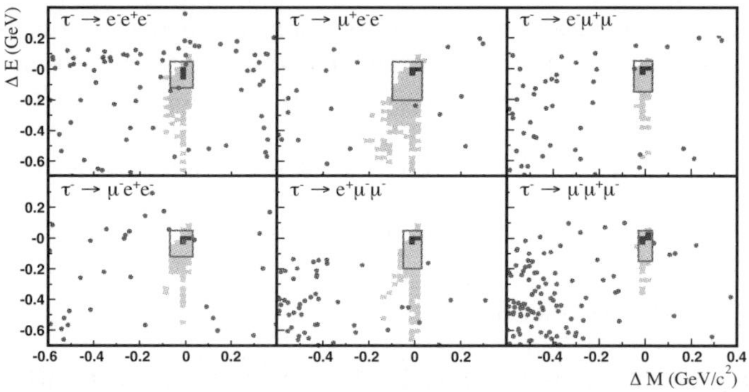

Figure 3: Observed data as dots and the boundaries of the signal region in $(\Delta M, \Delta E)$ plane for $\tau \to \ell\ell\ell$ searches. The dark and light shading indicates contours containing 50% and 90% of the selected MC signal events, respectively.

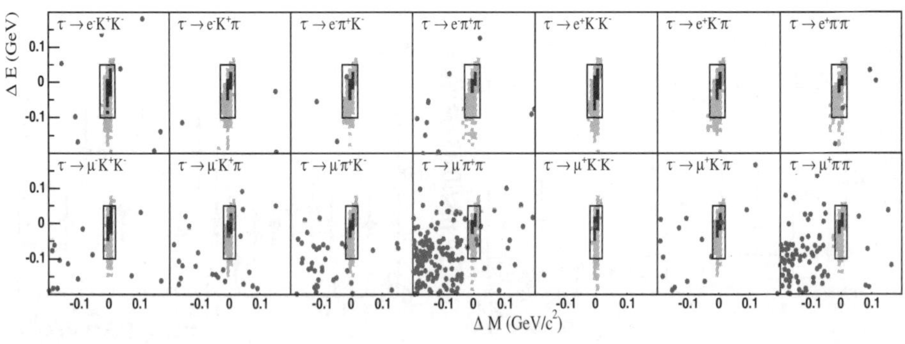

Figure 4: Observed data as dots and the boundaries of the signal region in $(\Delta M, \Delta E)$ plane for $\tau \to \ell hh$ searches. The dark and light shading indicates contours containing 50% and 90% of the selected MC signal events, respectively.

4 Results

No signal has been observed. Upper limits at 90% confidence level (C.L.) are set using: $\mathcal{B}^{90}_{UL} = N^{90}_{UL}/(2\varepsilon\mathcal{L}\sigma_{\tau\tau})$, where N^{90}_{UL} is the 90% C.L. upper limit on the number of signal events for $N_{\rm obs}$ events observed when $N_{\rm bgd}$ background events are expected, and ε is the signal efficiency. Efficiency estimates, the number of expected background events ($N_{\rm bgd}$) in the signal region (with total uncertainties), the number of observed events ($N_{\rm obs}$) in the signal region, and the 90% C.L. upper limit ($\mathcal{B}^{90}_{\rm UL}$) for each decay mode are tabulated below:

Mode	Efficiency [%]	$N_{\rm bgd}$	$N_{\rm obs}$	$\mathcal{B}^{90}_{\rm UL}(10^{-7})$
$e^-\gamma$	4.7 ± 0.3	1.9 ± 0.4	1	1.1
$\mu^-\gamma$	7.4 ± 0.7	6.2 ± 0.5	4	0.7
$e^-e^+e^-$	7.3 ± 0.2	1.5 ± 0.1	1	2.0
$\mu^+e^-e^-$	11.6 ± 0.4	0.4 ± 0.1	0	1.1
$\mu^-e^+e^-$	7.7 ± 0.3	0.6 ± 0.1	1	2.7
$e^+\mu^-\mu^-$	9.8 ± 0.5	0.2 ± 0.1	0	1.3
$e^-\mu^+\mu^-$	6.8 ± 0.4	0.4 ± 0.1	1	3.3
$\mu^-\mu^+\mu^-$	6.7 ± 0.5	0.3 ± 0.1	0	1.9
$e^-K^+K^-$	3.8 ± 0.2	0.2 ± 0.1	0	1.4
$e^-K^+\pi^-$	3.1 ± 0.1	0.3 ± 0.1	0	1.7
$e^-\pi^+K^-$	3.1 ± 0.1	0.1 ± 0.1	1	3.2
$e^-\pi^+\pi^-$	3.3 ± 0.2	0.8 ± 0.1	0	1.2
$\mu^-K^+K^-$	2.2 ± 0.1	0.2 ± 0.1	0	2.5
$\mu^-K^+\pi^-$	3.0 ± 0.2	1.7 ± 0.3	2	3.2
$\mu^-\pi^+K^-$	2.9 ± 0.2	1.0 ± 0.2	1	2.6
$\mu^-\pi^+\pi^-$	3.4 ± 0.2	3.0 ± 0.4	3	2.9
$e^+K^-K^-$	3.9 ± 0.2	0.0 ± 0.0	0	1.5
$e^+K^-\pi^-$	3.2 ± 0.1	0.2 ± 0.1	0	1.8
$e^+\pi^-\pi^-$	3.4 ± 0.2	0.4 ± 0.1	1	2.7
$\mu^+K^-K^-$	2.1 ± 0.1	0.1 ± 0.1	1	4.8
$\mu^+K^-\pi^-$	2.9 ± 0.2	1.5 ± 0.3	1	2.2
$\mu^+\pi^-\pi^-$	3.3 ± 0.2	1.5 ± 0.3	0	0.7

5 Summary

An improvement of five order of magnitude in the upper limits for 4 LFV τ decay modes over the past two decades is shown in Figure 5. The next five years promises to be most interesting phase in this evolution, when experimental results approach closer the predictions from different theoretical models.

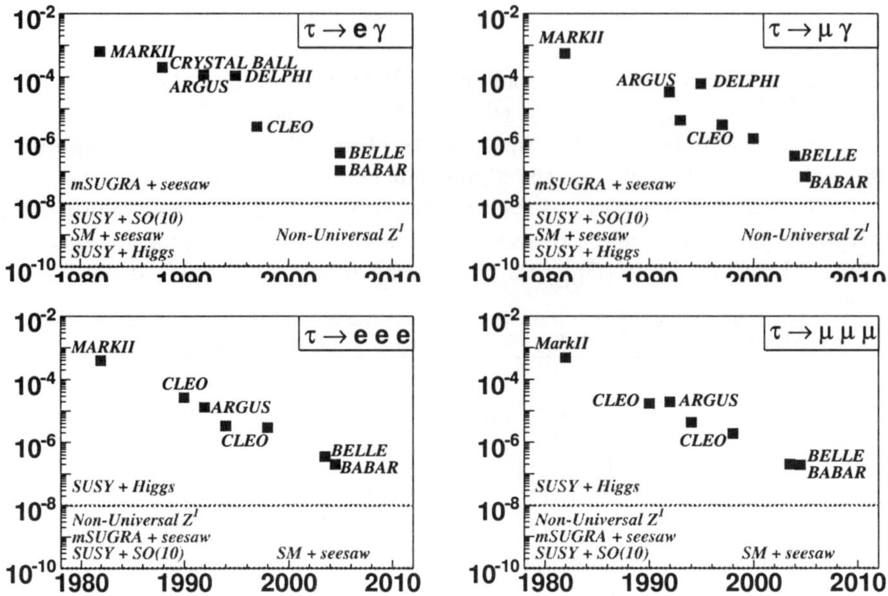

Figure 5: Evolution of experimental bounds ($\mathcal{B}^{90}_{\rm UL}$) and some predictions.

References

[1] B. Aubert *et al.* [BABAR Collaboration], Phys. Rev. Lett. **95**, 041802 (2005).

[2] J. R. Ellis, M. E. Gomez, G. K. Leontaris, S. Lola, D. V. Nanopoulos, Eur. Phys. J. C **14**, 319 (2000).

[3] J. R. Ellis, J. Hisano, M. Raidal, Y. Shimizu, Phys. Rev. D **66**, 115013 (2002).

[4] A. Masiero, S. K. Vempati, O. Vives, Nucl. Phys. B **649**, 189 (2003).

[5] T. Fukuyama, T. Kikuchi, N. Okada, Phys. Rev. D **68**, 033012 (2003).

[6] G. Cvetic, C. Dib, C. S. Kim, J. D. Kim, Phys. Rev. D **66**, 034008 (2002) [Erratum-ibid. D **68**, 059901 (2003)].

[7] C. x. Yue, Y. m. Zhang, L. j. Liu, Phys. Lett. B **547**, 252 (2002).

[8] A. Dedes, J. R. Ellis, M. Raidal, Phys. Lett. B **549**, 159 (2002).

[9] A. Brignole, A. Rossi, Phys. Lett. B **566**, 217 (2003).

[10] B. Aubert *et al.* [BABAR Collaboration], arXiv:hep-ex/0508012. Submitted to Phys. Rev. Lett. (2005).

[11] B. Aubert *et al.* [BABAR Collaboration], Phys. Rev. Lett. **92**, 121801 (2004).

[12] B. Aubert *et al.* [BABAR Collaboration], Phys. Rev. Lett. **95**, 191801 (2005).

ISR PHYSICS AT BABAR

V.Druzhinin [a]
Representing the BaBar Collaboration
Budker Institute of Nuclear Physics, 630090 Novosibirsk, Russia

Abstract. We present a review of BaBar results on e+e- annihilations into exclusive hadronic final states using the initial state radiation technique. Cross sections over the $\sqrt{s}$ range from threshold to 4.5 GeV, with very small point-to-point systematic errors, are presented for the 3π, $2(\pi^+\pi^-)$, $3(\pi^+\pi^-)$, $2(\pi^+\pi^-)2\pi^0$, $K^+K^-\pi^+\pi^-$, $2(K^+K^-)$ and $p\bar{p}$ final states. The proton form factor and the ratio of its electric and magnetic components are also presented.

1 Introduction

The BABAR detector [1] collect data at the PEP-II asymmetric-energy e^+e^- collider where 9-GeV electrons collide with 3.1-GeV positrons. The detector covers about 85% of the solid angle in the e^+e^- center-of-mass (c.m.) frame and provides excellent charged particles tracking and identification, and photon detection.

These detector features and high luminosity collected at the PEP-II allow to study the processes of e^+e^- annihilation into different hadronic states using initial state radiation (ISR) technique. In the ISR processes, $e^+e^- \to \gamma X$, a photon of energy E_γ is emitted by the initial electron or positron. As result the produced hadronic system has invariant mass $m = \sqrt{s(1-x)}$, where $\sqrt{s}$ is e^+e^- c.m. energy and $x = 2E_\gamma/\sqrt{s}$. The mass spectrum of the hadronic system X is related to the $e^+e^- \to X$ cross section as

$$\frac{d^2\sigma(e^+e^- \to \gamma X)}{dx d\cos\theta_\gamma} = W(s,x,\theta_\gamma)\sigma(m). \tag{1}$$

Here W is well-known photon radiator function describing ISR photon energy and angular distribution. The ISR photon is emitted predominantly along electron or positron direction. In our approach with detection of the ISR photon at large angle we use only about 10% of ISR events.

2 Selection of ISR events and background subtraction

For analysis we select events with all final particles (including hard ISR photon) detected. The charged particle identification is used to recognize the specific hadronic state and suppress background. For selected events we perform the kinematic fit with the requirement of energy and momentum balance. The two-photon mass for π^0 candidates is constrained to π^0 nominal mass. The cut on χ^2 of the kinematic fit provides additional background suppression.

[a] e-mail: druzhinin@inp.nsk.su

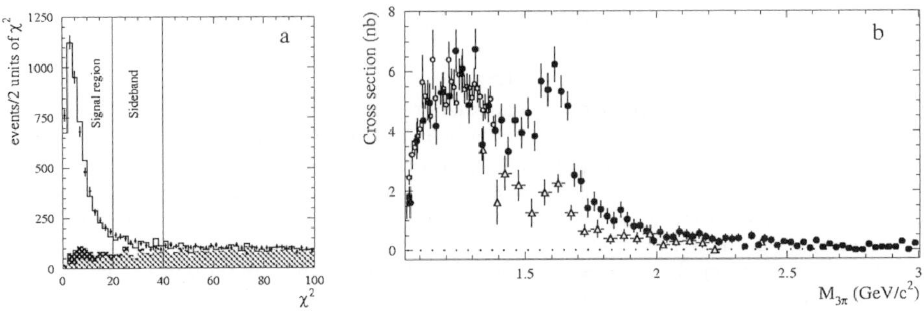

Figure 1: a). The χ_p^2 distribution for selected $3\pi\gamma$ candidates for data (points with error bars) and simulation (histogram). The dark and lightly shaded histograms shows the distributions for $e^+e^- \to \pi^+\pi^-\pi^0\pi^0$ and other background processes, respectively. b). The $e^+e^- \to \pi^+\pi^-\pi^0$ cross section measured by BABAR (filled circles), by SND (open circles), and DM2 (open triangles).

The main sources of remaining background are the ISR events with misidentified particles ($K^+K^-\pi^0\gamma$ for $3\pi\gamma$ process, $K^+K^-\gamma$ and $\pi^+\pi^-\gamma$ for $p\bar{p}\gamma$), events with high-energy π^0 instead of ISR photon ($p\bar{p}\pi^0$ for $p\bar{p}\gamma$). The third class of background processes is all other ISR processes and processes of continuum annihilation into quark-antiquark. For example, the χ^2 distribution for $e^+e^- \to \gamma\pi^+\pi^-\pi^0$ candidates is shown in Fig. 1a and seen to be well described by our simulation. Backgrounds from ISR events with misidentified particles ($K^+K^-\pi^0\gamma$) and events with a high-energy π^0 mimicing an ISR photon ($\pi^+\pi^-\pi^0\pi^0$) also peak at low chi^2 values (dark shaded histogram in Fig. 1a), but are a small fraction of the signal, and are measured from the data as a function of 3π mass and subtracted. Other sources of background involve missing or spurious particles and give a broad χ^2 distribution (light shaded histogram); they are subtracted using the χ^2 sideband region indicated.

3 Results

The $\pi^+\pi^-\pi^0$ final state [2]. The $e^+e^- \to 3\pi$ is a key process for study of the excited ω states. The measured $e^+e^- \to 3\pi$ cross section for the mass above 1.05 GeV is shown in Fig. 1b. Our results are consistent with data obtained by SND collaboration for energies below 1.4 GeV, but significantly exceed previous DM2 results for the mass region of ω'' resonance. From the fit of the cross section by a sum of the contributions of ω, ϕ, ω', and ω'' mesons we obtained the parameters of excited ω states: $M_{\omega'} = 1350 \pm 30$ GeV/c^2, $\Gamma_{\omega'} = 450 \pm 100$ GeV/c^2, $M_{\omega''} = 1660 \pm 10$ GeV/c^2, $\Gamma_{\omega''} = 230 \pm 40$ GeV/c^2.

The $\pi^+\pi^-\pi^+\pi^-$, $\pi^+\pi^-K^+K^-$, $K^+K^-K^+K^-$ final states [3]. Our results on the measurement of the cross sections of e^+e^- annihilation into four

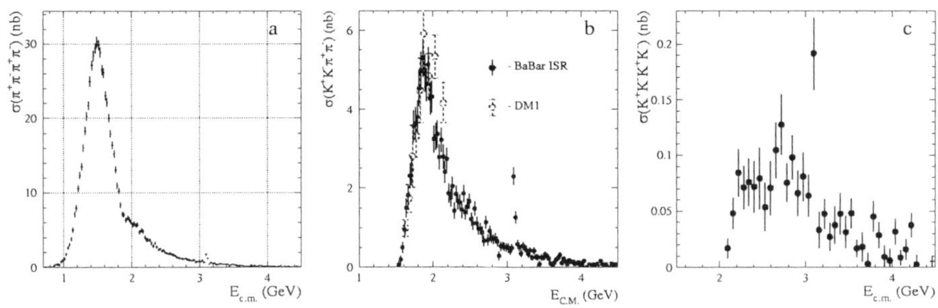

Figure 2: The measured cross sections for $e^+e^- \to \pi^+\pi^-\pi^+\pi^-$ (a), $e^+e^- \to \pi^+\pi^- K^+K^-$ (b), $e^+e^- \to K^+K^-K^+K^-$ (c).

charged particles are shown in Fig.2. This $e^+e^- \to \pi^+\pi^-\pi^+\pi^-$ process (Fig.2a) has large cross section, and its accurate measurement is important for calculation of the hadronic contribution into $(g-2)$ of muon. The BABAR data are in good agreement with previous direct e^+e^- measurements. For energies above 1.4 GeV they have world's best accuracy. From analysis of the 3-pion and 2-pion mass distributions we conclude that dominant intermediate state in this process is $a_1(1260)\pi$. For energies above 2 GeV the $f_0(1370)\rho$ intermediate state is also seen.

The cross section for $e^+e^- \to K^+K^-\pi^+\pi^-$ is shown in Fig. 2b. This process proceeds via $K^*K\pi$, $\phi\pi\pi$, and ρKK intermediate states. We also present the first measurement of the $e^+e^- \to K^+K^-K^+K^-$ cross section (Fig. 2c).

The $p\bar{p}$ final state [4]. The cross section for $e^+e^- \to p\bar{p}$ depends on two form factors: electric (G_E) and magnetic (G_M). From measuring the total cross section we can extract a combination of form factors. We define the effective form factor as $F(m) = \sqrt{(|G_M(m)|^2 + \tau|G_E(m)|^2)/(1+\tau)}$ where $\tau = 2m_p^2/m^2$. Such definition allows to compare our form-factor results with the data from previous e^+e^- and $p\bar{p}$ experiments which use the assumption that $|G_E| = |G_M|$. The ratio of the form factors can be extracted from the analysis of the distribution of the proton helicity angle (θ_p) in the $p\bar{p}$ rest frame. The ISR approach provides full θ_p coverage and hence high sensitivity to $|G_E/G_M|$. In contrast to previous e^+e^- and $p\bar{p}$ experiments, our measurement of cross section does not use the assumption that $|G_E| = |G_M|$. It should be noted that $p\bar{p}$ mass resolution near the $p\bar{p}$ threshold is less than 1 MeV and comparable with the energy spread of e^+e^- machines.

The $\cos\theta_p$ distribution was fitted in six bins of $m_{p\bar{p}}$ by a sum of two distributions corresponding G_E and G_M terms in the differential cross section. The G_E distribution is close to $\sin^2\theta$, the G_M distribution to $1+\cos^2\theta$. The obtained mass dependence of $|G_E/G_M|$ is shown in Fig. 3a. Our result disagrees significantly with previous measurement from LEAR [5].

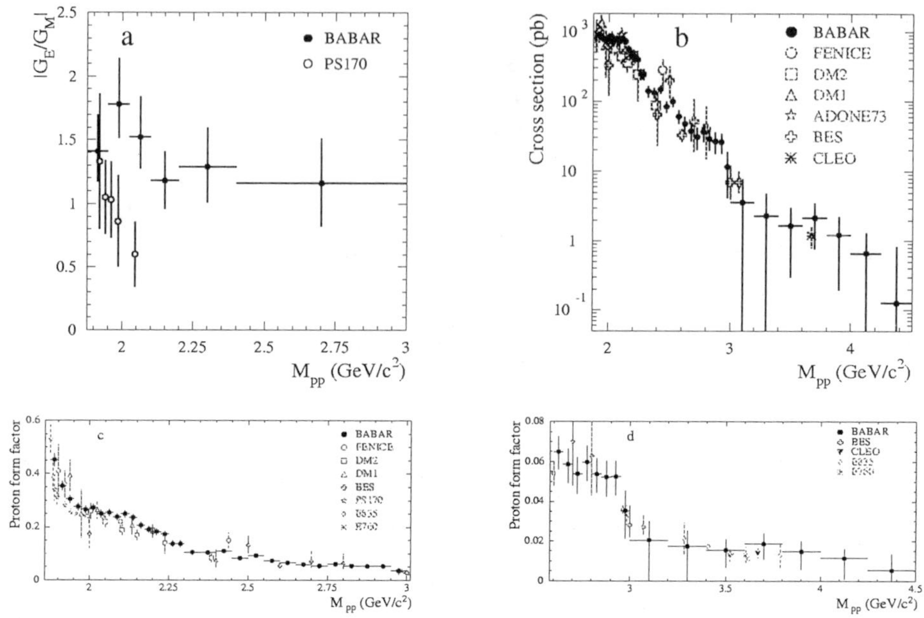

Figure 3: a). Mass dependence of the ratio $|G_E/G_M|$. b). The measured cross sections for $e^+e^- \to p\bar{p}$. c). and d).The effective proton form factor.

The obtained $e^+e^- \to p\bar{p}$ cross section shown in Fig. 3b (the contributions of J/ψ and $\psi(2S)$ decays are subtracted) is in reasonable agreement with previous e^+e^- measurements. Our data more full and accurate for energy range below 3 GeV. The effective form factor is shown in Fig. 3c,d. It has a complex mass dependence. We confirm the near-threshold enhancement in the form factor observed in the LEAR experiment. There are also two mass regions, near 2.25 GeV/c^2 and 3 GeV/c^2, that exhibit steep decreases in the form factor. This unusual mass dependence has never been discussed in the literature.

The 6π final states [6]. The near-threshold enhancement in the $p\bar{p}$ mass spectrum may be a manifestation of $p\bar{p}$ subthreshold resonance. Such resonance may be seen in $e^+e^- \to$ *hadrons* cross section near 1.9 GeV. A narrow dip in the $e^+e^- \to 6\pi$ cross section was observed in DM2 experiment and in the diffractive photoproduction of six pions in FOCUS experiment [7]. Our results on measurement of six pion cross sections are shown in Fig. 4. In both six-pion cross sections the dips near 1.9 GeV are clearly seen. The cross sections are fitted by a sum of resonance and continuum amplitudes. The fit results the resonant mass 1880 ± 30 MeV/c^2 (1860 ± 20 MeV/c^2) and width 130 ± 30 MeV/c^2 (160 ± 20 MeV/c^2) for $3(\pi^+\pi^-)$ $(2(\pi^+\pi^-)2\pi^0)$ mode. The obtained width is significantly differ from the FOCUS result of 29 ± 14 MeV/c^2.

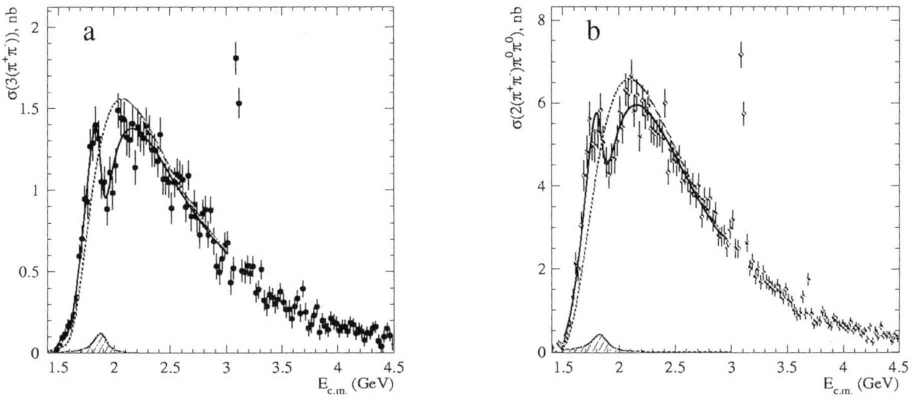

Figure 4: The measured cross sections for $e^+e^- \to 3(\pi^+\pi^-)$ (a), $e^+e^- \to 2(\pi^+\pi^-\pi^0)$ (b).

4 Summary

The program to measure low-energy hadronic cross sections using ISR is well underway at BaBar. We have presented the cross section measurements for $e^+e^- \to \pi^+\pi^-\pi^0$, $2(\pi^+\pi^-)$, $K^+K^-\pi^+\pi^-$, $2(K^+K^-)$, 6π. We significanly improved accuracy for these cross sections in the 1.4–4.5 GeV energy range.

We have also measured $e^+e^- \to p\bar{p}$ cross section and extracted the proton effective form factor. These are most full and accurate $e^+e^- \to p\bar{p}$ data in the energy range from the threshold up to 3 GeV. In contrast to previous experiment, our measurement does not use the assumption that $|G_E| = |G_M|$. From analysys of the proton angular distribution we have extracted the energy dependence of the $|G_E/G_M|$ ratio. Its value is found to be significanly greater than unity for energies below 2.1 GeV.

The analysys of $\pi^+\pi^-$, K^+K^-, $K\bar{K}\pi$, $\pi^+\pi^-\pi^0\pi^0$, 5π, and $D^{(*)}\bar{D}^{(*)}$ final states are in progress at BaBar.

References

[1] B.Aubert *et al.*, *Nucl. Instr. and Meth.* **A** 479, 1 (2002).
[2] B.Aubert *et al.*, *Phys.Rev.* **D** 70, 072004 (2004).
[3] B.Aubert *et al.*, *Phys.Rev.* **D** 71, 052001 (2005).
[4] B.Aubert *et al.*, hep-ex/0512023, *submitted to Phys.Rev.* **D**.
[5] G. Bardin *et al.*, *Nucl. Phys.* **B** 411, 3 (1994).
[6] B.Aubert *et al.*, *submitted to Phys.Rev.* **D**.
[7] P.L. Frabetti *et al.*, *Phys. Lett.* **B** 514, 240 (2001).

SEMILEPTONIC B DECAYS AT *BABAR*

V.Azzolini [a]

IFIC - Universitat de Valencia - CSIC , 46010 Valencia, Spain

*Abstract.*We report on studies of charmless semileptonic decays based on the data collected at the $\Upsilon(4S)$ resonance using the the *BABAR* detector [1] at the Stanford Linear Accelerator Center. We present a number of inclusive methods to isolate $b \to u\ell\bar{\nu}$ decays and suppress the more abundant $b \to c\ell\bar{\nu}$ process and show results of measurements based on: the lepton energy and squared lepton-neutrino invariant mass, E_ℓ and q^2, the hadron invariant mass and squared lepton-neutrino invariant mass, M_X and q^2. Exclusive charmless semileptonic decays have also been investigated. We report studies on $B \to \pi/\rho\ell\nu$ decay on untagged events, the $\bar{B} \to X_u\ell\bar{\nu}$ decay modes with hadronic tags, the $B^\pm \to \pi^0\ell^\pm\nu$ using semileptonic $B^- \to D^0\ell^- \bar{\nu}\,(X)$ tags and $B^0 \to \pi^- \ell^+ \nu$ with semileptonic $\overline{B}^0 \to D^{(*)+}\ell^- \bar{\nu}$ s tags. From the measurements of partial and total braching fractions, the magnitude of the CKM element $|V_{ub}|$ is derived using several theoretical predictions and frameworks.

The principal physics goal of the *BABAR* experiment is to establish *CP* violation in B mesons and to test whether the observed effects are consistent with the predictions of the Standard Model (SM). *CP* violating effects result in the SM from an irreducible phase in the Cabibbo-Kobayashi-Maskawa (CKM) matrix which describes the couplings of the charged weak current to quarks. An improved determination of the magnitude of the matrix element $|V_{ub}|$, the coupling strength of the b quark to the u quark, will contribute critically to tests of the consistency of the measured angles of the unitarity triangle of the CKM matrix.

The precise determination of $|V_{ub}|$ is very difficult as, due to the large charmed backgrounds, we are forced to measure partial branching fractions and extrapolate them to the full phase space by relying on QCD based theoretical calculations. It is therefore important to make redundant measurements by using several experimental techniques, and different theoretical frameworks.

1 Inclusive Measurements

In inclusive measurements, three kinematic variables are discussed in the literature, each having its own advantages: the lepton energy (E_ℓ), the hadronic invariant mass (M_X), and the leptonic invariant mass squared (q^2). The first measurements were restricted to the high end of the lepton spectrum where theoretical uncertainties are very large and therefore the extrapolation to the full spectrum becomes uncertain. Event selection based on M_X and q^2 allows us to select larger portions of phase space, but the underlying theoretical assumptions need to be carefully evaluated.

[a] e-mail: azzolini@slac.stanford.edu

In the first analysis here presented, semileptonic $b \to ue\bar{\nu}$ decays are selected using a novel approach based on simultaneous requirements for the electron energy, E_e, and the invariant mass squared of the $e\bar{\nu}$ pair, q^2 [b] [2]. The dominant charm background is suppressed by selecting a region of the q^2-E_e phase space where correctly reconstructed $b \to ce\bar{\nu}$ events are kinematically excluded.

Hadronic events containing an identified electron with energy $2.1\ GeV < E_e < 2.8\ GeV$ are selected. The neutrino 4-momentum is reconstructed from the visible 4-momentum and knowledge of the e^+e^- initial state.
The maximum kinematically allowed hadronic mass squared, for a given E_e and q^2, is $s_h^{max} = m_B^2 + q^2 - 2m_B(E_e + q^2/4E_e)$ for $\pm 2E_e > \pm\sqrt{q^2}$; we require $s_h^{max} \approx m_{D^0}^2$. We determine a partial branching fraction $\Delta B(\tilde{E}, \tilde{s}_{\mathrm{h}}^{\mathrm{max}}) = B(\bar{B} \to X_u e\bar{\nu})f_u$, unfolded for detector effects. The acceptance, f_u, is the fraction of $\bar{B} \to X_u e\bar{\nu}$ decays in the region of interest, $\tilde{E}_e > 2.0\ GeV$ and $\tilde{s}_{\mathrm{h}}^{\mathrm{max}} < 3.5\ GeV$ (shown in Fig. 2 - [3]), where $\tilde{E}_e$ and $\tilde{s}_{\mathrm{h}}^{\mathrm{max}}$ are the generated values in the B meson rest frame. We find $\Delta B(2.0, 3.5) = (3.54 \pm 0.33_{stat.} \pm 0.34_{syst.}) \times 10^{-4}$.

We extract $|V_{ub}| = [\Delta\mathcal{B}(\bar{B} \to X_u\ell\bar{\nu})/\Delta\zeta\ \tau_B]^{1/2}$ using $\tau_B = 1.604 \pm 0.023\ ps$ and the normalized partial rate,$\Delta\zeta$ as taken from Ref. [4]. The values used for the heavy quark parameters, $m_b = 4.61 \pm 0.08 GeV$ and $\mu_\pi^2 = 0.15 \pm 0.07 GeV$ are based on fits to *BABAR* $\bar{B} \to X_c\ell\bar{\nu}$ moments [5], translated to the Shape Function(SF) scheme of Ref. [6].

We find $|V_{ub}|(2.0, 3.5) = (3.95 \pm 0.26_{exp.} {}^{+0.58}_{-0.42}{}_{HQ} \pm 0.25_{theo.}) \times 10^{-3}$.

On the other hand, studies [7] indicate it is possible to reduce the theoretical error on the extrapolation by applying simultaneous cuts on M_X and q^2 in inclusive $\bar{B} \to X_u\ell\bar{\nu}$. In fact, while the M_X distribution has a large usable fraction of events, of the order of 70%, but depends on the SF, the q^2 distribution is less sensitive to non-perturbative effects but only a small fraction of events ($\sim 20\%$) is usable. We study the recoiled B candidate opposite of a fully reconstructed B in hadronic decay (B_{reco}), where are selected decays of the type $B \to \overline{D}\,Y$, where D refers to a charm meson, and Y is a collection of hadrons ($\pi^\pm, K^\pm, K_S^0, \pi^0$) with a total charge of ± 1 [c]. The kinematic consistency of a B_{reco} candidate with a B meson decay is checked using two variables, the beam-energy-substituted mass $m_{\mathrm{ES}} = \sqrt{s/4 - \vec{p}_B^{\,2}}$ and the energy difference, $\Delta E = E_B - \sqrt{s}/2$. Here $\sqrt{s}$ refers to the total energy in the $\Upsilon(4S)$ center of mass frame, and $\vec{p}_B$ and E_B denote the momentum and energy of the B_{reco} candidate in the same frame. In order to extract the partial charmless semileptonic branching ratio, $\Delta BR(\bar{B} \to X_u\ell\bar{\nu})$, in a given region of the M_X-q^2 plane, we define as signal the events with true values of the kine-

---

[b] hep-ex/0506036 - The data set used consists of 88.4 million $B\overline{B}$ pairs, corresponding to an integrated luminosity of $81.4\,\mathrm{fb}^{-1}$.

[c] hep-ex/0507017 - $211\,\mathrm{fb}^{-1}$.

matic variables in the chosen region, treating as background those that migrate from outside this region because of the resolution ((shown in Fig. 2 - [8]) and quote efficiences[d] only for those signal events. The partial branching fraction, in the signal region, is $\Delta BR(\bar{B} \to X_u \ell \bar{\nu}\ , M_X < 1.7 GeV/c^2\ , q^2 > 8 GeV^2/c^4\ ) = (0.87 \pm 0.13_{stat.+syst.} \pm 0.01_{theo}) \times 10^{-3}$.

To translate this into a measurement of $|V_{ub}|$,[e] we need the fraction of events inside the measurement region [f] as an external input; using the acceptance corrections calculated in the Bosch, Lange, Neubert and Paz [4, 10, 11] (BLNP) approach and by taking the SF parameters from [5], we find $|V_{ub}|^{BLNP}_{BABAR\ \bar{B}\to X_c \ell \bar{\nu}} = (4.65 \pm 0.34_{stat.+syst.}\ {}^{+0.46}_{-0.38}\ {}_{SF} \pm 0.23_{th}) \times 10^{-3}$, while with Bauer, Ligeti and Luke [12] (BLL) acceptances calculations, $|V_{ub}|^{BLL} = (4.82 \pm 0.36_{stat.+syst.} \pm 0.46_{theo.+SF}) \times 10^{-3}$.

In conclusion, the total error on $|V_{ub}|$ inclusive measurements is dominated by the experimental and theoretical uncertainties of the SF. The three approaches to measure inclusively $|V_{ub}|$ done at Babar show that no significantly changes occur changing the two sets of SF parameters coming from a fit to the photon energy spectrum and to the $\bar{B} \to X_c \ell \bar{\nu}$ moments. Being infact in good agreement, they give thus consistent results on $|V_{ub}|$.

2 Exclusive Measurements

Alternatively the measurements of the $\bar{B} \to X_u \ell \bar{\nu}$ decay rates can be done exclusively, because this allow for kinematic constrains and more efficient background suppression, but must rely on theoretical form-factor predictions. We distinguish between them through the tagging method used: the hadronic tag, the semileptonic tag, and finally the untagged one.

In events in which the decay of one B meson to a hadronic final state is fully reconstructed , the semileptonic decay, in $B^0 \to \pi^- \ \ell^+ \ \nu$ and $B^+ \to \pi^0 \ell^+ \ \nu$, of the second B is identified by the detection of a charged lepton and a pion [g]. The charmless meson in the semileptonic decay are reconstructed and the missing mass (see Fig. 3 - [13]) is calculated assuming that the pion and the charged lepton are the only particles present in the recoil except for the undetected neutrino. Evaluating the signal yield from a fit to $m_{\rm ES}$ in three regions of the invariant mass squared of the lepton pair, we obtain the total branching fractions $BR(B^0 \to \pi^- \ell^+ \nu\ ) = (1.14 \pm 0.27_{stat} \pm 0.17_{syst}) \times 10^{-4}$ and $BR(B^+ \to \pi^0 \ell^+ \nu\ ) = (0.86 \pm 0.22_{stat} \pm 0.11_{syst}) \times 10^{-4}$.

In the analyses using the semileptonic tag[h], we look for combinations of a D^+

[d] They are computed on simulations based on the DFN model. [9].

[e] The partial rate for $\bar{B} \to X_u \ell \bar{\nu}$ can be related to $|V_{ub}|$ though the relation previously defined.

[f] referred to as "acceptance"

[g] hep-ex/0507085 - 211 fb^{-1} .

[h] hep-ex/0506064 - 211 fb^{-1} (hep-ex/0506065 - 81 fb^{-1}).

or D^{*+} (D^0) meson and a charged lepton (e^- or μ^-) that are kinematically consistent with $\overline{B}^0 \rightarrow D^{(*)+}\ell^- \overline{\nu}$ ($B^- \rightarrow D^0\ell^- \overline{\nu}\,(X)$) decays. For each such B candidate, we define the recoil side as the tracks and calorimeter clusters that are not associated with the candidate and search for a signature of a $B^0 \rightarrow \pi^- \ell^+ \nu$ ($B^+ \rightarrow \pi^0\ell^+ \nu$) decay. We take advantage of the simple kinematics of the process to define discriminating variables like the cosine of θ_{BY} (see Fig. 2 - [14]) or of $\theta_{B\pi\ell}$ (Fig. 3 - [15]) [i] , and extract the signal yield from their distributions in three bins of q^2. Finally we calculate the total and the partial branching fractions using the signal efficiencies predicted by a Monte Carlo (MC) simulation. From the signal yields and the efficiencies evaluated, we extract the following results for the total branching fraction as $B(B^0 \rightarrow \pi^-\ell^+\nu) = (1.03 \pm 0.25_{stat.} \pm 0.13_{syst.}) \times 10^{-4}$ and $B\,(B^+ \rightarrow \pi^0\ell^+\,\nu) = (1.80 \pm 0.37_{stat.} \pm 0.23_{syst.}) \times 10^{-4}$.

For the untagged approach we present a determination of total branching fraction from charmless semileptonic B decays with exclusively reconstructed final states, $B \rightarrow h_u\ell\nu$, where the hadronic state h_u represents a $\pi^{\pm}$, π^0, $\rho^{\pm}$, or ρ^0 and ℓ represents e or μ [j].
The neutrino four-momentum, is inferred from the difference between the net four-momentum of the colliding-beam particles, $p_{\rm beams}$, and the sum of the four-momenta of all detected particles in the event. We discriminate against the remaining background using the variables $m_{\rm ES}$ and ΔE . Fig. 1 - [16] shows projections of the fitted $m_{\rm ES}$ vs ΔE distributions for each q^2 interval for $B \rightarrow \pi\ell\nu$ and $B \rightarrow \rho\ell\nu$, respectively. Integrated over the whole q^2 range, we observe 396 $\pi^-\ell^+\nu$, 137 $\pi^0\ell^+\nu$, 95 $\rho^-\ell^+\nu$, and 98 $\rho^0\ell^+\nu$ decays and the total branching fractions, $B(B^0 \rightarrow \pi^-\ell^+\nu) = (1.38 \pm 0.10_{stat.} \pm 0.16_{syst.} \pm 0.08_{FF}) \times 10^{-4}$ and $B(B^0 \rightarrow \rho^-\ell^+\nu) = (2.14 \pm 0.21_{stat.} \pm 0.48_{syst.} \pm 0.28_{FF}) \times 10^{-4}$.

In conclusion *BABAR* has produced in last months several and competitive total branching fraction exclusive measurements, that are consistent with previous measurements [17,18], but have higher statistical accuracy, are less dependent on theoretical form-factor predictions, and benefit from recent advances in theoretical calculations [19–21].

Acknowledgments

We are grateful for the extraordinary contributions of our PEP-II colleagues in achieving the excellent luminosity and machine conditions that have made this work possible. The success of this project also relies critically on the expertise and dedication of the computing organizations that support *BABAR*. The collaborating institutions wish to thank SLAC for its support and the

[i] Referring to the $D^{(*)+}\ell^-$ system as the "Y" system, we first calculate the cosine of θ_{BY}, the angle between p^*_B and p^*_Y and the cosine of $\theta_{B\pi\ell}$, the angle between p^*_B and $p^*_{\pi\ell}$.
[j] hep-ex/0507003 - 76 fb^{-1} .

kind hospitality extended to them. This work is supported by DOE and NFS (USA), NSERC (Canada), IHEP(China), CEA and CNRS-IN2P3 (France), BMBF and DFG (Germany), INFN (Italy), FOM (The Netherlands), NFR (Norway), MIST (Russia), and PPARC (United Kingdom). Individuals have received support from CONACyT (Mexico), the A. P. Sloan Foundation, the Research Corporation, and the Alexander von Humboldt Foundation.

References

[1] B. Aubert *et al.* (*BABAR* Collab.), Nucl. Instr. Methods Phys. Res., Sect. A **479**, 1 (2002).
[2] R. Kowalewski and S. Menke, Phys. Lett. B **541**, 29 (2002).
[3] *BABAR* Collaboration, B. Aubert *et al.* , hep-ex/0506036.
[4] B. Lange, M. Neubert, and G. Paz, hep-ph/0504071 (2005).
[5] B. Aubert *et al.* (*BABAR* Collab.), Phys. Lett. **93**, 011803 (2004).
[6] M. Neubert, Phys. Lett. **B612**, 13 (2005) and private communication.
[7] Bauer, Christian W. and Ligeti, Zoltan and Luke, Michael E., Phys. Lett. **D64**, 113004 (2001).
[8] *BABAR* Collaboration, B. Aubert *et al.* , hep-ex/0507017.
[9] F. DeFazio, and M. Neubert, JHEP **9906**, 017 (1999).
[10] S. W. Bosch, B. O. Lange, M. Neubert, and G. Paz, Nucl. Phys. B **699**, 335 (2004).
[11] S. W. Bosch, M. Neubert and G. Paz, JHEP **11**, 073 (2004), [hep-ph/0409115].
[12] C. W. Bauer, Z. Ligeti and M. Luke, hep-ph/0111387.
[13] *BABAR* Collaboration, B. Aubert *et al.* , hep-ex/0507085.
[14] *BABAR* Collaboration, B. Aubert *et al.* , hep-ex/0506064.
[15] *BABAR* Collaboration, B. Aubert *et al.* , hep-ex/0506065.
[16] *BABAR* Collaboration, B. Aubert *et al.* , hep-ex/0507003.
[17] CLEO Collaboration - S. B. Athar *et al.* .,*Phys.Rev.* D**68**, 072003 (2003).
[18] *BABAR* Collaboration - B. Aubert *et al.* .,*Phys.Rev.* **90**, 181801 (2003).
[19] M. Okamoto *et al.* . hep-lat/0409116.
[20] J. Shigemitsu *et al.* ., hep-lat/0408019.
[21] P. Ball and R. Zwicky, Phys. Rev. D**71** (2005),014015; - 0140029.

DALITZ PLOT ANALYSIS OF $D^0 \to \bar{K}^0 K^+ K^-$ AND D_{sJ} STATES AT BABAR

M.Pappagallo [a]
Representing the BaBar Collaboration
INFN and University of Bari, 70126 Bari, Italy

Abstract. A Dalitz plot analysis of D^0 events reconstructed in the hadronic decay $D^0 \to \bar{K}^0 K^+ K^-$ is presented here. The analysis is based on a data sample of 91.5 fb^{-1}. Herewith enclosed is an updated, preliminary analysis of D_{sJ} states using 125 fb^{-1}. All data have been collected with the BaBar detector at the PEP-II asymmetric-energy e^+e^- storage rings at SLAC running at center-of-mass energies on and 40 MeV below the $\Upsilon(4S)$ resonance.

1 Introduction

Although the BaBar project is mostly known as a B meson factory there is much more than B physics which can be done at this facility. The copious production of $c\bar{c}$ pairs from the continuum and high integrated luminosity, makes BaBar an excellent laboratory where to study the charm production and decays. In this paper we show results on a the study of $D^0 \to \bar{K}^0 K^+ K^-$ decay [1] and D_{sJ} states [2].

2 Dalitz plot analysis of $D^0 \to \bar{K}^0 K^+ K^-$

Dalitz plot analyses are useful in providing new information on resonances that contribute to three-body final states. They can help to enlighten old puzzles related to light meson spectroscopy, in particular on the structure of scalar mesons.

In order to obtain a measurement of the branching ratio $BR = \frac{\Gamma(D^0 \to \bar{K}^0 K^+ K^-)}{\Gamma(D^0 \to \bar{K}^0 \pi^+ \pi^-)}$, $D^0 \to \bar{K}^0 K^+ K^-$ and $D^0 \to \bar{K}^0 \pi^+ \pi^-$ decays are selected by requiring they come from a D^{*+}. The result is:

$$BR = \frac{\Gamma(D^0 \to \bar{K}^0 K^+ K^-)}{\Gamma(D^0 \to \bar{K}^0 \pi^+ \pi^-)} = (15.8 \pm 0.1(stat.) \pm 0.5(syst.)) \times 10^{-2}.$$

The Dalitz plot of the $D^0 \to \bar{K}^0 K^+ K^-$ is shown in Fig. 1. We observe a strong interference between the $\phi(1020)$ and a scalar meson which is identified as mostly due to the $a_0(980)$ resonance. The contribution of $a_0(980)^+$, in the right bottom corner, can also be observed. A partial wave analysis in the low mass K^+K^- region allows the K^+K^- scalar (S) and vector components (P) to be separated solving the following system of equations [3]:

[a] e-mail: marco.pappagallo@ba.infn.it

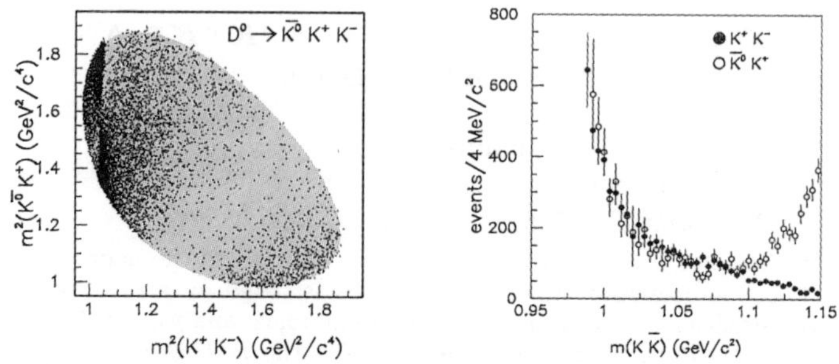

Figure 1: (Left) Dalitz plot of $D^0 \to \bar{K}^0 K^+ K^-$. (Right) Comparison between the scalar K^+K^- and the $\bar{K}^0K^+$ phase space corrected mass distributions.

$$\sqrt{4\pi}\left\langle Y_0^0 \right\rangle = S^2 + P^2$$

$$\sqrt{4\pi}\left\langle Y_1^0 \right\rangle = 2|S||P|cos\phi_{SP} \qquad \sqrt{4\pi}\left\langle Y_2^0 \right\rangle = \frac{2}{\sqrt{5}}P^2$$

where $\left\langle Y_L^0 \right\rangle_{L=0,1,2}$ are the efficiency corrected spherical harmonic moments. The resulting scalar K^+K^- and $\bar{K}^0K^+$ phase space corrected mass distributions are shown in Fig. 1 and show a good agreement. This supports the hypothesis that the $f_0(980)$ contribution is small, since $f_0(980)$ has isospin zero and therefore it cannot decay to $\bar{K}^0K^+$.

The K^+K^- S and P wave mass spectra, the $\bar{K}^0K^+$ mass spectrum and the phase difference ϕ_{SP} have been fitted simultaneously using a coupled channel Breit Wigner to describe the $a_0(980)$. The fit gives a measurement of $g_{\bar{K}K}$ which describe the $a_0(980)$ couplings to the $\bar{K}K$ system:

$$g_{\bar{K}K} = 464 \pm 29\,(\mathrm{MeV})^{1/2}.$$

An unbinned maximum likelihood fit has been performed for the decay $D^0 \to \bar{K}^0K^+K^-$ in order to use the distribution of events in the Dalitz plot to determine the relative amplitudes and phases of intermediate resonant and non-resonant states. The results of the Dalitz plot analysis can be summarised as follows (Table 1). *a)* The decay is dominated by $D^0 \to \bar{K}^0 a_0(980)^0$, $D^0 \to \bar{K}^0\phi(1020)$ and $D^0 \to K^- a_0(980)^+$. *b)* The $f_0(980)$ contribution is consistent with zero. *c)* The Double Cabibbo Suppressed contribution is consistent with zero. *d)* The remaining contribution is not consistent with being uniform, but can be described by the tail of a broad resonance, the $f_0(1400)$.

A search for CP asymmetries on the Dalitz plot has been performed. We have observed no statistically relevant asymmetries in fractions, amplitudes, or phases between D^0 and $\bar{D}^0$.

Final state	Amplitude	Phase(radians)	Fraction(%)
$\bar{K}^0 a_0(980)^0$	1.(fixed)	0.(fixed)	$66.4 \pm 1.6 \pm 7.0$
$\bar{K}^0 \phi(1020)$	$0.437 \pm 0.006 \pm 0.060$	$1.91 \pm 0.02 \pm 0.10$	$45.9 \pm 0.7 \pm 0.7$
$K^- a_0(980)^+$	$0.460 \pm 0.017 \pm 0.056$	$3.59 \pm 0.05 \pm 0.20$	$13.4 \pm 1.1 \pm 3.7$
$\bar{K}^0 f_0(1400)$	$0.435 \pm 0.033 \pm 0.162$	$-2.63 \pm 0.10 \pm 0.71$	$3.8 \pm 0.7 \pm 2.3$
$\bar{K}^0 f_0(980)$			$0.4 \pm 0.2 \pm 0.8$
$K^+ a_0(980)^-$			$0.8 \pm 0.3 \pm 0.8$
Sum			130.7 ± 2.2

Table 1: Results from the Dalitz plot analysis of $D^0 \to \bar{K}^0 K^+ K^-$.

3 D_{sJ} states

The unexpected discovery of narrow D_{sJ} states [4] [5] has raised new interest in the study of the charm spectroscopy. Here we report new preliminary results in the study of these D_{sJ} states.

3.1 $D^*_{sJ}(2317)^+$ and $D_{sJ}(2460)^+$ states

All the final states explored in this analysis involve one D_s^+ candidate decaying to $K^+K^-\pi^+$ where backgrounds are suppressed by selecting decays to $\bar{K}^{*0}K^+$ and $\phi\pi^+$.

$D_s^+\pi^0$ system (Fig. 2). In this selection, a cut on the π^0 momentum ($p_{\pi^0} > 400$ MeV/c) entirely eliminates the $D_s^*(2112)^+ \to D_s^+\pi^0$ signal. Estimates of the yield and the mass of the $D^*_{sJ}(2317)^+$ are obtained using an unbinned likelihood fit of the $D_s^+\pi^0$ mass spectrum. A reflection is produced by the $D_{sJ}(2460)^+ \to D_s^*(2112)^+\pi^0$ decay. Due to a kinematic coincidence, this reflection has a mean mass that is close to the $D^*_{sJ}(2317)^+$. These is no evidence of $D_{sJ}(2460)^+ \to D_s^+\pi^0$ decay for which we measure a 95% CL upper limit:

$$\frac{\mathcal{B}r(D_{sJ}(2460)^+ \to D_s^+\pi^0)}{\mathcal{B}r(D_{sJ}(2460)^+ \to D_s^+\pi^0\gamma)} < 0.11.$$

$D_s^{*+}\pi^0$ system (not shown). Estimates of the yield and mass of the $D^*_{sJ}(2317)^+$ are obtained using an unbinned likelihood fit of the $D_s^{*+}\pi^0$ mass spectrum. A reflection close to the $D_{sJ}(2460)^+$ mass is produced by $D^*_{sJ}(2317)^+ \to D_s^+\pi^0$ decays combined with unassociated γ particles.

$D_s^+\gamma$ system (not shown). There are two clearly visible structures in this spectrum on top of a gradually falling background distribution. The higher mass structure corresponds to the $D_{sJ}(2460)^+$ meson. The lower mass structure is a combination of reflections. The observation of $D_{sJ}(2460)^+ \to D_s^+\gamma$, prevents it from having $J^P = 0^\pm$. We measure a branching fraction:

$$\frac{\mathcal{B}r(D_{sJ}(2460)^+ \to D_s^+\gamma)}{\mathcal{B}r(D_{sJ}(2460)^+ \to D_s^+\pi^0\gamma)} = 0.375 \pm 0.054(stat.) \pm 0.057(syst.).$$

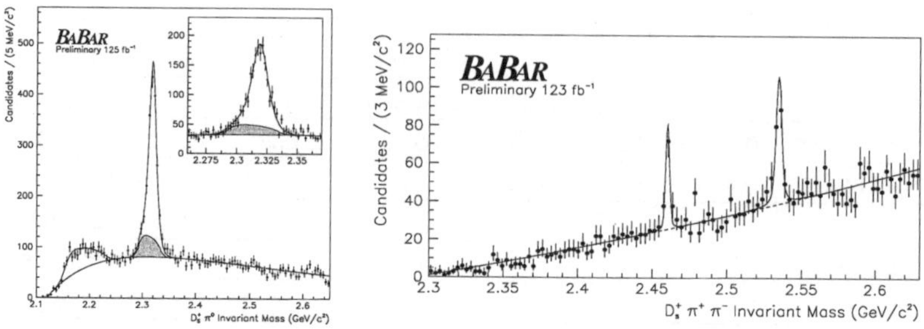

Figure 2: (Left)The $D_s^+\pi^0$ and (Right) the $D_s^+\pi^+\pi^-$ invariant mass distributions.

$D_s^+\pi^+\pi^-$ system (Fig. 2). To form $D_s^+\pi^+\pi^-$ candidates, each D_s^+ is combined with π^+ and π^- candidates having a momentum above 250 MeV/c. We observe two structures which correspond to $D_{sJ}(2460)^+$ and to $D_{s1}(2536)^+$ respectively. We measure a branching fraction:

$$\frac{\mathcal{B}r(D_{sJ}(2460)^+ \to D_s^+\pi^+\pi^-)}{\mathcal{B}r(D_{sJ}(2460)^+ \to D_s^+\pi^0\gamma)} = 0.082 \pm 0.018(stat.) \pm 0.011(syst.).$$

$D_s^+\pi^\pm$ systems (not shown). There is no evidence for isovector partners of $D_{sJ}^*(2317)^+$ in $D_s^+\pi^+$ and $D_s^+\pi^-$ systems [6].

From this studies, we have obtained new preliminary measurements of the $D_{sJ}^*(2317)^+$ and $D_{sJ}(2460)^+$ masses:

$$m(D_{sJ}^*(2317)^+) = 2318.9 \pm 0.3(stat.) \pm 0.9(syst.) \quad \text{MeV/c}^2$$

$$m(D_{sJ}^*(2460)^+) = 2459.4 \pm 0.3(stat.) \pm 1.0(syst.) \quad \text{MeV/c}^2$$

3.2 $D_{sJ}(2632)^+$ state

Recently, the SELEX collaboration [7] has reported a narrow state in the $D_s^+\eta$ and D^0K^+ mass distributions.

$D_s^+\eta$ system. After selecting $D_s^+ \to K^+K^-\pi^+$ and $\eta \to \gamma\gamma$ decays, a bi-dimensional background subtraction has been performed. The $D_s^+\eta$ mass distribution is shown in Fig. 3. There is no evidence for a $D_{sJ}(2632)^+$ state.

D^0K^+ system (Fig. 3). The D^0K^+ mass spectrum has been investigated using 3.7×10^6 $D^0 \to K^-\pi^+$ decays. There is no evidence for structure in the 2.632 GeV/c^2 mass region.

$D^{*+}K_s^0$ system (Fig. 3). The $D^{*+}K_s^0$ mass spectrum has been investigated using the $D^{*+} \to D^0\pi^+$ and $K_s^0 \to \pi^+\pi^-$ decay modes. The large, narrow peak

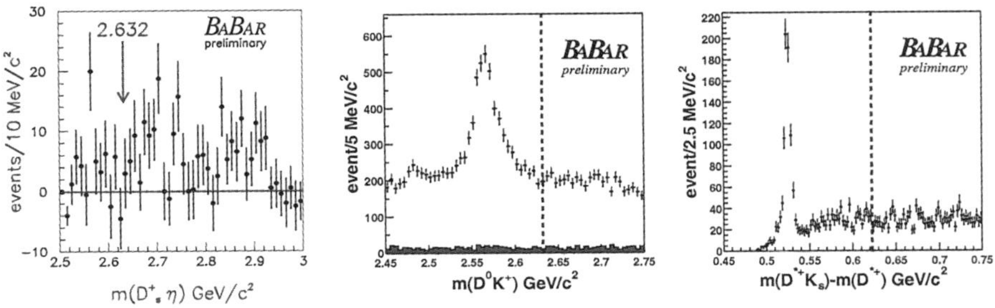

Figure 3: (Left) The $D_s^+\eta$ invariant mass distribution after background subtraction. (Middle) The D^0K^+ invariant mass distribution. The shaded histogram is the invariant mass distribution of wrong sign D^0K^- combinations. (Right) The distribution of the difference in invariant mass of the $D^{*+}K_s^0$ combination and D^{*+} candidate. The vertical lines indicate the mass location of the expected $D_{sJ}(2632)^+$ state.

just above the threshold results from the production of the $D_{s1}(2536)^+$. There is no evidence for production of $D_{sJ}(2632)^+$ state.

Acknowledgments

The author is grateful for the extraordinary contributions of our PEP-II colleagues in achieving the excellent luminosity and machine conditions that have made this work possible. The collaborating institutions wish to thank SLAC for its support and kind hospitality. This work is supported by DOE and NSF (USA), NSERC (Canada), IHEP (China), CEA and CNRS-IN2P3 (France), BMBF and DFG (Germany), INFN (Italy), FOM (The Netherlands), NFR (Norway), MIST (Russia), and PPARC (United Kingdom). Individuals have received support from CONACyT (Mexico), the A. P. Sloan Foundation, the Research Corporation, and the Alexander von Humboldt Foundation.

References

[1] BABAR Colaboration, B.Aubert *et al.*, *Phys.Rev.* **D** 72, 052008 (2005).
[2] BABAR Colaboration, B.Aubert *et al.*, *hep-ex/0408067*
BABAR Colaboration, B.Aubert *et al.*, *hep-ex/0408087.*
[3] S.U. Chung, *Phys.Rev.* **D** 56, 7299 (1997).
[4] BABAR Collaboration, B.Aubert *et al.*, *Phys.Rev.Lett.* **90**, 242001 (2003)
BABAR Collaboration, B.Aubert *et al.*, *Phys.Rev.* **D** 69, 031101 (2004).
[5] CLEO Collaboration, D. Besson *et al.*, *Phys.Rev.* **D** 68, 032002 (2003).
[6] T. Barnes, F.E. Close, H.J. Lipkin, *Phys.Rev.* **D** 68, 054006 (2003).
[7] SELEX Collaboration, A.V.Evdokimov *et al.*, *Phys.Rev.Lett.* **93**, 242001.

RARE MUONIC B-DECAYS AT ATLAS

K.Toms [a], N.Nikitine [b], S.Sivoklokov [c], L.Smirnova [d], D.Tlisov [e]
Skobeltsin Institute of Nuclear Physics, Moscow State University, 119992, Moscow, Russia

Abstract. We study the possibility of rare muonic B-mesons decays registration at the ATLAS LHC detector and their main backgrounds.

1 Introduction

ATLAS Detector (A Toroidal LHC Apparatus) is now under construction at CERN in the framework of the LHC (Large Hadron Collider) project with proton-proton 14 TeV center mass energy. This collider will be fully operational by 2007. Higgs boson and supersymmetry discovery are the prime ATLAS goals. The rare B-decays can make an essential contribution to the solution of the second problem.

Rare B-decays, produced by $b \to d, s$ quarks transitions, are forbidden at tree level in the Standard Model (SM). These decays can occur at the lowest order only through one-loop "penguin" and "box" diagrams. The branching ratios of these decays are very small: from 4×10^{-5} for rare radiative decay $B_d^0 \to K^*\gamma$ (discovered by CLEO at 1993 [1]) to 10^{-15} for rare Cabibbo–suppressed leptonic decay $B_d^0 \to e^+e^-$, that cannot be discovered even with LHC. In this report we will pay our attention to the rare muonic decays $B_{d,s}^0 \to \mu^+\mu^-$ with extremely small branching ratios:

$$Br(B_s^0 \to \mu^+\mu^-) = 3.5 \times \frac{|V_{ts}^* V_{tb}|^2}{2.2 \times 10^{-3}} \times 10^{-9}, \qquad (1)$$
$$Br(B_d^0 \to \mu^+\mu^-) = 0.9 \times \frac{|V_{td}^* V_{tb}|^2}{6.9 \times 10^{-5}} \times 10^{-10}$$

and main non-combinatoric backgrounds (BG) for them.

Rare muonic B-mesons decays are interesting from both theoretical and experimental points of view. In theory we have high precision SM prediction with minimal nonperturbative uncertanties and uncertanties tied with factorisation scale (see Section 2). Also, this branching ratios are sensitive to SM extentions. In experiment: only LHC will have enough statistics to detect rare muonic B-decays. ATLAS (and CMS) detectors will have some advantantages over LHCb detector studying rare muonic channels at nominal LHC luminosity. Finally, rare muonic decays have simple signature for experimental search.

[a] e-mail: ktoms@mail.cern.ch
[b] e-mail: nnikit@mail.cern.ch
[c] e-mail: sivokl@mail.cern.ch
[d] e-mail: lsmirnov@mail.cern.ch
[e] e-mail: tlisov@sinp.msu.ru

2 Theoretical description

From the theoretical point of view the $b \to q$ transitions $q = \{d, s\}$ are described using the effective Hamiltonian

$$H^{eff}(b \to q) = \frac{G_F}{\sqrt{2}} V_{tb} V_{tq}^* \sum_i C_i(\mu) O_i(\mu) \tag{2}$$

in the form of Wilson expansion. The set of Wilson coefficients $C_i(\mu)$ depends on the current model and contains the lowest order model contributions and perturbative QCD corrections [2]. The scale parameter μ is approximately equal to the mass of b-quark and is about 5 GeV. This parameter separates the perturbative and nonperturbative contributions of the strong interactions. The nonperturbative contribution is contained in the matrix elements of basis operators $O_i(\mu)$ between the initial and final hadronic states.

SM decay width of the rare muonic decays contain only one nonperturbative constant f_{B_q}. The value of this constant is known with high accuracy (about 5-10%), better than typical error of nonperturbatibe calculations. Furthermore the Wilson coefficient C_{10A} in NLO approach is not depend on the scale parameter μ, and doesn't give any new uncertanties in theoretical predictions. Main uncertanties (in order to 30%) are contained in numerical values of CKM-matrix elements.

3 $B^0_{d,s} \to \mu^+\mu^-$ decays simulation

Rare B-Physics simulations at ATLAS have a rich history. There were three important stages: the first, or TDR stage (1998–1999) used the Early ATLAS Detector layout based on GEANT3 and corresponding software [3]. The second stage (2002–2003) used the Final ATLAS Detector layout and intermediate software release [4]. And the last stage (2004–2005) used Final ATLAS Detector layout based on GEANT4 and new GRID-oriented software (so-called DC2 and Rome production). Below we introduce the first results of the Rome production and the comparison between these results and the previous ones.

In the framework of TDR stage the following results were obtained: after three years of LHC operation at initial luminosity (corresponding to 30 fb^{-1}) 4 signal events for $B^0_d \to \mu^+\mu^-$ decay and 27 signal events for $B^0_s \to \mu^+\mu^-$ decay, with 93 background events for the both decays were expected. After one year of LHC operation at nominal luminosity (corresponding to 100 fb^{-1}) 14 signal events for $B^0_d \to \mu^+\mu^-$ decay and 92 signal events for $B^0_s \to \mu^+\mu^-$ decay, with 660 common background events for both decays could be observed.

After three years of the LHC operation at initial luminosity $B^0_s \to \mu^+\mu^-$ decay with significance 2.8 σ could be detected. Adding one year with nominal luminosity we can set upper limit for $B_d^0 \to \mu^+\mu^-$ decay equal 3×10^{-10} with 95% CL, and detect $B_s^0 \to \mu^+\mu^-$ with 4.3σ significance [3].

In the framework of Rome production we have studied $B_s^0 \to \mu^+\mu^-$ with initial luminosity. For events reconstruction two vertexing strategies were used: CTMVFT vertex fit [5] and VKalVrt vertex fit [6]. We used a set of physical cuts to select signal events. Efficiencies of those cuts on signal and background events are presented in the Table 1, where p_T^{μ} - the transverse muon momentum, $\Delta R_{\mu\mu}$ - the distance between μ^+ and μ^- at(η, ϕ)-plane, $M_{\mu\mu}$ - the invariant dimuon mass, σ - the vertex resolution, L_{xy} - the distance between primary and secondary vertices of B_s^0-meson at transverse plane, α - the angle between momentum of $\mu^+\mu^-$-pair and direction of L_{xy} - distance.

Table 1: BG, signal events and selection cut efficiencies for $B_s^0 \to \mu^+\mu^-$ decay at 10 fb^{-1} integral LHC luminosity.

Cuts	BG: $b\bar{b}$ ($b\bar{b}b\bar{b}$, $b\bar{b}c\bar{c}$) $\to \mu^+\mu^- X$		Signal: $B_s^0 \to \mu^+\mu^-$	
Vertexing	CTMVFT	VKalVrt	CTMVFT	VKalVrt
$p_T^{\mu} > 6$ GeV, $\Delta R_{\mu\mu} < 0.9$	6.0×10^6 events		50 events	
$M_{\mu\mu}$ in $5.37^{+0.14}_{-0.07}$ GeV	2.0×10^{-2}	–	0.77	–
Isolation cut	5×10^{-2}		0.36	
$\sigma < 90\mu m$, $L_{xy} > 15\sigma$, $\alpha < 1^o$	2.8×10^{-3}	–	0.2	–
$L_{xy} > 11\sigma$, $\chi^2 < 15$	–	0.7×10^{-4}	–	0.4
Number of events	15 ± 10	20 ± 20	3	7

Founding on data from Table 1 it's possible to calculate the upper limit for $B_s^0 \to \mu^+\mu^-$ decay, for first year of ATLAS work at initial LHC luminocity. In order to do this, we extract cross section of B_s^0-meson from Rome Pythia sample, multiply it by acceptance of $B_s^0 \to \mu^+\mu^-$ decay, using LVL1 and LVL2 cuts: $\sigma(B_s^0)\times\alpha(B_s^0 \to \mu^+\mu^-) = 0.42\,\mu$b. Also we need to know the background cross section under the same conditions. From Rome Pythia samples we can obtain: $\sigma(BG) = 600$ pb. If we know full number of BG events for given luminocity, we can obtain the single event sensitivity (SES) using formula:

$$SES = \frac{1\,B_s^0 \to \mu 6\mu 6\,\text{event}}{\text{total number of BG events}} \times \frac{\sigma(BG)}{\sigma(B_s^0) \times \alpha(B_s^0 \to \mu^+\mu^-) \times \epsilon_\mu^2},$$

where ϵ_μ is the ATLAS single muon registration efficiency.

In Table 1 there is number of signal and background events for integral LHC luminosity 10 fb^{-1}. Using this data it's possible to get ATLAS upper limit at 90% CL for any initial integral LHC luminocity. These upper limits are presented in the Table 2. From this Table we can view, that even at 100 pb^{-1}, ATLAS exceed the upper limit for $B_s^0 \to \mu^+\mu^-$ branching ratio, obtained using combined data of CDF ans D0. For 30 fb^{-1} ATLAS upper limit is equal by the order of magnitude to SM predictions for $Br(B_s^0 \to \mu^+\mu^-)$.

Table 2: ATLAS sensitivity on $Br(B_s^0 \to \mu^+\mu^-)$ with Final detector layout.

Integral LHC Lumin.	BG ev. after $p_T^\mu > 6$ and $\Delta R_{\mu\mu} < 0.9$	Signal after all cuts	BG ev. after all cuts	ATLAS upp. lim. 90% CL	CDF&D0 upp. lim. 90% CL
100 pb^{-1}	$6.0 \cdot 10^4$	~ 0	~ 0.2	$6.4 \cdot 10^{-8}$	
10 fb^{-1}	$6.0 \cdot 10^6$	~ 7	~ 20	$7.0 \cdot 10^{-9}$	$1.2 \cdot 10^{-7}$
30 fb^{-1}	$1.8 \cdot 10^7$	~ 21	~ 60	$6.6 \cdot 10^{-9}$	

4 Some backgrounds for $B_{d,s}^0 \to \mu^+\mu^-$ decays

In order to find physics beyond SM in rare muonic decays, we need to know all possible SM backgrounds (BG).

It's well known that for ATLAS main contribution to the combinatoric background is coming from reactions $b\bar{b}$ $(b\bar{b}b\bar{b},\ b\bar{b}c\bar{c}) \to \mu^+\mu^-X$ with muons originating mainly from semileptonic b- and c-quark decays. This background always has been taken into account for studying the registration possibility of rare B-decays at ATLAS [3].

We will concentrate on BG from ecxlusive decays with small branching ratios and exotic decays, which are not included in standard MC-generators (like PYTHIA) and misidentification effects. Without cuts, background rate from exclusive decays and fake rate are both smaller than rate from combinatoric BG, but after applying necessary physical cuts, situation can change. Exclusive BG processes and fake rate can be potentially dangerous as they have signatures very similar to $B \to \mu^+\mu^-$. Below, will be considered some of backgrounds contributions.

4.1 Decays $B^{0,\pm} \to \pi^{0,\pm}\mu^+\mu^-$ as BG for $B_{d,s}^0 \to \mu^+\mu^-$

In the beginning, let us consider BG coming from $B \to \pi\mu^+\mu^-$ transitions. This BG appear because π-meson mass is equal to B-mass resolution for $B \to \mu^+\mu^-$ decays at ATLAS detector $\sigma_{B\to\mu\mu} \approx 80$ MeV (see the Table 8 in [4]).

Let us consider the origin of these BG. SM branching ratios of $B \to \pi\mu^+\mu^-$ decays predicted at the level [2, 7, 8]:

$$Br(B_d^0 \to \pi^0\mu^+\mu^-) = Br(B^\pm \to \pi^\pm\mu^+\mu^-) = 2.0 \times \frac{|V_{td}^* V_{tb}|^2}{6.9 \times 10^{-5}} \times 10^{-8}.$$

This is 10 times greater than SM prediction for $B_s^0 \to \mu^+\mu^-$ and 100 times greater than SM prediction for $B_d^0 \to \mu^+\mu^-$ (see Section 1). The BG appear when there is a soft pion in final state of $B \to \pi\mu^+\mu^-$ decay. In this case, invariant mass of muon pair is equal to

$$M_{\mu\mu} = M_B - M_\pi \approx M_B - \sigma_{B\to\mu\mu} \sim M_B.$$

We should mention that BG from $B^\pm \to \pi^\pm\mu^+\mu^-$ is less important than BG from $B_d^0 \to \pi^0\mu^+\mu^-$. In the case of $B^\pm \to \pi^\pm\mu^+\mu^-$ decay there are two charged tracks from B-meson vertex, only in the case of $p_T(\pi^\pm) < 0.5$ GeV In the same time for $B^0 \to \pi^0\mu^+\mu^-$ decay we always have two charged tracks from B-vertex. In this case cut on p_T of π^0 produced by energy resolution of electromagnetic calorimeter and can be up to 2-4 GeV.

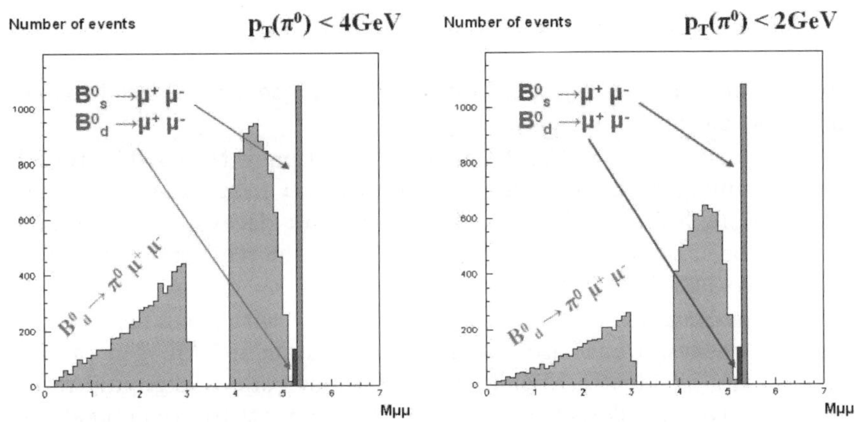

Figure 1: The invariant dimuon mass distribution for $B_d^0 \to \pi^0\mu^+\mu^-$ and $B_{d,s}^0 \to \mu^+\mu^-$ decays.

On the fig.1 there is shown the invariant dimuon mass distribution for $B_d^0 \to \pi^0\mu^+\mu^-$ and $B_{d,s}^0 \to \mu^+\mu^-$ decays. The distributions was obtained by MC-simulation on particle level. For muons there was applied LVL1 and LVL2 trigger cuts ($|\eta(\mu)| < 2.5$ and $p_T(\mu) > 6$ GeV). For $B_d^0 \to \pi^0\mu^+\mu^-$ we used theoretically calculated matrix element [2, 7, 8]. On both of the figures there were selected events where π^0 escapes identification at ATLAS. This selection is equivalent to $p_T < 4$ GeV for left fugure, or $p_T < 2$ GeV for right figure.

We can see on the fig.1 that end kinematical point of $B_d^0 \to \pi^0\mu^+\mu^-$ decay resides on the left side from the peaks of B_d^0- and B_s^0-mesons. So, it may seems that decay $B_d^0 \to \pi^0\mu^+\mu^-$ will not be BG for $B_{d,s}^0 \to \mu^+\mu^-$ decays. But, after full detector simulation all distributions will become more diffused by few $\sigma_{B\to\mu\mu}$ and $B_d^0 \to \pi^0\mu^+\mu^-$ mass distribution with high probability will overlap B_d^0-meson peak and, with a lesser degree, the B_s^0-meson peak. Full detector simulation is needed in order to get clear picture.

It's possible that vertex fit isolation and angle cuts [3, 9] after full detector simulation will reduce BG from $B \to \pi\mu^+\mu^-$. These cuts was absent in particle-level simulation due to their meaningless.

4.2 Misidentification and fake rates

Two-body hadronic decays of B-mesons can produce more important BG for $B \to \mu\mu$ than descibed in previous subsection. This BG are determined by hadron-muon misidentification at ATLAS detector. Typical value for misidentification approximately equal to 0.3-0.5% (see [10], Figure 8–23). Let's show importance of this BG with three examples.

First example is two-body hadronic decay $B_d^0 \to K^+\pi^-$. Branching of this decay $Br(B_d^0 \to K^+\pi^-) \approx 2 \times 10^{-5}$. So fake rate for $B \to \mu^+\mu^-$ is equal $Br(B_d^0 \to K^+\pi^-)\times(1/200)^2 \approx 0.5\times10^{-9}$, and nearly equal to $Br(B_s^0 \to \mu^+\mu^-)$ and exceed rate from $\mu^+\mu^-$ BG from previous subsection.

The second example of misidentification also tied with two-body hadronic decays, but differs from first example by origin of misidentification of muon pair. Let we consider the decay $B^+ \to (J/\psi \to \mu^+\mu^-)K^+$. Branching ratio of this decay $Br(B^+ \to (J/\psi \to \mu^+\mu^-)K^+) \sim 10^{-3} \times 6.0 \times 10^{-2}$. If we will take into account $K - \mu$ misidentification, we'll get 1/200. Let μ^+ from J/ψ decay has $p_T < 0.5$ GeV, i.e. is "soft" and is not detected by Inner Detector (it will add suppression factor of 1/10 into considered branching ratio). Then invariant mass μ^- from J/ψ and K^+ should be close to invariant mass of $B_{d,s}^0$-mesons. Effective fake rate for this case is approximately equal 3×10^{-8} and so $B^+ \to (J/\psi \to \mu^+\mu^-)K^+$ decay can produce the essential BG for $B_{d,s}^0 \to \mu^+\mu^-$.

Third example of misidentification importance is tied with semi-leptonic B-meson decays. Let we consider the decay $B_d^0 \to \pi^-\mu^+\nu_\mu$. Its branching is about 10^{-4}. This channel is included in MC-generators as background, but its contribution into BG is tied with decay $\pi^- \to \mu^-\bar{\nu}_\mu$. Fake rate for unphysical decay $B^0 \to \mu^-\mu^+\nu_\mu$ from $\pi - \mu$ misidentification isn't taken into account. Let us estimate this fake rate. Roughly it is equal to $Br(B_d^0 \to \pi^-\mu^+\nu_\mu) \times 1/200 \approx 0.5 \times 10^{-6}$. In order to estimate fake rate for $B^0 \to \mu^+\mu^-$ we need to decrease fake rate from $B^0 \to \mu^-\mu^+\nu_\mu$ by 10 times, which corresponds to soft-neutrino phase space.

So, fake rate from semileptonic decays of B-mesons is more important than fake rate from two-body hadronic B-meson decays but all of these fake rates needs detailed investigation using full-detector simulation.

4.3 Rare leptonic decays in external fields

Charged particles interactions with electromagnetic field are defined by invariant parameter $\chi = e\sqrt{(F_{\mu\nu}p_T^{\nu})^2}/m^3$, where $F_{\mu\nu}$ - the electromagnetic external field tensor, e - the charge of the particle. At ATLAS Detector there is external magnetic field with induction of $B = 2$ T and electric and magnetic field in detector matter. Density of electric field is $E \approx 10^{10}$ V/m, induction of magnetic field in matter is nearly equal to external field induction. Under such conditions for muons with different p_T paramerer $\chi_\mu \approx 10^{-9} - 10^{-12}$. So electromagnetic fields at ATLAS are weak.

Using the Volkov solution [11], Farry technique for external field calculations [12], and weak-field approximation for finding the correction for branching ratio from external field, we can obtain:

$$Br(B_q^0 \to \mu^+\mu^-, \chi_\mu) = Br(B_q^0 \to \mu^+\mu^-) \times \frac{1}{4\sqrt{\pi}} \frac{M_{B_q}}{E_{B_q}} \chi_\mu^{1/2} e^{-8/3\chi_\mu},$$

where E_{B_q} is the B_q-meson energy in the Laboratory System, M_{B_q} - the mass of B_q^0-meson and $Br(B_q^0 \to \mu^+\mu^-)$ - the branching ratios without external fields from equation (1). So, for ATLAS the external field contributions are fully negligible.

5 Conclusion

1. Already during the first year of LHC, with luminosity of 100 pb^{-1} ATLAS can exceed best CDF and D0 current upper limit on branching ratios of $B_s^0 \to \mu^+\mu^-$ decay.

2. After 30 fb^{-1} (equivalent to 3 years at initial luminosity) ATLAS will be able to achieve sensitivity at level of SM predictions for $B_s^0 \to \mu^+\mu^-$ decay.

3. Program of ATLAS rare muonic decays measurements can be continued from initial to nominal LHC luminosity.

4. The main contribution to the combinatorical background is coming from $b\bar{b}$ $(b\bar{b}b\bar{b},\, b\bar{b}c\bar{c}) \to \mu^+\mu^- X$ with muons originating mainly from semileptonic b- and c-quark decays.

5. It is very important to investigate the fake rate problem. It seems the most importang BG contribution is from $B^0 \to \pi^- \mu^+ \nu_\mu$. Also there is very important contributions from fake rate from two-body hadronic decays.

6. The rare exclusive decay $B_d^0 \to \pi^0 \mu^+ \mu^-$ can provide essential BG for decays $B_{d,s}^0 \to \mu^+ \mu^-$.

7. The external fields corrections for the decays $B_{d,s}^0 \to \mu^+ \mu^-$ gives negligible BG contribution.

Acknowledgments

Authors would like to express their gratitude for usefull discussion and help to J.Baines, A.Golutvin, D.Melikhov and M.Smizanska. This work was supported by Russian Ministry of Education and Science (Grants RI-111/002/079 and RI-111/002/080).

References

[1] CLEO Collaboration, *Phys.Rev.Lett.* 71, 674 (1993).
[2] A.Buras, M.Munz, *Phys.Rev.* **D** 52, 186 (1995); C.Bobeth et al., *JHEP* 0404, 071 (2004).
[3] CERN/LHCC/99-15, *ATLAS TDR* 15, 25 MAY 1999; *"1999 Workshop on SM Physics (and more) at the LHC"*, (CERN Yellow Reports CERN-2000-004).
[4] ATLAS B-Physics Group, ATL-PHYS-2005-002 (2005).
[5] F.Tartarelli, http://tarta.home.cern.ch/ tarta/vtx/docu.html
[6] V. Kostioukhine, ATL-PHYS-2003-031.
[7] D.Melikhov, N.Nikitin, S.Simula, *Phys.Rev.* **D** 57, 6814 (1998).
[8] D.Melikhov, B.Stech, *Phys.Rev.* **D** 62, 014006 (2000).
[9] N. Nikitine, F. Rizatdinova, L. Smirnova, *Phys.At.Nucl.* 62, 1697 (1999).
[10] *ATLAS TDR* 14, CERN/LHCC/99-14.
[11] D.M.Volkov, *Z. Phys* 94, 250 (1937).
[12] V.I.Ritus, *Lebedev Institute Rep.* 111, 5 (1979).

Z-SCALING AND STRANGE HIGH-P_T PARTICLE PRODUCTION IN P-P COLLISIONS AT RHIC

M.V.Tokarev [a]
Joint Institute for Nuclear Research,
141980 Dubna, Moscow region, Russia

Abstract. Results of analysis of new data on spectra of high-p_T strange particles produced in $p-p$ collisions at the Relativistic Heavy Ion Collider (RHIC) in the framework of z-scaling are presented. The scaling reflects symmetry properties such as self-similarity and fractality of hadron interactions at high energies. The properties of data z-presentation were established based on analysis of numerous high-p_T particle spectra obtained at the U70, ISR, SpS and Tevatron. These were used to obtain the scaling function $\psi(z)$ and predict spectra of strange particles over a wide p_T range at RHIC energies.

1 Introduction

Search for general features of strange particle production at high-p_T is of interest from theoretical and experimental point of view. Such features could clarify mechanisms of flavor transition and particle formation and understand origin of strangeness itself.

One of the scalings in high energy inclusive reactions is z-scaling observed in high-p_T particle production (see [1] and references therein). It is based on principles of locality, self-similarity and fractality of hadronic interactions. The scaling function ψ and scaling variable z used for data presentation are expressed via experimental quantities such as the inclusive cross section $Ed^3\sigma/dp^3$ and the multiplicity density of particles $dN/d\eta$. Data z-presentation reveals properties of energy and angular independence with power law at high z. The properties were established based on analysis of numerous high-p_T particle spectra obtained at the U70, ISR, SpS and Tevatron. The function $\psi(z)$ is interpreted as a probability density to produce a particle with a formation length z. The variable z reflects specific symmetry which relates to structure of space-time at small scales.

In the report we present results of analysis of new data on spectra of high-p_T strange particles produced in $p-p$ collisions at the Relativistic Heavy Ion Collider (RHIC) in the framework of z-scaling. Properties of z presentation are used to predict spectra of strange particles at RHIC energies over a wide p_T range. We consider that obtained results could give new constraints on distribution and fragmentation functions of strange quarks and more deep understanding of phenomenological aspects of QCD.

2 Z-scaling

The idea of z-scaling is based on the assumptions that gross feature of inclusive particle distribution of the process (1) at high energies can be described in terms of the corresponding kinematic characteristics

$$M_1 + M_2 \to m_1 + X \tag{1}$$

of the constituent subprocess written in the symbolic form (2)

$$(x_1M_1) + (x_2M_2) \to m_1 + (x_1M_1 + x_2M_2 + m_2) \tag{2}$$

satisfying the condition

$$(x_1P_1 + x_2P_2 - p)^2 = (x_1M_1 + x_2M_2 + m_2)^2. \tag{3}$$

[a] e-mail:tokarev@sunhe.jinr.ru

The equation is the expression of locality of hadron interaction at constituent level. The x_1 and x_2 are fractions of the incoming momenta P_1 and P_2 of the colliding objects with the masses M_1 and M_2. They determine the minimum energy, which is necessary for production of the secondary particle with the mass m_1 and the four-momentum p. The parameter m_2 is introduced to satisfy the internal conservation laws (for baryon number, isospin, strangeness, and so on).

The equation (3) reflects minimum recoil mass hypothesis in the elementary sub-process. To connect kinematic and structural characteristics of the interaction, the quantity Ω is introduced. It is chosen in the form

$$\Omega(x_1, x_2) = m(1 - x_1)^{\delta_1}(1 - x_2)^{\delta_2}, \tag{4}$$

where m is a mass constant and δ_1 and δ_2 are factors relating to the anomalous fractal dimensions of the colliding objects. The fractions x_1 and x_2 are determined to maximize the value of $\Omega(x_1, x_2)$, simultaneously fulfilling the condition (3)

$$d\Omega(x_1, x_2)/dx_1|_{x_2 = x_2(x_1)} = 0. \tag{5}$$

The fractions x_1 and x_2 are equal to unity along the phase space limit and cover the full phase space accessible at any energy.

Self-similarity is a scale-invariant property connected with dropping of certain dimensional quantities out of physical picture of the interactions. It means that dimensionless quantities for the description of physical processes are used. The scaling function $\psi(z)$ depends in a self-similar manner on the single dimensionless variable z. It is expressed via the invariant cross section $Ed^3\sigma/dp^3$ as follows

$$\psi(z) = -\frac{\pi s}{(dN/d\eta)\sigma_{in}} J^{-1} E\frac{d^3\sigma}{dp^3} \tag{6}$$

Here, s is the center-of-mass collision energy squared, σ_{in} is the inelastic cross section, J is the corresponding Jacobian. The factor J is the known function of the kinematic variables, the momenta and masses of the colliding and produced particles.

The function $\psi(z)$ is normalized as follows

$$\int_0^\infty \psi(z)dz = 1. \tag{7}$$

The relation allows us to interpret the function $\psi(z)$ as a probability density to produce a particle with the corresponding value of the variable z.

Principle of fractality states that variables used in the description of the process diverge in terms of the resolution. This property is characteristic for the scaling variable z written in the form

$$z = z_0\Omega^{-1}, \qquad \text{where} \qquad z_0 = \sqrt{\hat{s}_\perp}/(dN/d\eta). \tag{8}$$

The variable z has character of a fractal measure, $z \to \infty$ at $\Omega^{-1} \to \infty$. The quantity z_0 is finite part of z. It is the ratio of the transverse energy released in the binary collision of constituents and the average multiplicity density. The divergent part Ω^{-1} describes the resolution at which the collision of the constituents can be singled out of this process. The $\Omega(x_1, x_2)$ represents relative number of all initial configurations containing the constituents which carry fractions x_1 and x_2 of the incoming momenta. The momentum fractions x_1 and x_2 are determined in a way to minimize the resolution $\Omega^{-1}(x_1, x_2)$ of the fractal measure z with respect to all possible sub-processes (2) subjected to the condition (3). The variable z was interpreted as a particle formation length.

3 Results

Here we analyze the new data obtained by the STAR Collaboration [6] on p_T-spectra of neutral strange particles ($K_S^0, \bar{\Lambda}$) produced in $p-p$ collisions at $\sqrt{s}$ = 200 GeV.

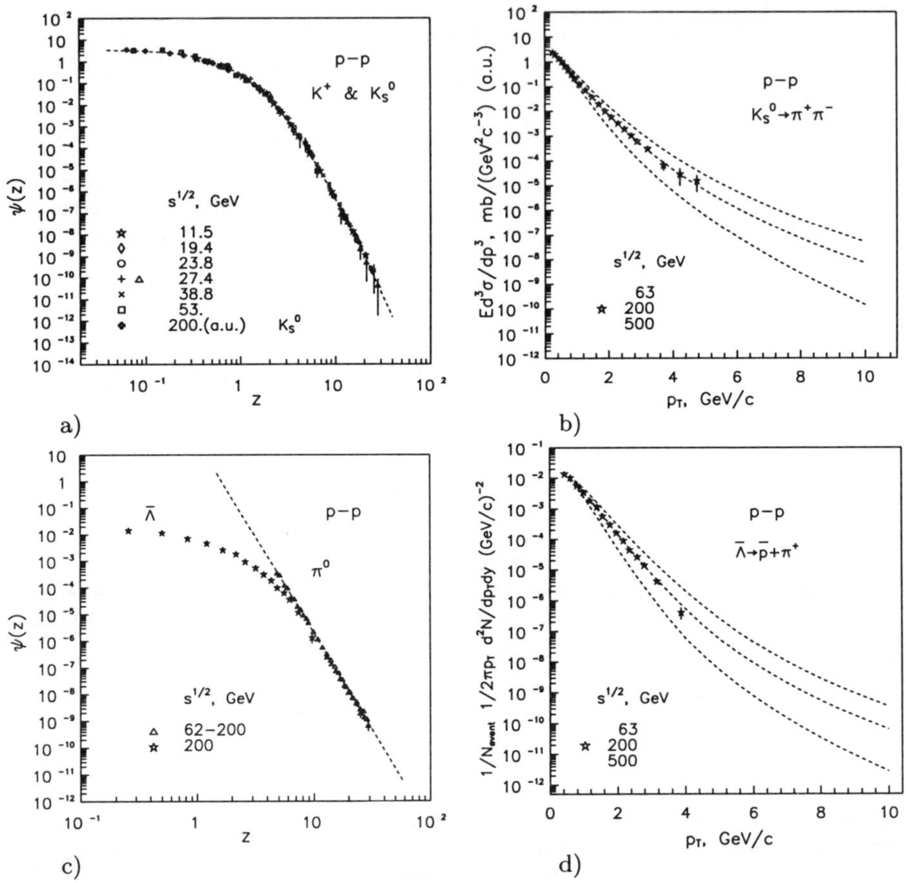

Figure 1: The z (a,c) and p_T (b,d) presentations of data on inclusive cross sections of K^+ and $K_S^0, \bar{\Lambda}$ produced in $p-p$ collisions at the U70, ISR, Tevatron and RHIC. Experimental data are taken from [2–5] and [6].

The scaling function $\psi(z)$ for data on K^+ and K_S^0-meson spectra obtained at the U70 [2], Tevatron [3, 4], ISR [5] and RHIC [6] are presented in Figire 1(a). The shape of the scaling function for K_S^0 is found to be in good agreement with $\psi(z)$ for K^+-mesons. Using parametrization of $\psi(z)$, the dependence of inclusive spectra of K_S^0-mesons on transverse momentum at $\sqrt{s}$ = 63, 200 and 500 GeV are plotted by the dashed line in Figure 1(b). The transverse momentum spectra of $\bar{\Lambda}$ produced in $p-p$ collisions are measured by the STAR Collaboration up to p_T = 4 GeV/c. The data are not enough to construct scaling function at high z. Therefore the F-dependence of z-presentation was used to determine the scaling function for $\bar{\Lambda}$. The

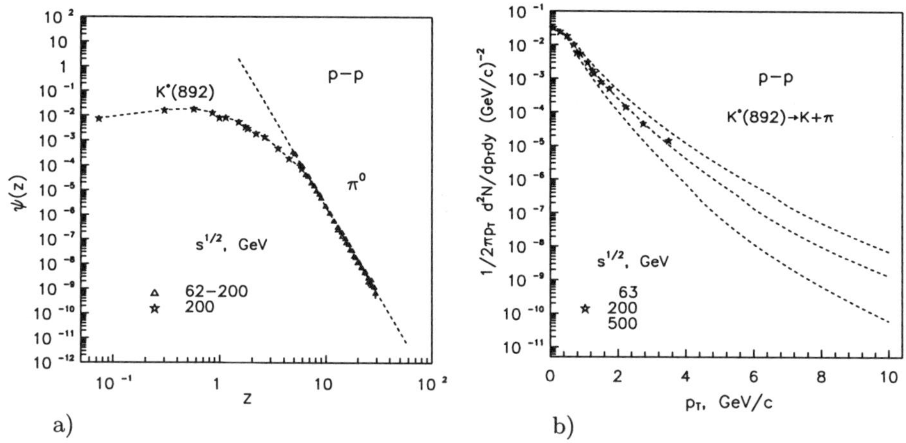

Figure 2: The z (a) and p_T (b) presentations of experimental data ($\star$) [7] on inclusive cross section of $K^\star$-mesons produced in $p-p$ collisions at the RHIC.

experimental data on inclusive cross section of π^0-mesons produced in $p-p$ collisions at $\sqrt{s} = 23-200$ GeV are used to construct the asymptotics of $\psi(z)$. The asymptotics is described by the power law, $\psi(z) \sim z^{-\beta}$, at high z. The dashed line shown in Figure 1(c) is the fit of the data on π^0-meson spectra. The parameterizations of $\psi(z)$ for $\bar{\Lambda}$ were used to predict spectra at $\sqrt{s} = 63, 200$ and 500 GeV at high p_T (see Figure 1(d)). Short-lived strange resonances are considered to be an efficient tool to probe properties of nuclear matter created in heavy-ion collisions. Vector resonance $K^{\star 0}$ are reconstructed via the hadronic decay channel $K^{\star,0} \to K^{\pm}\pi^{\mp}$. The transverse momentum $K^\star$ spectra for minimum bias $p-p$ collisions at $\sqrt{s} = 200$ GeV and $|y| < 0.5$ up to $p_T = 3.5$ GeV/c was obtained by the STAR Collaboration [7].

The F-independence of the scaling function at high z was used to construct the scaling function of $K^\star$ (Figure 2(a)) and predict transverse momentum spectra (Figure 2(b)) at $\sqrt{s} = 63, 200$ and 500 GeV up to 10 GeV/c.

4 Conclusions

The general regularities of high-p_T particle production described by z-scaling were found to be valid for strange hadrons produced in $p-p$ collisions in kinematical range accessible at the RHIC. Using the properties of z-scaling the spectra of $K_S^0, \bar{\Lambda}$ and $K^\star$ produced in $p-p$ collisions in high-p_T range at the RHIC energies were predicted. New evidence that mechanism of K_S^0 meson production reveals self-similar and fractal properties in high-p_T range was obtained. The obtained results are considered to be useful for understanding of strangeness origin in meson and baryons and verification of theory.

References

[1] M.Tokarev, I.Zborovský *Czech.J.Phys.*, **A** (Proc. Suppl.) 54, 445 (2004).
[2] V.V.Abramov *et al.*, *Sov. J. Nucl. Phys.* 41, 700 (1985).
[3] J.W.Cronin *et al.*, *Phys.Rev.* **D** 11, 3105 (1975).
D.Antreasyan *et al.*, *Phys.Rev.* **D** 19, 764 (1979).
[4] D.Jaffe *et al.*, *Phys.Rev.* **D** 40, 2777 (1989).
[5] B.Alper *et al.*, *Nucl.Phys.* **B** 87, 19 (1975).
[6] J.Adams, M.Heinz *et al.*, nucl-ex/0403020.
[7] J.Adams *et al.*, *Phys.Rev.* **C** 71, 064902 (2005).

GRID FOR EXPLORATION OF HIGH ENERGY PHYSICS: RDIG AND THE EGEE/LCG PROJECTS

V.Ilyin[a], A.Kryukov[b], A.Demichev[c]
D.V. Skobeltsyn Institute of Nuclear Physics, Moscow State University, 119992 Moscow, Russia

Abstract. We present current status of Russian Data-Intensive Grid and Russian participation in the global Grid EGEE/LCG projects.

1 Introduction

The EGEE (Enabling Grids for E-sciencE) project [1] brings together experts from over 27 countries with the common goal of creating global Grid infrastructure. The project aims to provide researchers in academia and industry with access to major computing resources, independent of their geographic location. In fact, any discipline which currently relies on advanced scientific computing resources may benefit from using the infrastructure. The EGEE project will also focus on attracting a wide range of new users to the Grid. The project concentrates on three core areas: (a) building a consistent, robust and secure Grid network that will attract additional computing resources, (b) continuously improving and maintaining the middleware in order to deliver a reliable service to users, (c) attracting new users from industry as well as science and providing them with high level of training and support they need. The EGEE Grid is based on the EU Research Network GEANT [2], a multi-gigabit pan-European data communications network for research and education, and exploits Grid expertise generated by many EU, national and international Grid projects to date such as the EU DataGrid project [3], other EU supported Grid projects and the national Grid initiatives [4] such as UK e-Science, INFN Grid, Nordugrid, US Open Science Grid and Russian Data-Intensive Grid (RDIG) [5]. The infrastructure will provide interoperability with other Grids around the globe, including the US and Asia, contributing to efforts to establish a world-wide Grid infrastructure. Currently the Grid system unifies 113 sites with more than 10,000 cpu and ~5 PB storage capacity in 30 countries (including 18 non-EGEE sites in 9 countries).

EGEE has always had close relations with other Grid initiatives falling into three broad categories: Infrastructure, Application and Support projects. Applications projects are important for EGEE both in making use of the infrastructure and helping to understand the requirements of user and feed these back into the development work of the project. Two applications projects, DILIGENT [6] (digital libraries) and GRIDCC [7] (remote instrumentation for

[a]e-mail: ilyin@sinp.msu.ru
[b]e-mail: kryukov@sinp.msu.ru
[c]e-mail: demichev@sinp.msu.ru

scientific measurements), have already been up and running for some time. They will soon be joined by BIOINFOGRID (bioinformatics) and Health-e-Child (computationally intensive research in paediatrics). Support projects include a variety of initiatives to contribute to Grid development, enhancing the impact and effectiveness of infrastructure and application projects. Efforts such as RDIG in Russia or SEE-GRID in the South-Eastern European region have already shown the success of these regional models for infrastructure extension. This strategy is now being adopted in other areas, with infrastructure projects covering China, the Mediterranean area, the Baltic States and Latin America.

A second two-year project (EGEE-2 in 2006-2007) is anticipated to follow on from EGEE, in which industry will progressively take up the operations and maintenance of a stable Grid infrastructure from the academic community. This is analogous to the way that GEANT has progressively transferred operations from the public to the private sector.

Implementation and certification of the performance of the evolving infrastructure initially will be guided by two pilot application domains. They are Biomedical Grids [8], where several communities are facing equally daunting challenges to cope with the flood of bioinformatics and healthcare data and LHC (Large Hadron Collider) [9] based experiments in high energy physics. The Large Hadron Collider (LHC), currently being built at CERN near Geneva, is the largest scientific instrument on the planet. When it begins operations in 2007, it will produce roughly 15 Petabytes (15 million Gigabytes) of data annually, which thousands of scientists around the world will access and analyze. The mission of the LHC Computing Project (LCG) [10] is to build and maintain a data storage and analysis infrastructure for the entire high energy physics community that will use the LHC. Among other applications are Earth Sciences and Astrophysics.

2 Russian Data-Intensive Grid (RDIG)

In order to ensure the full-scale participation of Russia in the EGEE project, the Consortium RDIG (Russian Data Intensive Grid) [5] has been formed in September of 2003. Memorandum has been signed by the heads of eight large institutes, including Institute for High Energy Physics (Protvino), Institute of Mathematical Problems of Biology (Pushchino), Institute for Theoretical and Experimental Physics (Moscow), Joint Institute for Nuclear Research (Dubna), Keldysh Institute for Applied Mathematics (Moscow), Skobeltsyn Institute of Nuclear Physics, Moscow State University (Moscow), Petersburg Nuclear Physics Institute (St.Petersburg) and RSC "Kurchatov Institute" (Moscow). Main goal of the consortium is to create a fully operated EGEE infrastructure in Russia including establishing a Russian Core Infrastructure Center (CIC), full scale participation in the EGEE evolution from the two pilot applications to an infrastructure serving multiple scientific and technological communities, outreach and dissemination activity targeting researchers from both scientific community and industry. For this aim RDIG enters EGEE structure as a regional federation "Russia". Today RDIG comprises more than 100 staff members, more than 10 sites, more than 500 CPU and ~100 Tbyte data storage capacity.

3 Russian Core Infrastructure Center (CIC)

The service activities in the EGEE Grid infrastructure is organized around three main functional organizations: Operations Management Center, Regional Operations Centers (ROCs) and Core Infrastructure Centers (CICs). The Operations Management Center (located at CERN) coordinates the operational activity of the infrastructure. The Regional Operations Centers provides a decentralized but efficiently coordinated support structure with front line user and deployment support. They have an important role as sources of expert advice and technical support in the process of building and operating the infrastructure. The ROCs have the expertise to assist the growing number of Resource Centers in joining the infrastructure, through the deployment of middleware releases and the development of procedures and capabilities to operate those resources.

In this paper we will mainly concentrate on a key part of the infrastructure - Core Infrastructure Centers (CICs). The CICs have two fundamental roles: to operate essential Grid infrastructure services that are not required at each Resource Center (RC, Grid sites where the very execution of jobs submitted to the Grid takes place) and to provide monitoring and operational troubleshooting services. In addition they provide support for management of Virtual Organizations (VOs), which are groups of individuals, often not bound to a single institution or enterprise, who, by reason of their common membership of the VO, and in sharing a common goal, are granted rights to use a set of resources on the Grid. There are five CICs in the EGEE infrastructure located at CERN, in UK, Italy, France and Russia. The CICs function as a single distributed entity by sharing operational information on a frequent and regular basis. They are in continuous contact and collectively manage day-to-day operation of the EGEE Grid infrastructure including the active monitoring of the infrastructure and the Resource Centers, and take appropriate action to protect the Grid from the effects of failing components and to recover from operational problems. For example: a Resource Center that is causing problems by generating invalid information may be removed from the Grid until the problem has been fixed; the operations support line of a Resource Center which has become inaccessible may be contacted to initiate recovery; a network operations center may be informed of suspected failures.

The Russian CIC provides the full scope Grid functionality for RDIG community so that users (in Russia and Europe) have access to Russian and European computer resources within the EGEE infrastructure. In particular, the Russian CIC provides all the services essential for operation of the Grid infrastructure as a whole. These services includes (but are not limited to): Resource Broker (Workload management), Information Service, Logging and Bookkeeping service, Replica and File Catalogs, Virtual Organization Management Services.

By now the complete set of Grid services has been put into operation in RDIG . Use of separate structure of the basic Grid services allows the Russian Grid segment to work both independently and within the framework of the Grid infrastructure of the global project EGEE. All-round testing and preliminary sets of jobs in the real conditions (on the basis of modeling events in upcoming experiments on Large Hadron Collider in the European Organization for Nuclear Research (CERN)) has been carried out.

Most of the Grid sites (Resource Centers, RC) contain two types of resources: computing (Working Nodes, WN) and storage (Storage Systems, SS) ones. In particular, all Russian RCs in EGEE/LCG/RDIG infrastructure have both types of the resources. A gateway to Grid for the resources provide the following services located at RCs: Computing Element (for WNs) and Storage Element (for SSs). Grid users communicate with the infrastructure services and, in particular, submit jobs to be executed at some RC via a set of clients combined in a package called User Interface (UI).

Currently, the LCG-2.6 middleware [11] is used for realization of these Grid services and the UI. The principal components of this middleware package are:

- the Globus Toolkit (GT2) developed by the Globus Project [12];
- the Condor system [13] developed at the University of Wisconsin, Madison;
- the Globus and Condor components and some other tools from US projects are integrated and packaged as the Virtual Data Toolkit by the VDT project [14];
- tools developed by the DataGrid Project (EDG) [3];
- new middleware components developed as part of the gLite toolkit [15] by the EGEE project; the current release of gLite provides improved tools for workload scheduling, Grid catalog, and a monitoring infrastructure; future releases will add additional functionality and provide re-engineered components with the aim of satisfying the requirements of the main EGEE application domains: high energy physics, biology and medicine.

The middleware will evolve towards a service oriented architecture (SOA) based on standards developed within the Web Services and related communities.

Among other services that the Russian CIC operates are the following:

- VO services: this includes the VO server itself (currently based on LDAP server, in the future we will use Virtual Organization Membership Service (VOMS) [16]), and as well includes the registration process web server [17] for the Russian regional VOs. Responsibility for managing the VO membership remains with the VO itself (VO manager). Instructions for a new users how to join the infrastructure are available at the SINP web-site [18].
- VO-specific services: This includes database services for replica services for a VO (or set of VOs), and may include other VO-specific or general information services depending on the needs of the VO (VO-box).
- The CIC performs monitoring and troubleshooting of the basic services and resources as well as performance monitoring, for example, of data transfers, job throughput etc.
- Operate middleware configuration management and change control procedures.
- Collect, store and make available as appropriate operational monitoring and authorization and accounting data from Resource Centers.
- The dissemination of EGEE middleware is an important subtask of the CIC; here the localization (translation of the documentation to Russian, etc.) is provided.

When EGEE was first launched, the infrastructure was managed centrally from the Operations Center at CERN. While this worked quite well, it also had disadvantages. Troubleshooting such a large a network is a tiring job, and the experience it generated was concentrated in one place. To lessen the work load and to make sure experience with Grid operations was more evenly spread out, EGEE came up with a scheme where the CICs shared the load. Dubbed CIC-on-duty, this new system began in October 2004. In this system, responsibility for managing the infrastructure passes around the globe on a weekly basis. Since April 2005, the Russian CIC joins the system and shares the responsibility with four other European CICs. The staff supervise the Grid sites, monitor problems, decide whether troubles are temporal or caused by a serious problems in a site's performance, and keep administrators of the sites informed about issues, contacting them by phone when necessary. An especially important part of a CIC-on-duty responsibility is providing the infrastructure security, examining its malfunctions, and taking measures against them, as well as summarizing experience and sharing it with other CICs. Already the CIC-on-duty scheme is proving worthwhile. While it seems that the number of bugs which the CICs have to deal with each day has stayed roughly constant, the number of sites they manage has grown immensely, as has the number of good sites. When the CIC-on-duty scheme began, only around 45% of sites passed their daily Site Functional Tests, but now this figure is closer to 80%. The approach taken in this scheme reflects the collaborative nature of the EGEE project in general, and the successes that such an approach can provide.

4 Participation of RDIG and SINP RC in Application Tasks

High Energy Physics and biomedical applications were of major importance for RDIG participation in the EGEE project at the initial stage. Tasks, directly concerning requests of LHC Computing, were major applications for us in 2004 - 2005. Biomedical applications are intended for needs of studying the biomacromolecules structures and dynamics, e-cell projects and computer aided drug-design, implementing new computational approaches and informational resources on biology. Starting from 2005, a new important application area for RDIG is plasma physics and nuclear fusion study in the framework of the ITER project [19]. In this area the features of computing are similar to that for the LHC needs: huge amount of data must be stored in different centers around the world and analyzed by numerous users from many sites. Therefore the EGEE middleware and infrastructure may be used for the needs of the ITER project with high efficiency and small re-engineering. One more application area supported by the EGEE/LCG/RDIG is Earth Science/Geophysics Research.

In accordance with the above mentioned priorities, the RDIG CIC and RCs support the following Virtual Organizations (VOs):

- VO supported in EGEE/LHC infrastructure:
  - LHC Affiliated VO for four LHC experiments and Accelerator/Beam Physics: ALICE, ATLAS, CMS, LHCB, DTEAM (technical VO)
  - non-LHC Affiliated VO: Biomed (Biomedical activity);
- Regional VOs supported only by RDIG:
  - Fusion (ITER Project, http://www.iter.org),
  - eEarth (eEarth Project, http://www.e-earth.ru),
  - PHOTON (PHOTON Project, http://egee.itep.ru/PHOTON),
  - AMS (AMS project, http://ams.cern.ch),
  - RGStest (new applications testing),

– RDTEAM (Grid infrastructure testing).

The support of the VOs implies that the users included in the corresponding VO LDAP or VOMS data bases have access to Grid services of the CIC and RCs. In other words, the users included in the VO data bases have ability to submit their jobs to the Grid computing resources, place/retrieve data to mass storage systems, monitor the resources and job execution, etc. The goal of the organization of the regional VOs is to support the national scientific projects and to test new application areas before including them into the global EGEE infrastructure. In the future RDIG plans to create and support VOs in the areas of chemical physics and bio-engineering.

Information about resources in RDIG as well as statistics on executed jobs, etc. is available at the RDIG monitoring page http://rocmon.jinr.ru:8080. For example, the RC at SINP MSU has a cluster of 60 WNs (Pentium IV CPUs, 2 GHz in average) and SS of the total capacity ~10 Tbyte. Detailed statistics on the RC operation provided by a few different monitoring tools is available at http://egee.sinp.msu.ru/rc/index.php . For instance, the resource center statistics from 2005-11-01 to 2006-01-30 (three months) read as follows:

Virtual Organization	Total Number of Jobs	Total Number of Users	Total CPU Time (hours:mins)	Total Elapsed Time (hours:mins)
ALICE	5	3	00 : 00	00 : 01
AMS	11	1	00 : 00	00 : 01
ATLAS	5665	25	30765 : 10	35802 : 29
BIOMED	2602	7	209 : 16	331 : 08
CMS	394	12	2803 : 19	3859 : 32
DTEAM	1221	29	38 : 50	154 : 30
FUSION	4871	2	653 : 37	1174 : 13
LHCB	1068	9	267 : 03	383 : 22
TOTAL:	**15837**	**88**	**34737 : 19**	**41705 : 19**

During the same period average cluster load was: CPU Time - 27.42%; Elapsed Time 32.92%. These figures are not very high so far because the EGEE/LCG/RDIG Grid infratructure is still at the stage of development (testing/debugging/updating of the middleware and services). At the production stage the average load of the RCs is expected to be close to 100%.

5 Conclusion

The RDIG infrastructure provides full-scale participation of Russian scientific institutes in the global project EGEE allowing access of Russian scientific communities to the huge distributed computing resources unified by EGEE Grid infrastructure. In addition, because of existence of all the basic Grid services in the Russian CIC, the Russian Grid segment may work both within the framework of the EGEE Grid infrastructure and independently providing an infrastructure for an access to distributed computing/storage facilities for national scientific projects.

Acknowledgments

This work was supported in different parts of the activity by the Federal Agency for Science and Innovations under contracts No. 02.434.11.1006 and No. 02.447.11.1002, INTAS-CERN grants No. 03-52-4297 and No. 05-103-7509, and RFBR grant No. 05-07-90292.

References

[1] EGEE Project: http://www.eu-egee.org
[2] GEANT Project: http://www.geant.net
[3] EU DataGrid Project: http://eu-datagrid.web.cern.ch/eu-datagrid
[4] e-Science project: http://www.nesc.ac.uk;
INFN Grid Project: http://grid.infn.it;
Nordugrid Project: http://www.nordugrid.org;
Open Science Grid: http://www.opensciencegrid.org
[5] RDIG Consortiium: http://egee-rdig.ru
[6] DILIGENT Project: http://www.diligentproject.org
[7] GRIDCC Project: http://www.gridcc.org
[8] Biomedical Virtual Organization:
http://public.eu-egee.org/applications/biomed.html
[9] LHC Project: http://www.cern.ch
[10] LCG Project: http://lcg.web.cern.ch/LCG
[11] LCG middleware: http://lcg.web.cern.ch/LCG/Sites/releases.html
[12] Globus Toolkit: http://www.globus.org
[13] Condor Project: http://www.cs.wisc.edu/condor
[14] Virtual Data Toolkit: http://vdt.cs.wisc.edu
[15] gLite Middleware: http://www.glite.org
[16] VOMS service: http://hep-project-grid-scg.web.cern.ch/hep-project-grid-scg/voms.html
[17] User registration in RDIG: http://rdig-registrar.sinp.msu.ru
[18] How to join EGEE-RDIG instruction:
http://egee.sinp.msu.ru/gateway/join_gw.php
[19] ITER Project: http://www.iter.org

PREDICTIONS FOR THE MUON g-2

M. Passera [a]
Dipartimento di Fisica, Università di Padova and INFN, 35131 Padova, Italy

Abstract. Standard Model predictions for the anomalous magnetic moment of the muon are reviewed and compared to the present experimental value.

1 Introduction

During the last few years, in a sequence of increasingly precise measurements, the E821 Collaboration at Brookhaven has determined $a_\mu = (g_\mu - 2)/2$ with a fabulous relative precision of 0.5 parts per million (ppm) [1, 2], allowing us to test all sectors of the Standard Model (SM) and to scrutinize viable alternatives to this theory [3]. The present world average experimental value is $a_\mu^{\text{EXP}} = 116\,592\,080\,(63) \times 10^{-11}$ (0.5 ppm) [2]. This impressive result is still limited by statistical errors. A new experiment, E969, has been approved (but not yet funded) at Brookhaven in 2004 [4]. Its goal is to reduce the present experimental uncertainty by a factor of 2.5 to about 0.2 ppm. A letter of intent for an even more precise $g-2$ experiment was submitted to J-PARC with the proposal to reach a precision below 0.1 ppm [5]. But how precise is the theoretical prediction?

The SM prediction a_μ^{SM} is usually split into three parts: QED, electroweak and hadronic (see [6] for recent reviews). This note provides a brief summary of the present status of these three contributions, and a comparison with the current experimental value.

2 QED and Electroweak Contributions

The QED contribution to a_μ arises from the subset of SM diagrams containing only leptons (e, μ, τ) and photons. First computed by Schwinger more than fifty years ago [7], it is now known up to terms of order $(\alpha/\pi)^4$, and leading five-loop contributions have been evaluated. The prediction currently stands at $a_\mu^{\text{QED}} = 116\,584\,719.4\,(1.4) \times 10^{-11}$ [8], where the error is due to the uncertainty of the $O(\alpha^4)$ and $O(\alpha^5)$ terms, and to the uncertainty of the fine-structure constant α. I refer the reader to Kataev's contribution to these proceedings for further details and very recent developments [9].

The electroweak (EW) contribution to a_μ is suppressed by a factor $(m_\mu/M_W)^2$ with respect to the QED effects. The one-loop part was computed by several authors [10]: $a_\mu^{\text{EW}}(1\text{ loop}) = \frac{5G_\mu m_\mu^2}{24\sqrt{2}\pi^2}\left[1 + \frac{1}{5}\left(1 - 4\sin^2\theta_W\right)^2 + O(m_\mu^2/M_{Z,W,H}^2)\right]$, where $G_\mu = 1.16637(1) \times 10^{-5}$ GeV^{-2} is the Fermi coupling constant. Closed analytic expressions for $a_\mu^{\text{EW}}(1\text{ loop})$ taking exactly into account the m_μ^2/M_B^2

[a] e-mail: passera@pd.infn.it

dependence ($B = Z, W$, Higgs, or other hypothetical bosons) can be found in refs. [11]. Employing the on-shell definition $\sin^2\theta_W = 1 - M_W^2/M_Z^2$ [12], where $M_Z = 91.1875(21)$ GeV and M_W is the SM prediction of the W mass (which can be derived, for example, from the simple formulae of [13] leading to $M_W = 80.383$ GeV for the Higgs mass $M_H = 150$ GeV), one obtains $a_\mu^{\rm EW}(1 \text{ loop}) = 194.8 \times 10^{-11}$.

The two-loop EW contribution to a_μ is not negligible because of large factors of $\ln(M_{Z,W}/m_f)$, where m_f is a fermion mass scale much smaller than M_W [14]. It was computed in 1995 [15]. The proper treatment of the contribution of the light quarks was addressed in [16,17]. These refinements significantly improved the reliability of the fermionic part (that containing closed fermion loops) of $a_\mu^{\rm EW}(\text{two loop})$ leading, for $M_H = 150$ GeV, to $a_\mu^{\rm EW} = 154(1)(2) \times 10^{-11}$ [17]. The first error is due to hadronic loop uncertainties, while the second one corresponds to an allowed range of $M_H \in [114, 250]$ GeV, to the current top mass uncertainty, and to unknown three-loop effects. The leading-logarithm three-loop contribution to $a_\mu^{\rm EW}$ is extremely small [17, 18]. The results of [19] for the two-loop bosonic part of $a_\mu^{\rm EW}$, performed without the large M_H approximation previously employed, agree with the previous evaluation [15] in the large Higgs mass limit.

3 The Hadronic Contribution

The evaluation of the hadronic leading-order contribution $a_\mu^{\rm HLO}$, due to the hadronic vacuum polarization correction to the one-loop diagram, involves long-distance QCD for which perturbation theory cannot be employed. However, using analyticity and unitarity, it was shown long ago that this term can be computed from hadronic e^+e^- annihilation data via the dispersion integral [20]

$$a_\mu^{\rm HLO} = \frac{1}{4\pi^3}\int_{4m_\pi^2}^{\infty} ds\, K(s)\sigma^{(0)}(s), \tag{1}$$

where $\sigma^{(0)}(s)$ is the total cross section for e^+e^- annihilation into any hadronic state, with extraneous QED corrections subtracted off. The kernel function $K(s)$ decreases monotonically for increasing s.

A prominent role among all e^+e^- annihilation measurements is played by the precise data collected in 1994-95 by the CMD-2 detector at the VEPP-2M collider in Novosibirsk for the $e^+e^- \to \pi^+\pi^-$ cross section at values of $\sqrt{s}$ between 0.61 and 0.96 GeV [21] (quoted systematic error 0.6%, dominated by the uncertainties in the radiative corrections). Recently [22] the CMD-2 Collaboration released its 1996-98 measurements for the same cross section in the full energy range $\sqrt{s} \in [0.37, 1.39]$ GeV. The part of these data for $\sqrt{s} \in [0.61, 0.96]$ GeV (quoted systematic error 0.8%) agrees with their earlier result published in [21]. In 2005, also the SND Collaboration (at the VEPP-2M collider as well) released its analysis of the $e^+e^- \to \pi^+\pi^-$ process for $\sqrt{s}$ between 0.39 and 0.98 GeV, with a systematic uncertainty of 1.3% (3.2%) for $\sqrt{s}$ larger (smaller) than 0.42 GeV [23]. However, a recent preliminary reanalysis of these data [24] uncovered an error in the treatment of the radiative corrections, reducing the value of the measured cross-section. The new preliminary SND result appears to be in good agreement with the corresponding one from CMD2.

In 2004 the KLOE experiment at the DAFNE collider in Frascati presented a precise measurement of $\sigma(e^+e^- \to \pi^+\pi^-)$ via the initial-state radiation (ISR) method at the ϕ resonance [25]. This cross section was extracted for $\sqrt{s}$ between 0.59 and 0.97 GeV with a systematic error of 1.3% and a negligible statistical one. There are some discrepancies between the KLOE and CMD-2 results, although their integrated contributions to a_μ^{HLO} are similar. The data of KLOE and SND [23] disagree above the ρ peak, where the latter are significantly higher. However, the value of the latter appears to be lower after the new preliminary reanalysis presented in [24]. The study of the $e^+e^- \to \pi^+\pi^-$ process via the ISR method is also in progress at BABAR [26] and Belle [27]. On the theoretical side, analyticity, unitarity and chiral symmetry provide strong constraints for the pion form factor in the low-energy region [28]. Recent evaluations of the dispersive integral based on the CMD-2 analysis of ref. [21] are in good agreement: $a_\mu^{\text{HLO}} = 6934\,(53)_{exp}(35)_{rad} \times 10^{-11}$ [29], $a_\mu^{\text{HLO}} = 6948\,(86) \times 10^{-11}$ [30], $a_\mu^{\text{HLO}} = 6924\,(59)_{exp}(24)_{rad} \times 10^{-11}$ [31], $a_\mu^{\text{HLO}} = 6944\,(48)_{exp}(10)_{rad} \times 10^{-11}$ [32]. The result of ref. [29] already includes KLOE's data and updates the one of ref. [33]. The recent data of CMD-2 [22] and SND [23, 24] are not yet included. Further significant progress is expected from the e^+e^- collider VEPP-2000 under construction in Novosibirsk [34] and, possibly, from DAFNE-2 [35].

The authors of [36] pioneered the idea of using vector spectral functions derived from the study of hadronic τ decays [37] to improve the evaluation of the dispersive integral. However, the latest analysis with ALEPH, CLEO, and OPAL data yields $a_\mu^{\text{HLO}} = 7110\,(50)_{exp}(8)_{rad}(28)_{SU(2)} \times 10^{-11}$ [33], a value significantly higher than those obtained with e^+e^- data (see [38] for recent preliminary results from Belle). Isospin-breaking corrections were applied [39]. Indeed, although the precise CMD-2 $e^+e^- \to \pi^+\pi^-$ data [21] are consistent with the corresponding τ ones for energies below ~ 0.85 GeV, they are significantly lower for larger energies. KLOE's $\pi^+\pi^-$ spectral function confirms this discrepancy with the τ data. SND's 2005 results [23] were compatible with the τ ones, but the very recent preliminary reanalysis presented in [24] seems to indicate that this is no longer the case. This puzzling discrepancy between the $\pi^+\pi^-$ spectral functions from e^+e^- and isospin-breaking-corrected τ data could be caused by inconsistencies in the e^+e^- or τ data, or in the isospin-breaking corrections which must be applied to the latter [40].

The hadronic higher-order (α^3) contribution a_μ^{HHO} can be divided into two

$a_\mu^{\mathrm{SM}} \times 10^{11}$		$\Delta \times 10^{11}$	σ
[29]	116 591 789 (76)	291 (98)	3.0 ⟨2.5⟩
[30]	116 591 803 (95)	277 (114)	2.4 ⟨2.0⟩
[31]	116 591 779 (76)	301 (98)	3.1 ⟨2.6⟩
[32]	116 591 799 (63)	281 (89)	3.1 ⟨2.7⟩
[33]	116 591 962 (70)	118 (95)	1.3 ⟨0.7⟩

Table 1: Standard Model vs. measurement.

parts: $a_\mu^{\mathrm{HHO}} = a_\mu^{\mathrm{HHO}}(\mathrm{vp}) + a_\mu^{\mathrm{HHO}}(\mathrm{lbl})$. The first one is the $O(\alpha^3)$ contribution of diagrams containing hadronic vacuum polarization insertions [41]. Its latest value is $a_\mu^{\mathrm{HHO}}(\mathrm{vp}) = -97.9\,(0.9)_{exp}(0.3)_{rad} \times 10^{-11}$ [31]. The second term, also of $O(\alpha^3)$, is the hadronic light-by-light contribution. As it cannot be directly determined via a dispersion relation approach using data (unlike the hadronic vacuum polarization contribution), its evaluation relies on specific models of low-energy hadronic interactions with electromagnetic currents. Three major components of $a_\mu^{\mathrm{HHO}}(\mathrm{lbl})$ can be identified: charged-pion loops, quark loops, and pseudoscalar (π^0, η, and η') pole diagrams. The latter ones dominate the final result and require information on the electromagnetic form factors of the pseudoscalars. In 2001 the authors of [42] uncovered a sign error in earlier evaluations of the dominating pion-pole part. Their estimate of $a_\mu^{\mathrm{HHO}}(\mathrm{lbl})$, based also on previous results for the quark and charged-pions loop parts [43], is $a_\mu^{\mathrm{HHO}}(\mathrm{lbl}) = 80\,(40) \times 10^{-11}$. A higher value was obtained in 2003 including short-distance QCD constraints: $a_\mu^{\mathrm{HHO}}(\mathrm{lbl}) = 136\,(25) \times 10^{-11}$ [44]. Further independent calculations would provide an important check of this result for $a_\mu^{\mathrm{HHO}}(\mathrm{lbl})$, a contribution whose uncertainty may become the ultimate limitation of the SM prediction of the muon $g-2$.

4 Standard Model vs. Measurement

The first column of Table 1 shows $a_\mu^{\mathrm{SM}} = a_\mu^{\mathrm{QED}} + a_\mu^{\mathrm{EW}} + a_\mu^{\mathrm{HLO}} + a_\mu^{\mathrm{HHO}}$. The values employed for a_μ^{HLO} are indicated by the reference on the left; all a_μ^{SM} values were derived with $a_\mu^{\mathrm{HHO}}(\mathrm{lbl}) = 80\,(40) \times 10^{-11}$ [42, 43]. Errors were added in quadrature. The differences $\Delta = a_\mu^{\mathrm{EXP}} - a_\mu^{\mathrm{SM}}$ are listed in the second column of Table 1, while the numbers of "standard deviations" (σ) appear in the third one. Lower discrepancies, shown in angle brackets, are obtained if $a_\mu^{\mathrm{HHO}}(\mathrm{lbl}) = 136\,(25) \times 10^{-11}$ [44] is used instead of $80\,(40) \times 10^{-11}$ [42,43].

5 Conclusions

The discrepancies between recent SM predictions of a_μ and the current experimental value vary in a very wide range, from 0.7 to 3.1 σ, according to the values chosen for the hadronic contributions. In particular, the leading-order hadronic contribution depends on which of the two sets of data, e^+e^- collisions or τ decays, are employed. The puzzling discrepancy between the $\pi^+\pi^-$ spectral functions from e^+e^- and isospin-breaking-corrected τ data could be caused by inconsistencies in the e^+e^- or τ data, or in the isospin-breaking corrections applied to the latter. Indeed, some disagreements occur between e^+e^- data sets, requiring further detailed investigations. On the other hand, the connection of τ data with the leading hadronic contribution to a_μ is less direct, and one wonders whether all possible isospin-breaking effects have been properly taken into account. Using e^+e^- data, the SM prediction of the muon $g-2$ deviates from the present experimental value by 2–3 σ.

As we mentioned in the Introduction, the impressive results of the E821 experiment are still limited by statistical errors, and there are proposals to reduce the present experimental uncertainty by a factor of 2.5 to about 0.2 ppm, or even less. While the QED and EW contributions appear to be ready to rival these precisions, much effort will be needed to reduce the hadronic uncertainty by a factor of two. This effort is challenging but possible, and certainly well motivated by the excellent opportunity the muon $g-2$ is providing us to unveil (or constrain) "new physics" effects.

Acknowledgments

Acknowledgments. I would like to thank the organizers for their excellent coordination of the 12th Lomonosov Conference on Elementary Particle Physics, and A. Studenikin for his kind invitation.

References

[1] H.N. Brown *et al.*, Phys. Rev. D **62** (2000) 91101; Phys. Rev. Lett. **86** (2001) 2227; G.W. Bennett *et al.*, ibid. **89** (2002) 101804; ibid. **89** (2002) 129903 (E).
[2] G.W. Bennett *et al.*, Phys. Rev. Lett. **92** (2004) 161802; hep-ex/0602035.
[3] A. Czarnecki and W.J. Marciano, Phys. Rev. D **64** (2001) 013014.
[4] R.M. Carey *et al.*, Proposal of the BNL Experiment E969, 2004; B.L. Roberts, hep-ex/0501012; hep-ex/0510056.
[5] J-PARC L.o.I. L17, B.L. Roberts contact person.
[6] M. Passera, J. Phys. G **31** (2005) R75; Nucl. Phys. Proc. Suppl. **155** (2006) 365; Proc. Sci. HEP2005 (2005) 305; M. Davier and W.J. Marciano, Ann. Rev. Nucl. Part. Sci. **54** (2004) 115; M. Knecht, hep-ph/0307239; A. Czarnecki and W.J. Marciano, Nucl. Phys. Proc. Suppl. **76** (1999) 245; V.W. Hughes and T. Kinoshita, Rev. Mod. Phys. **71** (1999) S133.

[7] J.S. Schwinger, Phys. Rev. **73** (1948) 416.
[8] T. Kinoshita and M. Nio, Phys. Rev. D **73** (2006) 053007.
[9] A.L. Kataev, these proceedings, hep-ph/0602098; hep-ph/0507174.
[10] R. Jackiw and S. Weinberg, Phys. Rev. D **5** (1972) 2396; I. Bars and M. Yoshimura, Phys. Rev. D**6** (1972) 374; G. Altarelli, N. Cabibbo and L. Maiani, Phys. Lett. B**40** (1972) 415; W.A. Bardeen, R. Gastmans and B. Lautrup, Nucl. Phys. B**46** (1972) 319; K. Fujikawa, B.W. Lee and A.I. Sanda, Phys. Rev. D **6** (1972) 2923.
[11] I.M. Ternov, V.N. Rodionov and A.I. Studenikin, Yad. Fiz. **37** (1983) 1270 [Sov. J. Nucl. Phys. **37** (1983) 755]; Annalen Phys. **46** (1989) 303; A.I. Studenikin, Zh. Eksp. Teor. Fiz. **97** (1990) 1407 [Sov. Phys. JETP **70** (1990) 795]; hep-ph/9808219; Yad. Fiz. **62** (1999) 2248 [Phys. Atom. Nucl. **62** (1999) 2071].
[12] A. Sirlin, Phys. Rev. D **22** (1980) 971.
[13] G. Degrassi, P. Gambino, M. Passera and A. Sirlin, Phys. Lett. B **418** (1998) 209; G. Degrassi and P. Gambino, Nucl. Phys. B **567** (2000) 3; A. Ferroglia, G. Ossola, M. Passera and A. Sirlin, Phys. Rev. D **65** (2002) 113002; M. Awramik, M. Czakon, A. Freitas and G. Weiglein, Phys. Rev. D **69** (2004) 053006; Phys. Rev. Lett. **93** (2004) 201805.
[14] T.V. Kukhto *et al.*, Nucl. Phys. B**371** (1992) 567.
[15] A. Czarnecki, B. Krause and W.J. Marciano, Phys. Rev. D **52** (1995) 2619; Phys. Rev. Lett. **76** (1996) 3267.
[16] S. Peris, M. Perrottet and E. de Rafael, Phys. Lett. B **355** (1995) 523; M. Knecht, S. Peris, M. Perrottet and E. de Rafael, JHEP **0211** (2002) 003.
[17] A. Czarnecki, W.J. Marciano and A. Vainshtein, Phys. Rev. D **67** (2003) 073006.
[18] G. Degrassi, G.F. Giudice, Phys. Rev. D **58** (1998) 053007.
[19] S. Heinemeyer, D. Stöckinger and G. Weiglein, Nucl. Phys. B**699** (2004) 103; T. Gribouk and A. Czarnecki, Phys. Rev. D **72** (2005) 53016.
[20] C. Bouchiat and L. Michel, J. Phys. Radium 22 (1961) 121; M. Gourdin and E. de Rafael, Nucl. Phys. B **10** (1969) 667.
[21] R.R. Akhmetshin *et al.* [CMD-2 Collaboration], Phys. Lett. B **578** (2004) 285.
[22] V.M. Aulchenko [CMD-2 Collaboration], JETP Lett. **82** (2005) 743; I. Logashenko, contribution to the HEP2005 International Europhysics Conference on High Energy Physics EPS, Lisboa, Portugal, July 21–27 2005, Proc. Sci. **HEP2005** (2005); contribution to the International Workshop "e^+e^- collisions from ϕ to ψ", February 27th – March 2nd 2006, Novosibirsk, Russia.
[23] M.N. Achasov *et al.*, J. Exp. Theor. Phys. **101** (2005) 1053.
[24] M.N. Achasov, contribution to the International Workshop "e^+e^- colli-

sions from ϕ to ψ", February 27th – March 2nd 2006, Novosibirsk, Russia.

[25] A. Aloisio *et al.* [KLOE Collaboration], Phys. Lett. B **606** (2005) 12.

[26] V. Druzhinin, these proceedings, hep-ex/0601020; M. Davier, Nucl. Phys. Proc. Suppl. **144** (2005) 238.

[27] B.A. Shwartz, Nucl. Phys. Proc. Suppl. **144** (2005) 245.

[28] G. Colangelo, Nucl. Phys. Proc. Suppl. **131** (2004) 185, and references therein.

[29] A. Höcker, hep-ph/0410081.

[30] F. Jegerlehner, Nucl. Phys. Proc. Suppl. **131** (2004) 213; J. Phys. G **29** (2003) 101.

[31] K. Hagiwara, A.D. Martin, D. Nomura and T. Teubner, Phys. Rev. D **69** (2004) 093003.

[32] J.F. de Trocóniz and F.J. Ynduráin, Phys. Rev. D **71** (2005) 073008.

[33] M. Davier, S. Eidelman, A. Höcker and Z. Zhang, Eur. Phys. J. C **31** (2003) 503.

[34] I.A. Koop *et al.*, Frascati Physics Series Vol. XVI (1999) 393; AIP Conf. Proc. **588** (2001) 319; B. Khazin and S. Eidelman, contributions to the International Workshop "e^+e^- collisions from ϕ to ψ", February 27th – March 2nd 2006, Novosibirsk, Russia.

[35] F. Ambrosino *et al.*, hep-ex/0603056.

[36] R. Alemany, M. Davier and A. Höcker, Eur. Phys. J. C **2** (1998) 123.

[37] M. Davier, A. Hocker and Z. Zhang, hep-ph/0507078 (Rev. Mod. Phys., to be published).

[38] K. Abe *et al.* [Belle Collaboration], arXiv:hep-ex/0512071; M. Fujikawa, contribution to the International Workshop "e^+e^- collisions from ϕ to ψ", February 27th – March 2nd 2006, Novosibirsk, Russia.

[39] W.J. Marciano and A. Sirlin, Phys. Rev. Lett. **61** (1988) 1815; V. Cirigliano, G. Ecker and H. Neufeld, Phys. Lett. B **513** (2001) 361; JHEP **0208** (2002) 002.

[40] S. Ghozzi and F. Jegerlehner, Phys. Lett. B **583**, 222 (2004); M. Davier, Nucl. Phys. Proc. Suppl. **131** (2004) 123; K. Maltman, Phys. Lett. B **633** (2006) 512; K. Maltman and C. E. Wolfe, Phys. Rev. D **73** (2006) 013004.

[41] B. Krause, Phys. Lett. B **390** (1997) 392.

[42] M. Knecht and A. Nyffeler, Phys. Rev. D **65** (2002) 73034; M. Knecht, A. Nyffeler, M. Perrottet and E. de Rafael, Phys. Rev. Lett. **88** (2002) 71802; A. Nyffeler, Acta Phys. Polon. B **34** (2003) 5197.

[43] M. Hayakawa and T. Kinoshita, Phys. Rev. D **57** (1998) 465; ibid. D **66** (2002) 019902(E); J. Bijnens, E. Pallante and J. Prades, Nucl. Phys. B **474** (1996) 379; ibid. B **626** (2002) 410.

[44] K. Melnikov and A. Vainshtein, Phys. Rev. D **70** (2004) 113006.

THE COMMENTS ON QED CONTRIBUTIONS TO $(G-2)_\mu$

A.L. Kataev [a], [b]
Institute for Nuclear Research, 117312 Moscow, Russia

*Abstract.*The comparison of definite numerical values of analytically evaluated asymptotic expressions for order α^4 and α^5 QED contributions to the muon anomalous magnetic moment with the results of numerical calculations are presented. It is stressed that observed agreement can be considered as the additional argument in favour of correctness of the recent direct numerical calculations.

1 Introduction

The calculations of perturbative QED contributions to the muon anomalous magnetic moments a_μ usually attract special interest. It is supported by the fact that for this low-energy quantity the interplay between pure QED contributions, electroweak (EW) and strong interactions effects plays important role. Moreover, the importance of taking into account strong interaction contributions to a_μ was clearly demonstrated by comparison of the most recent experimental data for a_μ [1] with available theoretical predictions (for a review see [2]). It reveals the appearance of definite, though not so clean, deviations of theoretical and experimental results. The existence of this deviation pushed ahead the desire to improve the knowledge about high order QED corrections to a_μ. As a result, the numerical calculations of the sum of 2958 most important five-loop diagrams, which depends on m_μ/m_e-ratio, were recently performed [3]

Here we concentrate ourselves on the comparison of the numerical results for order α^4 and α^5 contributions to a_μ, obtained in Refs. [4]- [9] from analytical high-loop contributions to the photon vacuum polarization function, with the existing results of direct numerical calculations.

2 The four-loop contributions

The renormalization group method plays an important role in the high-order perturbative calculations. As was shown in Ref. [10], the coefficients of the $\ln(m_\mu/m_e)$-terms in the expression for the vacuum polarization insertions into a_μ can be determined with the help of the renormalization group equations in the on-shell scheme. The general expression for the contributions to a_μ of the diagrams with dressed by electron loops internal photon line in the muon vertex reads

$$a_\mu = \frac{\alpha}{\pi}\int_0^1 dy(1-y)[d_R^\infty\left(\frac{-y^2}{1-y}\frac{m_\mu^2}{m_e^2},\alpha\right)-1] \tag{1}$$

where $d_R^\infty(\alpha,x)=[1+(\alpha/\pi)\Pi(\alpha,x)]^{-1}$ is the asymptotic photon propagator. Taking into account analytical expressions for the constant and logarithmic contributions to the photon vacuum polarization function in the on-shell scheme, one can obtain the number of asymptotic expressions for the diagrams, contributing to a_μ. The coefficients for the photon vacuum polarization function can be obtained using the renormalization group relations between QED β-function in the on-shell scheme and QED β-function in the momentum subtractions scheme which is known in the literature as the Ψ-function (see e.g. [11, 12]). The example of application of this formalism is the analytical calculation of the asymptotic limit of the subset of four-loop diagrams,

[a] e-mail: kataev@ms2.inr.ac.ru
[b] Supported in part by RFBR Grant N 05-01-00992

which contain in the photon line of the external muon vertex three-loop photon vacuum polarization contribution with two electron loops (**Subset 1**). The preliminary consideration of Ref. [4] (see [13]) was based on the application of analytical results for the four-loop contributions to the Ψ-function from the photon vacuum polarization diagram with three electron loops [14] (confirmed later on in [15]) and of the analogous four-loop corrections to the β-function in the on-shell scheme, taken from Ref. [16]. These considerations gave the following numerical contribution to a_μ [13] (see also [4])

$$a_\mu(\text{Subset 1}) = [0.923374... + O(\frac{m_e}{m_\mu})]\left(\frac{\alpha}{\pi}\right)^4 \tag{2}$$

which was in disagreement with the result of numerical calculations of Ref. [17], namely

$$a_\mu(\text{Subset 1}) = 1.4416(18)\left(\frac{\alpha}{\pi}\right)^4 \quad . \tag{3}$$

As was found in Ref. [5], this disagreement came from a theoretical error, which entered on-shell studies of Ref. [16]. It was corrected in Ref. [5] and the new asymptotic expression was obtained [4,5]:

$$a_\mu(\text{Subset 1}) = [1.452570... + O(\frac{m_e}{m_\mu})]\left(\frac{\alpha}{\pi}\right)^4 \quad . \tag{4}$$

It differs now from Eq.(3) in the third significant digit only. This discrepancy can be explained by the existence in Eq.(4) of the $O(m_\mu/m_e)$-corrections. The problem of comparing four-loop asymptotic and numerical contributions to a_μ appeared again in the process of consideration of the subset of diagrams, generated by the insertion into the internal photon line of the muon vertex three-loop vacuum polarization graphs in the "quenched" approximation, which do not contain internal loops. (**Subset 2**). Detailed calculational analysis of Ref. [8] gave the following numerical expression:

$$a_\mu(\text{Subset 2}) = [-0.290987... + O(\frac{m_e}{m_\mu})]\left(\frac{\alpha}{\pi}\right)^4 \quad . \tag{5}$$

while the numerical results of Ref. [17] were

$$a_\mu(\text{Subset 2}) = -0.7945(202)\left(\frac{\alpha}{\pi}\right)^4 \quad . \tag{6}$$

The disagreement was mostly removed in Ref. [18] after re-evaluating these diagrams using the integration VEGAS routine (the result of Eq.(6) was obtained using the integration routine RIWIAD). It was explained in Ref. [18] that the new number

$$a_\mu(\text{Subset 2}) = -0.2415(19)\left(\frac{\alpha}{\pi}\right)^4 \tag{7}$$

differed from the one of Eq.(6) due to severe underestimation of errors, which appeared in the process of the calculations of Ref [17]. Moreover, it is closer to the asymptotic expression of Eq.(5), though definite difference of order $O(m_e/m_\mu)$ still remained. This difference stimulated the authors of Ref. [9] to improve the precision of Eq.(5), combining the asymptotic [19,20] and threshold [21] results and applying the developed in Ref. [22] variant of the Padé resummation technique. After these improvements Eq.(5) moved in the direction of numerical value of Eq.(7). Indeed, the delicate calculation of Ref. [9] gave

$$a_\mu(\text{Subset 2}) = -0.230362(5)\left(\frac{\alpha}{\pi}\right)^4 . \tag{8}$$

Finally, in the work of Ref. [23] the following new result was obtained:

$$a_\mu(\text{Subset 2}) = -0.230596(416)\left(\frac{\alpha}{\pi}\right)^4 . \tag{9}$$

It is in perfect agreement with Eq.(8). The remaining difference between the results of Eq.(9) and Eq.(7) was traced to a problem discovered in Ref. [23] of rounding-off errors caused by keeping an insufficient number of effective digits in carrying out renormalization by numerical means. Thus the discovered in Ref. [8] discrepancy between the results of analytically-oriented and numerical calculations was finally eliminated.

3 The five-loop contributions

The calculations of the asymptotic contributions to a_μ from the diagrams with single dressed internal photon line was continued at the five-loop level in Refs. [6]- [8]. In particular, Ref. [6] shows that the coefficient of the subset of diagrams with internal four-loop light-by-light scattering graphs, composed from diagrams with two electron loops (**Subset 3**), is small

$$a_\mu(\text{Subset 3}) = [-\frac{\mathrm{a}_4^{[2,\mathrm{l-l}]}}{2} - 2.0237 + \mathrm{O}(\frac{\mathrm{m_e}}{\mathrm{m}_\mu})]\left(\frac{\alpha}{\pi}\right)^5 \tag{10}$$

where $a_4^{[2,l-l]}$ is the constant term of the logarithmically divergent sum of the four-loop light-by-light scattering contributions to the photon vacuum polarization function. The subset of the five-loop contributions to a_μ, formed by insertion into photon line of the sum of the four-loop "quenched" vacuum polarization graphs (**Subset 4**), gives the similar five-loop correction [7]

$$a_\mu(\text{Subset 4}) = [-\frac{\mathrm{a}_4^{[1]}}{2} - 0.7334 + \mathrm{O}(\frac{\mathrm{m_e}}{\mathrm{m}_\mu})]\left(\frac{\alpha}{\pi}\right)^5 \tag{11}$$

where $a_4^{[1]}$ is the constant term of the sum of the four-loop "quenched" vacuum polarization diagrams. The contribution to a_μ from the five-loop diagrams of **Subset 5**, which is generated by inserting into photon line of the four-loop vacuum polarization graphs with two electron loops, also contain still unknown constant terms [7]:

$$a_\mu(\text{Subset 5}) = [-\frac{\mathrm{a}_4^{[2]}}{2} + 0.6642 + \mathrm{O}(\frac{\mathrm{m_e}}{\mathrm{m}_\mu})]\left(\frac{\alpha}{\pi}\right)^5 . \tag{12}$$

The numerical calculations of the diagrams contributing to Eq.(10)-Eq.(12) are still unavailable. However, there are subsets of five-loop diagrams contributing to a_μ, which can be approximated by their asymptotic expressions and were numerically calculated recently in the process of the work of Ref. [3]. These are the diagrams of six concrete subsets. **Subset 6** is formed by insertion into photon line of the muon vertex four-loop vacuum polarization graphs with three electron loops. **Subset 7** consists of diagrams, generated by dressing this photon line by two two-loop vacuum polarization insertions with electron loops. The photon line of the diagrams of **Subset 8** contains three-loop photon vacuum polarization insertions with two electron loops and the additional one-loop bubble. **Subset 9** represents the sum of the diagrams, which have "quenched" three-loop and one-loop subsequent vacuum polarization insertions into photon line. Thus, **Subset 8** and **Subset 9** can be generated from the four-loop **Subset 1** and **Subset 2** by inserting into their dressed photon line the additional one-loop electron bubble. **Subset 10** consists from the five-loop diagrams with the photon line, dressed by two-loop photon vacuum polarization contribution and two electron loops. **Subset 11** is formed by inserting into photon line of four subsequent one-loop vacuum polarization contributions. It is necessary to emphasize, that all vacuum polarization subgraphs of the diagrams, considered in this report, are containing **electron** loops **only**. Let us compare the numerical analogs of analytical asymptotic five-loop contributions to a_μ, calculated in Ref. [7], with the concrete results [24] of direct numerical calculations of Ref. [3]. The value of the constant, related to **Subset 9**, was derived in the process of this work by more careful numerical studies of the analytical result of Ref. [7]. The asymptotic analytical expression of the **Subset 11** was confirmed in [25]. **Table 1.** The coefficients for several order $(\alpha/\pi)^5$ contributions to a_μ.

Subset of diagrams	asymptotic result	numerical result
a_μ (Subset 6)	$[-\frac{a_4^{[3]}}{2} + 2.8523 + \mathrm{O}(\frac{m_e}{m_\mu})]$ [7]	2.88598(9) [24]
a_μ (Subset 7)	$[4.8176... + \mathrm{O}(\frac{m_e}{m_\mu})]$ [7]	4.74212(14) [24]
a_μ (Subset 8)	$[7.4491... + \mathrm{O}(\frac{m_e}{m_\mu})]$	7.45270(88) [24]
a_μ (Subset 9)	$[-1.3314... + \mathrm{O}(\frac{m_e}{m_\mu})]$ [8]	-1.20841(70) [24]
a_μ (Subset 10)	$[27.7188... + \mathrm{O}(\frac{m_e}{m_\mu})]$ [7]	27.69038(30) [24]
a_μ (Subset 11)	$[20.1832... + \mathrm{O}(\frac{m_e}{m_\mu})]$ [7]	20.14293(23) [24]

We observe satisfactory agreement between the entries to the second and third column of Table 1. The existing difference can be explained by the effects of the $\mathrm{O}(m_e/m_\mu)$-corrections. Indeed, for the several subsets the five-loop numerical calculations Ref. [3] are in agreement with the related analytical expressions obtained in Ref. [26] both for the asymptotic and $\mathrm{O}(m_e/m_\mu)$-contributions. Among these subsets are **Subsets 7,10,11**. Thus, their asymptotic analytical results, calculated in Ref. [7], and confirmed in Ref. [26], really differ from the numerical ones by the leading $\mathrm{O}(m_e/m_\mu)$-terms. In the case of the five-loop diagrams from **Subset 9**, the $\mathrm{O}(m_e/m_\mu)$- terms should also be responsible for elimination of minor differences between asymptotic and numerical results. Indeed, since this statement is correct for the four-loop diagrams of **Subset 2** (compare Eq.(6) and Eq.(9) with Eq.(10)), it should also be correct in the case of five-loop diagrams from **Subset 9**, which have similar topological structure. A similar argument may also hold for the five-loop diagrams from **Subset 8**. The discussions presented in this summary provide additional support for the results of the numerical calculations [3] of the definite five-loop contributions to a_μ. However, the dominant contribution of evaluated 2958 diagrams is coming from the subset with light-by-light scattering electron loop internal insertion, which at present is difficult to check. Moreover, other 6122 diagrams still remained uncalculated. In spite of this we think that that the agreement of the new tenth-order QED contribution to a_μ, namely $663(20)(\alpha/\pi)^5$ [3], with the *preliminary* results of improved renormalization-group inspired estimates of the high-order corrections to a_μ [27], namely $\approx 658(\alpha/\pi)^5$ [28], is of **theoretical interest** and deserve further attention. However, the current difference between phenomenological and experimental results for a_μ can not be described by the five-loop corrections (for a review see [2]). They may become really important for the more precise a_μ experiments.

In conclusion I would like to to thank T. Kinoshita for useful discussions from 1990 to the present. I am grateful to S.L. Adler for his invitation to visit Institute for Advanced Study in Princeton, where this summary was written.

References

[1] G. W. Bennett *et al.*, Phys. Rev. Lett. **92**, 161802 (2004).
[2] M. Passera, J. Phys. G **31**, R75 (2005).
[3] T. Kinoshita and M. Nio, Phys. Rev. D **73** , 05053007 (2006).
[4] R. N. Faustov, A. L. Kataev, S. A. Larin and V. V. Starshenko, Phys. Lett. B **254**, 241 (1991).
[5] T. Kinoshita, H. Kawai and Y. Okamoto, Phys. Lett. B **254**, 235 (1991).
[6] A. L. Kataev, JETP Lett. **54**, 602 (1991)
[7] A. L. Kataev, Phys. Lett. B **284**, 401 (1992).
[8] D. J. Broadhurst, A. L. Kataev and O. V. Tarasov, Phys. Lett. B **298**, 445 (1993).
[9] P. A. Baikov and D. J. Broadhurst, in Proc. of 4th Int. Workshop AIHENP95, Pisa, Italy, 1995, p.167, hep-ph/9504398.
[10] B. Lautrup and E. De Rafael, Nucl. Phys. B **70**, 317 (1974).
[11] M. Gell-Mann and F. E. Low, Phys. Rev. **95**, 1300 (1954).
[12] N. N. Bogolyubov and D. V. Shirkov, "Introduction To The Theory Of Quantized Fields," Intersci. Monogr. Phys. Astron. **3**, 1 (1959).
[13] R. N. Faustov, A. L. Kataev, S. A. Larin and V. V. Starshenko, Preprint TIFR/TH/90-8.
[14] S. G. Gorishny, A. L. Kataev and S. A. Larin, Phys. Lett. B **194**, 429 (1987).
[15] S. G. Gorishny, A. L. Kataev, S. A. Larin and L. R. Surguladze, Phys. Lett. B **256** 81 (1991).
[16] J. Calmet and E. de Rafael, Phys. Lett. B **56**, 181 (1975).
[17] T. Kinoshita, B. Nizic and Y. Okamoto, Phys. Rev. D **41**, 593 (1990).
[18] T. Kinoshita, Phys. Rev. D **47**, 5013 (1993).
[19] S. G. Gorishny, A. L. Kataev and S. A. Larin, Nuovo Cim. A **92**, 119 (1986).
[20] S. G. Gorishny, A. L. Kataev and S. A. Larin, Phys. Lett. B **273**, 141 (1991) [Erratum-ibid. B **275**, 512 (1992); B **341**, 448 (1995)].
[21] B. H. Smith and M. B. Voloshin, Phys. Lett. B **324**, 117 (1994) [Erratum-ibid. B **333**, 564 (1994)].
[22] D. J. Broadhurst, J. Fleischer and O. V. Tarasov, Z. Phys. C **60**, 287 (1993).
[23] T. Kinoshita and M. Nio, Phys. Rev. D **60**, 053008 (1999).
[24] T. Kinoshita, private communication (2005).
[25] D. J. Broadhurst, Z. Phys. C **58**, 339 (1993).
[26] S. Laporta, Phys. Lett. B **328**, 522 (1994).
[27] A. L. Kataev and V. V. Starshenko, Phys. Rev. D **52**, 402 (1995).
[28] A. L. Kataev, Nucl. Phys. Proc. Suppl. B **155**, 369 (2006), hep-ph/0507174.

LOCALIZATION OF SCALAR AND FERMIONIC EIGENMODES IN SU(2) LATTICE GAUGE THEORY

M.I.Polikarpov [a], F.V.Gubarev [b], S.M.Morozov [c], S.V.Syritsyn [d], V.I.Zakharov [e]
Institute of Theoretical and Experimental Physics, B.Cheremushkinskaya 25, Moscow 117218, Russia

Abstract. We discuss the localization of eigenmodes of the covariant lattice Laplacian and overlap Dirac operator in lattice SU(2) gluodynamics. It occurs that the localization properties of the eigenmodes for fundamental scalars and fermions are different in confining configurations and in those from which confining disorder was eliminated by center-vortex removal. Localization of eigenmodes on subspaces of dimension less than four indicates the existence of the low dimensional structures in the confining vacuum.

1 Introduction

Study of the lattice Yang-Mills vacuum using eigenmodes of overlap Dirac operator and of covariant Laplacian operator is rather popular now; the very recent publications on this subject are [1–3]. Eigenvalues (eigenmodes) are gauge invariant (covariant) characteristics of the vacuum thus they can provide a useful information about confinement, topology and fermion condensate. Moreover it has been found that some eigenmodes are localized on low dimensional subspaces, thus indicating the existence of low dimensional objects in the confining field theory. The localization phenomenon is directly related to solid state Anderson localization [4] in a disordered medium. The material of this talk is mainly based on publications [5,6].

A natural measure of the localization of scalar field is provided by the inverse participation ratio (IPR) I_λ which is defined as follows (for review see for example Ref. [7]). Let $\rho_\lambda(x) = |\phi_\lambda(x)|^2$, $\sum_x \rho_\lambda(x) = 1$, where $\phi_\lambda(x)$ is an eigenmode of the covariant Laplacian in a given gauge field background, $-\Delta(A)\phi_\lambda = \lambda\,\phi_\lambda$. Then for any finite volume V the IPR is defined by

$$I_\lambda = V \sum_x \rho_\lambda^2(x)\,, \tag{1}$$

and characterizes the inverse fraction of sites contributing to the support of $\rho_\lambda(x)$. Note that for delocalized modes $\rho_\lambda(x) = 1/V$ and hence $I_\lambda = 1$, while an extremely localized mode, $\rho_\lambda(x) = \delta_{x,x_0}$ and $I_\lambda = V$. Moreover, for eigenmodes localized on a fraction f of sites (so that sup $\rho_\lambda = \mathrm{V_f} = \mathrm{f\ V}$) we have $I_\lambda = V/V_f = 1/f$. Thus if IPR for a given eigenmode is proportional to the system volume then this eigenmode is localized.

[a] e-mail: polykarp@itep.ru
[b] e-mail: gubarev@itep.ru
[c] e-mail: smoroz@itep.ru
[d] e-mail: syritsyn@itep.ru
[e] e-mail: xxz@mppmu.mpg.de

If we average IPR over gauge field configurations we have to allow for a mixture of both localized and extended modes and the average value of IPR is given by

$$\langle I_\lambda(\beta)\rangle = A(\beta) + B(\beta)V/V_f\,. \tag{2}$$

Usually [7], one considers localization only in terms of the total volume. Recently, there appeared data that the localization volume can depend on the lattice spacing as well. Namely, such effect was observed first in [8] for fermionic modes (for so called Asqtad fermions). The localization properties of scalar probe particles were investigated in Ref. [6] and, again, strong dependence on a was observed. For SU(3) gluodynamics with improved action and overlap Dirac operator the a dependence was found recently in [2]. Observed dependence on the lattice spacing seems specific for dynamics of Yang-Mills theories and allows to find the dimension, d, of the subspace where eigenmode is localized. At fixed physical volume we have $I_\lambda = V/\,V_f = Va^4/\,(V_f a^4) = a^{d-4}\;V^{phys}/\,V_f^{phys}$ and therefore the lattice spacing dependence is given by (*cf.* eq. (2)):

$$\langle I_\lambda\rangle \;=\; A + B\cdot a^{d-4}\,. \tag{3}$$

Below we discuss the eigenmodes in the confining and in the deconfining vacuum. The deconfining gauge field configuration can be obtained introducing finite temperature, instead we study the so-called vortex-removed configurations. Recall that the vortex-removed configurations [9] are obtained by multiplying the lattice configuration in maximal center gauge by the center-projected configuration. This is known to remove the confinement properties of the lattice configuration (the asymptotic string tension is zero), but is otherwise a rather minimal disturbance of the original lattice, since plaquette values away from P-vortex locations are left unchanged. Locally, the vortex-removed configuration is actually somewhat more disordered than the original configuration.

2 Fundamental scalars

2.1 Localization volume

The covariant lattice Laplacian in the fundamental representation of the gauge group is given by

$$\triangle^{ab}_{xy} = \sum_\mu \left[U^{ab}_\mu(x)\delta_{y,x+\hat\mu} + U^{\dagger ab}_\mu(x-\hat\mu)\delta_{y,x-\hat\mu} - 2\delta^{ab}\delta_{xy}\right], \tag{4}$$

where a, b are color indices. We have calculated the average IPR of the lowest-lying eigenmode at several lattice couplings β, and plot in Fig. 1 (left) the subtracted IPR data vs. the lattice volume V in physical units, here $V[\mathrm{fm}^4] = Va^4(\beta) = (La(\beta))^4$ is the lattice volume in physical units, $a = a(\beta)$ is the lattice

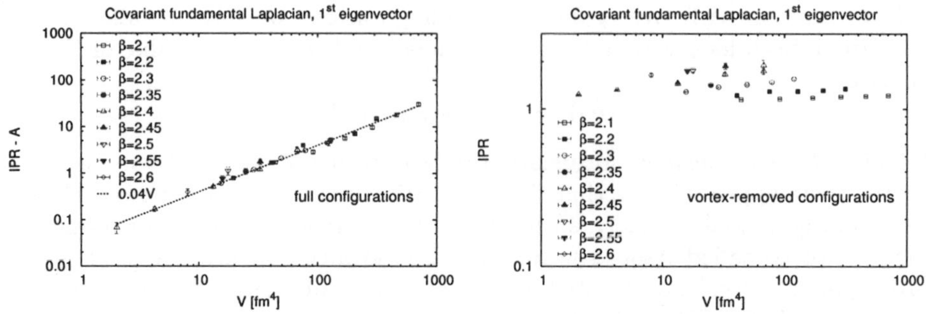

Figure 1: (left) Log-log plot of $IPR - A$ vs. lattice volume in physical units (fm^4). The dotted line is a best fit to linear dependence on volume. (right) Same as in the left figure for the configurations with vortices removed. In this case the data is consistent with extended, rather than localized, eigenmodes.

spacing (in fm) at coupling β. By the "subtracted" IPR data we refer to our data for $I_0(\beta)$ minus the constant $A(\beta)$ obtained from the best fit to eq. (2). We find that the IPR values at different β fall roughly on the same straight line, which implies that the localization 4-volume is constant in physical units, $V_f \approx (2fm)^4$, cf. eq. (2).

2.2 Removing of center vortices

The next question is whether the localization we have found is due specifically to the confining disorder of the lattice configuration, or whether it is due simply to local stochastic fluctuations in the link variables. To address this question, we have computed the low-lying eigenmode spectrum in the center-projected, and vortex-removed configurations. Fig. 1 (right) shows that the lowest eigenmode computed in the center-projected (or "vortex-only") configuration is localized much as in the original lattice. The vortex-removed data, however, is drastically different. In Fig. 1 (right), we see that that the IPRs of the vortex-removed configurations remain essentially constant as L increases, and therefore localization is completely absent. The obvious conclusion is that localization is associated with large-scale, confining disorder provided by center vortices, rather than local, non-confining fluctuations of the link variables.

2.3 Mobility edge

So far we have considered only the lowest eigenmode. For the higher modes, we find at any fixed volume that the I_n decreases (b increases) as the eigenvalue λ_n increases, until there is no localization at all. For the non-localized states, we found that the IPR are small (O(1)) even on large lattices, and do not increase with lattice size. This situation is familiar in the Anderson model [4], where eigenmodes at the upper and lower ends of the spectrum are localized, and eigenmodes in the bulk of the spectrum are extended. The boundaries between localized and extended states are known as "mobility edges" [10].

2.4 Higher representations

We have also calculated the low-lying eigenmode spectrum of the covariant Laplacian for the adjoint and $j = 3/2$ representation of the SU(2) gauge group. We find that the low-lying modes are strongly localized for adjoint representation than for fundamental, and they are stronger localized for the $j = 3/2$ representation than than for adjoint. The data for IPR falls on a single line when plotting $IPR \times a^2$, $(IPR \times a^4)$ vs. lattice volume $(La)^4$ in physical units for adjoint $(j = 3/2)$ representation. Additional calculations show that the localization region for adjoint representation is in fact point-like rather than surface-like, with a localization volume that shrinks to zero in physical units in the continuum limit. The data for $j = 3/2$ representation suggests that the localization region is always on the scale of the lattice spacing, regardless of the length of the lattice spacing in physical units.

Adjoint representation is by definition unaffected by the vortex removing procedure, thus IPR for all modes in adjoint representation for vortex removed configurations is the same as in the original vacuum. The $j = 3/2$ representation of SU(2) has the same n-ality as the fundamental representation. However, unlike the fundamental representation, vortex removal has essentially no effect on localization of the low-lying $j = 3/2$ eigenmodes [6].

3 Fermions

3.1 Localization properties

We use the massless overlap Dirac operator[f] [11]:

$$D = \frac{\rho}{a}\left(1 + \frac{A}{\sqrt{AA^\dagger}}\right), \quad A = D_W - \frac{\rho}{a}, \tag{5}$$

where A is the Wilson Dirac operator with negative mass term. Anti-periodic (periodic) boundary conditions in time (space) directions were employed. It turns out that for SU(2) gluodynamics the optimal value of ρ parameter is 1.4. Furthermore, we have used the minmax polynomial approximation [12] to compute the sign function $\mathrm{sign}(A) = A/\sqrt{AA^\dagger} \equiv \gamma_5\,\mathrm{sign}(H)$, where $H = \gamma_5 A$ is hermitian Wilson Dirac operator. In order to improve the accuracy and

[f] The invaluable assistance of G. Schierholz and T. Streuer in the overlap operator implementation is kindly acknowledged.

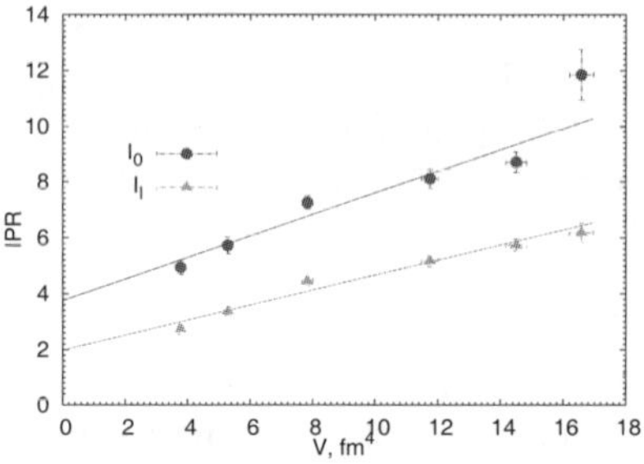

Figure 2: Volume dependence of I_0 and I_l at fixed lattice spacing $a = 0.1394(8)$ ($\beta = 2.35$).

performance about one hundred lowest eigenmodes of H were projected out. Note that the eigenvalues of (5) lie on the circle of radius ρ centered at $(\rho, 0)$ in the complex plane. In order to relate them with continuous eigenvalues of the Dirac operator the circle was stereographically projected onto the imaginary axis [13]. The information about statistics and number of gauge field configurations used is given in [5].

The localization properties of eigenmodes of the overlap Dirac operator are characterized by IPR (1), where $\rho_\lambda(x) = \psi_\lambda^\dagger(x)\psi_\lambda(x)$, $\sum_x \rho_\lambda(x) = 1$, and $\psi_\lambda(x)$ is an eigenmode of the overlap Dirac operator in a given gauge field background with virtuality λ, $D\,\psi_\lambda = \lambda\,\psi_\lambda$. In Fig. 2 we plot IPR of zero eigenmodes (I_0) and of low-lying eigenmodes, the later is defined as: $I_l = \sum_\lambda \langle I_\lambda \rangle / N_l$ where the sum is over $0 < \lambda \leq 50$ MeV, N_l is the total number of modes in this interval. Fitting the data to Eq. (2) we found for exact zero modes $A = 3.7(5)$, $B = 0.38(5)$ with $\chi/n.d.f = 3$, and for low-lying modes $A = 1.9(2)$, $B = 2.26(3)$ with $\chi/n.d.f = 2$. The proportionality of I_0 and I_l to lattice volume, V, indicates that the low lying and zero mode are localized.

3.2 Removing of center vortices

As in case of scalars the vortex-removing procedure [9] drastically changes the properties of the vacuum: confinement and chiral condensate disappear. Moreover topological susceptibility, $< Q^2/V >$, vanishes after removing of the center vortices. The topological charge of the configuration was defined by the index of overlap Dirac operator $Q = n_+ - n_-$, and $n_\pm$ is the number of exact zero modes with positive (negative) chirality. On Fig. 3 we show the quantity $< \pi\rho(\lambda) >^{\frac{1}{3}}$ which is related to chiral condensate through the Banks-Casher relation [14], $\langle \bar{\psi}\psi \rangle = -\pi\rho(\lambda_n \to 0)$, where $\rho(\lambda_n \to 0)$ is the density of the (delocalized) zero modes in the limit of infinite volume V. For finite lattice volumes these modes are near-zero and this is the reason of the limit $\lambda_n \to 0$. It is seen that for the vortex removed configurations there exist the gap in the

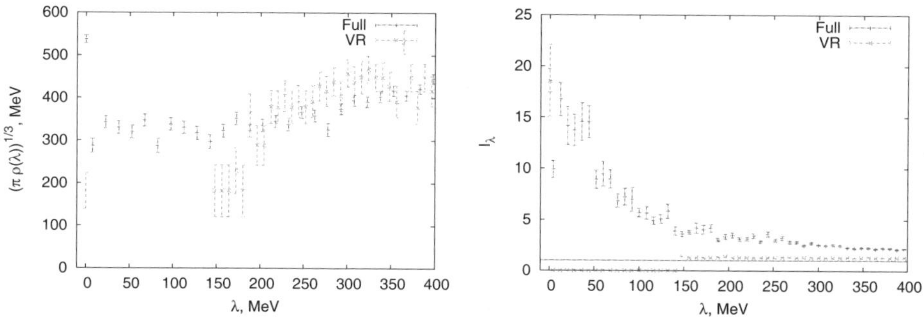

Figure 3: (left) $< \pi\rho(\lambda) >^{\frac{1}{3}}$ for full and vortex removed (VR) configurations. (right) IPR for full and vortex removed (VR) configurations. The averages are taken over 200 gauge field configurations at $\beta = 2.5$ on lattice $16^3 \cdot 18$.

distribution of the near-zero eigenvalues, thus chiral condensate vanishes for vortex removed gauge fields.

On Fig. 3 (right) IPR for full and vortex removed configurations is shown. After vortex removing all modes (except of a few surviving zero modes) become delocalized, $(IPR) \approx 1$. Thus, we can conclude that the non-trivial localization properties of the eigenfunctions of the Dirac operator are directly related to the center vortices. Note that similar observations were made by using other types of lattice fermions, see in particular [15] and references therein.

Another qualitative feature seen in Fig. 3 (right) is the existence of a kind of a 'merging point', $\lambda_{merge} \approx 150 - 200\,\text{MeV}$. Namely, for $\lambda_n > \lambda_{merge}$ the values of IPR practically do not depend on the eigenvalue λ_n. The data for various values of the lattice spacing indicate scaling in physical units [5], at least an approximate one, of the position of the merging point. Just the similar effect was observed in the data on localization of a test fundamental color scalar. If one approaches the mobility edge from above, then the IPR blows up as $(\lambda_n - \lambda_{mob})^{-\alpha}$ where $\alpha > 0$. Note that this simple dependence holds for a fixed, large total volume V_{tot} and the singularity is smoothened due to finiteness of the total volume. Now, for chiral fermions one expects that the mobility edge is $\lambda_{mob} = 0$. Then, the growth of the IPR with $\lambda_n \to 0$ indicated by our data is reminiscent of existence of the critical exponent α in the non-relativistic case.

3.3 Lattice spacing dependence of IPR and fine tuning

The most striking feature exhibited by the data is a strong dependence of the IPR at fixed physical volume, V^{phys}, on the lattice spacing a. The data were fitted [5] by eq. (3). The IPR data for low lying modes strongly suggest that the dimensionality of localization region is zero, $d = 0$; the relevant $\chi^2/n.d.f.$

is 2.3 which should be compared with its values 3.4 ($d = 1$) and 4.9 ($d = 2$). In other words, the volume occupied by the low lying eigenfunctions tends to zero in physical units as a^4. The exact zero modes seems to be localized in one dimensional ($d = 1$) regions; $\chi^2/n.d.f.$ in this case is 0.38 while it is 0.95 for $d = 0$ and 0.85 for $d = 2$.

Note that the a dependence of the IPR was suggested first by the data of Ref. [8] where dimension d between 2 and 3 was found for lowest-lying fermionic modes for Asqtad fermions[g]. For SU(3) gluodynamics for overlap fermions the dimension of the localization region was found in Ref. [2], the result is $d = 2$ for zero modes and $d = 3$ for the lowest-lying modes. At this moment, it is difficult to clarify the reason for the difference between these and our results. On the other hand the most important fact is the a dependence of IPR which indicates that there exist lower dimensional submanifolds in QCD vacuum. The same statement was done in Refs. [16] on the base of analysis of the topological charge density distribution. The strong a dependence of the IPR is in sharp contrast with scaling, in physical units, of the density of states and of the chiral condensate (defined through the Casher-Banks relation). Thus, the emerging picture is that the size of the eigenfunctions shrinks strongly with $a \to 0$ while the eigenvalues remain stable. In other words, the data suggest that the phenomenon of fine tuning is quite a common feature of the Yang-Mills dynamics [h]. Note that the phenomenon of the fine tuning was observed first by studying the properties of the lattice monopoles and vortices, see, e.g., [17]. However the definition of the monopoles and vortices involves projected fields which makes theoretical interpretation more difficult. The fine tuning seen in the data on the fermion localization is defined in explicitly gauge invariant terms.

Acknowledgments

Numerous discussions with J. Greensite and S. Olejnik on the topics of this review are gratefully acknowledged. This work was partially supported by grants RFBR-05-02-16306a, RFBR-0402-16079 and EU Integrated Infrastructure Initiative Hadron Physics (I3HP) under contract RII3-CT-2004-506078 (MIP); RFBR-05-02-16306a and RFBR-05-02-17642 (S.M.M.).

References

[1] S. Schaefer and T. DeGrand, [hep-lat/0508025]; B. Svetitsky, [hep-lat/0511003]; B. Svetitsky, Y. Shamir and M. Golterman, [hep-

[g] The recent calculations [3] give dimension 3 or 3.5.

[h] The standard definition of the fine tuning, which goes back to the Higgs physics, is as follows. If a relativistic system has a typical size r_0 then the energy levels of order $\lambda \sim 1/r_0$ are "natural". If, on the other hand, $\lambda \ll 1/r_0$, the level is "fine tuned".

lat/0508015];
F. Bruckmann and E.-M. Ilgenfritz, [hep-lat/0509020];
S. Solbrig et al., [hep-lat/0509052];
J. Greensite, Š. Olejník, M.I. Polikarpov, S.N. Syritsyn and V.I. Zakharov, Proc.Sci.LAT2005, 325 (2005);
C. Gattringer and S. Solbrig, Proc.Sci.LAT2005 (2005) 127;
F. Bruckmann and E-M. Ilgenfritz, Proc.Sci.LAT2005, 305 (2005);
M.I. Polikarpov, F.V. Gubarev, S.M. Morozov and V.I. Zakharov, [hep-lat/0510098].

[2] Y. Koma, E.-M. Ilgenfritz, K. Koller, G. Schierholz, T. Streuer, V. Weinberg, Proc.Sci.LAT2005, 300 (2005).

[3] C. Bernard et al., [hep-lat/0510025].

[4] P. Anderson, Phys. Rev. 109, 1492 (1958).

[5] F.V. Gubarev, S.M. Morozov, M.I. Polikarpov and V.I. Zakharov, JETP Lett. 82, 343 (2005).

[6] J. Greensite, Š. Olejník, M.I. Polikarpov, S.N. Syritsyn and V.I. Zakharov, Phys. Rev. D71, 114507 (2005).

[7] B. Kramer, A. MacKinnon, Rep.Prog.Phys. 56, 1469 (1993);
C. Gattringer et al., Nucl. Phys. B617, 101 (2001);
T. Kovacs, Phys. Rev. D67, 094501 (2003).

[8] C. Aubin *et al.* [MILC Collaboration], [hep-lat/0410024].

[9] P. de Forcrand and M. D'Elia, Phys. Rev. Lett. 82, 4582 (1999).

[10] N. Mott, J. Non-Cryst. Solids 1, 1 (1968).

[11] H. Neuberger, Phys. Lett. B417, 141 (1998); Phys. Lett. B427, 353 (1998).

[12] L. Giusti, C. Hoelbling, M. Luscher and H. Wittig, Comput. Phys. Commun. 153, 31 (2003).

[13] T.-W. Chiu, S.V. Zenkin, Phys.Rev. D59, 074501 (1999); T.-W. Chiu, C.-W. Wang, S.V. Zenkin, Phys.Lett. B438, 321 (1998).

[14] T. Banks and A. Casher, Nucl. Phys. B169, 103 (1980).

[15] J. Gattnar et al., Nucl. Phys. B716, 105 (2005).

[16] A. Alexandru, I. Horvath and J.B. Zhang, Phys.Rev. D72 (2005) 034506; I. Horvath et al., Phys.Lett. B617, 49 (2005); I. Horvath et al., Phys.Lett. B612, 21 (2005); I. Horvath Nucl.Phys. B710, 464 (2005), Erratum-ibid. B714, 175 (2005); I. Horvath et al., Phys.Rev. D68, 114505 (2003); I. Horvath et al., Phys.Rev. D67, 011501 (2003).

[17] F.V. Gubarev, A.V. Kovalenko, M.I. Polikarpov, S.N. Syritsyn and V.I. Zakharov, Phys. Lett. B574, 136 (2003); A.V. Kovalenko, M.I. Polikarpov, S.N. Syritsyn and V.I. Zakharov, Phys. Rev. D71, 054511 (2005);
V. I. Zakharov, Phys. Usp. 47, 37 (2004), Usp. Fiz. Nauk 47, 39 (2004).

RESUMMATION OF LARGE LOGARITHMS WITHIN THE METHOD OF EFFECTIVE CHARGES

C.J. Maxwell

Institute for Particle Physics Phenomenology (IPPP) ,University of Durham, Durham, DH1 3LE UK

*Abstract.*We show how the resummation of large logarithms can be incorporated into the method of effective charges. As an example this approach is applied to the event shape variables thrust and heavy-jet mass in e^+e^- annihilation. Whilst there are some problems with the behaviour of the effective charge in the two-jet region, smaller power corrections are required to fit the data compared to the standard approach.

In this talk I will describe some recent work together with Michael Dinsdale concerning the relative size of non-perturbative power corrections for QCD event shape observables [1]. For e^+e^- event shape *means* the DELPHI collaboration have found in a recent analysis that if the next-to-leading order (NLO) perturbative corrections are evaluated using the method of Effective Charges [2] then one can obtain excellent fits to data without including any power corrections [3]. In contrast fits based on the use of standard fixed-order perturbation theory in the $\overline{MS}$ scheme with a physical choice of renormalization scale equal to the c.m. energy, require additional power corrections C_1/Q with $C_1 \sim 1$ GeV. Power corrections of this size are also predicted in a model based on an infra-red finite coupling [4] , which is able to fit the data reasonably well in terms of a single parameter. Given the DELPHI result it is interesting to consider how to extend the method of Effective Charges to event shape *distributions* rather than means.

Consider an e^+e^- observable $\mathcal{R}(Q)$, e.g. an event shape observable- thrust or heavy-jet mass, Q being the c.m. energy.

$$\begin{aligned}\mathcal{R}(Q) &= a(\mu, \mathrm{RS}) + \mathrm{r}_1(\mu/\mathrm{Q}, \mathrm{RS})\mathrm{a}^2(\mu, \mathrm{RS}) + \\ &+ r_2(\mu/Q, \mathrm{RS})\mathrm{a}^3(\mu, \mathrm{RS}) + \cdots, \end{aligned} \tag{1}$$

Here $a \equiv \alpha_s/\pi$. Normalised with the leading coefficient unity, such an observable is called an *effective charge.* The couplant $a(\mu, \mathrm{RS})$ satisfies the beta-function equation

$$\frac{da(\mu, \mathrm{RS})}{d\ln(\mu)} = \beta(a) = -ba^2(1 + ca + c_2a^2 + c_3a^3 + \cdots) . \tag{2}$$

Here $b = (33 - 2N_f)/6$ and $c = (153 - 19N_f)/12b$ are universal, the higher coefficients c_i, $i \geq 2$, are RS-dependent and may be used to label the scheme, together with dimensional transmutation parameter Λ [5]. The *effective charge*

$\mathcal{R}$ satisfies the equation

$$\frac{d\mathcal{R}(Q)}{d\ln(Q)} = \rho(\mathcal{R}(Q)) = -b\mathcal{R}^2(1 + c\mathcal{R} + \rho_2\mathcal{R}^2 + \rho_3\mathcal{R}^3 + \cdots \,. \tag{3}$$

This corresponds to the beta-function equation in an RS where the higher-order corrections vanish and $\mathcal{R} = a$, the beta-function coefficients in this scheme are the RS-invariant combinations

$$\begin{aligned} \rho_2 &= c_2 + r_2 - r_1 c - r_1^2 \\ \rho_3 &= c_3 + 2r_3 - 4r_1r_2 - 2r_1\rho_2 - r_1^2 c + 2r_1^3 \,.\,. \end{aligned} \tag{4}$$

Eqn.(3) for $d\mathcal{R}/d\ln Q$ can be integrated to give

$$\begin{aligned} b\ln\frac{Q}{\Lambda_{\mathcal{R}}} &= \frac{1}{\mathcal{R}} + c\ln\left[\frac{c\mathcal{R}}{1+c\mathcal{R}}\right] + \\ &+ \int_0^{\mathcal{R}(Q)} dx \left[\frac{b}{\rho(x)} + \frac{1}{x^2(1+cx)}\right] \end{aligned} \tag{5}$$

The dimensionful constant $\Lambda_{\mathcal{R}}$ arises as a constant of integration. It is related to the dimensional transmutation parameter $\tilde{\Lambda}_{\overline{MS}}$ by the exact relation,

$$\Lambda_{\mathcal{R}} = e^{r/b}\tilde{\Lambda}_{\overline{MS}} \,, \tag{6}$$

Here $r \equiv r_1(1, \overline{MS})$ with $\mu = Q$, is the NLO perturbative coefficient.

Eq.(5) can be used to convert the measured data for the observable $\mathcal{R}$ into a value of $\Lambda_{\overline{MS}}$ bin-by-bin. Such an analysis was carried out in Ref. [6] for a number of e^+e^- event shape observables, including thrust and heavy jet mass which we shall focus on here. It was found that the fitted Λ values exhibited a clear plateau region, away from the two-jet region, and the region approaching $T = 1/3$ where the NLO thrust distribution vanishes. Event shape distributions for thrust (T) or heavy-jet mass (ρ_h) contain large kinematical logarithms, $L = \ln(1/y)$, where $y = (1 - T)$, $\rho_h, \cdots$. These must be resummed to all-orders in the two-jet region.

$$\frac{1}{\sigma}\frac{d\sigma}{dy} = A_{LL}(aL^2) + L^{-1}A_{NLL}(aL^2) + \cdots \,. \tag{7}$$

Here LL, NLL, denote leading logarithms, next-to-leading logarithms, etc. For thrust and heavy-jet mass the distributions *exponentiate* [7]

$$\begin{aligned} R_y(y') &\equiv \int_0^{y'} dy \frac{1}{\sigma}\frac{d\sigma}{dy} = C(a\pi)\exp(Lg_1(a\pi L) \\ &+ g_2(a\pi L) + ag_3(a\pi L) + + \cdots) + D(a\pi, y) \end{aligned} \tag{8}$$

Here g_1 contains the LL and g_2 the NLL.$C = 1 + O(a)$ is independent of y and D contains terms that vanish as $y \to 0$. It is natural to define an effective charge $\mathcal{R}(y')$ so that

$$R_y(y') = \exp(r_0(y')\mathcal{R}(y')) . \tag{9}$$

This effective charge will have the expansion

$$r_0(L)\mathcal{R}(L) = r_0(L)(a + r_1(L)a^2 + r_2(L)a^3 + \cdots) \tag{10}$$

Here $r_0(L) \sim L^2$, and the higher coefficients $r_n(L)$ have the structure

$$r_n = r_n^{\rm LL} L^n + r_n^{\rm NLL} L^{n-1} + \cdots \tag{11}$$

Usually one resums these logarithms to all-orders using the known closed-form expressions for $g_1(aL)$ and $g_2(aL)$, where a is taken to be the $\overline{MS}$ coupling with a "physical" scale choice $\mu = Q$ ($\overline{MS}$PS). Instead we want to resum logarithms to all-orders in the $\rho(\mathcal{R})$ function (ECH). The form of the ρ_n RS-invariants (Eq.(4)) means that the ρ_n have the structure

$$\rho_n = \rho_n^{\rm LL} L^n + \rho_n^{\rm NLL} L^{n-1} + \cdots . \tag{12}$$

One can then define all-orders RS-invariant LL and NLL approximations to $\rho(\mathcal{R})$,

$$\begin{aligned} \rho_{\rm LL}(\mathcal{R}) &= -b\mathcal{R}^2(1 + c\mathcal{R} + \sum_{n=2}^{\infty} \rho_n^{\rm LL} L^n \mathcal{R}^n) \\ \rho_{\rm NLL}(\mathcal{R}) &= -b\mathcal{R}^2(1 + c\mathcal{R} + \sum_{n=2}^{\infty} (\rho_n^{\rm LL} L^n \\ &+ \rho_n^{\rm NLL} L^{n-1})\mathcal{R}^n) . \end{aligned} \tag{13}$$

The resummed $\rho_{\rm NLL}(\mathcal{R})$ can then be used to solve for $\mathcal{R}_{\rm NLL}$ by inserting it in Eq.(5). Notice that since $\Lambda_{\mathcal{R}}$ involves the *exact* value of $r_1(1, \overline{MS})$ there is no matching problem as in the standard $\overline{MS}$PS approach. The resummed $\rho_{LL}(\mathcal{R})$ can be straightforwardly numerically computed using

$$\rho_{\rm LL}(x) = \beta(a)\frac{d\mathcal{R}_{\rm LL}}{da} = -ba^2\frac{d\mathcal{R}_{\rm LL}}{da} , \tag{14}$$

with a chosen so that $\mathcal{R}_{\rm LL}(a) = x$. The same relation with $\beta(a) = -ba^2(1+ca)$ suffices for $\rho_{NLL}(\mathcal{R})$, although in this case one needs to remove $NNLL$ terms, e.g. an L^0 term which would otherwise be included in ρ_2. This can be accomplished by numerically taking limits $L \to \infty$ with $L\mathcal{R}$ fixed.

A crucial feature of the effective charge approach is that it resums to all-orders *RG-Predictable* pieces of the higher-order coefficients, thus the NLO

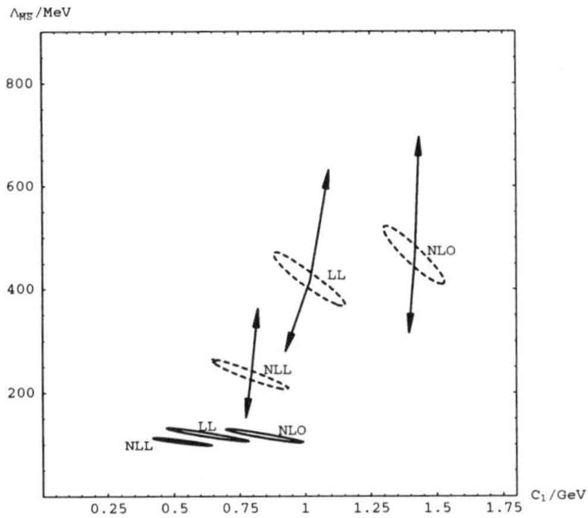

Figure 1: Fits to the thrust distribution for $\Lambda_{\overline{MS}}$ and C_1. Solid ellipses use ECH, dashed ellipses $\overline{MS}$PS (with the arrows showing the effect on the central value of varying $Q/2 < \mu < 2Q$). The ellipses indicate 2σ errors generated by allowing χ^2 to vary within 4 of its minimum.

ECH result (assuming $c = 0$ for simplicity) corresponds to an RS-invariant resummation

$$a + r_1 a^2 + r_1^2 a^3 + \cdots + r_1^n a^{n+1} + \cdots \,. \tag{15}$$

Thus even at fixed-order without any resummation of large logs in $\rho(\mathcal{R})$ a *partial* resummation of large logs is automatically performed. Furthermore one might expect that the LL ECH result contains already NLL pieces of the standard $\overline{MS}$PS result.

We now turn to fits simultaneously extracting $\Lambda_{\overline{MS}}$ and the size of power corrections C_1/Q from the data. To facilitate this we use the result that inclusion of power corrections effectively shifts the event shape distributions, which can be motivated by considering simple models of hadronization, or through a renormalon analysis [8]. Thus we define

$$R_{PC}(y) = R_{PT}(y - C_1/Q) \,. \tag{16}$$

This shifted result is then fitted to the data for 1-thrust and heavy jet mass. e^+e^- data spanning the c.m. energy range from $44 - 189$ GeV was used (see [1] for the complete list of references). The resulting fit for 1-thrust is shown in Figure 1.

The ECH fits for thrust and heavy jet mass show great stability going from NLO to LL to NLL, presumably because at each stage a partial resummation of higher logs is automatically performed. The power corrections required with ECH are somewhat smaller than those found with $\overline{MS}$PS, but we do not find as dramatic a reduction as DELPHI find for the means. This may be because their analysis corrects the data for bottom quark mass effects which we have ignored. The fitted value of $\Lambda_{\overline{MS}}$ for ECH is much smaller than that found with $\overline{MS}$PS, ($\alpha_s(M_Z) = 0.106$ (thrust) and 0.109 (heavy-jet mass)). Similarly small values are found with the Dressed Gluon Exponentiation (DGE) approach [9]. A problem with the Effective Charge resummations is that the $\rho(\mathcal{R})$ function contains a branch cut which limits how far into the 2-jet region one can go. We are limited to $1 - T > 0.05M_Z/Q$ in the fits we have performed. This branch cut mirrors a corresponding branch cut in the resummed $g_1(aL)$ function. Similarly as $1 - T$ approaches $2/3$ the leading coefficient $r_0(L)$ vanishes and the Effective Charge formalism breaks down. We need to restrict the fits to $1-T < 0.18$. From the "RG-predictability" arguments we might expect that these difficulties would also become apparent for a NNLL $\overline{MS}$PS resummation.

Acknowledgements

Alexander Studenikin is thanked for his painstaking organization of this stimulating and enjoyable meeting.

References

[1] M.J. Dinsdale and C.J. Maxwell, Nucl. Phys. **B713** (2005) 465.
[2] G. Grunberg, Phys. Lett. **B95** (1980) 70; Phys. Rev. **D29** (1984) 2315.
[3] DELPHI Collaboration (J. Abdallah et. al.) Eur. Phys. J. **C29** (2003) 285.
[4] Y.L. Dokshitzer and B.R. Webber, Phys. Lett. **B404** (1997) 321.
[5] P.M. Stevenson, Phys. Rev. **D23** (1981) 2916
[6] S.J. Burby and C.J. Maxwell, Nucl. Phys. **B609** (2001) 193.
[7] S. Catani, G. Turnock, B.R. Webber and L. Trentadue, Phys. Lett. **B263** (1991) 491.
[8] B.R. Webber [hep-ph/9411384].
[9] E. Gardi and J. Rathsman, Nucl. Phys. **B638** (2002) 243.

GENERALIZED DUAL SYMMETRY OF NONABELIAN THEORIES, MONOPOLES AND DYONS

C.R.Das [a]
The Institute of Mathematical Sciences, Chennai, India
L.V.Laperashvili [b]
The Institute of Theoretical and Experimental Physics, Moscow, Russia
H.B.Nielsen [c]
The Niels Bohr Institute, Copenhagen, Denmark

Abstract. We consider nonabelian $SU(N) \times \widetilde{SU(N)}$ theories, revealing the generalized dual symmetry. Beta-functions are investigated. It is shown that they are antisymmetric under the interchange $\alpha \leftrightarrow 1/\alpha$, and have zero at $\alpha = 1$. Monopoles, or dyons, are responsible for the phase transition. Considering critical points at $\alpha_1 \approx 0.4$ and $\alpha_2 \approx 2.5$, we give an explanation of the freezing of QCD α_s.

1 Renormalization group equations (RGEs) and duality

In the last years gauge theories essentially operate with the fundamental idea of duality. In the present investigation we considered $SU(N) \times \widetilde{SU(N)}$ nonabelian theories, describing a system of fields with non-dual g and dual $\tilde{g}$ charges, in terms of loop variables.

Following the idea by Zwanziger [1] to describe symmetrically non–dual and dual abelian fields A_μ and $\tilde{A}_\mu$, covariantly interacting with electric $j_\mu^{(e)}$ and magnetic currents $j_\mu^{(m)}$, respectively, we suggested to construct the generalized Zwanziger–type action for the pure nonabelian gauge theories. Such a theory shows the invariance under the generalized dual operation [2], that is, has a dual symmetry under the interchange:

$$\alpha \leftrightarrow \tilde{\alpha}, \qquad \text{or} \qquad \alpha \leftrightarrow \frac{1}{\alpha}, \tag{1}$$

which follows from the charge quantization condition:

$$g\tilde{g} = 4\pi n, \quad n \in Z, \quad \text{or} \quad \alpha\tilde{\alpha} = 1, \tag{2}$$

where

$$\alpha = \frac{g^2}{4\pi}, \qquad \tilde{\alpha} = \frac{\tilde{g}^2}{4\pi}. \tag{3}$$

In general case, we write the following RGEs [3]:

$$\frac{\mathrm{d}\ln\alpha(t)}{\mathrm{dt}} = -\frac{\mathrm{d}\ln\tilde{\alpha}(t)}{\mathrm{dt}} = \beta(\alpha) - \beta(\tilde{\alpha}) = \beta^{(total)}(\alpha), \tag{4}$$

which comes from the dual symmetry and charge quantization condition, valid for arbitrary t: $\alpha(t)\tilde{\alpha}(t) = 1$. In Eq. (4) $\beta(\alpha)$ is the well–known perturbative beta–function. Here we see that the total beta–function (4) for the pure nonabelian theory is antisymmetric under the interchange (1), what means that $\beta^{(total)}(\alpha)$ has zero ("fixed point") at $\alpha = \tilde{\alpha} = 1$: $\beta^{(total)}(\alpha = \tilde{\alpha} = 1) = 0$.

---

[a] e-mail: crdas@imsc.res.in
[b] e-mail: laper@itep.ru
[c] e-mail: hbech@nbi.dk

2 An example of beta–function in the case of the pure $SU(3)$ colour gauge group

In the gluondynamics (pure $SU(3)$ colour gauge theory) β–function is given by the following series over $\alpha/4\pi$ [4]:

$$\beta(\alpha) = -\left[\beta_0 \frac{\alpha}{4\pi} + \beta_1 \left(\frac{\alpha}{4\pi}\right)^2 + \beta_2 \left(\frac{\alpha}{4\pi}\right)^3 + ...\right] = -\sum_{n=0}^{\infty} \beta_n \left(\frac{\alpha}{4\pi}\right)^{n+1} \quad (5)$$

In the 3–loop approximation:

$$\beta_0 = 11, \qquad \beta_1 = 102, \qquad \beta_2 = 1428.5. \quad (6)$$

QCD shows that α_s is freezing at the value $\alpha_s \approx 0.4$ [5]. Then we have:

$$\beta(\alpha) = 0 \qquad \text{for} \qquad 0.4 < \alpha, \tilde{\alpha} < \frac{1}{0.4} = 2.5.$$

The behaviours of $\beta(\alpha)$ and $\beta(\tilde{\alpha}) = \beta(1/\alpha)$ are given by curves 1 and 2 of Figure 1.

In the light of contemporary ideas of the abelization of $SU(N)$ gauge theories [6], it is attractive to investigate the behaviour of $\beta^{(total)}(\alpha)$ in the vicinity of the point $\alpha = 1$.

Lattice investigations of the pure $SU(3)$ theories show that in some region $\alpha > \alpha_{conf}$ non–perturbative effects make up composite configurations of monopoles, which form a monopole condensate. Such a condensate creates strings between chromo–electric charges, which confine these charges. It is natural to think that the same configurations are created in the local $SU(3)$ gauge theory and imagine them as the Higgs fields $\tilde{\phi}(x)$ of scalar monopoles. 'T Hooft [6] suggested to consider such a gauge, in which monopole degrees of freedom, hidden in composite monopole configurations, become explicit and abelian. He developed *a method of the Maximal Abelian Projections (MAP)*. According to this method, in the non–perturbative region scalar monopoles are created only by the diagonal $SU(3)$ components $(A_\mu)^i_i$ of gauge fields $(A_\mu)^i_j$, where $i, j = 1, 2, 3$ are colour indices, and interact only with the diagonal $\widetilde{SU(3)}$ components of gauge fields $(\tilde{A}_\mu)^i_j$. In the non–perturbative region, the non–diagonal $SU(3)$ and $\widetilde{SU(3)}$ components of gauge fields are suppressed and the interaction of monopoles with dual gluons is described by $\widetilde{U(1)} \otimes \widetilde{U(1)}$ (Cartain) subgroup of $\widetilde{SU(3)}$ group. These monopoles can be approximately described by the Higgs fields $\tilde{\phi}(x)$ of scalar chromo–magnetic monopoles, interacting with gauge fields $\tilde{A}_\mu$. Recalling the generalized dual symmetry, we assume that the similar composite configurations have to be produced by dual gauge fields $\tilde{A}_\mu$, and described by the Higgs fields $\phi(x)$ of scalar chromo–electric "monopoles", interacting with gauge fields A_μ. The interaction of "monopoles" with gluons also has to be described by $U(1) \otimes U(1)$ subgroup of $SU(3)$ group. In general, $U(1)$–subgroups are arbitrary embedded into the $SU(N)$ gauge group: $U(1)^{N-1} \subset SU(N)$, and in the non–perturbative region $SU(N)$ gauge theory is reduced to the abelian $U(1)^{N-1}$ theory with $N-1$ different types of abelian monopoles.

Using notations: $f_{\mu\nu,i} \equiv (F_{\mu\nu})^i_i$, $a_{\mu,i} \equiv (A_\mu)^i_i$ and $\tilde{a}_{\mu,i} \equiv (\tilde{A}_\mu)^i_i$, we have the following equations, valid into the non–perturbative region of QCD ($i = 1, 2$):

$$\partial_\nu f_{\mu\nu,i} = \frac{i}{2} g_{(MAP)} \left[\phi_i^+ \mathcal{D}_\mu \phi_i - (\mathcal{D}_\mu \phi_i)^+ \phi_i\right], \quad (7)$$

$$\partial_\nu f^*_{\mu\nu,i} = \frac{i}{2}\tilde{g}_{(MAP)}\left[\tilde{\phi}_i^+\tilde{\mathcal{D}}_\mu\tilde{\phi}_i - (\tilde{\mathcal{D}}_\mu\tilde{\phi}_i)^+\tilde{\phi}_i\right], \tag{8}$$

where the covariant derivatives $\mathcal{D}_\mu$ and $\tilde{\mathcal{D}}_\mu$ are: $\mathcal{D}_\mu = \partial_\mu - ig\,a_\mu$, $\tilde{\mathcal{D}}_\mu = \partial_\mu - i\tilde{g}\,\tilde{a}_\mu$. Abelian monopoles $\tilde{\phi}_{1,2}$ and "monopoles" $\phi_{1,2}$ in Eqs. (7) and (8) have the charges $g_{(MAP)}$ and $\tilde{g}_{(MAP)}$: $\alpha_{(MAP)} = \alpha/2$, $\tilde{\alpha}_{(MAP)} = \tilde{\alpha}/2$.

In the nonberturbative region, the curve 3 of Figure 1 represents beta-function given by the contribution of the abelian theory (abelian dominance) of two scalar monopole fields $\tilde{\phi}_{1,2}$ and two scalar electric fields $\phi_{1,2}$:

$$\frac{\mathrm{d}\ln\alpha_{(MAP)}(t)}{\mathrm{d}t} = -\frac{\mathrm{d}\ln\tilde{\alpha}_{(MAP)}(t)}{\mathrm{d}t}$$

$$= 2\left[\frac{\alpha_{(MAP)} - \tilde{\alpha}_{(MAP)}}{12\pi}\left(1 + 3\frac{\alpha_{(MAP)} + \tilde{\alpha}_{(MAP)}}{4\pi} + ...\right)\right], \tag{9}$$

what gives the following beta–functions:

$$\frac{\mathrm{d}\ln\alpha(t)}{\mathrm{d}t} = -\frac{\mathrm{d}\ln\tilde{\alpha}(t)}{\mathrm{d}t} = \frac{\alpha - \tilde{\alpha}}{12\pi}\left(1 + 3\frac{\alpha + \tilde{\alpha}}{8\pi} + ...\right). \tag{10}$$

As it was shown in Ref. [3], the abelian RGEs can be considered by perturbation theory simultaneously only in the small region (approximately):

$$0.2 \lesssim \alpha^{(em)}, \tilde{\alpha}^{(em)} \lesssim 1. \tag{11}$$

The behaviour of the effective fine structure constants was investigated in the vicinity of the phase transition point in compact lattice QED by the Monte Carlo simulation method [7]. The following result was obtained:

$$\alpha_{crit}^{lat.QED} \approx 0.20 \pm 0.015, \qquad \tilde{\alpha}_{crit}^{lat.QED} \approx 1.25 \pm 0.10, \tag{12}$$

which is very close to the perturbative region (11) for parameters $\alpha^{(em)}$ and $\tilde{\alpha}^{(em)}$. Using the two–loop approximation for the effective potential in the Higgs model of dual scalar electrodynamics, we have obtained almost the same result in Ref. [8]. Then our abelian monopoles, arising in QCD as a result of MAP, have the following critical constant value: $\tilde{\alpha}_{(MAP)}^{(crit)} \approx 1.25$, what gives the beginning of the confinement region for QCD: $\alpha_1 = \alpha_{conf} = 1/(\tilde{\alpha}^{(crit)}) = 1/(2\tilde{\alpha}_{(MAP)}^{(crit)}) \approx 1/2.5 = 0.4$. We have received an explanation of the value of freezing α. By dual symmetry, the end of the perturbative region for the scalar field ϕ is: $\tilde{\alpha}_{conf} \approx 0.4$, what corresponds to $\alpha_2 = 1/\alpha_1 \approx 2.5$.

The behaviour of the total beta–function for the pure $SU(3)$ colour gauge theory is given by curves 1,2,3 in Figure 1 (curve 1' corresponds to QCD), where the regions of the formation of strings also are shown.

3 Dyons

The dual symmetry of the pure nonabelian theories leads to the natural assumption that in the non–perturbative region, not monopoles and "monopoles", but dyons are responsible for the confinement [9]. Then we have united Higgs abelian scalar dyon fields $\phi_{1,2,...N-1}$, having simultaneously electric and magnetic charges, and the following equations for each components $i = 1, 2, ..., N-1$ can be considered:

$$\partial_\lambda f_{\lambda\mu,i} = ig_{(MAP)}\left[\phi_i^+\mathcal{D}_\mu\phi_i - (\mathcal{D}_\mu\phi_i)^+\phi_i\right], \tag{13}$$

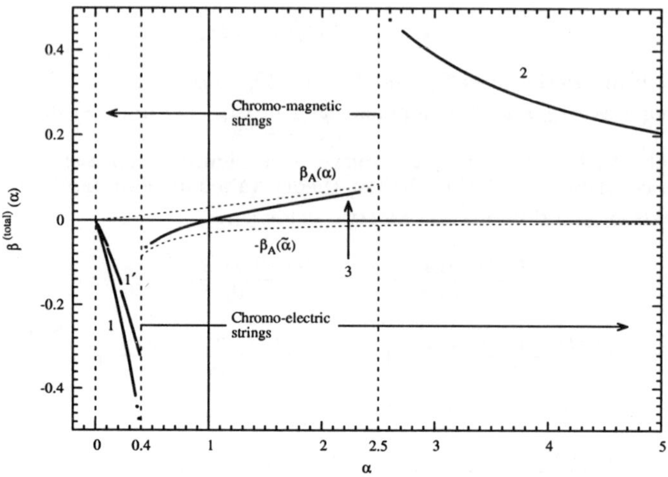

Figure 1: The behaviour of the total beta-function for the pure $SU(3)$.

$$\partial_\lambda f^*_{\lambda\mu,i} = i\tilde{g}_{(MAP)} \left[\phi_i^+ \tilde{\mathcal{D}}_\mu \phi_i - (\tilde{\mathcal{D}}_\mu \phi_i)^+ \phi_i\right], \tag{14}$$

where

$$\alpha_{(MAP)} = \frac{\alpha}{N-1} \qquad \text{and} \qquad \tilde{\alpha}_{(MAP)} = \frac{\tilde{\alpha}}{N-1}. \tag{15}$$

The behaviour of the $SU(3) \times \widetilde{SU(3)}$ total beta–function, given by Figure 1, also is valid for the case of dyons.

Acknowledgments

This work was supported by the Russian Foundation for Basic Research (RFBR), project No. 05–02–17642.

References

[1] D. Zwanziger, *Phys.Rev.* **D** 3, 343 (1971); ibid., **D** 3, 880 (1971).
[2] Chan Hong-Mo, Tsou Sheung Tsun, *Phys.Rev.* **D** 52, 6134 (1995); ibid., **D** 56, 3646 (1997); ibid., **D** 57, 2507 (1998).
[3] L.V. Laperashvili, H.B. Nielsen, *Mod.Phys.Lett.* **A** 14, 2797 (1999).
[4] O.V. Tarasov, A.A. Vladimirov, A.Yu. Zharkov, *Phys.Lett.* **B** 93, 429 (1980).
[5] Yu.A. Simonov, *Yad.Fiz.* **58**, 113 (1995); A.M. Badalian, Yu.A. Simonov, *Yad.Fiz.* **60**, 714 (1997).
[6] G. 't Hooft, *Nucl.Phys.* **B** 190, 455 (1981).
[7] J. Jersak, T. Neuhaus and P.M. Zerwas, *Phys.Lett.* **B** 133, 103 (1983); *Nucl.Phys.* **B** 251, 299 (1985).
[8] L.V. Laperashvili, D.A. Ryzhikh, H.B. Nielsen, *Int.J.Mod.Phys.* **A** 16, 3989 (2001); *Phys.Atom.Nucl.* **65**, 353 (2002).
[9] M.N. Chernodub, F.V. Gubarev, M.I. Polikarpov, JETP Lett. **69**, 169 (1999); arXiv: hep-lat/9801010.

SUMMATION OF FEYNMAN DIAGRAMS IN N=1 SUPERSYMMETRIC ELECTRODYNAMICS

K.Stepanyantz

Faculty of Physics, Moscow State University, 119992 Moscow, Russia

*Abstract.*Summation of Feynman diagrams is made in N=1 supersymmetric electrodynamics, regularized by higher derivatives, using Schwinger-Dyson equations and Ward identities. The terms, which can not be summed by this way disappear due to a special new identity.

1 Introduction

The supersymmetry essentially improves the ultraviolet behavior of a theory. Thanks to that it is possible to suggest the form of the β-function exactly to all orders of the perturbation theory even in theories with unextended supersymmetry. The form of the exact β-function was proposed first in Ref. [1] from the investigation of the instanton contributions structure. For $N = 1$ supersymmetric electrodynamics, which will be considered here, such β-function (that is called the exact Novikov, Shifman, Vainshtein and Zakharov (NSVZ) β-function) is

$$\beta(\alpha) = \frac{\alpha^2}{\pi}\Big(1 - \gamma(\alpha)\Big), \tag{1}$$

where $\gamma(\alpha)$ is the anomalous dimension of the matter superfield.

Explicit calculations, made with the dimensional reduction [2], confirm this proposal, but require a special choice of the subtraction scheme [3,4]. Explicit calculations in two- [5,6], three- [7] and partially four-loop [8] approximations for the $N = 1$ supersymmetric electrodynamics with the higher derivative regularization [9,10] reveal that renormalization of the operator $W_a C^{ab} W_b$ is exhausted at the one-loop and the Gell-Mann-Low function coincides with the exact NSVZ β-function and has corrections in all orders of the perturbation theory.

Here we consider the problem how to extend results of these papers to all orders of the perturbation theory [11,12].

2 $N = 1$ supersymmetric electrodynamics and its regularization by higher derivatives

The massless $N = 1$ supersymmetric electrodynamics with the higher derivatives term in the superspace is described by the following gauge invariant action:

$$S = \frac{1}{4e^2}\text{Re}\int d^4x\, d^2\theta\, W_a C^{ab}\Big(1+\frac{\partial^{2n}}{\Lambda^{2n}}\Big)W_b + \frac{1}{4}\int d^4x\, d^4\theta\, \Big(\phi^* e^{2V}\phi + \tilde\phi^* e^{-2V}\tilde\phi\Big). \tag{2}$$

It is well known, that adding of the higher derivative term does not remove divergences in one-loop diagrams. In order to regularize them, it is necessary to insert the Pauli-Villars determinants $\det PV(V, M_i)$ [13]. Then the generating functional can be written as

$$Z = \int DV\, D\phi\, D\tilde{\phi} \prod_i \Big(\det PV(V, M_i)\Big)^{c_i} \exp\Big(i(S_{ren} + S_{gf} + S_S)\Big), \qquad (3)$$

where S_{ren} is the renormalized action, S_{gf} is the gauge fixing action, S_S denotes terms with sources and the coefficients c_i satisfy conditions

$$\sum_i c_i = 1; \qquad \sum_i c_i M_i^2 = 0. \qquad (4)$$

Below we will assume, that $M_i = a_i\Lambda$, where a_i are some constants. Insertion of Pauli-Villars determinants allows to cancel remaining divergences in all one-loop diagrams, including diagrams, containing insertions of counterterms.

In our notations the generating functional for the connected Green functions is written as $W = -i\ln Z$. The effective action Γ is obtained by making a Legendre transformation.

3 Schwinger-Dyson equation

From generating functional (3) it is possible to obtain [11] the Schwinger-Dyson equation for the two point Green function of the gauge field by the standard methods [11]. Substituting into this equation solutions of the Ward identities we obtain

$$\frac{d}{d\ln\Lambda}\Gamma^{(2)}_V\Big|_{p=0} = \frac{d}{d\ln\Lambda}\int \frac{d^4p}{(2\pi)^4}\frac{d^4q}{(2\pi)^4}\, d^4\theta\, V\partial^2\Pi_{1/2}V\left(\frac{1}{2q^2}\frac{d}{dq^2}\ln(q^2G^2) - \frac{8f}{q^2G}\right) - \text{similar terms with the Pauli-Villars fields}, \qquad (5)$$

where the function $G(q^2)$ is defined by the two-point Green function as follows:

$$\frac{\delta^2\Gamma}{\delta\phi_x^*\delta\phi_y} = \frac{D_x^2\bar{D}_x^2}{16}G(\partial^2)\delta^8_{xy} \qquad (6)$$

and the other functions are defined by

$$\frac{1}{4}\frac{\delta}{\delta V_x}\frac{\delta}{\delta\phi_y}\exp\Big(\frac{2}{i}\frac{\delta}{\delta J_z}+2V_z\Big)\phi_z\Big|_{p=0} = -2\partial^2\Pi_{1/2x}\Big(\bar{D}_x^2\delta^8_{xy}\delta^8_{xz}\Big)F(q^2)$$
$$+\frac{1}{8}D^a C_{ab}\bar{D}_x^2\Big(\bar{D}_x^2\delta^8_{xy}D^b_x\delta^8_{xz}\Big)f(q^2) - \frac{1}{16}q^\mu G'(q^2)\bar{D}\gamma^\mu\gamma_5 D_x\Big(\bar{D}_x^2\delta^8_{xy}\delta^8_{xz}\Big)$$
$$-\frac{1}{4}\bar{D}_x^2\delta^8_{xy}\delta^8_{xz}\,G(q^2). \tag{7}$$

Note, that the functions $F(q^2)$ and $f(q^2)$ can not be defined from the Ward identities. The terms with the Pauli-Villars fields are given in Ref. [11].

We see, that the second term in equation (5) contains the function f, while the terms, containing the unknown function F, are completely cancelled. However, explicit three- [7] and four-loop [8] calculations show, that the term, containing the unknown function f is equal to zero. Therefore, it is possible to propose, that the following identity take place:

$$\frac{d}{d\ln\Lambda}\int\frac{d^4q}{(2\pi)^2}\frac{f(q^2)}{q^2G(q^2)} = 0, \tag{8}$$

(Similar identity in the massive case is given in Ref. [11].) This identity was proved for a special class of diagrams in Ref. [12]. The method, proposed in it, possibly can be generalized for the other diagrams.

Due to identity (8) and a similar identity for the massive case, the integrals, defining the two-point Green function of the gauge field, are reduced to integrals of total derivatives in the four-dimensional spherical coordinates and can be easily calculated. As a result of this calculation [11] we obtain:

1. The renormalization constant Z_3 should be chosen so that

$$\frac{1}{e^2}Z_3\Big(e,\Lambda/\mu\Big) = \frac{1}{e^2} - \frac{1}{4\pi^2}\ln\frac{\Lambda}{\mu}, \tag{9}$$

where μ is a normalization point. This means, that the divergences in the two-point function of the gauge field exist only in the one-loop approximation.

2. The final expression for the corresponding part of the effective action (without the gauge fixing terms) in the massless case can be written as

$$\Gamma^{(2)}_V = \int\frac{d^4p}{(2\pi)^4}\,d^2\theta\,W_a(-p)\,C^{ab}W_b(p)\left[\frac{1}{4e^2}+\frac{1}{16\pi^2}\ln\frac{\mu}{p}-\frac{1}{16\pi^2}\ln(ZG)+\text{const}\right]. \tag{10}$$

According to this formula the Gell-Mann-Low function is given by (1), has corrections in all orders of the perturbation theory and coincides with the exact NSVZ β-function.

4 Conclusion

Summation of Feynman diagrams, defining the two-point Green function of the gauge field in the limit $p \to 0$ in $N = 1$ supersymmetric electrodynamics, can be partially made by Schwinger-Dyson equations and Ward identities. Diagrams, which can not be summed by this way, give a nontrivial contribution starting from the three-loop approximation. Calculation of their sum exactly to all orders of the perturbation theory is possible, but appears to be rather complicated technically. A summation algorithm was constructed in Ref. [12].

Note, that Eq. (8), which can be easily guessed from explicit calculations, is a new identity, which is not reduced to the Ward identities. This identity is closely connected with relation (10) for the two-point Green functions, from which the exact NSVZ β-function is obtained. Then there are some questions: What is a true reason of this identity? Is it a consequence of some symmetry of the theory? So far we have no answers to these questions.

Using identity (8) it is possible to prove, that the Gell-Mann-Low function coincides with the exact NSVZ β-function. However, the renormalization of the operator $W_a C^{ab} W_b$ is exhausted at the one-loop, in accordance with the structure of the anomaly supermultiplet. This means, that the anomaly puzzle is naturally solved, if the higher derivatives regularization is used for the calculations.

Acknowledgments

This work was supported by RFBR grant No 02-01-00126.

References

[1] V.Novikov, M.Shifman, A.Vainstein, V.Zakharov, *Phys.Lett.* 166 **B**, 329, (1985).
[2] W.Siegel, *Phys.Lett.* 84 **B**, 193, (1979).
[3] O.V.Tarasov, V.A.Vladimirov, *Phys.Lett.* 96 **B**, 94, (1980).
[4] I.Jack, D.R.T.Jones, C.G.North, *Nucl.Phys.* **B** 486, 479, (1997).
[5] A.Soloshenko, K.Stepanyantz, hep-th/0203118.
[6] A.Soloshenko, K.Stepanyantz, *Theor.Math.Phys.* 134, 377, (2003).
[7] A.Soloshenko, K.Stepanyantz, *Theor.Math.Phys.* 140, 1264, (2004).
[8] A.Pimenov, K.Stepanyantz, *Theor.Math.Phys.* to appear.
[9] A.Slavnov, *Theor.Math.Phys.* 23, 3, (1975).
[10] T.Bakeyev, A.Slavnov, *Mod.Phys.Lett.* **A** 11, 1539, (1996).
[11] K.Stepanyantz, *Theor.Math.Phys.* 142, 29, (2005).
[12] K.Stepanyantz, hep-th/0511012.
[13] L.Faddeev, A.Slavnov, *"Gauge fields, introduction to quantum theory"*, (Benjamin Reading), 1990.

SPECTRUM OF HIGGS PARTICLES IN THE EXCEPTIONAL SUPERSYMMETRIC STANDARD MODEL

S. F. King[a], S. Moretti[b], R. Nevzorov[c d]
School of Physics and Astronomy, University of Southampton, Southampton, SO17 1BJ, U.K.

Abstract. We discuss the spectrum of Higgs bosons in the framework of the exceptional supersymmetric standard model (ESSM). The presence of a Z' and exotic particles predicted by the ESSM allows the lightest Higgs particle to be significantly heavier than in the MSSM and NMSSM. When the mass of the lightest Higgs boson is larger than $135-140\,\mathrm{GeV}$ the heaviest scalar, pseudoscalar and charged Higgs states lie beyond the TeV range.

One of the strongest theoretical arguments in favour of softly broken supersymmetry (SUSY) is associated with the partial unification of the standard model (SM) gauge interactions with gravity. However, the minimal supersymmetric standard model (MSSM) being incorporated in supergravity suffers from the μ problem. Indeed the superpotential of the MSSM contains one bilinear term $\mu\hat{H}_d\hat{H}_u$, so that μ is expected to be either zero or of the order of the Planck scale. At the same time, in order to get the correct pattern of electroweak symmetry breaking (EWSB), μ is required to be in the EW or TeV range. In the framework of the simplest extension of the MSSM — the Next–to–Minimal Supersymmetric Standard Model (NMSSM) — an 'effective' μ–term is generated dynamically. But the NMSSM superpotential is invariant under a discrete Z_3 symmetry which leads to the formation of domain walls in the early universe (For a review of the MSSM and NMSSM see e.g. [1].).

The μ problem is solved automatically within superstring inspired models based on the E_6 gauge group or its rank–6 subgroup $SU(3)_C \times SU(2)_W \times U(1)_Y \times U(1)_\psi \times U(1)_\chi$. Two anomaly-free $U(1)_\psi$ and $U(1)_\chi$ symmetries of the rank-6 model are defined by [2]: $E_6 \to SO(10)\times U(1)_\psi$, $SO(10) \to SU(5)\times U(1)_\chi$. Near the string scale the rank-6 model can be reduced to an effective rank–5 model with only one extra gauge symmetry $U(1)'$:

$$U(1)' = U(1)_\chi \cos\theta + U(1)_\psi \sin\theta\,. \tag{1}$$

If $\theta \neq 0$ or π the extra $U(1)'$ gauge symmetry forbids an elementary μ term but allows interaction $\lambda S H_d H_u$ in the superpotential. After EWSB the scalar component of the SM singlet superfield S acquires a non-zero vacuum expectation value (VEV) breaking $U(1)'$ and an effective μ term of the required size is automatically generated. Clearly there are no domain wall problems in such model since there is no additional Z_3 symmetry.

[a]E-mail: sfk@hep.phys.soton.ac.uk.
[b]E-mail: stefano@hep.phys.soton.ac.uk.
[c]E-mail: nevzorov@phys.soton.ac.uk.
[d]On leave of absence from the Theory Department, ITEP, Moscow, Russia.

In this article we explore the Higgs sector in the framework of a particular E_6 inspired supersymmetric model with an extra $U(1)_N$ gauge symmetry in which right handed neutrinos do not participate in the gauge interactions. ($\theta = \arctan\sqrt{15}$). Only in this exceptional supersymmetric standard model (ESSM) right–handed neutrinos may be superheavy, sheding light on the origin of the mass hierarchy in the lepton sector [3]- [4]. The particle content of the ESSM involves three complete fundamental 27 representations of E_6. It ensures anomaly cancellation within each generation. In addition to the complete 27_i representations doublet and anti-doublet from extra $27'$ and $\overline{27'}$ can and must survive to low energies to preserve gauge coupling unification. Thus in addition to a Z' the ESSM involves extra matter that forms three $5+5^*$ representations of $SU(5)$ plus three $SU(5)$ singlets which carry $U(1)_N$ charges.

The ESSM Higgs sector includes two Higgs doublets H_u and H_d as well as a SM–like singlet field S [3]- [4]. The Higgs effective potential can be written as

$$V = V_F + V_D + V_{soft} + \Delta V\,, \tag{2}$$

$$\begin{aligned}
V_F &= \lambda^2|S|^2(|H_d|^2+|H_u|^2)+\lambda^2|(H_dH_u)|^2\,,\\
V_D &= \tfrac{g_2^2}{8}\left(H_d^\dagger\sigma_a H_d + H_u^\dagger\sigma_a H_u\right)^2 + \tfrac{g'^2}{8}\left(|H_d|^2-|H_u|^2\right)^2 +\\
&\quad + \tfrac{g_1'^2}{2}\left(\tilde{Q}_1|H_d|^2+\tilde{Q}_2|H_u|^2+\tilde{Q}_S|S|^2\right)^2\,,\\
V_{soft} &= m_S^2|S|^2+m_1^2|H_d|^2+m_2^2|H_u|^2+\Big[\lambda A_\lambda S(H_uH_d)+h.c.\Big]\,,
\end{aligned} \tag{3}$$

where $g' = \sqrt{3/5}g_1$; g_2, g_1 and g_1' are $SU(2)_W$, $U(1)_Y$ and $U(1)_N$ gauge couplings; while $\tilde{Q}_1$, $\tilde{Q}_2$ and $\tilde{Q}_S$ are effective $U(1)_N$ charges of H_d, H_u and S respectively [3]- [4]. The couplings g_2 and g' are known precisely. Assuming gauge coupling unification one finds that $g_1'(Q) \simeq g_1(Q)$ for any $Q < M_{\mathrm{GUT}}$ [4].

At tree–level the Higgs potential is described by the sum of the first three terms in Eq. (2). The last term ΔV represents the contribution of loop corrections to the Higgs effective potential. In Eq. (2) V_F and V_D are the F and D terms. The soft SUSY breaking terms are collected in V_{soft}. A simple counting shows that the ESSM Higgs sector contains only one additional singlet field and one extra parameter compared to the MSSM. Therefore it can be regarded as the simplest extension of the Higgs sector of the MSSM.

At the physical vacuum the ESSM Higgs fields develop the VEVs $\langle H_d\rangle = \frac{v_d}{\sqrt{2}}$, $\langle H_u\rangle = \frac{v_u}{\sqrt{2}}$ and $\langle S\rangle = \frac{s}{\sqrt{2}}$, thus breaking the $SU(2)_W\times U(1)_Y\times U(1)_N$ symmetry to $U(1)_{\mathrm{EM}}$, associated with electromagnetism. Instead of v_d and v_u it is more convenient to use $\tan\beta = \frac{v_u}{v_d}$ and $v=\sqrt{v_d^2+v_u^2}$, where $v = 246\,\mathrm{GeV}$. After EWSB two CP-odd and two charged Goldstone modes in the Higgs sector are absorbed by the Z, Z' and $W^\pm$ gauge bosons so that only six physical degrees of freedom are left. They form one CP-odd and two charged Higgs states (as in the MSSM) with masses

$$m_A^2 \simeq \frac{2\lambda^2 s^2 x}{\sin^2 2\beta} + O(M_Z^2)\,, \qquad m_{H^\pm}^2 \simeq m_A^2 + O(M_Z^2) \tag{4}$$

respectively and three CP-even states (as in the NMSSM) with masses

$$m_{h_3}^2 \simeq m_A^2 + O(M_Z^2)\,, \qquad m_{h_2}^2 \simeq g_1'^2 \tilde{Q}_S^2 s^2 + O(M_Z^2)\,, \tag{5}$$

$$\begin{aligned} m_{h_1}^2 \simeq & \tfrac{\lambda^2}{2} v^2 \sin^2 2\beta + \tfrac{\bar{g}^2}{4} v^2 \cos^2 2\beta + g_1'^2 v^2 \Big(\tilde{Q}_1 \cos^2\beta + \tilde{Q}_2 \sin^2\beta\Big)^2 - \\ & - \tfrac{\lambda^4 v^2}{g_1'^2 Q_S^2} \Big(1 - x + \tfrac{g_1'^2}{\lambda^2}\Big(\tilde{Q}_1 \cos^2\beta + Q_2 \sin^2\beta\Big) Q_S\Big)^2 + \ldots\,, \end{aligned} \tag{6}$$

where $x = \frac{A_\lambda}{\sqrt{2}\lambda s} \sin 2\beta$.

At least one CP–even Higgs boson is always heavy preventing the distinction between the ESSM and MSSM Higgs sectors. Indeed the mass of the singlet dominated Higgs scalar particle m_{h_2} is always close to the mass of Z' boson $M_{Z'} \simeq g_1' \tilde{Q}_S s \sim g_1 s$ that has to be heavier than $500 - 600\,\mathrm{GeV}$ [5]. The masses of the charged, CP–odd and one CP–even Higgs states are governed by m_A. The mass of the SM–like Higgs boson given by *Eq.*(6) is set by M_Z. The last term in Eq.(6) must not be allowed to dominate since it is negative. This constrains x around unity for $\lambda > g_1'$. As a consequence m_A is confined in the vicinity of $\frac{\lambda s}{\sqrt{2}} \tan\beta$ and is much larger than the masses of the Z' and Z bosons. At so large values of m_A the masses of the heaviest CP–even, CP–odd and charged states are almost degenerate around m_A.

The qualitative pattern of the Higgs spectrum obtained for $\lambda > g_1'$ in the leading one–loop approximation is shown in Fig. 1. The numerical analysis reveals that the heaviest CP–even, CP–odd and charged Higgs states lie beyond the TeV range in the considered case. The second lightest CP–even Higgs boson is predominantly a singlet field so that it will be quite difficult to observe this particle at future colliders. With decreasing λ the allowed range of m_A enlarges so that charged, CP–odd and second lightest CP–even Higgs states may have masses in the $200 - 300\,\mathrm{GeV}$ range when $\lambda < g_1'$. But for $m_A < 500\,\mathrm{GeV}$ and $\lambda < g_1'$ we get a MSSM–type Higgs spectrum with the lightest SM–like Higgs boson below $130 - 135\,\mathrm{GeV}$ and with the heaviest scalar above $600\,\mathrm{GeV}$ being singlet dominated and irrelevant. The non–observation of Higgs particle at LEP rules out most parts of the ESSM parameter space in this case.

From Fig. 1 and Eq. (6) it becomes clear that at some value of m_A (or x) the lightest CP–even Higgs boson mass m_{h_1} attains its maximum value. At tree–level the upper bound on m_{h_1} is given by the sum of the first three terms in Eq. (6). The inclusion of loop corrections increase the bound on the lightest CP-even Higgs boson mass in SUSY models substantially. In Fig. 2 we plot

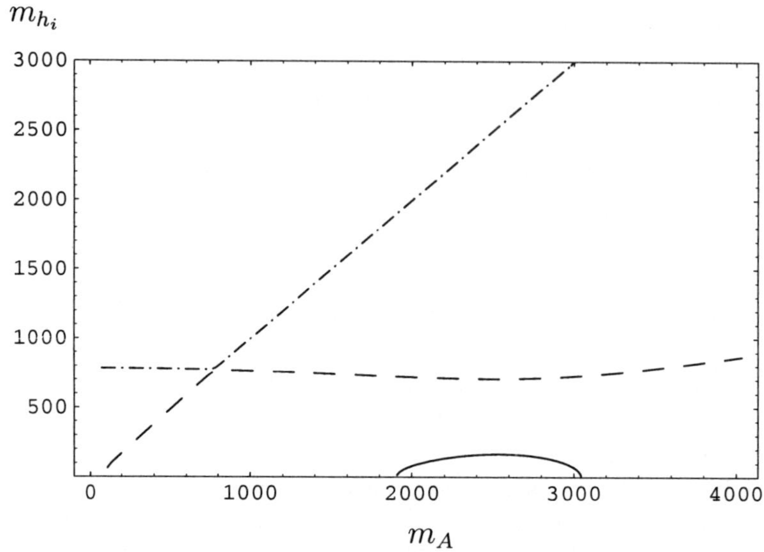

Figure 1: The one–loop masses of the CP–even Higgs bosons versus m_A for $\lambda(M_t) = 0.794$, $\tan\beta = 2$, $M_{Z'} = M_S = 700\,\text{GeV}$ and $X_t = \sqrt{6}M_S$. Solid, dashed and dashed–dotted lines correspond to the masses of the lightest, second lightest and heaviest Higgs scalars.

the two-loop upper bounds on the mass of the lightest Higgs particle in the MSSM, NMSSM and ESSM as a function of $\tan\beta$. At moderate values of $\tan\beta$ ($\tan\beta = 1.6 - 3.5$) the upper limit on the lightest Higgs boson mass in the ESSM is considerably higher than in the MSSM and NMSSM. It reaches the maximum value $\sim 150 - 155\,\text{GeV}$ at $\tan\beta = 1.5 - 2$. In the considered part of the parameter space the theoretical restriction on the mass of the lightest CP-even Higgs boson in the NMSSM exceeds the corresponding limit in the MSSM because of the extra contribution to $m^2_{h_1}$ induced by the additional F-term in the Higgs scalar potential of the NMSSM. The size of this contribution, which is described by the first term in Eq. (6) is determined by the Yukawa coupling λ. The upper limit on the coupling λ caused by the validity of perturbation theory in the NMSSM is more stringent than in the ESSM due to the presence of exotic $5 + \overline{5}$-plets of matter in the particle spectrum of the ESSM. This is the reason why the upper limit of m_{h_1} in the NMSSM is considerably less than in the ESSM at moderate values of $\tan\beta$.

At large $\tan\beta > 10$ the contribution of the F-term of the SM–type singlet field to $m^2_{h_1}$ vanishes. Therefore with increasing $\tan\beta$ the upper bound on the lightest Higgs mass in the NMSSM approaches the corresponding limit in the MSSM. In the ESSM the theoretical restriction on m_{h_1} also diminishes when $\tan\beta$ rises but it is still $4 - 5\,\text{GeV}$ larger than the one in the MSSM because of the $U(1)_N$ D-term contribution to $m^2_{h_1}$ (the third term in Eq. (6)).

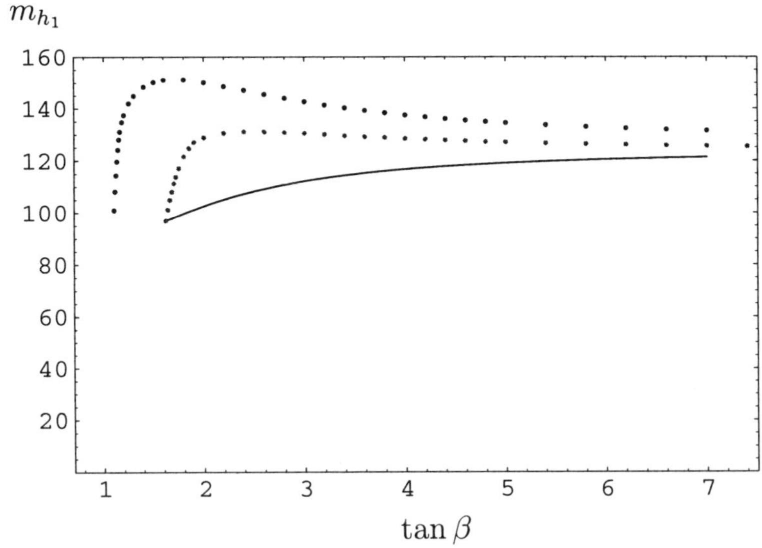

Figure 2: The dependence of the two-loop upper bound on the lightest Higgs boson mass on $\tan\beta$ for $m_t(m_t) = 165\,\text{GeV}$, $m_Q^2 = m_U^2 = M_S^2$, $X_t = \sqrt{6}M_S$ and $M_S = 700\,\text{GeV}$. The solid, lower and upper dotted lines represent the limit on m_{h_1} in the MSSM, NMSSM and ESSM.

The discovery at future colliders of superpartners of observed quarks and leptons as well as a relatively heavy SM–like Higgs boson with mass 140 – 155 GeV, that corresponds to $\lambda > g_1'$ in the ESSM, will permit to distinguish the ESSM from the simplest supersymmetric (like MSSM and NMSSM) and other extensions of the SM. Another possible manifestations of the exceptional SUSY model at the LHC are related with the enhanced production of l^+l^-, $t\bar{t}$ and/or $b\bar{b}$ pairs coming from either a Z' boson or exotic particle decays.

Acknowledgements: The authors would like to to acknowledge support from the PPARC grant PPA/G/S/2003/00096, the NATO grant PST.CLG.980066 and the EU network MRTN 2004-503369.

References

[1] D.J.H. Chung, L.L. Everett, G.L. Kane, S.F. King, J.D. Lykken, L.T. Wang, *Phys. Rept.* 407, 1 (2005).

[2] J.L. Hewett, T.G. Rizzo, *Phys. Rept.* 183, 193 (1989).

[3] S.F. King, S. Moretti, R. Nevzorov, to be published in *Phys. Lett.* **B**, *hep-ph/0511256.*

[4] S.F. King, S. Moretti, R. Nevzorov, to be published in *Phys. Rev.* **D**, *hep-ph/0510419.*

[5] F. Abe et al., *Phys. Rev. Lett.* 79, 2192 (1997).

QUANTIZATION of NONLINEAR FIELDS on CLASSICAL BACKGROUND [a]

M.Chichikina [b]

Faculty of Physics, Moscow State University, 119992 Moscow, Russia

*Abstract.*In this presentation we offer scheme of quantization of boson fields on the classical background by means of the Bogoliubov group variables.

In this work we propose quantizations, new a way, which will allow to use the Bogoliubov group variables for systems having arbitrary symmetry group (provided that we we know representation of this group), including for non-stationary systems.

The development of a scheme of quantization occurs in three basic stages.

I. At the first stage we pass to new variable - Bogoliubov group variables - and all considered quantities we express in the terms new variable.

Let variable $x^{'}$ are related to x by group transformation:

$$x' = F(x, a), \qquad x'' = F\left(F(x, a), b\right) = F(x, c), \qquad c = \varphi\left(a, b\right).$$

Variations of coordinates at a variation of parameters of group a are:

$$\left(\delta x'\right)^{i} = \xi_s^i(x')B_p^s(a)\delta a^p,$$

where $i = 0, 1, 2, 3$- number of coordinate, $p = 1..., r$, and k - number of generators of group. The group properties of transformation are defined by tensor $B_p^s(a)$.

Let's define Bogoliubov transformation as follows:

$$f(x) = gv(x^{'}) + u(x^{'}),$$

dimensionless parameter g is supposed to be large, and $(u(x^{'}), a)$ are independent new variable (Bogoliubov group variable). Explisitly separated large component depends on $x^{'}$ as well as a quantum part. Thus we restore invariancy with respect to transformations group, which was broken by explisit separation of classical component, as it was pointed out in introduction. However consideration of variable τ as independent leads to the fact that right-hand side of our equation contains now on k variable more than left one. In order to equalize the number of variables we shall impose on $u(x^{'})$ some invariant conditions.

Let's make it as follows: we choose, some functions $N^a(x^{'})$ (the number of functions is equal to number of group parameters)and we shall impose on function $u(x)$ the following conditions:

$$\omega(N^a, u) = 0.$$

[a]Supported by Russian President Grant no. 1450.2003.2

[b]e-mail: chich@goa.bog.msu.ru

$\omega(N^a, u)$ is invariant symplectic form.

Use of this condition allows to obtain the equations, which determine group variable as functionals of initial $f(x)$ and $f_n(x)$ on D.

II. At the second stage we develop the perturbation theory. All integrals of motions are expressed in new variables. Analyzing received expressions, we obtain conditions, at which application of perturbation theory is correct, namely: the classical part owes to satisfy to some equations. These equations turn out coinside with Eileur-Lagrange equations. However this result is not trivial: it was received not as a consequence of variational principle, but as a condition of development of a perturbative scheme for our system in the terms new variables, while for reception of these equations we did not need to know Hamiltonian structure as generator of transaltions on time.

So, at the second stage we obtain equations of motion for classical component.

It is made as follows: Having in explisit form of expression for new variable as functionals from old variable on D it is possible to define the operators $\hat{q}(x)$ and $\hat{p}(x)$:

$$\hat{p}(x) = \frac{1}{\sqrt{2}}\Big(f_n(x) + i\frac{\delta}{\delta f(x)}\Big), \qquad \hat{q}(x) = \frac{1}{\sqrt{2}}\Big(f(x) - i\frac{\delta}{\delta f_n(x)}\Big).$$

We can treat $\hat{q}(x)$ and $\hat{p}(x)$ as operators of coordinate and momentum of a field oscillators and to develop a scheme of secondary quantization. However there is another pair of selfconjugated operators, constructed from $f_n(x)$, $f(x)$, $\frac{\delta}{\delta f(x)}$ and $\frac{\delta}{\delta f_n(x)}$, which satisfy the same commutation relation. Thus, number of possible field states appears to be doubled and requires a reduction.

Therefore we shall act as follows:

we shall develop a scheme of the perturbation theory in the terms of Bogoliubov variables despite of appearance of extra states. After that we make a reduction of states number, that is this reduction will be as a matter of fact the dynamical scheme.

Now we have quantum operators of coordinate and momentum (and they are are expressed in the terms of new variable). Now we can perform quantization of system by the following replacement:

$$f(x^{'}) \longrightarrow \hat{q}(x), \qquad f_n(x^{'}) \longrightarrow \hat{p}(x).$$

In the terms of new variable $\hat{q}(x)$ and $\hat{p}(x)$ are series on inverted powers of coupling constant. Hence, and the integrals of a motion of system are series:

$$O = g^2 O_{-2} + gO_{-1} + O_0 + \frac{1}{g}O_1 + \ldots$$

Here O_{-2} are C-number, the operators O_{-1} are linear with respect to $u(x^{'})$, $u_n(x^{'})$, $\frac{\partial}{\partial u(x^{'})}$, $\frac{\partial}{\partial u_n(x^{'})}$. Since there are no normalizible eigenvectors of such operators, so for correct development of perturbation theory it is necessary that operators were identically equal to zero. This requirement are performed, if classical component satisfies equations of a motion on D. Then the operators O_{-1} are equal to zero.

Therefore further we shall consider classical part to be solution of Cauchy problem with boundary conditions on D.

III. The third stage - construction of system state space .

Above we have mentioned, that number of possible field states is double on comparison with a real situation, and consequently the reduction of number of states is necessary.

At this stage of construction of our scheme we achieve correct number of degrees of freedom which will be equal to its real number. During realization of a reduction we obtain equations of motion for quantum correction for our field (the equations appear nonlinear). Also at this stage the explisit expressions for creation-annihilation operators of quantums of a field are obtained as well as and spectrum of frequencies, that is actually, energy spectrum. Besides we obtain expressions for integrals of motion as generators of corresponding symmetry transformations:

$$O_0 = i\Im^{\alpha}\frac{\partial}{\partial\tau^{\alpha}},$$

In particular, Hamiltonian is generator of shift on time. Let's note, that equations of motion of a classical part and explisit form of hamiltonian are received independently, that is is overcome basic difficulty of the description of non-stationary systems, which was described in introduction.

Also in zero order we shall obtain Heizenberg equation in the terms new variables.

The scheme proposed above has been applied to the different systems. The most complicated example is interacting gravitational and scalar field. We obtain that classiacl part of gravitational field has to be solution of the Einstein equiation.

Quantum corrections w_{st} for gravity have to satisfy to the nonlinear evolition equations.

As example we consider quantization on the background on flat metric

$$\begin{pmatrix} -1 & 0 & 0 & 0 \\ 0 & 1 & 0 & 0 \\ 0 & 0 & 1 & 0 \\ 0 & 0 & 0 & 1 \end{pmatrix}$$

than

$$F_{st} = \delta_{st}.$$

Quantum addend for gravity w_{st} looks like:

$$w_{st} = \begin{pmatrix} w_{22} & w_{23} \\ w_{23} & -w_{22} \end{pmatrix}$$

Suppose quantum addend depends on one space coordinate (say, x) only. Equations of motion for quantum part of gravitational filed lookis like :

$$\frac{\partial^2 w_{st}}{\partial t^2} = \frac{\partial^2 w_{st}}{\partial x^2}.$$

It means that quantum part of gravitational filed is flate doublepolarized waves. All the other important rezults it is possible to find in [3] .

So we can state that totality of the received resultsis the following in general case:

-Equation of motion of a classical part;

-Equation of motion of the quantum additive;

-Energy spectrum;

-Expressions for integrals of a motion;

-Heisenberg eqations;

-Explisit form of the field operator in the terms of new variables;

allows to assert, that our theory gives the complete description of system of boson fields, quatized on classical background. While theory guarantees exact performance of the conservation laws in any order of perturbation theory and also allows to avoid a zero-mode problem.

Acknowledgments

Acknowledgments for Prof O.A.Khrustalev for discussions

References

[1] N.N.Bogoliubov *the Ukrainian Mathematical Journal.* v.2. p. 3-24. (1950).

[2] O.A.Khrustalev, M.V.Chichikina *Teor.Mat.Fiz.* T. 111. N2.p. 242-251 (1997).

[3] O.D.Timofeevskaya O.A.Khrustalev, M.V.Chichikina *Moscow Univeristy Bulleten* N4 p.3, N5 p.3, N6 p.11-14 (2003).

COLLAPSE OF POSITRONIUM AND VACUUM INSTABILITY

A.E. Shabad, [a]
P.N. Lebedev Physics Institure, Leninsky prospect 53, Moscow 119991, Russia
V.V. Usov[b]
Center for Astrophysics, Weizmann Institute of Science, Rehovot 76100, Israel

*Abstract.*A hypercritical value for the magnetic field is determined, which provides the full compensation of the positronium rest mass by the binding energy in the maximum symmetry state and disappearance of the energy gap separating the electron-positron system from the vacuum. The compensation becomes possible owing to the falling to the center phenomenon. The structure of the vacuum is described in terms of strongly localized states of tightly mutually bound (or confined) pairs. Their delocalization for still higher magnetic field, capable of screening its further growth, is discussed.

It is accepted that magnetic fields are stable in pure quantum electrodynamics (QED), and other interactions (weak or strong) or magnetic monopoles have to be involved to make the magnetized vacuum unstable [1]. In this Talk (see Ref. [2] for a detailed version) we establish, within the frame of QED, that there exists a hypercritical value of the magnetic field

$$B_{\rm hpcr}^{(1)} \simeq \frac{m^2}{4e}\exp\left\{\frac{\pi^{3/2}}{\sqrt{\alpha}} + 2C_{\rm E}\right\} \simeq 10^{28}B_0 \simeq 10^{42}{\rm G}\,. \qquad (1)$$

Here $\alpha = e^2/4\pi \simeq 1/137$, $B_0 = m^2/e = 4.4 \times 10^{13}$G, m is the electron mass. The value (1) is less than the magnetic field $\sim (10^{47} - 10^{48})$ G expected to be present near superconductive cosmic strings [3] and the one ($\sim 10^{47}$G) produced at the beginning of the inflation [4]. The hypercritical magnetic field leads to the shrinking of the energy gap between an electron and positron that takes place due to the falling to the center. The latter is caused by the ultraviolet singularity of the photon propagator (or Coulomb potential) [5] plus the dimensional reduction in the magnetic field [6] - [8]. We discuss the vacuum structure around the hypercritical magnetic field and its possible decay under a further growth of the magnetic field - that may cause its screening.

We rely on the theory of the falling to the center developed in [9] that implies deviations from the standard quantum theory manifesting themselves when extremely large electric fields near the singularity become important. In that theory the singularity in the Schrödinger-like equation yields a singular measure in the scalar product and hence the geometry of a black-hole-like object. (Stress, that the geometry is induced: no interaction of gravitational origin is included.)

We proceed from the (3+1)-dimensional Bethe-Salpeter (BS) equation in an approximation, which is the ladder approximation once the photon propagator

[a]e-mail: shabad@lpi.ru
[b]e-mail: fnusov@wicc.weizman.ac.il

(in the coordinate space) is taken in the Feynman gauge: $D_{ij}(x) = g_{ij}D(x^2)$, $x^2 \equiv x_0^2 - \mathbf{x}^2$, $g_{ii} = (1, -1 - 1 - 1)$. In an asymptotically strong magnetic field this equation may be written [2] in the following (1+1)-form covariant under the Lorentz transformations along the axis 3:

$$(\mathrm{i}\overrightarrow{\hat{\partial}_{\|}} - \frac{\hat{P}_{\|}}{2} - m)\Theta(t,z)(-\mathrm{i}\overleftarrow{\hat{\partial}_{\|}} - \frac{\hat{P}_{\|}}{2} - m)$$
$$= \mathrm{i}8\pi\alpha \sum_{i=0,3} D\left(t^2 - z^2 - \frac{P_\perp^2}{(eB)^2}\right) g_{ii}\gamma_i\Theta(t,z)\gamma_i, \qquad (2)$$

where $\Theta(t,z)$ is the 4×4 (Ritus transform of) BS amplitude, $t = x_0^{\mathrm{e}} - x_0^{\mathrm{p}}$ and $z = x_3^{\mathrm{e}} - x_3^{\mathrm{p}}$ are the differences of the coordinates of the electron (e) and positron (p) along the time x_0 and along the magnetic field $\mathbf{B} = (0, 0, B_3 = B)$. $P_{\|}$ and $P_\perp$ are projections of the total (generalized) momentum of the positronium onto the (0,3)- subspace and the (1,2)-subspace. Only two Dirac gamma-matrices, $\gamma_{0,3}$, are involved, $\hat{\partial}_{\|} = \partial_t\gamma_0 - \partial_z\gamma_3$, $\hat{P}_{\|} = P_0\gamma_0 - P_3\gamma_3$.

Equation (2) is valid in the domain, where the argument of D is greater than the electron Larmour radius squared $(L_B)^2 = (eB)^{-1}$. When $B = \infty$, this domain covers the whole exterior of the light cone $z^2 - t^2 \geq 0$.

In deriving equation (2) the expansion over the complete set of Ritus matrix eigenfunctions [10] was used in [2] that accumulate the dependence on the transversal spacial and spinorial degrees of freedom. This expansion yields an infinite set of equations, where different pairs of Landau quantum numbers n^{e}, n^{p} are entangled, Eq. (2) being the equation for the $(n^{\mathrm{e}} = n^{\mathrm{p}} = 0)$-component that decouples from this set in the limit $B = \infty$.

In the ultra-relativistic limit $P_0 = P_3 = P_\perp = 0$ equation (2) is solved by the most symmetric Ansatz $\Theta = I\Phi$, where I is the unit matrix.

The ultraviolet singularity on the light cone $(x^2 = 0)$ of the free photon propagator, $D(x^2) = -(\mathrm{i}/4\pi^2)(1/x^2)$, after this expression is used in eq. (2), leads to falling to the center in the Schrödinger-like differential equation

$$-\frac{\mathrm{d}^2\Psi(s)}{\mathrm{d}s^2} + \left(m^2 - \frac{1}{4s^2}\right)\Psi(s) = \frac{4\alpha}{\pi}\frac{1}{s^2}\Psi(s), \quad (eB)^{-1/2} \ll s \leq \infty, \qquad (3)$$

to which the radial part of equation (2) is reduced in the most symmetrical case, when the wave function $\Phi(x) = s^{-1/2}\Psi(s)$ does not depend on the hyperbolic angle ϕ in the space-like region of the two-dimensional Minkowsky space, $t = s\sinh\phi$, $z = s\cosh\phi$, $s = \sqrt{z^2 - t^2}$.

The solution that decreases at $s \to \infty$ is given by the McDonald function with imaginary index: $\Psi(s) = \sqrt{s}\ K_\nu(ms)$, $\nu = \mathrm{i}\,2\sqrt{\alpha/\pi} \simeq 0.096\,\mathrm{i}$, oscillating when $s \to 0$. The falling to the center [11] holds for any positive α.

According to [9], the singular equation (3) should be considered as the generalized eigenvalue problem with respect to α. The operator in the left-hand side is self-adjoint provided the standing wave condition is imposed,

$$\Psi(s)|_{s=L_B} = 0\,, \qquad (4)$$

that treats the Larmour radius as the lower edge of the normalization box. The discrete eigenvalues $\alpha_n(L_B)$ condense in the limit $B = \infty$ to become a continuum of states that form the (rigged) Hilbert space of vectors orthogonal with the singular measure $s^{-2}ds$. The latter fact allows to normalize them to δ-functions and interpret as free particles emitted and absorbed by the singular center. As long as the Larmour radius is much smaller than the only characteristic length in eq.(3), electron Compton length, $L_B \ll m^{-1} \simeq 3.9 \times 10^{-11}$ cm., the small-distance asymptotic regime is reached, and nothing "behind the horizon," $s < L_B$, - where the two-dimensional equations (2) and (3) are not valid - may affect the problem. In this way the existence of the limit of vanishing regularization length, impossible in the standard theory, is achieved.

For sufficiently large magnetic fields L_B becomes so small that the region where $K_\nu(ms)$ oscillates gets inside the domain of validity of equation (3). The value of the magnetic field when this happens for the first node is just (1). The corresponding Larmour radius is about fourteen orders of magnitude smaller than m^{-1} and makes $\sim 10^{-25}$ cm. Eq. (1) tells how large the magnetic field should be in order that the boundary problem (3), (4) might have a solution, in other words, that the point $P_0 = \mathbf{P} = 0$ might belong to the spectrum. Therefore, if the magnetic field exceeds the first hypercritical value the positronium ground state exists [c] with its rest energy compensated for by the mass defect.

The ultra-relativistic state $P_\mu = 0$ has the internal structure of what was called a "confined state", belonging to kinematical domain called "sector III" in [9], i.e. the one whose wave function behaves as a standing wave combination of free particles near the lower edge of the normalization box and decreases as $\exp(-ms)$ at large distances. The effective "Bohr radius", i.e. the value of s that provides the maximum to the wave function makes $s_{\max} = 0.17m^{-1}$. This is certainly much less than the standard Bohr radius $(e^2m)^{-1}$. Taken at the level of 1/2 of its maximum value, the wave function is concentrated within the limits $0.006\ m^{-1} < s < 1.1\ m^{-1}$. But the effective region occupied by the confined state is still much closer to $s = 0$, since - in accord with the aforesaid - the probability density of the confined state is the wave function squared *weighted with the measure* $s^{-2}ds$ *singular in the origin* [9] and is hence concentrated near the edge of the normalization box $s \simeq 10^{-25}$cm, and not in the vicinity of the maximum of the wave function. The electric fields

[c] A relation like (1) is present in [8]. There, however, a different problem is studied and, correspondingly, a different meaning is attributed to that relation: it expresses the mass gained dynamically - in the course of spontaneous breakdown of the chiral invariance in massless QED - by a massless Fermion in terms of the magnetic field applied to it.

at such distances are about 10^{43} Volt/cm. Certainly, there is no evidence that the standard quantum theory should be valid under such conditions. This fact encourages the use of a theory that admits deviations from the standard quantum theory that close to the singularity point.

At $B = B^{(1)}_{\mathrm{hpcr}}$ the total energy and momentum of a positronium in the ground state are zero. This state is not separated from the vacuum by an energy gap, and has maximum symmetry in the coordinate and spin space. Hence, it may be related to the vacuum and responsible for its structure.

At $B > B^{(1)}_{\mathrm{hpcr}}$ the eigenvalues of the BS equation (2) for the total 2-momentum components $P_{0,3}$ of the e^+-e^- system are expected to shift into the space-like region (we keep $P_\perp = 0$), whereas for $B < B^{(1)}_{\mathrm{hpcr}}$ the c.m. 2-momentum of the real pair was, naturally, time-like. If $P_{0,3} \neq 0$, at least for far space-like $P_\parallel$, the situation can be modelled by the same equation as (3), but with the large negative quantity $m^2 + P_\parallel^2/4$ substituted for m^2. Then the wave function would contain two oscillating exponentials for large space-like intervals,

$$\exp\left\{\pm \mathrm{i}s\sqrt{\left|m^2 + \frac{P_\parallel^2}{4}\right|}\right\}\exp\left\{\mathrm{i}P_\parallel\frac{x_\mathrm{e}+x_\mathrm{p}}{2}\right\}, \quad s \gg (-P_\parallel^2)^{-1/2}, \tag{5}$$

and two oscillating exponentials $\exp(\pm 2\mathrm{i}\ln s\,\sqrt{\alpha/\pi})$ for small ones, $s \sim L_\mathrm{B}$. In the Lorentz frame, where $P_0 = 0, P_3 \neq 0$, and with the time arguments in the two-time BS amplitude equal to one another: $x_0^\mathrm{e} = x_0^\mathrm{p}$, the solution oscillates along the magnetic field with respect to the relative coordinate $x_3^\mathrm{e} - x_3^\mathrm{p}$ (mutually free particles) and with respect to the c.m. coordinate $x_3^\mathrm{e} + x_3^\mathrm{p}$ (vacuum lattice).

We are now in the kinematical domain called sector IV, or deconfinement sector in Refs. [9]. Here the constituents are free at large intervals and near the singular point $s = 0$. The wave incoming from infinity is partially reflected, and partially penetrates to the singular point, the probability of creation of the delocalized (free) states being determined by the barrier transmission coefficient [9]. Such states may exist if one succeeds to satisfy e.g. periodic conditions, to be imposed on the lower and upper boundaries of the normalization volume, instead of condition (4), appropriate in sector III. The possibility to obey them is provided again by the falling to the center. Now, the delocalized states in two-dimensional Minkowsky space correspond to electron and positron that circle along Larmour orbits with vanishing radii in the plane orthogonal to the magnetic field and simultaneously perform, when the interval between them is large, a free motion along the magnetic field. They have magnetic moments and seem to be capable of screening the magnetic field. This provides the mechanism that may prevent the classical magnetic field from being larger than a *second hypercritical field*, for which the delocalization first appears. No sooner than the delocalized states are found in our present problem one may definitely claim the instability of the vacuum with the second hypercritical magnetic field or - which is the same - the instability of such field under the pair creation that might provide the mechanism for its diminishing. For the present, we state that the first hypervalue (1) is such a value of the magnetic field, the exceeding of which would already cause restructuring of the vacuum.

Acknowledgments

This work was supported by the Russian Foundation for Basic Research (project no 05-02-17217) and the President of Russia Programme (LSS-1578.2003.2), as well as by the Israel Science Foundation of the Israel Academy of Sciences and Humanities.

References

[1] M. Bander and H.R. Rubinstein, *Phys.Lett.* **B** 280, 121 (1992); **289**, 385 (1992); R.C. Duncan, astro-ph/0002442;
[2] A.E. Shabad and V.V. Usov, hep-th/0512236.
[3] E. Witten, *Nucl.Phys.* **B** 249, 557 (1985); E.M. Chudnovsky, G.B. Field, and D.N. Spergel, *Phys.Rev.* **D** 34, 944 (1986).
[4] S. Kawati and A. Kokado, *Phys.Rev.* **D** 39, 3612 (1989); D. Grasso and H.R. Rubinstein, *Phys.Rep.* **348**, 163 (2001).
[5] J.S. Goldstein, *Phys.Rev.* **91**, 1516 (1953); P.I. Fomin, V.P. Gusynin, V.A. Miransky and Yu.A. Sitenko, *Rivista Nuovo Cimento* **6**, numero 5 (1983); N. Setô, *Progr.Theor.Phys.Suppl.* **95**, 25 (1988).
[6] R. Loudon, *Am.J.Phys.* **27**, 649 (1959); Yu.M. Loskutov and V.V. Skobelev, *Phys.Lett.* **A** 56, 151 (1976); V.N. Oraevskii, A.I. Rez and V.B. Semikoz, *Zh.Eksp.Teor.Fiz.* **72**, 821 (1977); L.B. Leinson and V.N. Oraevskii, *Sov.J.Nucl.Phys.* **42**, 254 (1985).
[7] A.E. Shabad and V.V. Usov, *Astrophys. and Space Sci.* **117**, 309 (1985); *ibid* **128**, 377 (1986).
[8] V.P. Gusynin, V.A.Miransky and I.A. Shovkovy, *Phys.Rev.* **D** 52, 4747 (1995); *Nucl.Phys.* **B** 462, 249 (1996); C.N. Leung, Y.J. Ng and A.W. Ackley, *Phys.Rev.* **D** 54, 4181 (1996);
[9] A.E. Shabad, hep-th/0403177; *J.Phys.A: Math.Gen.* **38**, 7419 (2005).
[10] V.I. Ritus, *Sov.Phys.JETP* **48**, 778 (1978).
[11] L.D. Landau and E.M. Lifshits, "*Quantum Mechanics*", (Pergamon Press, Oxford) 1991.

ON INFLUENCE OF INTENSIVE STATIONARY ELECTROMAGNETIC FIELD ON THE BEHAVIOR OF FERMIONIC SYSTEMS

V.N.Rodionov,[a] A.M.Mandel, E.V.Arbuzova

Russian State Geological Prospecting University, 118873, Moscow, Russia,

International University "Dubna", 141980, Dubna, Moscow region, Russia

*Abstract.*Exact solutions for Schrödinger and Pauli equations for charged particles in an external stationary electromagnetic field of an arbitrary configuration are constructed. The corresponding equations for complex energy of particles bounded by short range potential are deduced. Explicit analytical expressions are given for the shift and width of a quasistationary level for different configurations of the external field. It is shown that the common view on the stabilizing role of magnetic field concerns only scalar particles.

An influence of external electromagnetic fields on nonrelativistic reactions with charged particles and on the behavior of bound systems has being investigated systematically for a long time [1–4]. However, a set of problems, in our opinion, requires additional studying. First, it is the appearance of physically meaningless divergences in equations for the complex energy of quasistationary bound states. Second, the question about the behavior of fermionic systems in super strong electromagnetic fields is not enough elucidated. It is important, that the field can become intensive not only due to increasing of the strength, but also due to decreasing of binding energy of the system. In other words, in vicinity of the reaction threshold an arbitrary external field is strong. Third, the common view on the stabilizing role of magnetic field in decays of quasistationary states is inadequate for spinor particles [5]. In this paper we treat a part of these deficiencies.

Let us consider a charged particle bounded by a short range potential like a δ-potential and located in an external stationary electromagnetic field of an arbitrary configuration. The process of particle coming out of a δ-well leads to the transition from an isolated energy level to a level in a continuous spectrum, *i.e.* to a decay of the bound state. The external electromagnetic field influences on this transition in two ways. First, it forms wave functions of the "free" state and, second, shifts and broads the bound level in a δ-well. As the result the initial state of the particle in the external field becomes quasistationary. The most adequate instrument for the investigation of such states is the complex energy formalism [2].

In order to receive the wave function of a free charged spinless particle in a stationary electromagnetic field of a general configuration, it is necessary to solve the Schrödinger equation with the following Hamiltonian:

$$\widehat{H}_{Sh} = -\frac{\hbar^2}{2m}\Delta - i\hbar\omega_H x\frac{\partial}{\partial y} + \frac{1}{2}m\omega_H^2 x^2 + eEx\sin\varphi + eEz\cos\varphi, \qquad (1)$$

[a]e-mail: physics@msgpa.ru

where e and m are the charge and the mass of the particle; E and H are the strengths of the electric and the magnetic fields, respectively; φ is the angle between the corresponding vectors; $\omega_H = eH/mc$ is the cyclotron frequency. We suppose that the magnetic field has the orientation along axis z, and the vector of the electric field strength lies in the plane xz.

The moving of the particle in a plane normal to the magnetic field has the same character as the one in the absence of electric field. Corresponding eigenfunctions are Hermit functions $u_n(\rho)$, each of them is related to a certain Landau level n . The movement of the particle along axis z is similar to the one in the uniform electric field, and it can be described by Airy functions. It is not difficult to see [6] that the wave function of the charged particle in the field of the considered configuration has the form

$$\psi_n(\vec{r},t) = N\,u_n(\rho)\,\mathrm{B}(\xi)\exp\left[\frac{i}{\hbar}\left(yp_y - t\tilde{W}\right)\right], \tag{2}$$

where N is a normalization coefficient, function $B(\xi)$ is proportional to the Airy function Ai regular at infinity.

The total energy $\tilde{W}$ of the particle in the given field is

$$\tilde{W} = \hbar\omega_H(n+1/2) + eEz_0\cos\varphi. \tag{3}$$

The solution of the Schrödinger equation constructed in that way is valid for a scalar particle. In order to take into account spin states of the particle in the concerned non-relativistic approximation it is necessary to pass on to the Pauli equation for the spinor wave function $\bar{\psi}_{n\sigma}(\vec{r},t)$.

The Hamiltonian of the Pauli equation involves the energy of the spin interaction with the external stationary magnetic field $\vec{H}$:

$$\widehat{H}_P = \widehat{H}_{Sh} + \frac{e\hbar}{2mc}\vec{\sigma}\vec{H}.$$

The spinor wave function of the mixed state with an arbitrary time dependence of the coefficients has the form

$$\bar{\psi}_{n\sigma}(\vec{r},t) = \psi_n(\vec{r},t)\begin{pmatrix} a_1(t)\exp(-i\omega_H t/2) \\ a_2(t)\exp(i\omega_H t/2) \end{pmatrix}, \tag{4}$$

where the scalar function $\psi_n(\vec{r},t)$ is defined by the formula (2).

In this case the spin projection on the axis z is conserved. This conservation is provided by the homogeneity and the stationarity of the magnetic field. The energy of the particle is defined, instead of the expression (3), by the more common formula

$$\tilde{W} = \hbar\omega_H(n+1/2+\sigma) + eEz_0\cos\varphi, \tag{5}$$

where $\sigma = 0$ for the scalar particle and $\sigma = \pm 1/2$ for the spinor one.

The scalar part of the time depending Green function of a charged particle in the given field can be expressed in terms of elementary functions. The expression for the probability P of the decay of a particle bound state in a unit of time (or for the width Γ of the bound level) in the considered field is received in [6] with help of the Green function, using the procedure developed in papers [7]. The final equation for the complex energy of a bound level in the stationary external field of a general configuration looks like

$$(-W)^{1/2} - (-W_0)^{1/2} = \frac{1}{2}\left(\frac{\hbar}{i\pi}\right)^{1/2} \int\limits_0^\infty \frac{dt}{t^{3/2}} \exp\left(\frac{i}{\hbar}W_0 t\right) \times \left\{\frac{\omega_H t}{2}\frac{f_\sigma(\omega_H t/2)}{\sin(\omega_H t/2)} \exp\left(\frac{iS}{\hbar}\right) - 1\right\}, \tag{6}$$

where

$$S = -\frac{(eE\cos\varphi)^2 t^3}{24m} + \frac{(eE\sin\varphi)^2 t}{2m\omega_H^2}\left[\frac{\omega_H t}{2}\cot\left(\frac{\omega_H t}{2}\right) - 1\right]. \tag{7}$$

Function f_σ is identically equal to unity for scalar particles, but for spinors it depends on the polarization of the bound particles in the initial state and of free particles in the final one.

We see that the received equation is resolved explicitly with respect to W. Note, that all integrals in intermediate expressions used in the derivation as well as in the final formula (6) *are finite*.

Let us stop now on some variants of the solution of the received equations for simple configurations of an external field. The first one is the external electrostatic field. At that our equation (6) takes the form [8]

$$(-\varepsilon)^{1/2} - 1 = (-\varepsilon_0)^{1/2} + \frac{F^{1/3}}{\pi}\left[Ai^{'}\left(Bi^{'} + iAi^{'}\right) + \varepsilon_0 F^{-2/3} Ai\,(Bi + iAi)\right],$$

where the common argument of the Airy functions Ai, Bi and of their derivatives Ai', Bi' is $-\varepsilon_0 F^{-2/3}$, and the zero approximation gives $\varepsilon_0 = -1$. Here the energy ε and the strength of the electrostatic field F are measured in units of $|W_0|$ and in characteristic atom units [4].

In a weak field ($F \ll 1$) we get the well-known asymptotic expression [2]

$$-\varepsilon = 1 + \frac{F^2}{16} + \frac{i}{4}F\exp\left(-\frac{4}{3F}\right).$$

In the strong field $F = 1$ (i.e. comparable with intra-atomic one) it is easy to get from (6) $-\varepsilon \approx 1{,}0411 + 0{,}0451 \cdot i$. One more configuration of an external field which we consider is essentially important to show the role of the magnetic field. For simplicity we put the angle $\varphi = 0$ and average over polarizations of bound particles. It is convenient to measure the magnetic field h like the electric one in natural atom units [4].

If $h \le F \ll 1$ it is possible to receive [5] from our equation (6)

$$(-\varepsilon)^{1/2} - 1 = \frac{F^2}{32} + \frac{h^2}{48a} + i\frac{F}{8}\left(1 + \frac{h}{3aF^2}\right)\exp\left(-\frac{4b^{3/2}}{3F}\right).$$

Here the dependent on spin factors a and b take values $a = -2, b = 1 + h/2$ for a scalar particle and values $a = b = 1$ for a spinor one. Therefore, the corrections to the complex energy due to a magnetic field essentially depend on spins of particles. The magnetic field decreases the width of a scalar particle bound level and by that stabilizes it. On the contrary, the width of the level of a particle with spin 1/2 increases in a magnetic field, *i.e.* the latter strengthens the destabilizing action of an electric field. That difference is still more noticeable in a strong magnetic field [5]. This field suppresses exponentially the withdrawal of scalar particles from the δ-well, but intensifies it linearly for particles with spin:

$$\mathrm{Im}(-\varepsilon)^{1/2}_{\sigma=0} \sim h^{1/2} \exp\left(-\frac{2^{1/2}h^{3/2}}{3F}\right); \qquad \mathrm{Im}(-\varepsilon)^{1/2}_{\sigma=1/2} \sim \frac{h}{F^{1/3}} Ai^2(F^{-3/2}).$$

If we distinguish particle polarizations after ionization it is not difficult to see with help of equation (6) that electrons with spins directed along a magnetic field in the initially non-polarized beam behave like scalar particles. Thus, the increasing of the level width in a magnetic field is provided only by electrons with spins oriented against the field.

So, for an adequate description of the influence of a magnetic field on the behavior of fermionic quasistationary systems it is necessary to take into account spin states of fermions even if the considered problem is not relativistic.

This work was supported in part by the grant of President of Russian Federation for leading scientific schools (Grants SS-2027.2003.2, SS-5332.2006.2)

References

[1] L. V. Keldysh, *Zh.Exsp.Teor.Fiz.* **34**, 1138 (1958), (*Sov.Phys.–JETP* **34**, 788 (1959)); W. Franz, *Z.Naturf.* A. **13.** 484 (1958); Ya. B. Zel'dovich, *Zh. Exsp. Teor. Fiz.* **39**, 776 (1961), (*Sov.Phys.–JETP* **12.**, 542 (1961)).

[2] Yu. N. Demkov and G. F. Drukarev, *Zh. Exsp. Teor. Fiz.* **47**, 918 (1964),(*Sov. Phys.–JETP* **20**, 614 (1965)).

[3] A. I. Baz, Ya. B. Zel'dovich, and A. M. Perelomov, *"Scattering, Reactions and Decays in Nonrelativistic Quantum Mechanics"*, (1971), (US Department of Commerce, Natural Technical Information Service, N69, 25848)

[4] V. S. Popov, B. M. Karnakov, and V. D. Mur, (1998) *Zh. Exsp. Teor. Fiz.* **113**, 1579 (1998), (*Sov.Phys.–JETP* **86**, 860 (1998)).

[5] V. N. Rodionov, G. A. Kravtzova, and A. M. Mandel, *Zh.Exsp.Teor.Fiz. Lett.* **75**, 435 (2002), (*Sov.Phys.–JETP Letters* **75**, 363 (2002)); V. N. Rodionov, G. A. Kravtzova, and A. M. Mandel, *Zh.Exsp.Teor.Fiz.Lett.* **78**, 253 (2003), (*Sov.Phys.–JETP Letters* **78**, 218 (2003)).

[6] Rodionov V.N., Mandel A.M., Arbuzova E.V. hep-ph/0510357.

[7] V. N. Rodionov, *Zh.Exsp.Teor.Fiz.* **111**, 3 (1997), (*Sov.Phys.–JETP* **84**, 1 (1998)); A. I. Nikishov and V. I. Ritus *Zh.Exsp.Teor.Fiz.* **85**, 1544 (1983), (*Sov.Phys.–JETP* **58**, 14 (1983)).

[8] V. G. Kadyshevskii and V. N. Rodionov, *Theor.Math.Phys.* **125**, 1668 (2000), (*Theor.Math.Phys.* **125**, 1654 (2000)).

RADIATIVE EFFECTS IN THE STANDARD MODEL EXTENSION

V.Ch. Zhukovsky[a], A.E. Lobanov[b], E.M. Murchikova[c]
Faculty of Physics, Moscow State University, 119992 Moscow, Russia

*Abstract.*The possibility of radiative effects that are due to interaction of fermions with the constant axial-vector background in the standard model extension is investigated. Electron-positron photo-production and photon emission by electrons and positrons were studied. The rates of these processes were calculated in the Furry picture. It was demonstrated that the rates obtained strongly depend on the polarization states of the particles involved. In consequence of this fact ultra-relativistic particles should occupy states with a preferred spin orientation, i.e., photons have the sign of polarization opposite to the sign of the effective potential, while charged particle are preferably in the state with the helicity coinciding with the sign of the effective potential. This leads to evident spatial asymmetries.

Physical laws have been confirmed to be CPT and Lorentz invariant with high accuracy in numerous experiments [1]. Nevertheless, for the last few years there have been made a number of assumptions that these symmetries are only approximate. The modern quantum field theoretical viewpoint admits a possibility of Lorentz invariance violation through spontaneous symmetry breaking mechanism in more fundamental theories, such as string theories, noncommutative geometry, etc. The resulting theory can be effectively described in the framework of the Standard Model Extension (SME) [2], which was recently used in calculations of various effects of Lorentz violation in the low energy range (see e.g. [3,4]).

In the present work we investigate the possibility of e^+e^- pair production by a photon and photon radiation by electrons induced by the interaction of fermions with the so called constant axial-vector Lorentz breaking background b^μ, — the name generally excepted in the framework of the SME. We assume that the photon dispersion law remains unchanged, and hence we neglect Cherenkov radiation. Our calculations are made in the framework of the Furry picture, i.e., without making any expansion in powers of the parameter b^μ.

The Dirac equation that takes into account interaction with the axial-vector term is

$$\left(i\gamma^\mu\partial_\mu - \gamma^\mu b_\mu\gamma^5 - m\right)\psi = 0. \tag{1}$$

For the sake of simplicity, we choose $b^\mu = \{b, 0, 0, 0\}$. Then the orthonormalized system of solutions for equation (1) can be written as follows [5]:

$$\psi(x) = \frac{|\Delta_{q\zeta}|}{\sqrt{2q^0}}\, e^{-iq^0x^0} e^{i\mathbf{q}\mathbf{x}\Delta_{q\zeta}}(\gamma^\mu q_\mu + m)(1 - \zeta\gamma^5\gamma_\nu S^\nu_{tp})\psi^0, \tag{2}$$

---

[a]e-mail: zhukovsk@phys.msu.ru
[b]e-mail: lobanov@phys.msu.ru
[c]e-mail: murchikova@yahoo.com

where $\Delta_{q\zeta} = 1 + \zeta b/|\mathbf{q}|$, $S^{\mu}_{tp} = \{|\mathbf{q}|/m, q^0\mathbf{q}/|\mathbf{q}|m\}$, and ψ^0 is a constant bispinor. Particle kinetic momentum components, related to group 4-velocity by the relation $q^{\mu} = mu^{\mu}$, $q^2 = m^2$, are used as quantum numbers of this problem. In this case the eigenvalues $\zeta = \pm 1$ determine the particle helicity.

It is clear that the relation of the canonical momentum $\mathbf{P}$ to the kinetic momentum $\mathbf{q}$ is determined by the formula $\mathbf{P} = \mathbf{q}\Delta_{q\zeta}$, and the particle energy is given by the same relation $\varepsilon = \sqrt{\mathbf{q}^2 + m^2}$, as in the case of a free particle in the Lorentz invariant theory.

Consider the electron-positron pair photo-production process. The probability of pair production by a polarized photon is defined by the relation

$$P = \frac{e^2}{k^0}\int d^4x\, d^4y \int \frac{d^4p\, d^4q}{(2\pi)^5}\,\delta(p^2 - m^2)\delta(q^2 - m^2) \times \mathrm{Sp}\{\gamma_{\mu}(x)\varrho_{\bar{e}}(x, y; p, \zeta_p)\gamma_{\nu}(y)\varrho_e(y, x; q, \zeta_q)\}\varrho^{\mu\nu}_{ph}(y, x; k). \tag{3}$$

Here, $\varrho_e(y, x; q, \zeta_q)$, $\varrho_{\bar{e}}(x, y; p, \zeta_p)$ are the electron and positron density matrices respectively, $\varrho^{\mu\nu}_{ph}(y, x; k)$ is the initial photon density matrix.

It is convenient to express the results of integrating with the use of dimensionless quantities. Introducing the notations

$$\begin{array}{lll} \eta = k^0/2m, & p^0/m = \eta + y, & q^0/m = \eta - y, \\ d = |b|/m, & \bar{\zeta}_{\bar{e},e} = \zeta_{p,q}\,\mathrm{sign}(b), & \bar{g} = g\,\mathrm{sign}(b), \end{array} \tag{4}$$

where g is the sign of helicity of the photon, we have

$$W_{\bar{g}\bar{\zeta}_{\bar{e}}\bar{\zeta}_e} = \frac{e^2 m}{32\eta^2}\int_{y_-}^{y_+} \frac{dy}{\sqrt{(\eta + y)^2 - 1}\sqrt{(\eta - y)^2 - 1}} \times \left[\bar{\zeta}_{\bar{e}}\sqrt{(\eta + y)^2 - 1} + \bar{\zeta}_e\sqrt{(\eta - y)^2 - 1} + 2d + 2\bar{g}\eta\right]^2. \tag{5}$$

The integration limits $y_{\pm}$ in the formula (5) are different for different values of parameter d. In the case $d < 1/2$, we have $y_{\pm} = \pm y_0$, if $\bar{\zeta}_{e,\bar{e}} = 1$. For other values of $\bar{\zeta}_{e,\bar{e}}$ the integral is equal to zero. Here

$$\eta_1 = \frac{1 + d^2}{2d}, \qquad y_0 = \frac{|\eta - d|\sqrt{\eta - \eta_1}}{\sqrt{\eta - d/2}}. \tag{6}$$

The result of integration can be expressed in terms of elliptic integrals.

In the case of interest $d \ll 1$, the following approximate formula is valid:

$$W_{\bar{g}\bar{\zeta}_{\bar{e}}\bar{\zeta}_e} \approx \frac{e^2 m}{16\eta}\ln\frac{1 + \sqrt{1 - \eta_1/\eta}}{1 - \sqrt{1 - \eta_1/\eta}}(1 + \bar{\zeta}_{\bar{e}})(1 + \bar{\zeta}_e)(1 + \bar{g}). \tag{7}$$

Near the reaction threshold defined by the relation $\eta = \eta_1$, the transition rate is described by the formula

$$W_{\bar{g}\bar{\zeta}_{\bar{e}}\bar{\zeta}_e} \approx \frac{e^2 m}{8\eta}\sqrt{1 - \eta_1/\eta}(1 + \bar{\zeta}_{\bar{e}})(1 + \bar{\zeta}_e)(1 + \bar{g}), \tag{8}$$

and for $d\eta \gg 1$, by the formula

$$W_{\bar{g}\bar{\zeta}_{\bar{e}}\bar{\zeta}_e} \approx \frac{e^2 m}{16\eta} \ln(8d\eta)\,(1+\bar{\zeta}_{\bar{e}})(1+\bar{\zeta}_e)(1+\bar{g}). \tag{9}$$

It is clear therefore that only the photons with the helicity sign coinciding with the sign of the effective potential can effectively produce pairs. For photons with the opposite helicity sign, the pair production is suppressed. The rate of the process as a function of the photon inverse energy is depicted in Fig.1.

Consider now the cross-channel to the above process that corresponds to photon radiation by a charged particle. It should be pointed out that the subsequent formulas are valid both for an electron and for a positron.

Introduce dimensionless variables

$$x = q^0/m, \quad \gamma = p^0/m, \quad d = |b|/m, \quad \bar{\zeta}_{i,f} = \zeta_{p,q}\,\mathrm{sign}(b), \tag{10}$$

then the radiative transition rate is defined as

$$W_{\bar{g}\bar{\zeta}_i\bar{\zeta}_f} = \frac{e^2 m}{8\gamma(\gamma^2-1)} \int_{x_-}^{x_+} \frac{dx}{\sqrt{x^2-1}} \times \left[\bar{\zeta}_i\sqrt{\gamma^2-1} + \bar{\zeta}_f\sqrt{x^2-1} + 2d + \bar{g}(\gamma - x)\right]^2. \tag{11}$$

The integration limits $x_\pm$ in the formula (11) are the same as for calculation of probability of neutrino spin light [5]. The integration is carried out elementary, therefore we show only approximate formulas for the case $d \ll 1$.

If $d\gamma \ll 1$, expression (11) leads to the formula

$$W_{\bar{g}\bar{\zeta}_i\bar{\zeta}_f} \approx \frac{2e^2 m d^3}{3}\gamma(\gamma^2-1)^{1/2}(1-\bar{g}\beta)^2(1-\bar{\zeta}_i)(1+\bar{\zeta}_f), \tag{12}$$

where $\beta = \sqrt{\gamma^2-1}/\gamma$ is the charged particle velocity. In the relativistic limit $(\gamma \gg 1)$, the transition rate is given by the expression

$$W_{\bar{g}\bar{\zeta}_i\bar{\zeta}_f} \approx \frac{e^2 m}{16\gamma}\left[\ln(1+4d\gamma) - \frac{4d\gamma(1+6d\gamma)}{(1+4d\gamma)^2}\right](1-\bar{\zeta}_i)(1+\bar{\zeta}_f)(1-\bar{g}). \tag{13}$$

Thus, we see that charged particles at high energies radiate primarily photons with the helicity sign opposite to the sign of the effective potential. Moreover, the radiation process is accompanied by the particle helicity flip. The rate of the process as a function of the electron inverse energy is depicted in Fig.2.

Let us consider now the radiation power. It can be verified that if $d\gamma \ll 1$, the radiation power becomes

$$I_{\bar{g}\bar{\zeta}_i\bar{\zeta}_f} \approx \frac{2e^2 m^2 d^4}{3}(\gamma^2-1)(1-\beta\bar{g})(3\gamma^2(1-\beta\bar{g})-1)(1-\bar{\zeta}_i)(1+\bar{\zeta}_f). \tag{14}$$

In the relativistic limit, the radiation power is equal to

$$I_{\bar{g}\bar{\zeta}_i\bar{\zeta}_f} \approx \frac{e^2 m^2}{16}\left[\ln(1+4d\gamma) - \frac{2d\gamma(6+15(4d\gamma)+11(4d\gamma)^2)}{3(1+4d\gamma)^3}\right](1-\bar{\zeta}_i)(1+\bar{\zeta}_f)(1-\bar{g}). \tag{15}$$

Hence, the total power of radiation increases logarithmically.

It can be seen from equations (13) and (15) that in the ultra-relativistic limit ($d\gamma \gg 1$) the average energy of emitted photons is $\langle\varepsilon_\gamma\rangle/\varepsilon_e \approx 1-(3\ln(1+4d\gamma)-9/2)^{-1}$. It is evident that initial particles loose almost all their energy through radiation. It should be emphasized that radiation by relativistic particles is completely circulary polarized, the polarization sign is opposite to the sign of the effective potential.

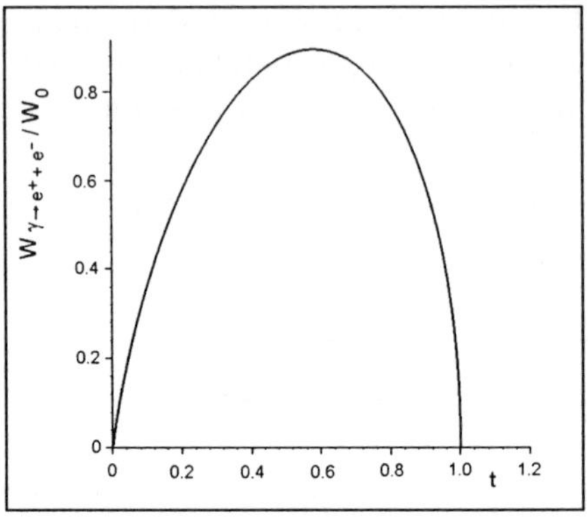

Figure 1: The rate of e^-e^+ pair production. $W_0 = e^2m/2\eta_1 \approx 10^{13}\,(|b|/1\text{eV})\,\text{sec}^{-1}$; $t = \eta_1/\eta \approx 2.5\cdot10^{11}(1\text{eV})^2/|b|\varepsilon_\gamma$.

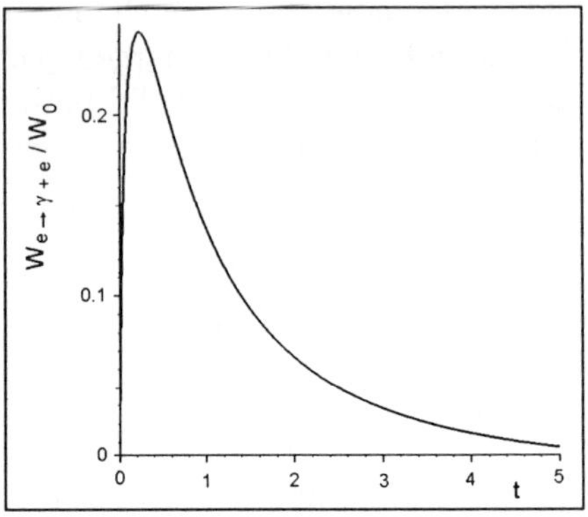

Figure 2: The rate of the radiative transition. $W_0 = 2e^2md \approx 2\cdot10^{13}\,(|b|/1\text{eV})\,\text{sec}^{-1}$; $t = (4d\gamma)^{-1} \approx 0.6\cdot10^{11}(1\text{eV})^2/|b|\varepsilon_e$.

It was demonstrated that the rates obtained strongly depend on the polarization states of the particles involved. In consequence of this fact ultra-relativistic particles should occupy states with a preferred spin orientation, i.e., photons have the sign of polarization opposite to the sign of the effective potential, while charged particle are preferably in the state with the helicity coinciding with the sign of the effective potential. This leads to evident spatial asymmetries in the reaction products, which may have certain consequences observable in astrophysical and cosmological studies.

This work was supported in part by the grant of President of Russian Federation for leading scientific schools (Grant SS — 2027.2003.2).

References

[1] K. Hagiwara et al., *Phys.Rev.D* **66**, 010001 (2002).

[2] S. M. Carroll, G. B. Field, and R. Jackiw, *Phys.Rev.D* **41**, 1231 (1990); D. Colladay and V. A. Kostelecký, *Phys.Rev.D* **55**, 6760 (1997); *Phys.Rev.D* **58**, 116002 (1998).

[3] R. Jackiw and V. A. Kostelecký, *Phys.Rev.Lett.* **82**, 3572 (1999).

[4] D. Ebert, V. Ch. Zhukovsky, and A. S. Razumovsky, *Phys.Rev.D*, **70**, 025003 (2004), hep-th/0401241.

[5] A. E. Lobanov, *Phys.Lett.B* **619**, 136 (2005), hep-ph/0506007.

TWO-FREQUENCY UNDULATOR AND HARMONIC GENERATION OF AN ULTRARELATIVISTIC ELECTRON

V.Mikhailin, K.Zhukovsky [a]

Department of Optics and Spectroscopy, Faculty of Physics, Moscow State University, 119992 Moscow, Russia

*Abstract.*The radiation spectrum of the relativistic electron in the two-frequency magnetic field, oscillating in both transversal directions is analytically investigated. The spectral properties of the planar bi-harmonic undulator radiation are discussed. The effect of the field and the undulator parameters is elucidated. It is demonstrated how the bi-harmonic undulator can be exploited to regulate the emission of certain selected harmonics and hence contribute to the development of the efficient devices with high extraction and narrow spectrum.

Since the first successful amplification experiment with the Free Electron Laser (FEL), conducted 30 years ago, significant changes in physics have affected the role of electron accelerators. Nowadays, they are used for the production of high brightness X-ray bursts with the ultimate goal of reaching the hard X-ray region. The improvement of the undulators technology during the past years [1-5] have given rise to new undulator design concepts, for example, the exotic undulator devices, capable of producing radiation with arbitrary polarization and enhancing the emission on a particular harmonic. Such devices employ special schemes like two-frequency and bi-harmonic undulators, where a relativistic electron crosses the undulator with the on-axis field, oscillating along the same axis with the relatively integer multiplies spatial periods (2, 3, 4, etc.). These devices are known for large extraction efficiency and narrow spectrum features (see, e.g., [6]). In this work, we consider some of possible solutions for the design of undulators that can also be employed for Self Amplified Spontaneous Emission in Free Electron Lasers (SASE FELs).

Consider the following undulator configuration:

$$\vec{H} = \left(0,\, H_y \cos\frac{2\pi x}{\lambda_1},\, H_z \sin\frac{2\pi x}{\lambda_2}\right),\ \lambda_1 \neq \lambda_2 \text{ and } H_y \neq H_z. \tag{1}$$

Relativistic electrons propagate in the magnetic undulator with N periods of length λ_1 along the x-axis. In this case, the following inequalities are assumed: $\gamma \gg 1, \beta_\perp \ll 1, \beta_\perp H_\parallel \ll H_\perp, \vec{E} = 0$. The key parameters of the undulator are, as follows:

$$k_1 = \frac{eH_y}{mc^2}\frac{\lambda_1}{2\pi},\ k_2 = \frac{eH_z}{mc^2}\frac{\lambda_1}{2\pi},\ k_1 \neq k_2, \tag{2}$$

and we assume without any loss of generality that the undulator frequencies $\omega_1 = 2\pi\beta_x^0 c/\lambda_1$, $\omega_2 = 2\pi\beta_x^0 c/\lambda_2$ are related with an integer factor p: $\omega_2 = p\omega_1$. The undulator brightness on the axis at $\theta = 0$ is expressed in terms of the generalized Bessel functions of four arguments [1,7-9] The resonance frequencies

[a] e-mail: zhukovsk@phys.msu.ru

ω_n are determined by the zeros in the phase $\phi_n = \frac{\omega}{2\gamma^2}(1+\frac{1}{2}(k_1^2+k_2^2)) - n\omega_1$, dependent on the unique harmonic number n and the undulator harmonic number p:

$$\omega_n = n\phi,\ n = 0, 1, 2, \ldots,\ \phi = \frac{2\omega_1\gamma^2}{1+k^2},\ k^2 = \frac{1}{2}(k_1^2+k_2^2)). \tag{3}$$

The intensity of the on-axis radiation is determined by the series over the resonance frequencies. The polarization of the emitted radiation is orthogonal to the undulator axis. For even undulator harmonic coefficient ($\omega_2 = p\omega_1,\ p$ even) the odd harmonics will only have a Z component and even harmonics will be Y polarized. For odd undulator number $p = 3, 5, 7\ldots$, even harmonics will be totally suppressed, and the remaining odd ones will have both Y and Z components. Examples of the radiation spectrum of the two-frequency undulator with $k_1 = k_2 = 1$ and undulator harmonic numbers $p = 2$ and $p = 3$ are demonstrated in Figs. 1.1(a) and 1.1(b) respectively with contributions of the Y (light grey columns) and Z (dark grey columns) components of intensity. The dominant harmonic is radiated with $n = p$, i.e. its frequency is $\omega = p\omega_1$. For far-distant frequencies ω_1 and ω_2 the radiation of low harmonics is suppressed. The increase of the harmonic undulator number p intensifies the radiation of the main and the higher harmonics, the reasonable increase of the parameter k has the same effect. With $k \sim 1 (k > 1)$, the spectrum widens and high harmonics suppress the main one. For $p > 2$ and $k_1 \sim k_2$, the spectrum becomes a complicated sequence of maximums and minimums with the general tendency to fade out with further growth of n (Fig. 1.2 with $p = 3$: Y and Z components of the harmonic brightness for the harmonic number n, generated on the double frequency undulators with $k_1 = k_2 = 2$ and $\omega_2 = 3\omega_1$). The radiation of higher harmonics may have very complex character with several local extremums for Y and Z components. The examples for the Y and Z components of the seventh harmonic $n = 7$ intensity, generated in the double frequency undulator with harmonic undulator number $p = 3$, are plotted in Figs. 1.3(a) and 1.3(b) respectively as functions of field parameters k_1 and k_2. Two local maximums of comparable intensity are easily seen in these plots. For $\omega_1 \neq \omega_2$, the minimum of the intensity of the 7-th harmonic no more follows the line $k_1 = k_2$ as was the case for $\omega_1 = \omega_2$. It can be found numerically. Evaluation of the intensity of the main harmonic radiation in the background of the near 10 harmonics and far harmonics indicates the best range of $k_{1,2} \approx 0.6 - 0.8$. Another example of the two frequency undulator is given by the planar bi-harmonic undulator (PBHU) [10, 11] — the magnetic device with linear polarization and with on-axis field, consisting of two parts, oscillating at different wave-lengths, one with period λ_u and the other with period λ_u/h, where h is an integer:

$$H_y = H_1[\sin(k_u z) + d\sin(hk_u z)],\ \ h \equiv \text{integer}. \tag{4}$$

In our work we employ the transparent mathematical formalism, based on the use of generalized Bessel functions [1,7-9] and allowing deep physical insight. The on axis undulator brightness can be obtained, following the usual procedure, illustrated in the previous section of our work. In this way, we obtain the integral representation for the on-line intensity that is essentially determined by the factor S_n. Fig. 1.4 gives an idea of the relative amount of the power, radiated by different harmonics and represents the S_n factor in the radiation integral for $h = 3, d = 0.1$ for the first (n=1, solid line) and third (n=3, dotted line) harmonic of the PBHU as functions of the undulator parameter k. In the case of even values of h, the S_n factor, plotted in Fig. 1.5 as a function of the d parameter for $k = 1.7, h = 2; n = 1$, – solid line, $n = 2$ – dotted line, and $n = 3$ – dashed line. The second harmonic gives a weak on-axis signal for the small value of d. For more substantive values of d, the harmonic is more intense and even stronger than the first, for $|d| > 0.4$. The first harmonic is not particularly sensitive to variations of d, while the third one is strongly influenced and can be even eliminated by the appropriate choice of parameter d (in this case $|d| = 0.5$). In the case of odd h: no even harmonic are radiated on axis. For example, for $h = 3, k = 1$, as is seen in Fig. 2.1 for $n = 1$ (solid line) and $n = 3$ (dotted line), the third harmonic is strongly suppressed for $d = 0.5$. In Fig. 2.2 ($d = 0.05, h = 3$), the main harmonic ($n = 1$, solid line) is dominant for $k < 1$, while the others ($n = 3$–dotted, $n = 5$–dashed, $n = 7$–dash-dotted) contribute only when $k > 1$. In Fig.2.3 with $h = 3$ the choice of $d = -0.5$ keeps the $n \geq 3$ harmonics ($n = 3$ – dotted, $n = 5$–dashed, $n = 7$–dash-dotted) significantly suppressed as compared with $n = 1$ (solid line) harmonic, whereas the $n = 5$ harmonic is enhanced for $k \sim 2$. Fig.2.4 shows that for the same $k \sim 2$ and $d = 0.5$, the fifth and the seventh harmonics are enhanced without any significant suppression of the first and third ones. This fact may have practical consequences as the reduction of FEL mirror degradation effects due to hard component of undulator radiation. To conclude the theoretical investigation of the harmonic radiation by the PBHU, we note that it is possible to extend and generalize this approach, considering the three component, as well as the multi-component cases. The systems considered so far have been conceived to enhance the effects associated with the e-beam bunching, consequent to FEL type interactions and can therefore be exploited in many flexible ways, including bunching induced by seeding. The analysis developed here can also be exploited in FEL oscillators as well as in SASE FEL. Thus, a bi-harmonic device can produce a large amount of higher harmonic intra-cavity power along with very short pulses using the mechanism of non-linear harmonic generation combined with pulse propagation and slippage effects. The use of "exotic" undulator in FEL oscillators may provide a new element to be added to a complex FEL device of the type considered so far in order to provide, e.g., a seed with high peak power and short time duration.

Fig. 1.1 (a)

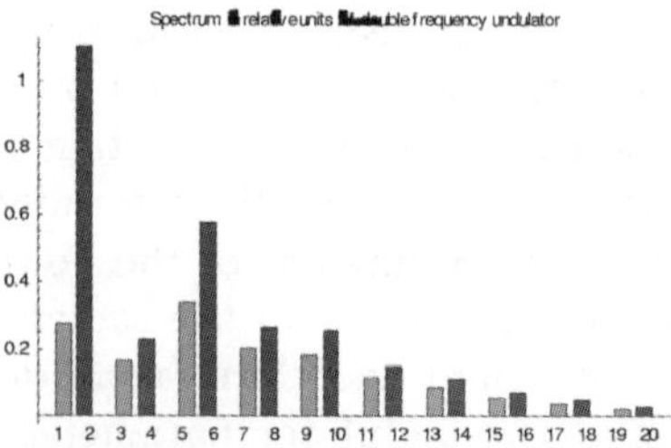

Fig. 1.1(b)

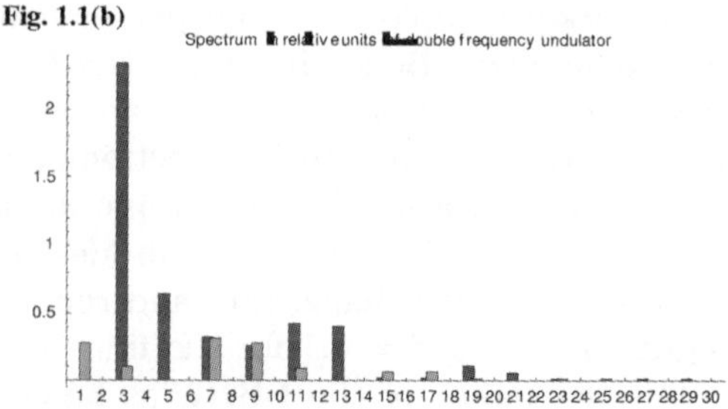

Fig. 1.2

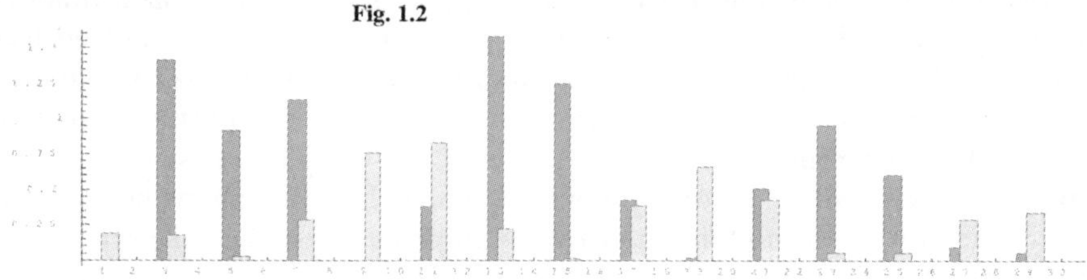

Fig. 1.3(a)

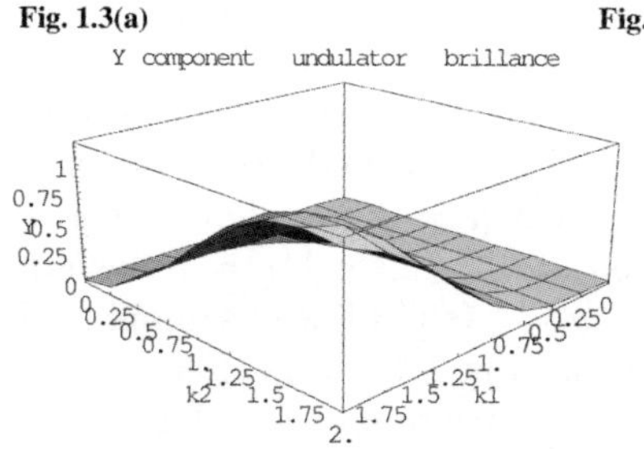

Fig. 1.3(b)

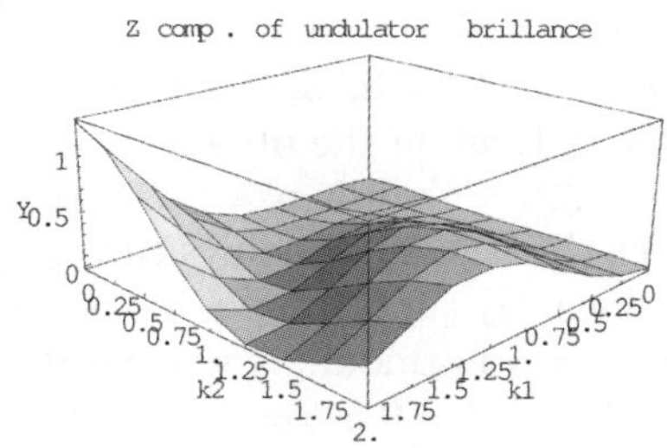

Fig. 1.4

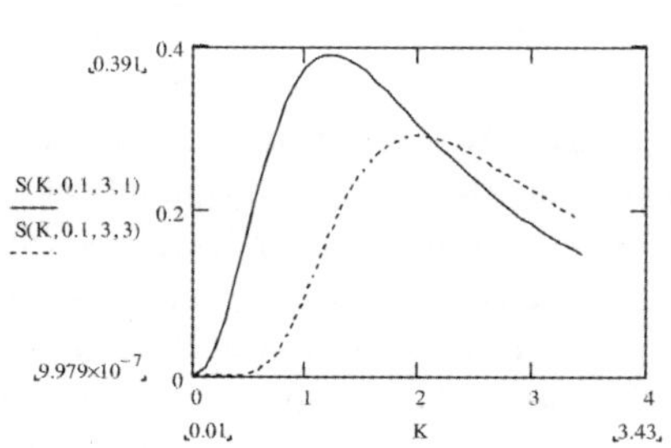

Fig. 1.5

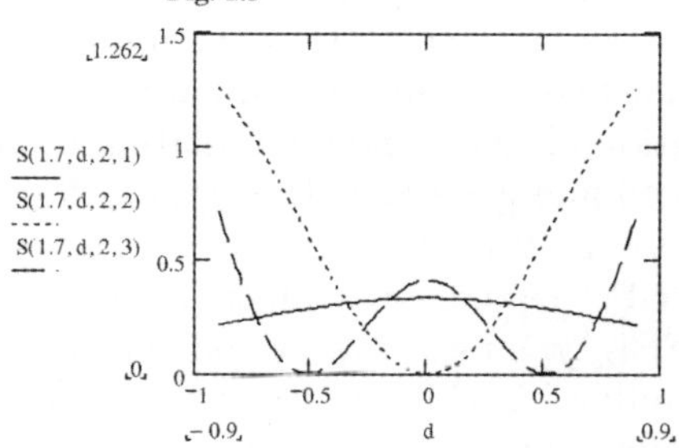

Fig. 2.1

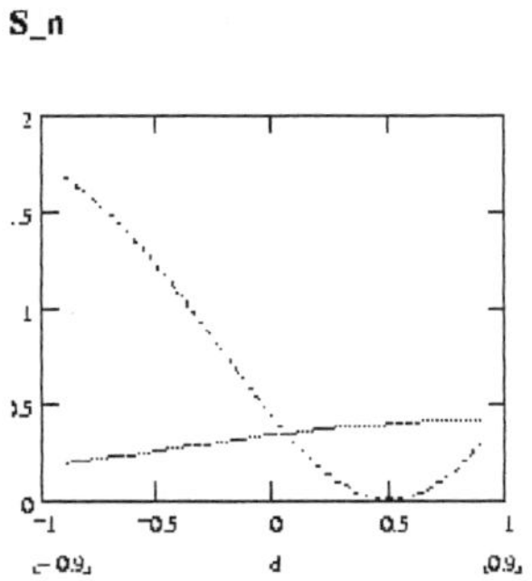

Fig. 2.2

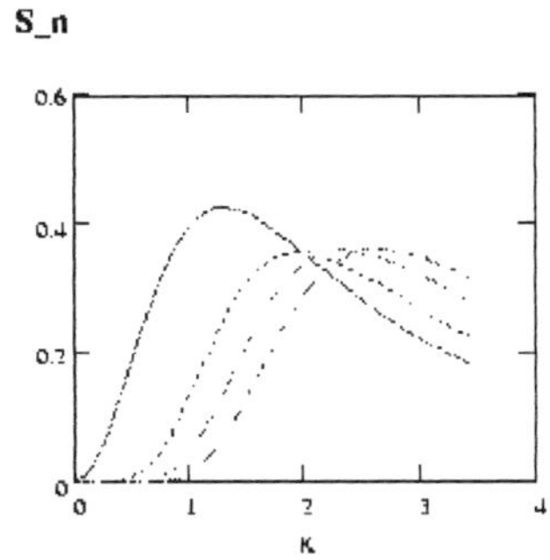

Fig. 2.3

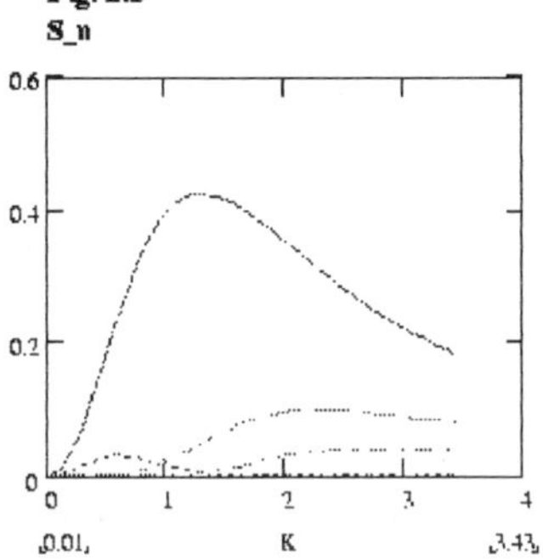

Fig. 2.4

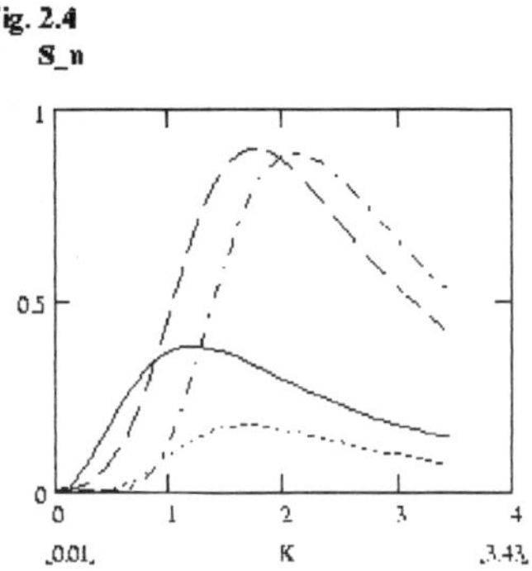

References

[1] I.M. Ternov, V.V. Mikhailin, V.R. Khalilov. *"Synchrotron Radiation and its Applications"* (Moscow) 1980.

[2] *"Synchrotron Radiation theory and its development"*, V.A Bordovitsyn ed. (World Scientific Publishing, Singapore), 1999.

[3] G.Dattoli, L.Gianessi, P.L.Ottaviani, H.P.Freund, S.G.Biedron, S.Milton. *Nuclear Instruments and Methods in Physics research* **A** 495, 48 (2002).

[4] A.A. Sokolov, D.V. Gal'tsov, V.Ch. Zhukovsky *Zh.Tekhn. Fiz.*, 43, 682 (1973)

[5] K-J. Kim, *Nucl. Instr. and Meth.*A 246, 67 (1986)

[6] W.B.Colson, C.Pellegrini, A.Renieri, Eds. *"Laser Handbook"*, vol.VI (North Holland, Amsterdam) 1993.

[7] G.Dattoli, A.Renieri, A.Torre, *"Lectures on the free electron Laser Theory and Related Topics"* (World Scientific) 1993.

[8] A.I.Nikishov and V.I.Ritus, *Zh. Eksp. Teor. Fiz.* 46, 776 (1964)

[9] G.Dattoli, L.Gianessi, L.Mezzi, A.Torre, *Nuovo Cim.* 105B, 327 (1990)

[10] G. Dattoli, A. Doria, L. Giannessi and P. L. Ottaviani, *Nucl. Instrum. Meth.* **A** 507, 388 (2003).

[11] G. Dattoli, P. L. Ottaviani, S. Pagnutti, *J. Appl. Phys.* 97, 11, 3102 (2005)

INFLUENCE OF THE ELECTRON SPIN ON THE ANGULAR DISTRIBUTION OF THE PHOTOELECTRONS

I. Andrievskiy[a], V. Tlyachev[b]

Faculty of Physics, Adygheya State University, 385000, Maykop, Russia

*Abstract.*The process of photoeffect for K- and L-shell is considered. Expressions for differential cross section correct to first order in αZ are obtained. It is shown that received expressions are depend on spin states

1 Introduction

Interest to a photoeffect does not disappear, and last years the new theoretical models describing process of photoionization (see, for example, [1] and the literature quoted there) are offered.

In the most of theoretical works given in [1, 2] the angular distribution of DCS (differential cross section) was considered post summation over electron's initial and final spin states. Therefore the opportunity to observe influence of electron spin states on angular distribution of DCS was lost.

In the given work we offer research of influence of electron spin states and photons' polarization on angular distribution of outgoing electrons.

We'll consider photons with linear or right circular polarization. For the last one, as it shown in [2], dependence of DCS for photoeffect on spin states is the most strong.

We expand upon the Sauter-Sommerfeld theory by taking into account the spin states of the initial and final electron. The photoeffect is concidered for K- and L-shell. The wave function of the final electron state is represented in Sommerfeld-Maue approximation. The initial state of the electron can be described by the relativistic wave function of a hydrogen-like atom.

Two physically allocated cases connected with a direction of a vector of polarization, a vector of velocity of a photoelectron and a direction of a falling photon are comprehensible to the further research: when the vector of polarization coincides with a direction of vector of photoelectron velocity and when the vector of polarization coincides with a direction of falling of a photon.

Let's characterize electron spin states by two-component spinor $w_\zeta(\vec{l})$, which satisfy an equation on eigenvalues [2]:

$$(\vec{\sigma}\vec{l})w_\zeta(\vec{l}) = \zeta w_\zeta(\vec{l}), \quad \zeta = \pm 1.$$

Here $\vec{\sigma}$ represents the Pauli spin operators; the vector $\vec{l}$ characterizes the spin direction: for $\zeta = 1$ the spin oriented parallel to $\vec{l}$, for $\zeta = -1$ it is antiparallel to $\vec{l}$.

[a]e-mail:JohnAndr@yandex.ru
[b]e-mail: tim1988@maykop.ru

The DCS for photoeffect can be presented in the same form for all cases:

$$\frac{d\sigma}{d\Omega} = N[(1+\zeta\zeta^{'})(S_1+\zeta S_2)+(1-\zeta\zeta^{'})(F_1+\zeta F_2)];$$
$$d\Omega = \sin(\theta)d\theta d\varphi, \quad 0\le\theta\le\pi, \quad 0\le\varphi\le 2\pi.$$

Here θ is polar emission angle, ζ describes initial electron spin state, $\zeta^{'}$ describes final electron spin state. S_1, S_2, F_1 and F_2 are the functions, which describe the angular distribution of photoelectrons.

2 Transition from a state $nS_{1/2}$

Actually the diagrams of dependence of the angular distribution on the angle θ have been analyzed in most works. But determining of the angle θ^{max} which has the maximum of DCS is very important point. Therefore it is useful to analyze the dynamics of dependence of θ^{max} from the photoelectrons velocity. All the diagrams in given work show this dependence.

For circular polarization of a photon at $\vec{l}=\vec{n}$ the curve 1 on Fig.1 is constructed for the basic maximum for $\zeta=-\zeta'=-1$, the curve 2 corresponds to values $\zeta=\zeta'=1$, and also coincides with a curve constructed for the differential section averaged on initial and final spin states. Besides that the curve 2 coincides with a picture of movement of a maximum for linear polarization at $\zeta=\zeta'=1$ and at $\zeta=\zeta'=-1$. Precisely the same picture is observed and at $\vec{l}=\vec{v}/v$. That is obvious with the physical point of view. Thus the curve 1 on Fig. 1 corresponds to values $\zeta=-\zeta'=1$, the curve 2 – corresponds to values $\zeta=\zeta'=-1$.

On Fig.2 the curve 1 for circular polarization of a photon in case when $\vec{l}=\vec{n}$ is constructed for the basic maximum and corresponds to values $\zeta=\zeta'=-1$. Except for that since $v=0.8$ there is a second (weakened) maximum to which the curve 2 is corresponded. For $\vec{l}=\vec{v}/v$ curves 1 and 2 on Fig.2 correspond according to the basic and the second maximum at $\zeta=\zeta'=1$.

Thus, it is necessary to note, that influence of spin characteristics at various transitions for different orientations of a vector of polarization exists. In case of transitions without changing direction of spin ($\zeta=\zeta'=1$ and $\zeta=\zeta'=-1$) both for $\vec{l}=\vec{v}/v$, and for $\vec{l}=\vec{n}$ (and for linear, and circular polarization of a photon) the plot of dependence of an angle θ^{max} for which it is maximum of the differential cross section as a function of the photoelectron velocity v is the same as the plot constructed for differential section averaged on initial and final spin states. It is shown that the influence of spin effects is the strongest in transitions with reorientation of spin: $\zeta=-\zeta'=-1$ for $\vec{l}=\vec{n}$; $\zeta=-\zeta'=1$ for $\vec{l}=\vec{v}/v$. Besides for circular polarization of photons in cases when $\zeta=\zeta'=-1$ at $\vec{l}=\vec{n}$ and $\zeta=\zeta'=-1$ at $\vec{l}=\vec{v}/v$ since $v=0.8$ appear the second (weakened) maximum of the differential cross section.

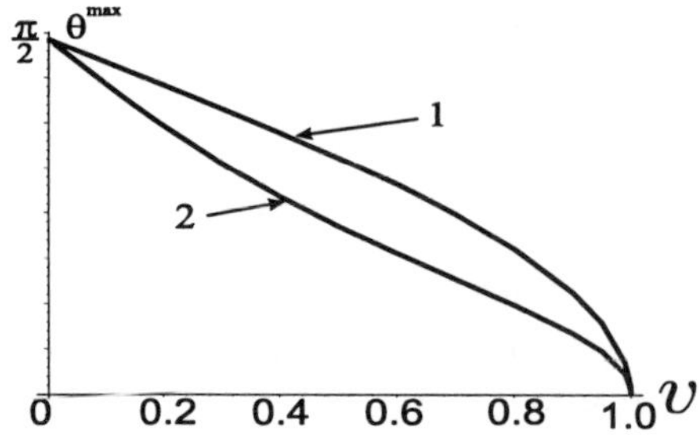

Figure 1: Distribution for polar emission angle for which it is maximum of the DCS as a function of the photoelectron velocity, K-shell. The curve 1 is constructed for the basic maximum for $\zeta = -\zeta' = -1$, the curve 2 corresponds to values $\zeta = \zeta' = 1$

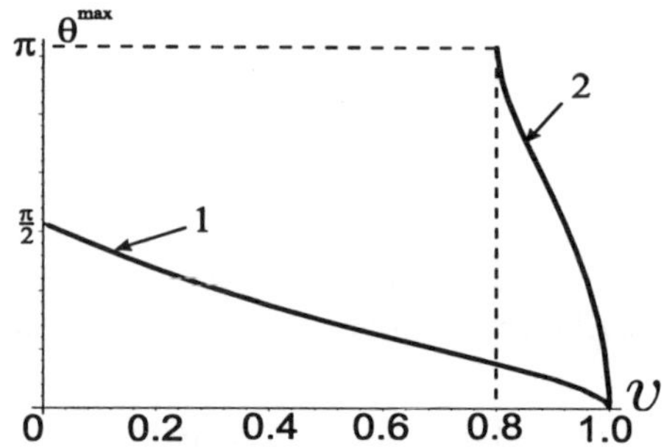

Figure 2: The curves 1 and 2 correspond according to the basic and the second maximum at $\zeta = \zeta' = 1$. $\vec{l} = \vec{v}/v$

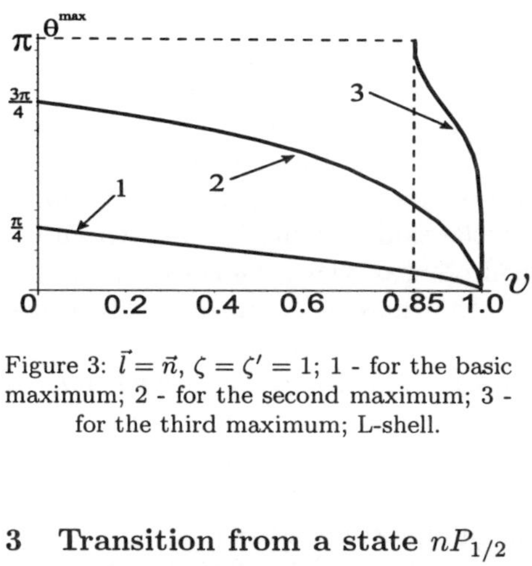

Figure 3: $\vec{l} = \vec{n}$, $\zeta = \zeta' = 1$; 1 - for the basic maximum; 2 - for the second maximum; 3 - for the third maximum; L-shell.

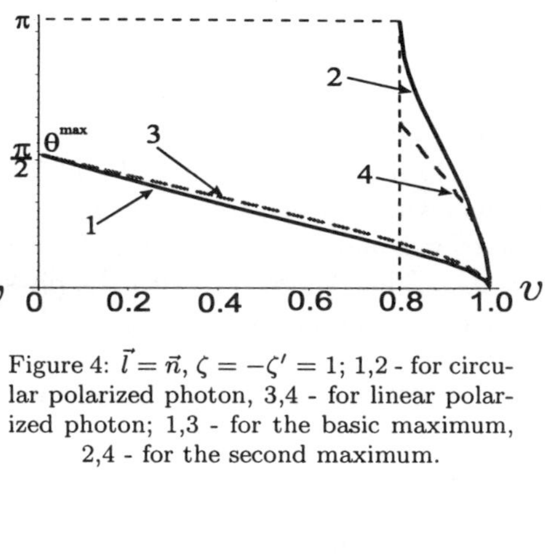

Figure 4: $\vec{l} = \vec{n}$, $\zeta = -\zeta' = 1$; 1,2 - for circular polarized photon, 3,4 - for linear polarized photon; 1,3 - for the basic maximum, 2,4 - for the second maximum.

3 Transition from a state $nP_{1/2}$

For a photoeffect from a L-shell when $\vec{l} = \vec{n}$ and $\zeta = \zeta' = 1$ on all an interval of values of v is present two maxima of cross section (the basic and the weakened) both for circular, and for linear polarization (and corresponding curves coincide). However for linear polarization since beta=0.85 there is a third maximum (see Fig.3). Other picture arises at $\zeta = \zeta' = 1$. There are only two maxima both for linear, and for circular polarization, thus corresponding curves do not coincide (see Fig.4).

For the case when $\vec{l} = \vec{v}/v$ the curve for the basic maximum is the same for $\zeta = \zeta' = 1$ and for $\zeta = \zeta' = -1$ both for linear, and for circular polarization of the photon. Since $v = 0,375$ there is a second maximum for the case of circular polarization of the photon at $\zeta = \zeta' = 1$. Since $v = 0,4$ the second

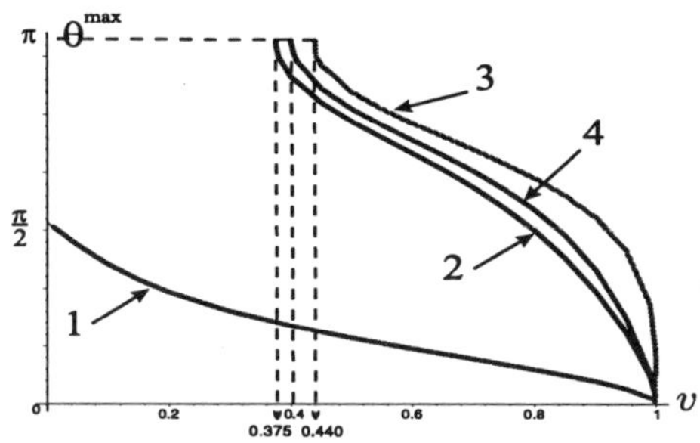

Figure 5: $\vec{l} = \vec{v}/v$; 1 - for the basic maximum; 2 - the second maximum when $\zeta = \zeta' = 1$, 3 - the second maximum when $\zeta = \zeta' = -1$ (for circular polarized photon); 4 - for the second maximum for linear polarized photon when $\zeta = \zeta' = 1$, $\zeta = \zeta' = -1$.

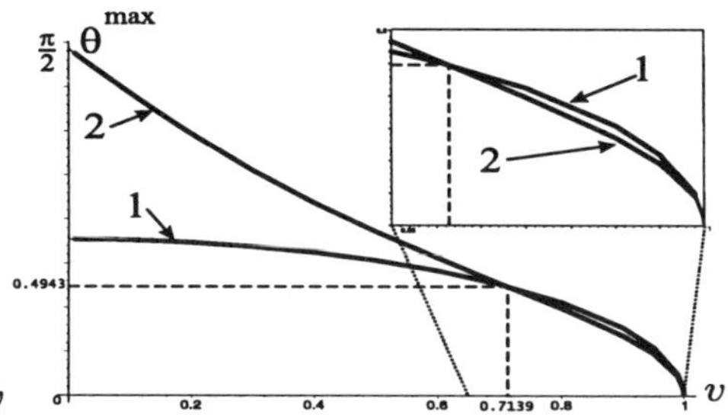

Figure 6: Distribution for the polar emission angle θ^{max} for which it is maximum of the differential cross section as a function of the photoelectron velocity v. 1 - for the cross section (10) from [3]; 2 - for the DCS for K-shell after averaging on initial and final spin states.

maximum arises for the case of linear polarization both for $\zeta = \zeta' = 1$ and $\zeta = \zeta' = -1$. Since $v = 0,440$ there is the second maximum for the case of circular polarization of photon at $\zeta = \zeta' = -1$ (see Fig.5).

Averaged differential section on initial and final spin states we can compare results of the given work to results of work [3] in which the differential cross section of photoeffect for K-shell correct to third order in αZ is received. As wave function for electron in final state the in [3] high-energy limit of the exact Coulomb wave function is used. In [3] is considered the relativistic photoeffect, therefore, apparently from Fig. 6, the received results not completely coincide with the results received by us (since in our work it is considered both relativistic, and not relativistic area). Nevertheless, for enough high velocities of photons ($v \geq 0,7$) the good consent is observed.

4 Conclusion

Thus, we obtained expressions for photoeffect differential cross section for K and L-shells for two cases of orientation of the vector of polarization $\vec{l} = \vec{v}/v$ and $\vec{l} = \vec{n}$ for linear and circular polarized photon. We analysed the influence of spin orientation on angular distribution of DCS. The dependence of the received expressions on spin states is shown.

References

[1] A.H. Sorensen, Phys. Rev. A **64**, 012703 (2001).

[2] V.G. Bagrov et al., Sov. Phys. JETP. **67(6)**, 1085 (1988).

[3] T.A. Weber, C.J. Mullin, Phys Rev. **126**, 615 (1962).

FROM RADIATION THEORY TO THE DYNAMICS OF RELATIVISTIC CHARGED PARTICLES

V.Bordovitsyn [a],T.O.Pozdeeva [b]

Faculty of Physics, Tomsk State University, 634050 Tomsk, Russia

*Abstract.*The most general differential conservation low of energy and momentum densities generated by an arbitrary moving relativistic charge is formulated. The radiation damping is obtained via integration over an extremely small hypersurface, which surrounds the world line of a point charge between two near proper time moments.

The damping force is one of the most important problems in quantum and classical electrodynamics of charged particles. Starting with H. A. Lorentz [1] and M. Abragam [2] many of the most eminent authors were engaged in this area (see reviewing works [3, 4]). Unfortunately part of these works was fulfilled in the noncovariant form that initially led to the hard confusion in the determination of the basic radiation characteristics [5] and even to incorrect physical summaries [5, 6] (for detailly see [4]).

The first entirely covariant formulation of the radiation theory and damping force was given by P. A. M. Dirac [7]. In many respects the further discussion of the moot points favoured to the clarification of this situation (see [3, 4] and others). Nevertheless the method suggested by Dirac for derivation of the damping force on the basis of the half-difference limit of the retarded and advanced field potentials and developed by his followers is to our mind not quite satisfactory in respect to physical point of view (see also [8]) and need in a more precise definition.

Note that the exotics for the damping force phenomenon repeatedly to reduced to the series of perfectly uncommon physical suppositions and hypotheses [9–14] deserving careful study until now. So the most simple and physically valid target setting of a single relativistic charged particle gains a special significance.

In the most general form of the conservation low of energy-momentum density in the differential looks like follows:

$$D_\nu \mathcal{P}^{\mu\nu}_e + D_\nu \mathcal{P}^{\mu\nu}_{ext} + D_\nu \mathcal{P}^{\mu\nu}_{self} = 0, \qquad (1)$$

Here $\mathcal{P}^{\mu\nu}_e$ and $\mathcal{P}^{\mu\nu}_{ext}$ are correspondingly the density tensors of the charge energy-momentum and the external electromagnetic field and

$$\mathcal{P}^{\mu\nu}_{self} = \overline{\mathcal{P}}^{\mu\nu} + \widetilde{\overline{\mathcal{P}}}^{\mu\nu} + \widetilde{\mathcal{P}}^{\mu\nu}$$

is the sum of density tensors of the energy-momentum of the electromagnetic fields connected wit a charge $\overline{\mathcal{P}}^{\mu\nu}$, the radiation field $\widetilde{\mathcal{P}}^{\mu\nu}$ and the mixed one

[a]e-mail: varbord@sibmail.com

[b]e-mail: pozdeeva@inbox.ru

$\widetilde{\overline{\mathcal{P}}}^{\mu\nu}$ (see e. g. [3, 15] and others).

As a result the equation (1) provides the balance of all the forces acting on a charged particle. To find the total force acting on a charge one ought to fulfill the limit transition from the field presentation to world line point of a point particle. As is well known in integrating over four-dimensional space the first two terms in (1) give the Lorentz force [16] [c]

$$F_e^\mu = \frac{dP_e^\mu}{d\tau} = m_0 w^\mu = \frac{e}{c} H_{ext}^{\mu\nu} v_\nu \tag{2}$$

We present the rest third term in such a way

$$\frac{dP_{self}^\mu}{d\tau} = \frac{d\overline{P}^\mu}{d\tau} + \frac{d\widetilde{\overline{P}}^\mu}{d\tau} + \frac{d\widetilde{P}^\mu}{d\tau} = \frac{1}{c}\int D_\nu \mathcal{P}_{self}^{\mu\nu} d\Upsilon.$$

here $d\Upsilon = cdTdV$ is the element of four-space volume. Let us proceed to integration over the closed hypersurface which element in the most general case can be presented as a number of elements of the light con and Bha-bha tube [17] by means of Ostragradsky-Gauss' law in the right hand of this formula:

$$d\sigma_\nu = \varepsilon^2 e_\nu d\Omega c d\tau. \tag{3}$$

Here

$$e_\nu = \frac{\tilde{r}_\nu}{\varepsilon} - \frac{1}{c} v_\nu \tag{4}$$

is a unit space-like vector satisfying the conditions

$$e_\nu e^\nu = 1, \; v_\nu e^\nu = 0.$$

The invariant quantity $\varepsilon = -\tilde{r}_\rho v^\rho / c$ plays the role of the spherical wave radius with the solid angle $d\Omega$ in the rest frame.

In integrating over the solid angle (see also [3]) with a primary element of the closed hypersurface in (2) we get

we get

$$\frac{dP_{self}^\nu}{d\tau} = \frac{1}{c}\frac{d}{d\tau}\int D_\nu \mathcal{P}_{self}^{\mu\nu} d\Upsilon = \frac{2}{3}\frac{e^2}{c^5}(w_\rho w^\rho v^\mu + \frac{c^3}{\varepsilon} w^\mu). \tag{5}$$

It is necessary to take in mind that all the kinematic variables are taken at the trajectory time moment t in the right hand of this identity. However as a whole all the expressions still corresponds to the observation time moment $\tilde{t}$. Thus, it is not adequate to use formula (5) at this step for the determining of the total balance of all the forces acting on a particle

[c] Here we use the metric in which $v^\mu v_\mu = -c^2$, here $\mu, \nu, \rho, ...0, 1, 2, 3$.

One ought to carry out the limit transition $\varepsilon \to 0$ that corresponds the substitution [d]

$$v^\mu \to v^\mu - w^\mu \Delta\tau,\ w^\mu \to w^\mu - \dot{w}^\mu \Delta\tau \tag{6}$$

where

$$\dot{w}^\mu = d^2 v^\mu / d\tau^2,\ \Delta\tau = \varepsilon / c \to 0.$$

Now in place of (5) we have

$$\frac{dP^\mu_{self}}{d\tau} = \frac{1}{c}\frac{d}{d\tau}\int D_\nu \mathcal{P}^{\mu\nu}_{self} d\Upsilon = \frac{2}{3}\frac{e^2}{c^3}\left(\frac{1}{c^2} w_\rho w^\rho v^\mu - \dot{w}^\mu + \frac{c}{\varepsilon} w^\mu\right).$$

Due to the conversation low of the total momentum

$$\frac{dP^\mu_e}{d\tau} + \frac{dP^\mu_{ext}}{d\tau} + \frac{dP^\mu_{self}}{d\tau} = 0.$$

following from (1) with account (2) we obtain

$$F^\mu_e = \left(m_0 + \frac{2}{3}\frac{e^2}{\varepsilon c^2}\right) w^\mu = \frac{e}{c} H^{\mu\nu}_{ext} v_\nu + \frac{2}{3}\frac{e^2}{c^3}\left(\dot{w}^\mu - \frac{1}{c^2} w_\rho w^\rho v^\mu\right). \tag{7}$$

To avoid the divergence of the second term in front of w^μ it should be taken in mind that the minimum possible value of ε coincides with the meaning of the classical electron radius r_0. Then the value

$$\tilde{m}_{dyn} = \frac{2}{3}\frac{e^2}{r_0 c^2} \tag{8}$$

is in good agreement with the well known expression for the dynamic electromagnetic electron mass (see f. e. [18] and others). Supposing the observed electron mass coincides with the magnitude $m_e = m_0 + \tilde{m}_{dyn}$ the identity (7) passes into the Lorentz-Abraham-Dirac (LAD) equation

$$F^\mu_e = m_e w^\mu = \frac{e}{c} H^{\mu\nu}_{ext} v_\nu + \Gamma^\mu, \tag{9}$$

which considers damping with the Abraham vector

$$\Gamma^\mu = \frac{2}{3}\frac{e^2}{c^3}\left(\dot{w}^\mu - \frac{1}{c^2} w_\rho w^\rho v^\mu\right).$$

If we use only the first formula for v_μ in the expansion (6) and assume that acceleration is determined only with the external electromagnetic field in the limit of the minimal time $\Delta\tau = \varepsilon / c$ we come to the Mo-Papas equation [13]

$$F^\mu_e = m_e w^\mu = \frac{e}{c} H^{\mu\nu}_{ext} + \frac{2}{3}\frac{e^2}{c^3}\left(\frac{e}{c} H^{\mu\nu} w_\nu - \frac{1}{c^2} w_\rho w^\rho v^\mu\right) \tag{10}$$

[d] It is easy to show that the use of these relations before and after integration gives the same result.

instead of the LAD equation (8).

In contrast to the LAD equation (10) does not have self-accelerating solutions. It follows that the problem of non-physical self-accelerating solutions arise only in the time intervals less than

$$\Delta\tau = \frac{\varepsilon}{c} = \frac{2}{3}\frac{e^2}{\tilde{m}_0 c^3}.$$

This conclusion is in perfect accordance with the rigorous quantum theory of radiation damping [19].

This work was supported by RFBR grant 06-02-16719.

References

[1] H.A.Lorentz, *Arch. Neer. Sci. Ex.* **25** 363 (1982);*"The Theory of Electrons"*, (Teubner , Leipzig), 1909; 2- nd ed.: 1915.

[2] M.Abragam, *Zs. Phys.* **14** 236 (1904); *"Theorie der Strahlung"*. Bd.2, 13-15, (Teubner , Leipzig),1905.

[3] C.Teilelboim, D.Villaroel, Ch. G van Weert, *Rivista Nuovo Cimento* **3** 1 (1980).

[4] V.L.Ginzburg, *"Theoretical Physics and Astrophysics"*, Ch.3.(Pergamon, Oxford) 1969.

[5] M.Born, *Ann. der Phys.* **30** (1909).

[6] W.Pauli, *" Theory of Relativity"*, Pergamon,Oxford) 1958.

[7] P.A.M.Dirac, *Proc. Roy. Soc.* **A167** (1938) .

[8] P.A.Hogan, *Nuovo Cimento* **B15** (1937).

[9] P.Caldirola, *Nuovo Cimento* **3** 297 (1956).

[10] L.Lanz, *Nuovo Cimento* **23** 196 (1962).

[11] P.F.Browne, *Ann. of Phys.* **59** 254 (1970).

[12] B.P.Kosyakov, *Usp. Fiz. Nauk* **162** 161 (1992); *"Fiz element. chastic i yadra"* 34, 2002, in Russian.

[13] Tse Chin Mo, C. H Papas, *Phys. Rev.* **D4** 3566 (1971).

[14] A.V.Sermyagin, *Preprint*, (JINR, Dubna) 1978, in Russian.

[15] V.G.Bagrov, V.A.Bordovitsyn, *"Synchrotron Radiation Theory and its Development"*. In memory of I.M. Ternov. Ed.: V.A.Bordovitsyn , (World Scientific, Singapore), 1999.

[16] L.D.Landau and E.M,Lifshitz, *"The Classical Theory of Fields"*, (Pergamon Press, Oxford), 1980.

[17] H.J Bha-bha, *Proc. Roy. Soc.* **A172** 384 (1939).

[18] J.Schwinger, *Found. Phys.* **13** 373 (1939).

[19] V.S.Krivitkii, V.N. Tsytovich *Usp. Fiz. Nauk* **161** 126 (1991), in Russian.

ON THE WAVE ZONE OF UNIFORMLY ACCELERATED CHARGE

V.A. Bordovitsyn [a], B.G. Bulenok [b], T.O. Pozdeeva

Tomsk State University, Lenin Ave. 36, 634050 Tomsk, Russia

Abstract. At first, a space structure of the wave zone of radiation from a single arbitrarily moving charge is investigated. It is shown that in case of hyperbolic acceleration the wave zone is spherically symmetric. This explains the well-known paradox of the absence of radiation damping force in this case.

1 Introduction

Let us remember at the beginning that the hyperbolic motion of uniformly accelerated charge occurs at constant acceleration a_0 in the rest (proper) frame and in this case the world line of straight line moving particle represents the equilateral hyperbola in the space-time plane.

Many long years the radiation problem of a uniformly accelerated charge was one of the most knotty problem of electrodynamics (see more detailed in [1,2]). This circumstance led to the series of widely discussed contradictions which started from the negation even of the existence of radiation a hyperbolically accelerated charge and in such a way the presence of the wave zone (Born paradox) correspondingly [3,4]. The basis of our work underlies the idea borrowed from the book by V. L. Ginzburg [1] which consists in states that in principle the radiation of uniformly accelerated charge does not differ from that one of an arbitrary accelerated charge.

2 The wave zone of radiation of an arbitrary moving charge

In the general case of an arbitrary moving charge the spatial structure of the radiation wave zone was investigated due to the general formulas of the radiation theory in the work [5]. According to the conclusions of this work (see also [6]) the wave zone border, where the power of pure electromagnetic wave field begins to prevail over the power of the convective-wave electromagnetic fields, must be determined taking into account the configuration of the angular distribution of power of all the electromagnetic fields generated by a charge.

In accordance with [5, 6] the analytic criterion of the wave zone formation can be presented in the following form:

$$\tilde{r} \gg \frac{c^2}{a} \frac{\beta(1-\beta^2)|(\beta-\cos\theta)\cos\varphi\sin\alpha+\sin\varphi\cos\alpha|\sin\theta}{(1-\beta\cos\theta)^2\sin^2\varphi\sin^2\alpha+[(\beta-\cos\theta)\cos\varphi\sin\alpha+\sin\theta\cos\alpha]^2}. \quad (1)$$

[a] e-mail: vabord@sibmail.com

[b] e-mail: vadim@tspu.edu.ru

Here α is an angle between the directions of velocity and acceleration vectors, the angles θ and φ define the direction from the charge to observer by means of the unit vector

$$\mathbf{n} = (\sin\theta\ \cos\varphi, \sin\theta\ \sin\varphi, \cos\theta)$$

in the instantaneous coordinate system with the velocity directed along the axis- Z and the acceleration lying in the plane XZ.

According to (1) the radiation wave zone of an arbitrary moving charge has very complicated anisotropic character. At first time this fact was checked in [7] by the example of synchrotron radiation (see also [8]).

The space anisotropy of the wave zone is mostly pronounced in the in the plain$(\beta, \mathbf{a})$ for which $\varphi = 0, \pi$. In this case the angle criterion of the way zone formation (1) takes the simplest form:

$$\tilde{r}|_{\varphi=0,\pi} \gg \frac{c^2}{a}\beta(1-\beta^2)\frac{\sin\theta}{|\sin\theta\cos\alpha \pm (\beta - \cos\theta)\sin\alpha|}. \tag{2}$$

It is easy to find from (2) the two directions

$$\theta_{1,2} = \arccos(\beta\sin^2\alpha \pm \cos\alpha\sqrt{1-\beta^2\cos^2\alpha}),$$

along which the wave zone tends to infinity. Obviously these are the two directions along which the radiation is absolutely absent. In the ultrarelativistic case of wave zone formation the wave zone border approaches very closely to

$$\theta_{1,2} = \arccos(\sin^2\alpha \pm \cos\alpha\sin\alpha)$$

practically without changing of their cross sizes orthogonal to directions $\theta_{1,2}$.

The above mentioned reasoning are illustrated in Fig.1. The conventional wave zone imagination corresponds to the equality sign in (1). Outside the surface shown in the figures the power of the radiation fields prevails, so it is the wave zone, but the inside space corresponds to the convective zone with the prevailing of the mixed convective wave fields. In such a way in the general case of an arbitrary moving charge the wave zone appears not immediately and not for any distances from a charge $\tilde{r}$. The field emitted at a moment t becomes a radiation field at different distances from the charge $\tilde{r} = c(\tilde{t} - t)$ and correspondingly at different time of observation $\tilde{t}$.

3 The wave zone of a hyperbolically accelerated charge

The criterion of a wave zone of a hyperbolic accelerated charge can be received from the above mentioned formulae by the limiting transition $\alpha \to 0$. It is a special case, when the criterion of a formation of a wave zone (1) loses dependence from angles

$$\tilde{r}_{\alpha=0} \gg \frac{c^2}{a}\beta(1-\beta^2),$$

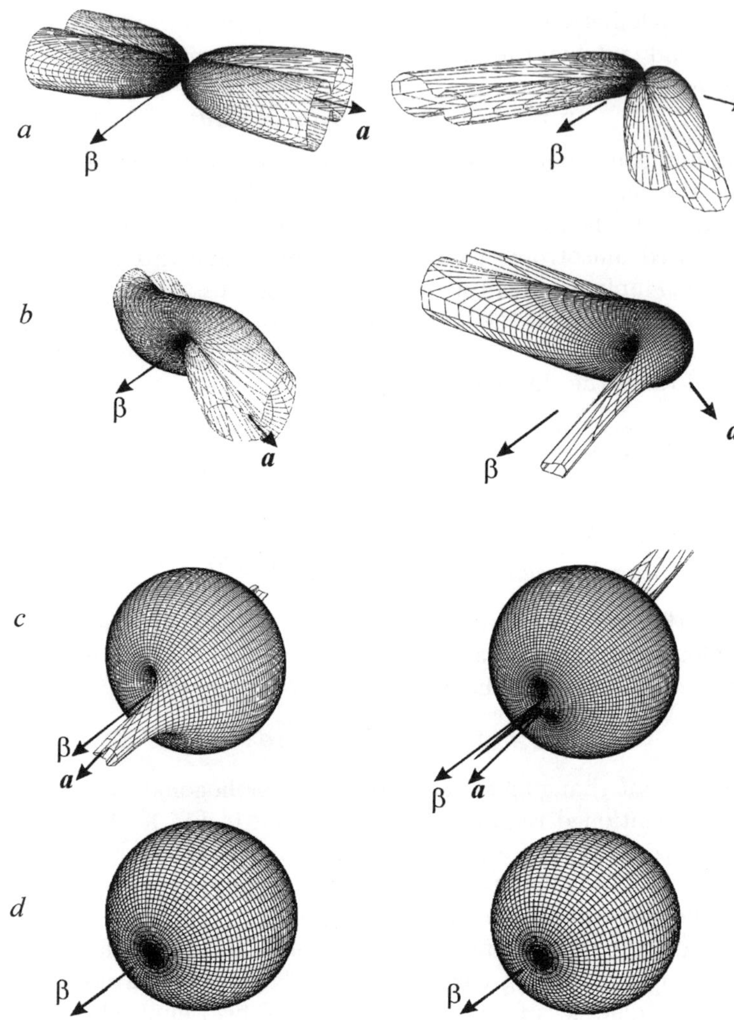

Figure 1: Evolution of a wave zone of radiation from $\beta = 0.001$ (left) to $\beta = 0.9$ (right) for $\tilde{r}|_{\alpha=\pi/2}$ (a), $\tilde{r}|_{\alpha=\pi/4}$ (b), $\tilde{r}|_{\alpha=\pi/30}$ (c), and $\tilde{r}|_{\alpha\simeq 0}$ (d).

which corresponds to spherical symmetry of a wave zone. In accordance with the equations of motion of a charge (see [1]) with the initial condition $\beta|_{t=0} = 0$ we shall have

$$\tilde{r}|_{\alpha=0} \gg ct. \tag{3}$$

Hence the wave zone of hyperbolically accelerated charge always exists and as it is shown in the pictures appears in the spherical symmetry manner at the same distances from a charge, and it must go to infinity when $\beta \to 1$.

As to the spatial directions along which the wave zone tends to infinity (here they correspond to the values $\theta = 0, \pi$) is should be noted that firstly the convective zone vanishes at the limit along these directions, and secondly as along the directions $\theta = 0, \pi$ the radiation is absolutely absent so there is no corresponding projection of the radiation damping. Along all the other directions the radiation of the hyperbolically accelerated charge is not zero, but the radiation damping is absent as before, because the energy loss are identical in all the directions and completely compensate each other. That's why the paradoxical effect of the radiation damping absence for the hyperbolically accelerated charge in spite of none zero self-radiation power can be considered as a consequence of the absolutely spherical symmetry of the wave zone of radiation.

Acknowledgments

This work was supported by RFBR grant 06-02-16719, and partially by grant of "Dynasty" fund.

References

[1] V.L. Ginzburg, *"Theoretical Physics and Astrophysics"*, Ch.3 (Pergamon, Oxford) 1969.
[2] F. Rohrlich, *"Classical Charged Particles"*, (Addison-Wesley, New York) Ch.5.3, 1990.
[3] M. Born, *Ann. der Phys.* **30**, 1 (1909).
[4] W. Pauli. *"Theory of Relativity"*, (Pergamon, Oxford) 1958.
[5] V.A. Bordovitsyn, V.G. Bulenok, T.O. Pozdeeva. *NIMB* **3**, 201 (2003)
[6] V.G. Bagrov, V.A. Bordovitsyn. In the book *"Radiation theory of relativstic particles"*,(Fizmatlit, Moscow, in Russian) 23-26, 2002.
[7] V.G. Bagrov. *Optika i Spectroskopiya* (in Russian) **18**, 541 (1965).
[8] V.G. Bagrov, V.A. Bordovitsyn, G.F. Kopytov. *Izv. wz. Fiz.* (in Russian) 15, N3, 30 (1972).

DYNAMICAL METHODS OF INVESTIGATION IN APPLICATION TO QUANTUM SYSTEMS

Yuri A. Rylov [a]

Abstract. One suggests to investigate the quantum systems simply as dynamic systems, ignoring the quantum principles and constraints imposed by them. Such dynamic methods of investigation appear to be more adequate and effective, than the conventional quantum methods of investigation. Using these methods, one can overcome the principal problem of quantum field theory: join of nonrelativistic quantum principles with the relativity principles. Investigation of dynamic system $\mathcal{S}_S$, described by the Schrödinger equation, admits one to prove that the Copenhagen interpretation is incompatible with quantum mechanics formalism. Besides, it is shown that sometimes the application of quantum principles leads to incorrect results.

The action for the Schrödinger particle $\mathcal{S}_S$ has the form

$$\mathcal{A}_S[\psi] = \int \left\{ \frac{i\hbar}{2}\left(\psi^* \partial_0 \psi - \partial_0 \psi^* \cdot \psi\right) - \frac{\hbar^2}{2m} \nabla \psi^* \nabla \psi \right\} d^4 x \qquad (1)$$

where $\psi = \psi(t, \mathbf{x})$ is a complex wave function. The action carries out the complete description of the Schrödinger particle $\mathcal{S}_S$.

Connection between the particle and the wave function is described by the relations.

$$\langle F(\mathbf{x}, \mathbf{p})\rangle = B \int \mathrm{Re}\left\{\psi^* F(\mathbf{x}, \hat{\mathbf{p}})\psi\right\} d\mathbf{x}, \qquad \hat{\mathbf{p}} = -i\hbar\nabla, \qquad B = \left(\int \psi^* \psi d\mathbf{x}\right)^{-1} \qquad (2)$$

which define the mean value $\langle F(\mathbf{x}, \mathbf{p})\rangle$ of any function $F(\mathbf{x}, \mathbf{p})$ of position $\mathbf{x}$ and momentum $\mathbf{p}$. We shall refer to these relations together with the restrictions imposed on its applications as the quantum principles, because von Neumann [1] has shown, that all proposition of quantum mechanics can be deduced from relations of this type.

Setting $\hbar = 0$ in the action (1), we hope to obtain a description of classical particle. Instead, the description disappears, and in this sense the description (1) is artificial. To obtain the natural description of the Schrödinger particle, we produce the change of variables

$$\psi \to \Psi_b : \quad \psi = |\Psi_b| \exp\left(\frac{b}{\hbar} \log \frac{\Psi_b}{|\Psi_b|}\right) \qquad b = \text{const} \neq 0 \qquad (3)$$

in the action (1). We obtain

$$\mathcal{A}_S[\Psi_b] = \int \left\{ \frac{ib}{2}\left(\Psi_b^* \partial_0 \Psi_b - \partial_0 \Psi_b^* \cdot \Psi_b\right) - \frac{b^2}{2m} \nabla \Psi_b^* \nabla \Psi_b + \frac{b^2}{2m}\left(\nabla|\Psi_b|\right)^2 - \frac{\hbar^2}{2m}\left(\nabla|\Psi_b|\right)^2 \right\} dt d\mathbf{x} \qquad (4)$$

The dynamic equation becomes to be nonlinear, if $b^2 \neq \hbar^2$

$$ib\partial_0 \Psi_b = -\frac{b^2}{2m}\nabla^2 \Psi_b - \frac{\hbar^2 - b^2}{8m}\left(\frac{(\nabla \rho)^2}{\rho^2} + 2\nabla \frac{\nabla \rho}{\rho}\right)\Psi_b, \tag{5}$$

However, the description becomes to be natural in the sense, that after setting $\hbar = 0$, the action $\mathcal{A}_S [\Psi_b]$ turns into the action

$$\mathcal{A}_{\mathrm{Scl}} [\Psi_b] = \int \left\{ \frac{ib}{2}\left(\Psi_b^* \partial_0 \Psi_b - \partial_0 \Psi_b^* \cdot \Psi_b\right) - \frac{b^2}{2m}\nabla \Psi_b^* \nabla \Psi + \frac{b^2}{2m}\left(\nabla |\Psi_b|\right)^2 \right\} dt d\mathbf{x} \tag{6}$$

which describes the statistical ensemble $\mathcal{E}[\mathcal{S}_{\mathrm{d}}]$ of free classical particles $\mathcal{S}_{\mathrm{d}}$. The action $\mathcal{A}_{\mathcal{E}[\mathcal{S}_{\mathrm{d}}]}$ for this statistical ensemble can be represented in the form

$$\mathcal{A}_{\mathcal{E}[\mathcal{S}_{\mathrm{d}}]}[\mathbf{x}] = \int \frac{m}{2}\left(\frac{d\mathbf{x}}{dt}\right)^2 dt d\boldsymbol{\xi} \tag{7}$$

where $\mathbf{x} = \mathbf{x}(t, \boldsymbol{\xi})$ is a 3-vector function of independent variables $t, \boldsymbol{\xi} = \{\xi_1, \xi_2, \xi_3\}$. The variables (Lagrangian coordinates) $\boldsymbol{\xi}$ label the particles $\mathcal{S}_{\mathrm{d}}$ of the statistical ensemble $\mathcal{E}[\mathcal{S}_{\mathrm{d}}]$. The statistical ensemble $\mathcal{E}[\mathcal{S}_{\mathrm{d}}]$ is a dynamic system of the hydrodynamic type. One can show that the dynamic system, described by the action $\mathcal{A}_{\mathrm{Scl}}[\Psi_b]$ (6) is a partial case (irrotational flow) of the dynamic system $\mathcal{E}[\mathcal{S}_{\mathrm{d}}]$ [2].

Change of variables, leading from the action $\mathcal{A}_{\mathcal{E}[\mathcal{S}_{\mathrm{d}}]}[\mathbf{x}]$ to the action $\mathcal{A}_{\mathrm{Scl}}[\Psi_b]$ contains integration (see [2] or mathematical appendices to [4, 5]). Because of this integration this change of variables was not known for a long time. The constant b in the action $\mathcal{A}_{\mathrm{Scl}}[\Psi_b]$ is an arbitrary constant of integration (gauge constant). Arbitrary integration functions are "hidden" inside the wave function Ψ_b. Thus, the limit of Schrödinger particle (4) at $\hbar \to 0$ is a statistical ensemble $\mathcal{E}[\mathcal{S}_{\mathrm{d}}]$, but not an individual particle $\mathcal{S}_{\mathrm{d}}$. It means, that the *wave function describes a statistical ensemble of particles, but not an individual particle*, and Copenhagen interpretation, where the wave function describes an individual particle, is *incompatible* with the quantum mechanics formalism.

One can show, that the Schrödinger particle $\mathcal{S}_{\mathrm{S}}$ is a special case of the statistical ensemble $\mathcal{E}[\mathcal{S}_{\mathrm{st}}]$ of stochastically moving free particles $\mathcal{S}_{\mathrm{st}}$, i.e. the action (4) for $\mathcal{S}_{\mathrm{S}}$ can be obtained from the action for the statistical ensemble $\mathcal{E}[\mathcal{S}_{\mathrm{st}}]$ by means of a proper change of variables [2].

The statistical ensemble $\mathcal{E}[\mathcal{S}_{\mathrm{st}}]$ of *stochastic* particles $\mathcal{S}_{\mathrm{st}}$ is a dynamic system, described by the action

$$\mathcal{A}_{\mathcal{E}[\mathcal{S}_{\mathrm{st}}]}[\mathbf{x}, \mathbf{u}_{\mathrm{df}}] = \int \left\{ \frac{m}{2}\left(\frac{d\mathbf{x}}{dt}\right)^2 + \frac{m}{2}\mathbf{u}_{\mathrm{df}}^2 - \frac{\hbar}{2}\nabla \mathbf{u}_{\mathrm{df}} \right\} dt d\boldsymbol{\xi} \tag{8}$$

where $\mathbf{u}_{\rm df} = \mathbf{u}_{\rm df}(t, \mathbf{x})$ is the diffusion velocity, describing the mean value of the stochastic component of the velocity, whereas $\frac{d\mathbf{x}}{dt}(t, \boldsymbol{\xi})$ describes the regular component of the particle velocity, and $\mathbf{x} = \mathbf{x}(t, \boldsymbol{\xi})$ is a 3-vector function of independent variables $t, \boldsymbol{\xi} = \{\xi_1, \xi_2, \xi_3\}$. The variables $\boldsymbol{\xi}$ label stochastic particles $\mathcal{S}_{\rm st}$, constituting the statistical ensemble. The operator $\boldsymbol{\nabla}$ is defined in the space of coordinates $\mathbf{x}$. Resolving dynamic equation $\delta\mathcal{A}/\delta\mathbf{u}_{\rm df} = 0$ with respect to $\mathbf{u}_{\rm df}$ and eliminating $\mathbf{u}_{\rm df}$, from dynamic equations, we obtain the dynamic equations of the hydrodynamic type

$$m\frac{d^2\mathbf{x}}{dt^2} = -\boldsymbol{\nabla}U(\rho, \boldsymbol{\nabla}\rho), \qquad U(\rho, \boldsymbol{\nabla}\rho) = \frac{\hbar^2}{8m}\left(\frac{(\boldsymbol{\nabla}\rho)^2}{\rho^2} - 2\frac{\boldsymbol{\nabla}^2\rho}{\rho}\right) \tag{9}$$

$$\rho = \left[\frac{\partial(x^1, x^2, x^3)}{\partial(\xi_1, \xi_2, \xi_3)}\right]^{-1} = \frac{\partial(\xi_1, \xi_2, \xi_3)}{\partial(x^1, x^2, x^3)} \tag{10}$$

Hydrodynamic equations (9) may be written in terms of the wave function [2,4,5]. The proper change of the variables together with *integration* turns the action (8) for $\mathcal{E}[\mathcal{S}_{\rm st}]$ into the action containing the quantum constant $\hbar$ and an arbitrary integration constant b.

The wave function is a method of description of *any ideal fluid,* but not a specific quantum quantity. Quantum and classical dynamic systems distinguish *dynamically* (by the form of their actions). The form of description (in terms of wave function, or in terms of position and momentum of the particle) is of no importance.

Considering the statistical ensemble of stochastic particles as a starting point of the quantum mechanics, *one does not need quantum principles*, because the interpretation is carried out directly via the particles of the statistical ensemble. Besides, the quantum theory turns into a consistent statistical theory, where there are two sorts of particles: (1) individual stochastic particle $\mathcal{S}_{\rm st}$ and (2) statistical average particle $\langle\mathcal{S}_{\rm st}\rangle$, which is the statistical ensemble, normalized to one particle. Formalism of quantum mechanics deals only with the statistical average particle $\langle\mathcal{S}_{\rm st}\rangle$. All predictions of QM relate only to $\langle\mathcal{S}_{\rm st}\rangle$. In accordance with two sorts of particles we have two sorts of measurements: (1) S-measurement produced over $\mathcal{S}_{\rm st}$ and (2) M-measurement produced over $\langle\mathcal{S}_{\rm st}\rangle$. The two kinds of measurements have different properties and one may not confuse them (see for details [3]).

Using statistical ensemble as a starting point, we can test validity the quantum principles, which appear to be wrong in some cases. For instance, the momentum distribution $w(\mathbf{p}) = |\langle\mathbf{p}|\psi\rangle|^2$ is in reality a distribution over mean momenta $\langle\mathbf{p}\rangle$ [6].

Description of "quantum particle" $\mathcal{S}_{\rm S} = \mathcal{E}[\mathcal{S}_{\rm st}]$ in the "classical language" is realized by equations

$$\frac{d\mathbf{p}}{dt} = -\boldsymbol{\nabla} U\left(\rho, \boldsymbol{\nabla}\rho\right), \qquad \frac{d\mathbf{x}}{dt} = \frac{\mathbf{p}}{m}, \tag{11}$$

where U is defined by the second relation (9).

Dynamic equations (11) are the *partial* differential equations, because ρ contains derivatives with respect to ξ_α, $\alpha = 1, 2, 3$.

Description of "classical particle" $\mathcal{S}_{\rm Scl} = \mathcal{E}\left[\mathcal{S}_{\rm d}\right]$ in the "classical language" has the form

$$\frac{d\mathbf{p}}{dt} = 0, \qquad \frac{d\mathbf{x}}{dt} = \frac{\mathbf{p}}{m} \tag{12}$$

where $\mathbf{x} = \mathbf{x}\left(t, \boldsymbol{\xi}\right)$, $\mathbf{p} = \mathbf{p}\left(t, \boldsymbol{\xi}\right)$. Dynamic equations (12) are *ordinary* differential equations.

Describing quantum system $\mathcal{S}$ and corresponding classical system $\mathcal{S}_{\rm cl}$ in the classical language, we recognize that dynamic equations for the quantum system $\mathcal{S}$ are *partial* differential equations, whereas dynamic equations for its classical approximation $\mathcal{S}_{\rm cl}$ are *ordinary* differential equations. The difference between the ordinary and partial differential equations is dynamical (but not quantum). This invariant difference is a foundation of the procedure of the dynamical disquantization (transition to classical approximation).

Dynamic disquantization is the procedure of projecting of derivatives onto the direction of 4-current j^k

$$\partial^l \rightarrow \partial_{||}^l = \frac{j^l j^k}{j^s j_s} \partial_k, \qquad l = 0, 1, 2, 3,$$

To obtain ordinary differential equations, one needs to project derivatives onto one direction.

The dynamic disquantization admits one to *formalize* the transition to the classical description. Being applied to Schrödinger particle $\mathcal{S}_{\rm S}$, the dynamic disquantization transform $\mathcal{S}_{\rm S} = \mathcal{E}\left[\mathcal{S}_{\rm st}\right]$ into the statistical ensemble $\mathcal{E}\left[\mathcal{S}_{\rm d}\right]$ of free classical particles $\mathcal{S}_{\rm d}$. The procedure of dynamic disquantization is determined by the dynamic system *completely* and does not refer to the quantum constant and the quantum principles.

[1] J.V. Neumann, *Mathematische Grundlagen der Quantenmechanik.* Berlin, Springer, 1932. chp. 4.
[2] Yu.A. Rylov, *J. Math. Phys.* **40**, 256, (1999)
[3] Yu.A. Rylov, e-print, http://arXiv.org/abs/physics/0210003
[4] Yu. A. Rylov, e-print, http://arXiv.org/abs/physics/0410045
[5] Yu. A. Rylov, e-print, http:// arXiv.org/abs/physics/0412032
[6] Yu. A. Rylov, e-print, http://arXiv.org/abs/physics/0402068
[7] Yu.A. Rylov, *J. Math. Phys.* **32**, 2092, (1991).
[8] Yu. A. Rylov e-print, http://arXiv.org/abs/physics/0504031

SOME ASPECTS OF MULTITRACE MATRIX MODELS

A.O.Shishanin [a]

Moscow State Industrial University, 109280 Moscow, Russia

*Abstract.*To examine solutions with two segments for hermitian Goldstone matrix model, specific parameter ξ described phase is defined. Existence conditions of phases are discussed with $\xi = 0$ or 1 and $\xi = 1/2$. Propagator and vacuum energy are calculated in a symmetric case with $\xi = 1/2$.

It is known from the work of 't Hooft [1] that the only planar diagrams survive the large N limit of SU(N) gauge field theory. Zero-dimensional matrix models are important since they count number of planar diagrams for many-dimensional field theories. The counting of planar diagrams was obtained by combinatoric [2] and analytical [3] methods. The solution [3] is unique but model with negative square of mass has some phases [4], [5], [6], [7].

Note that more famous application of zero-dimensional matrix models consists to fit the two-dimensional quantum gravity [8]. In particular, usual hermitian matrix models is a lattice analog of quantum gravity without matter [9]. It is possible to consider two-dimensional quantum gravity with major terms of curvature. More simple matrix model with multitrace terms for this case was solved in ref. [10]. If we leave in action only free term with twotrace summand at large N, this model behaves as vector model at $N \to \infty$ [11].

In this work we consider model united both [4] and [10] cases with action

$$S(\Phi) = -\frac{1}{2} tr\Phi^2 + \frac{g}{N} tr\Phi^4 + \frac{h}{N^2}(tr\Phi^2)^2. \tag{1}$$

Partition function Z and vacuum energy E_0 are related by the ratio

$$Z = e^{-N^2 E_0} = \int D\Phi e^{-S(\Phi)}. \tag{2}$$

Here Φ is a $N \times N$ hermitian matrix, g and h are the coupling constants.

Let us introduce scale the eigenvalue of Φ as $\lambda(x)$, and the eigenvalue distribution function $u(\lambda)$. Then the saddle point equation at the large N limit becomes

$$-\frac{1}{2}\lambda + 2g\lambda^3 + 2hD\lambda = \fint_L d\mu \frac{u(\mu)}{\lambda - \mu} \tag{3}$$

with condition for $u(\lambda)$

$$\int_L u(\lambda)\, d\lambda = 1,$$

$$u(\lambda) \geq 0, \lambda \in L, \qquad u(\lambda) = 0, \lambda \notin L.$$

[a] e-mail: shishandr@rambler.ru

Here D is the propagator of same model described by the following expression

$$D =< \frac{1}{N} tr\Phi^2 >= \int_L u(\lambda)\, \lambda^2\, d\lambda.$$

Assume that L consists from two segment [a,b] and [c,d]. To close the system of equations and find $u(\lambda)$ we enter new parameter of a model [5]:

$$\xi = \int_a^b u(\lambda)\, d\lambda. \tag{4}$$

Let us remind the features of theory when $h = 0$ [4]. Here it was found:
A. BIPZ solution [3],

$$u_A = \frac{1}{\pi}(2g\lambda^2 - \frac{1}{2} + 4ga^2)\sqrt{4a^2 - \lambda^2},$$

$$12ga^4 - a^2 - 1 = 0,$$

$$E^0(g) = -\frac{1}{24}(a^4 + 10a^2) - \frac{1}{2}loga^2 - \frac{3}{8};$$

B. two-cut symmetric solution with $\xi = \frac{1}{2}$(weak coupling solution) [4] - [6], $g < \frac{1}{16}$,

$$u_B = \frac{2g|\lambda|}{\pi}\sqrt{(\delta^2 - \lambda^2)(\lambda^2 - \gamma^2)}, \qquad \gamma \leq \lambda \leq \delta,$$

$$\delta^2 = \frac{1}{4g} + \frac{1}{\sqrt{g}}, \quad \gamma^2 = -\frac{1}{4g} + \frac{1}{\sqrt{g}}, \quad D = \frac{1}{4g},$$

the vacuum energy is defined as

$$E^0(g) = -\frac{1}{16g} + \frac{1}{4}\ln(4g) - \frac{3}{8}; \tag{5}$$

C. solution with $\xi = 0$ or $1(U(N)$ like solution) ,

$$u_C(\lambda) = \frac{1}{\pi}(2g\lambda^2 + 2g\sigma\lambda + 2g\sigma^2 + g\tau^2 - \frac{1}{2})\sqrt{(\sigma + \tau - \lambda)(\lambda - \sigma + \tau)},$$

$$\tau^2 = \frac{1 - \sqrt{1 - 60g}}{15g}, \quad \sigma^2 = \frac{3 + 2\sqrt{1 - 60g}}{20g}.$$

Let us return to more general theory with $h \neq 0$. In this case we obtain:
A. DDSW solution [10],

$$u_A(\lambda) = \frac{1}{\pi}(-\frac{1}{2} + 4ga^2 + 2hD + 2g\lambda^2)\sqrt{4a^2 - \lambda^2},$$

$$16gha^8 + (12g + 4h)a^4 - a^2 - 1 = 0, \qquad D = \frac{16ga^6 - a^4}{1 - 4ha^4},$$

$$E^0(g,h) = \frac{3}{4} - 4ga^4 - 6g^2a^8 - \frac{1}{2}loga^2;$$

B. two-cut symmetric solution with $\xi = \frac{1}{2}$,

$$u_B(\lambda) = \frac{2g|\lambda|}{\pi}\sqrt{(\delta^2 - \lambda^2)(\lambda^2 - \gamma^2)}, \quad \gamma \le \lambda \le \delta, \tag{6}$$

$$\delta^2 = \frac{1}{4(g+h)} + \frac{1}{\sqrt{g}}, \qquad \gamma^2 = -\frac{1}{4(g+h)} + \frac{1}{\sqrt{g}},$$

$$D = \frac{1}{4(g+h)};$$

the condition for exitence of phase is $\gamma^2 \ge 0$; this gives

$$g \ge 16(g+h)^2; \tag{7}$$

if $h = 0$, we have $g < \frac{1}{16}$; the vacuum energy can be expressed as

$$E^0(g,h) = -\frac{g}{16(g+h)^2} + \frac{1}{4}\ln(4g) - \frac{3}{8}; \tag{8}$$

C. solution with $\xi = 0$ or 1,

$$u_C(\lambda) = \frac{1}{\pi}(2g\lambda^2 + 2g\sigma\lambda + 2hD + 2g\sigma^2 + g\tau^2 - \frac{1}{2})\sqrt{(\sigma + \tau - \lambda)(\lambda - \sigma + \tau)}.$$

Equations, which define τ, σ, D, have the form:

$$3g\tau^2 + 2g\sigma^2 - \frac{1}{2} + 2hD = 0,$$

$$3g\sigma^2\tau^2 - \frac{1}{4}\tau^2 + \frac{3}{4}g\tau^4 + hD\tau^2 = 1,$$

$$D = \frac{\tau^2}{2}(6g\sigma^4 + \frac{1}{2}g\tau^4 + 6g\sigma^2\tau^2 - \frac{1}{2}\sigma^2 - \frac{1}{8}\tau^2 + hD(2\sigma^2 + \frac{1}{2}\tau^2)).$$

Excepting here σ and D, one can derive a single equation for τ

$$5g^2h\tau^8 + (11gh + 15g^2)\tau^4 - 2g\tau^2 + 4(g+h) = 0, \tag{9}$$

where σ and D connect with τ as

$$\sigma^2 = \frac{3}{8}\tau^2 + \frac{1}{2g\tau^2}, \quad D = \frac{2\tau^2 - 15g\tau^4 - 4}{8h\tau^2}.$$

To find a range of (g, h) for the definition of physical values we need compose the discriminant of given algebraic equation (9) for τ^2. From the multiplier of discriminant for this equation we find inequality which defines structure of corner's bifurcation as

$$225g \geq 4(45g + 41h)^2. \qquad (10)$$

A comparison between (7) and (10) shows that domain $\xi = 0$ or 1 is situated inside other parabola when $\xi = \frac{1}{2}$. Thus, we extended the formula (8) for vacuum energy in symmetric case [4], [6] and found the range of definition for two phases.

References

[1] 't Hooft G., *Nucl.Phys.* **B**72, 461(1974).

[2] Tutte W.T., *Can.J.Math.* **14**, 21(1962);
Koplik J., Neveu A., Nussinov S., *Nucl.Phys.* **B**123, 109(1977).

[3] Brezin E., Itzykson C., Parisi G. and Zuber J.B., *Comm.Math.Phys.* **59**, 35(1978).

[4] Shimamune Y., *Phys.Lett.* **B**108, 407(1982).

[5] Arefeva I.Ya., Ilchev A.S., Mitrjushkin, in Proc.*"III International Symposium on Selected Topics in Statistical Mechanics"*, Dubna, vol.1, p.20(1984).

[6] Cicuta G.M., Molinari L., Montaldi E., *Mod.Phys.Lett.* **A**1, 125(1986);

[7] Cicuta G.M., Molinari L., Montaldi E., *J.Phys. A: Math.Gen.* **20**, L67(1987);
Jurkiewicz J., *Phys.Lett.* **B**261, 260(1991);
Sasaki M. and Suzuki H., *Phys.Rev.* **D**43, 4015(1991);
Lechtenfeld O., *Int.J.Mod.Phys.* **A**7, 2335 and 7097(1992);
Senechal D., *Int.J.Mod.Phys.* **A**7, 1491(1992);
Akemann G,, *Nucl.Phys.* **B**482, 403(1996), hep-th/9606004.

[8] Di Francesco P., Ginsparg P., Zinn-Jystin J., *Phys.Rep.* **254**, 1(1995);
Ginsparg P. and Moore G. *Lectures on 2D string theory*,Cambridge University Press, 1993.

[9] David F., *Nucl.Phys.* **B**257, p.45 and p.543(1985);
Kazakov V.A., *Phys.Lett.* **B**159, 282(1985);
Ambjorn J., Durhuus B., Frolich J., *Nucl.Phys.* **B**257, 433(1985).

[10] Das S., Dhar A., Sengupta A. and Wadia S., *Mod.Phys.Lett.* **A**5, 1041(1990).

[11] Brezin E. and Hikami S., *J.Phys.* **A**12, 759(1979).

RESISTERS AND PACIFISTS, THE INTELLIGENTSIA FIGHTS BACK

by John Kuhn Bleimaier[a]
15 Witherspoon Street, Princeton, New Jersey 08542 USA

The intelligentsia has always been a bastion of idealism, a hotbed of activity in the quest for social justice. Members of the intelligentsia have traditionally been identified with opposition to repression. This has taken the form of passive resistance, political engagement, revolutionary activity and even terrorism. The latter, extreme response to repression, has, in itself, evoked a complex response from the intelligentsia. The use of terror as an implement for political change has elicited reactions ranging from sympathy to revulsion. While the intelligentsia may be unified by a shared idealism, members of the intelligentsia are often divided on the issue of how to appropriately respond to repression.

I

Members of the intelligentsia contemplating involvement in the political life of the community are faced with two radically different stereotypes. Shakespeare's Hamlet embodies the debilitating effect of intellectual activity. "To be or not to be." At the most fundamental level the thoughtful individual can find himself in a quandary leading to inaction. On the other hand the Plato's Philosopher King represents the ideal of the societal leader who sees the big picture, possesses a world view and functions decisively on the basis of disinterested comprehension of the social good. Both the Prince of Denmark and the Philosopher King would be likely to engender different responses from their fellow intellectuals who happen to be members of the society at large. Hamlet's indecisiveness may inspire a coup. Whereas intellectual citizens of the land of the Philosopher King who do not share his particular vision for society might plot revolution. In either instance, the repression of indecision or the repression of the visionary may evoke a response from the intelligentsia.

II

The modern world has its intellectual roots in the Protestant Reformation. The intellectual elite of Western Europe at the time responded to the tyranny of the Roman Church much as members of the intelligentsia have done to repression ever since. One group violently broke with Rome and established the freedom of conscience by the force of arms. Another group espoused the Christian ideal of nonviolence and founded sects of pacifists standing outside and in opposition to the social and religious order. The third group of intellectuals worked from within the Roman Catholic Church, using the political channels of its structure in order to effectuate the Counter Reformation.

The Reformation represents a pivotal moment in the development of the intelligentsia because it established the right, nay the obligation, of thinking people to confront the ideological monopoly of the powers that be. The Reformation placed the world of ideas at the center of the political arena. Prior to the Reformation princes fought over territory and for economic advantage. The

---

[a]B.A.; Master of International Affairs; Juris Doctor; member of the New York, New Jersey and US Supreme Court bars; Fellow of Mathey College, Princeton University

Reformation ushered in a world in which competing systems of ideas challenged one another in the political arena.

III

We may be appalled by the concept of conflicts based on differing theological doctrines. However, the intelligentsia has significance in the life of the body politic only when opposing ideas compete with one another. The intelligentsia was irrelevant during the Dark Ages when personal sovereigns contended with each other on the basis of self interest alone. Are wars waged because of dynastic rivalries less violent that those fought because of opposed ideologies? Indeed the clash of idea systems is potentially less frightful because such a clash can be resolved by nonviolent means. The winner and the looser in the sphere of ideas can be determined by the ballot box.

It should come as no surprise that the great philosophers of democracy burst on the world scene on the heals of the Reformation. When religious freedom of conscience has once been asserted can personal self determination be far behind? The French and English philosophers of the state formed the intellectual framework within which the debate on the future of society would be conducted. Suddenly intellectual issues became central to the life of the state and the intelligentsia attained a position of substantial importance in society.

Thomas Jefferson announced the supremacy of conscience in his ringing Declaration of Independence of the American colonies. With nothing more than the strength of an intellectual argument, he promulgated the doctrine that the laws of any sovereign are a nullity when they contravene the laws of God and nature. The Jeffersonian intelligentsia in the New World of the 18th Century established once and for all the pivotal nature of the intelligentsia's reaction to repression. When the intellectual leaders of the thirteen North American colonies determined that the sovereign of the British Empire had crossed the line of acceptable conduct in relation to his subjects, those American intellectuals could completely reject His Britannic Majesty's authority over their society. Our concept of democracy is itself an embodiment of the intelligentsia's reaction to repression.

Geography made the American Revolution possible. The North American intelligentsia was located in situo thousands of miles distant from the sovereign who sought to exercise tyrannical powers. The intellectuals and the population at large rose up against their oppressor and successfully asserted their authority. The intelligentsia's response to repression on the European continent was substantially different.

The French Revolution represented the assertion of the rights of man which had been articulated by members of the intelligentsia. However, the excesses of the mob as played out on the streets and squares of France took the form of the Terror. The social and political concepts developed by intellectuals were lost in the chaos of blood and smoke. Unlike its American predecessor, the French Revolution spun out of the control of the intelligentsia who were its theoretical progenitors.

IV

We can see Marxism as the response to the failure of the intelligentsia to follow through and come to power during the French Revolution. In the final analysis the French Revolution represented a failed response to autocratic repression. The principles of liberty equality and brotherhood were vociferously proclaimed. But after the maelstrom of the Napoleonic wars the old Bourbon monarchy was restored. In order for the intelligentsia to maintain control over the revolutionary response to repression it was necessary to install a transitional authoritarian polity, the dictatorship of the proletariat. The Marxist dictatorship of the proletariat represents nothing more than a misnomer. For this dictatorship to have any meaning or any practical chance of success it had to be a dictatorship of the Marxist intellectuals in the name of the proletariat.

The French Revolution had spun out of the control of its intellectual instigators and monarchical repression had thus not been successfully confronted. The Marxist model postulated the temporary establishment of an alternative authoritarian structure to channel the energies of the revolution and maintain it within the grasp of the Marxist intelligentsia.

V

Of course, Marxism was not the only intellectual response to repression in the 19th Century. There were utopian socialists and liberal democrats representing an infinite variety of ideological hues and shadings. Perhaps the most interesting of the political movements to have swept through the ranks of the intellectuals of the period was anarchism. Anarchism represents the classic response to repression. If the state is repressive then the answer to state tyranny would be the destruction of the state system of social control.

In a sense, anarchism and pacifism are two sides of the same coin. They could be viewed as two forms of extreme idealism which hold that the systems of societal control are themselves the causes of social injustice. Both the anarchist and the pacifist must seem to be naive from the perspective of the pragmatic social activist who believes that a society of laws, and of law enforcement, represents the best hope for the attainment of justice. On the contrary anarchists and pacifists believe that man can thrive in the absence of the state and its inescapable reliance on the threat or use of force.

Historically the 19th Century anarchists will probably remain most memorable because they introduced the concept of terrorism. This tool for social action represents the destabilization of authority by the application of frightful violence. The terrorist destroys the society's confidence in the government by demonstrating that the government is unable to deliver on its social contract to maintain security for the individual and for social institutions. A terrorist not only destroys a target but he makes every aspect of social existence seem frail and vulnerable. The terrorist destroys confidence in the system. The terrorist confronts the tyranny of state repression with the tyranny of fear. The ever present threat of bombings and assassinations kept more or less repressive regimes on the defensive for much of the last decades of the 19th Century.

Within the intelligentsia the terrorism of the anarchists helped to stimulate the reaction of the conservative intelligentsia. The apologists of monarchy, the proponents of gradualism and the advocates of socially conservative agendas gained support by pointing to the seemingly mindlessness violence advocated by the anarchist architects of terror.

Indeed one of the products of the cycle of state repression and anarchist terrorism was the development of the right wing intelligentsia. The ideologues of the extreme right took sustenance from the manifestations of anarchist violence during the period prior to the First World War. Absolute monarchists pointed to the apparent irrationality of the opponents of the state as providing justification for repression. While the majority of the intelligentsia consisted of opponents of state repression of various stripes, there was a distinct right wing intelligentsia which attempted to justify all the excesses of even the most repressive regimes.

After the First World War with the virtual collapse of monarchy in Europe, the right wing intelligentsia became allied with conservative agrarian populists with fascism as the result. The rapid rise of fascist movements in Italy, Germany and Spain came as a response to perceptions of Bolshevik repression in the Soviet Union. The architects of fascism were not, as is now widely supposed, merely thugs and hooligans. There was a fascist intelligentsia which articulated the theoretical basis of the nascent ideology. Interestingly there were even fascist manifestations in the arts and literature. The tragic events surrounding the Second World War have so thoroughly discredited the fascist ideology that it is easy to forget that this too was a movement within the intelligentsia.

VI

Indeed the Second World War can be seen as the true European Civil War. In virtually every nation of Europe there were members of the intelligentsia who supported variously liberal democracy, Marxism, socialism, fascism, chauvinism and a literal rainbow of subordinate ideologies. Far from a clear cut battle of good against evil, the Second World War was a massive bloodletting based on the release of a myriad of animosities compounded by the violent interaction of a plethora of antagonistic movements. Each of the ideological enemies had their systematic ideologues. The allegiances of the intelligentsia throughout the whole of Europe were hopelessly divided. At the beginning of the 1930s, in each major power, Britain, France and Germany, there were liberal democratic, communist and fascist movements of substantial significance.

From the dawn of mass culture propagandists have attempted to characterize wars as definitive conflicts between good and evil. In societies which claim to be democratic it is important to "sell" a war to the populace by representing it as an epic battle between the forces of light and darkness. In actual fact all historic wars have represented international struggles between complex competing interests. The complexity of the underlying motivations for conflict are even greater in the case of "world wars" involving multiple state entities.

For more than half a century we have become accustomed to viewing the Second World War as the quintessential struggle between good and evil. So much is this the case that the word "fascist," referring to one of the competing ideologies of the 20th century, has become a simple term of abuse and has lost its meaning in political science. However, a dispassionate view of World War II will reveal that it was a complicated struggle between diverse interests with neither side enjoying a monopoly of virtue or vice. This is most simply illustrated by asking the question: Can we really count Joseph Stalin among the angels while relegating Carl Gustav Mannerheim to the camp of the demons?

The Second World War was at one level a contest between incompatible national interests. Both Germany and France sought military primacy on the

European continent. In the sphere of economics British and German industry were in competition in world markets. Nationalistic aspirations were also a factor as in the frictions between Germanic and Slavic peoples in eastern Europe. The competing ideologies of the right and the left played an enormous part it setting the stage for the great international cataclysm.

The divergent interests of the contending state entities forms an important part of the picture. German policy was geared toward righting the wrongs of the Versailles Peace Treaty of 1920. England sought to insure the integrity of its colonial empire in an ever more threatening environment. Italy wished to reassert herself as a world power. Hungary wanted to reclaim territories seceded after the First World War. The Soviet Union strove to break out of the "cordon sanitaire" which had isolated it as an ideological pariah in the international community. Beyond the boundaries of Europe the United States was struggling to extricate itself from the grip of a devastating economic depression with its capitalistic economic system intact. Japan intended to establish a geopolitical footprint which would be consistent with its potential economic strength. China was motivated by a desire to reassert its nationhood and step out of the shadow of quasi-colonialism.

The foregoing enumeration of motivations which contributed to the conflict we have called the Second World War is inherently simplistic. Its purpose is, however, to illustrate that each contending nation state was preoccupied with its own policy goals. These objectives were always inherently selfish, but not categorically evil. In the year 1939, just as today, no sovereign was motivated exclusively by idealism or wickedness in the formation of its policy.

In the mid-twentieth century throughout Europe political awareness had become universal. National policies affecting war or peace had to take account of the new concept of "public opinion." National policy was no longer formulated by personal sovereigns and their courtier advisors without regard to the masses. With universal literacy and mass media, including radio, international politics had already become the politics of popular mobilization.

It is interesting to note that, on the eve of World War II, in virtually every nation state there were dissonant voices speaking out against the established national policy. In Britain Sir Oswald Mosley and his black shirts sought to realign their country's foreign policy with that of the fascist states. In France the anti war elements coined the phrase, "Mourier pour Dantzig?" ("Is it worth dying for Danzig?") Meanwhile the French right wing campaigned with the slogan, "Better Hitler than Blum!" The exploits of the German resistance to Hitler and of the Italian anti-fascists are well known. Not a single of the European belligerent nations represented a genuinely unified monolith prior the outbreak of war in 1939.

Once the Second World War actually began the disunity within each of the combatant powers became even more marked. Collaboration with occupying powers was widespread and voluntary. Vichy France constituted an important and willing economic component of the axis war machine. The Nazi SS successfully recruited volunteers in Belgium, Denmark and Norway. An entire axis army corps under the command of General Andrei Vlassov was raised from among Soviet citizens. Even in England no less a figure than former prime minister David Lloyd-George advocated making peace with Germany after the fall of France. On the other hand within Germany conservative monarchists plotted the overthrow of Adolf Hitler. In Italy local communists carried on a large scale guerrilla war against the Mussolini government.

The disparate interests of the combatant sovereigns and the substantial disunity of the belligerent peoples leads me to the conclusion that the Second World War as fought on the European continent could more accurately be called the "European Civil War." Within the warring societies, quite literally, brother rose up against brother and father against son. The individual partisans were variously motivated by ideology, nationalism, historic animosity, and class antagonism.

It is certainly an enormous misconception to believe that this European Civil War was fought for or against the Jews from the perspective of any of the belligerent states. The so called "Jewish question" was a lively subject of debate and contention throughout Europe from the time of the French Revolution. However, no country or alliance, with the possible exception of the Soviet Union, espoused the cause of the Jews as a matter of explicit policy. On the other hand in virtually all European societies prior to World War II attitudes which are now called antisemitic were widely held.

Looking at the side of the Allies, negative attitudes toward Jews were prevalent among French conservative nationalists both before and after the celebrated Dreyfuss affair. British Tories were renowned for their low opinion of the "Laventine race." Antisemitism in Poland was all pervasive. None of these allies went to war against the axis in order to support the Jews.

Among the axis powers, no European nation traditionally exhibited less antisemitism than did Italy. Mussolini's ideology was based on extreme nationalism and economic fascism, not on opposition to the Jews as a people. In Germany, on the other hand, there is no denying that opposition to the position of Jews in politics, the economy and culture was a cornerstone of Nazi doctrine. However, it was not the Nazi antisemitism but rather larger geopolitical considerations which caused the formation of the alliance to defeat the axis. No European leader told his people that they were going to war against the fascists in order to protect the interests of the Jews.

The European Civil War was also not principally a conflict of left versus right or liberalism versus conservatism. Within Nazi Germany the most vociferous opponents of Adolf Hitler were the conservative monarchist military class, not the trade unionists. Britain fought the war ostensibly in order to protect the integrity of its far-flung and admittedly racist empire. Liberal distaste for fascist totalitarianism was hardly a consideration. Certainly concern for fundamental human rights was not among the policy motivations for the Stalinist Soviet Union. Truly the contending powers were united against one another in marriages of convenience, not idealistic crusades.

Once we realize that World War II was indeed a European Civil War and not a conflict between light and darkness, we can understand our contemporary world more clearly. Despite the harangues of demagogues the conflicts in our time are not struggles between angels and devils. When we look back on the historic past we seriously err if we adopt the truisms of present day propagandists without serious, dispassionate consideration

VII

During the 1940s most of Europe experienced the nightmare of military occupation. The intelligentsia's reaction to this phenomenon was widely varied.

Passive resisters, freedom fighters and collaborationists were to be found in every society. This fact compounds our impression that World War Two was really a civil war. Certainly at its onset and during much of the war the various national intelligentsias were divided and confounded.

It was only after the conclusion of hostilities that a sense of a unified intelligentsia began to coalesce. The horrors of war had left an almost universal sense of contempt for the intellectual constructs of the past which had allowed the calamity to take place. It was in the back wash of the experiences of the war that the passive resistance of Mahatma Ghandi in India and later of Martin Luther King in the United States fired the imaginations of the intelligentsia in the second half of the 20th Century.

However, elsewhere in the world the ideologies of violence continued to have their adherents. The end of European colonialism in Africa came in no small measure as a result of popular insurrections like the Mau Mau rebellion in Kenya. Furthermore, the state of Israel was created with the strategic use of terror, first to force an end to the British mandate in the area and then to stimulate the mass exodus of Arab refugees from Palestine. It was this thoughtful application of terrorist tactics which made the establishment of the Jewish state possible. This successful strategy was conceived and implemented by members of the Jewish intelligentsia.

As we have seen there is a logical continuum when we study the response of the intelligentsia to repression. Action and reaction are laws in the social sciences no less than in physics, The creation of Israel led to the adoption of terrorist tactics by Arabs, both Muslims and Christians. The escalation of violence in the Middle East and the occupation of Islamic holy shrines by Israel resulted in the questioning of the effectiveness of secular Islam. Repression, foreign and domestic, in Middle Eastern countries resulted in a movement of the Islamic intelligentsia known as fundamentalism.

Terrorism is the intellectual response to powerlessness. In a democratic society intellectuals will attempt to effect change be means of logical persuasion. However, when faced with an all pervasive tyranny, members of the intelligentsia may take up arms in unconventional ways.

It is a truism that one man's freedom fighters are another man's terrorists. What appears as rightful authority from one perspective is viewed as intolerable despotism from another. If members of the intelligentsia can refrain from the rhetoric of confrontation they can contribute to the de-escalation of violence in the world community. The watchword of western Christendom has been progress. The advancement of the human condition can be most effectively obtained when thoughtful men and women step back from the extremes of repression and terrorism. The intelligentsia can perform its function best when it advocates that activity for which it is most abundantly qualified: dialogue.

VIII

In a world of mass media one of the victims of a press oligarchy is language. Buzz words eliminate the concept of nuance. The shock impact of words constitute the means and the end of a journalism fueled by advertising revenue and the economic self interest of the dominant class.

In the 21st Century the word "terrorism" has been widely used but seldom defined. Indeed the word "terrorist" has become a simple term of abuse, devoid of precise meaning in the field of political science. It is widely recognized that, in the rhetoric of the media, your enemy is a "terrorist" while your friend is a "freedom fighter." This fact was most forcefully demonstrated when the self same, fierce, muhajadeen tribesmen in Afghanistan shifted from status as "freedom fighters," opposing the Soviet occupation of their country, to "terrorists," opposing the US lead occupation of their homeland.

It is one of the duties of the intelligentsia to uphold the integrity of language and to combat the debasement of the meaning of words. It is the "word," after all, which constitutes the divine spark which makes us human. It is the ability to articulate abstractions by means of language which sets humans apart from the animal kingdom. If the intelligentsia is to constitute an agency for progress in the world at large, it must vigorously defend the accurate use of words to convey universally accepted meanings.

"Terrorism" is defined as a mode of governing, or a mode of opposing government, by means of unlawful intimidation. The use of fear as an agent of social control constitutes terrorism. For this definition of terrorism to be useful we must qualify that the "fear" utilized by terrorism is the fear of unjustified harm, as opposed to the fear of the just consequences of a violation of law. Thus terrorism is by definition the use of lawless intimidation. A policeman or a soldier acting within the constraints of his duty and in accordance with accepted legal authority is not a terrorist. However, that same policeman or soldier could be viewed as a terrorist if he sought to intimidate outside the confines of his legal duty. Thus "state terrorism" constitutes the use of fear, in violation of the norms of international law, in order to coerce the behavior of another state, an individual societal group or the domestic population at large.

A private citizen may be a terrorist when he seeks to oppose government by means of unlawful intimidation. However, the use of lawful means to oppose government is obviously not terrorism. Organizing a demonstration or a strike does not constitute terrorism, although it may serve to intimidate the government. The exercise of the lawful right of self defense is not terrorism. Resistance to foreign occupation does not equate to terrorism. Furthermore, organized opposition to tyranny in any form is not terrorism.

A terrorist seeks to control society by means of unlawful intimidation. It is the extralegal and unjustified use of fear as an instrument of coercion which defines terrorism.

CLINICAL APPROACH TO INVESTIGATION AND SOLUTION OF INTERNATIONAL CONFLICTS OR PARANOIA IN THE AGE OF MASS-MEDIA [a]

Mikhail Reshetnikov
East European Psychoanalytic Institute
18-A, Bolshoy pr., St.-Petersburg, 197198, Russia [b]

1 Introduction

In mid-90th I was invited to participate in the mission of the US President Carter Fund for resolution of Russian-Estonian conflict in Narva. That time my friend and colleague, eminent American psychiatrist and psychoanalyst professor Vamic Volcan told me that "every international conflict develops like paranoia".

This idea stuck in my head, but then I could neither elaborate it, nor find its implications in Volcan's works. But, originally it was his idea.

Let me remind you some aspects of clinical picture of paranoia. I am sure that these aspects are very important for understanding of contemporary situation in the world.

2 Paranoia

As we know paranoia is "single-symptom" psychic disorder. Typically, patient's delusional thoughts and ideas have ordinary contents: in patient's opinion someone persecutes him, deceives, betrays, tries to humiliate, to stress his inferiority, to poison or infect or even destroy him.

Outside of this "cognitive mistake", such patients usually haven't any behavioral disorders or strange features, and they often demonstrate high social achievements. Now it is generally accepted that delusional disorders have mainly psycho-social roots, and their main triggers are: psychic traumata in early life, especially experiences of humiliation, physical or psychic violence, parental abuse and over-demanding expectations from the child.

In result feeling of basic trust doesn't develop, and such person from the very beginning perceives his close environment or the whole world as hostile. But in most cases of prominent pathology there would be "the most dangerous" figures or groups and types of relationships.

3 Historical review

Let me remind you as well, that in XIX century it was already known that paranoia develops gradually, "like normal character develops in other people". Kraepelin wrote, that such patients, develop "a tendency to evaluate and interpret life experiences extremely subjectively and connect them to one's own

---

[a] Short version

[b] e-mail: www.oedipus.ru veip@yndex.ru

wishes and fears". Then Kraepelin stresses, that if such people are deeply religious, they easily form an idea that they are chosen by God and demonstrate "tendency to preach and to seek followers, which often could be successful".

Kraepelin was the first who spoke about transmission of illness from one person to another. He called it "induced madness" that happens mostly in cases of paranoia. Suspicions and guesses gradually transmute into sureness and then into unshakable belief. You can understand how it happens in the age of mass-media.

4 Case histories

Case histories of paranoia belong to the most tragic and impressive chapter of psychiatry, which in modern times is multiplied by cinema and became an element of mass-culture (and it is very symptomatic).

This is not a history of psychiatry, so I will only refer to a few classic cases, described by Kraepelin, Bleuler and Freud. In all these cases we can see the ideas of persecution, injustice, social humiliation with further transformation in seeking truth and revenge, which sometimes take form of serial murder without any guilt, remorse and even fear of punishment. There are also ideas of messiah's mission, self-sacrifice, martyrdom and death for redemption as a way of approaching God. It is about individual cases. And what is about a group?

5 Social pathology

In his paper "Group Psychology and the Analysis of the Ego" Freud expresses revolutionary idea that a contrast between individual and group psychic phenomenon "loses a great deal of its sharpness when it is examined more closely", so the psychology of personality is "at the same time social psychology as well". Freud adds that mass psychology have to consider every person not as individual subject but rather as "a member of race, of a nation, of a cast, of a profession, of an institution" and he stresses that "a group is always impulsive, changeable and irritable The impulses which a group obeys are always imperious It (a group) has a sense of omnipotence". So, in a group we see the same social phenomena, as in individual life, but with greater feeling of omnipotence.

6 The content of hypothesis

I have nothing to add to this, I only expand this idea from group psychology to group pathology, and formulate the following idea: if there has been a deep psychic trauma in group background connected to (national) humiliation, then delusional ideas (or relationships) could develop (even after a long time).

Under certain conditions (negative economical, social or political factors) such ideas can transmute into unshakable belief of these people (or large ethnic group) that they are righteous, chosen by God and have special messiah mission, together with the ideas of pride, greatness and self-sacrifice in name of redemption or revenge. And such messiah mission can take the cruelest form.

I understand that this hypothesis is quite vulnerable, but I dare to expand it to all ethnic groups that went through military occupation, repression and colonization, then got freedom or possibility to return back to their historical lands and soon felt themselves at the backyard of history and civilization.

7 Intergenerational transmission

Now I will turn to Vamic Volcan's works that were mentioned above, in particular, to his paper "Traumatized societies" (2003). Volcan pays attention to the fact that investigations of national affects and massive psychic trauma (inflicted by the hostile group) showed the special significance of intergenerational transmission.

Investigation of concentration camps survivals, conducted after Second World War confirmed that parents transmit to their children something more than just an anxiety, hatred to enemies or other negative feelings.

Children of survivals (after national tragedies) develop deeper identification with their parents and manifest symptoms connected to psychic contents of parental past - and past in general (which they didn't and couldn't witness). This concept of "identification" is well-known. The most important here is the fact that adults who went through massive psychic trauma can "insert" their traumatized self-images into developing identity of their children. In result the children become carriers of damaged parental images.

After massive trauma (inflicted by hostile group) hundreds, thousands or even millions of individuals insert their traumatized images into their children, and all these "inserted images" are connected to the same traumatic event.

In result "common task" of the next generation is to save a "memory" of parental trauma, to mourn their losses, to act upon their humiliation or (if it isn't possible) - to take revenge on their enemies. In any case the main task is: to save mental representation of parental trauma which gradually (during decades and centuries) supports especial identity of particular large group.

Vamic Volcan called such mental representation "the chosen trauma" of a large group. When the large group is threatened by a new economical, political or religious crisis, its leaders (intuitively or intentionally) appeal to that "chosen trauma" which has a special potential for group's emotional consolidation.

8 Practical example

We can find practical example in Yugoslavia. In times of instability official authorities (mostly Serbs) began to exploit the memory of Kosovo battle, which has been 600 years ago and memory of legendary Serb Prince Lasar, who had been captured and killed in this battle by Moslems.

In result Bosnian Moslems were perceived as guilty in all historical troubles of Serbs, so it was a "legitimate" reason for all forms of revenge: Serbs began to kill, robber and rape their enemies with Middle-age cruelty.

I won't give other examples or draw any parallels. They are quite obvious.

9 Translation of criminal experience

In conclusion I should say that I am not trying directly or indirectly (by explanation) to approve terrorism as an extreme manifestation of international conflict, even taking in account that it is "a weapon of weak". My task is essentially different - to understand the roots of violence and terrorism from the deep psychological perspective and to find ways to minimize terrorism and to deprive it of psychological supplies and social basis, where it finds its forces and followers.

My goal is - to allow boys and girls who are born today or will be born tomorrow in any ethnic group to be protected from the transmission of criminal and semi-criminal traumatic experience of previous generations and prevent them from joining terrorists. I am sure that this goal can be reached, but not in result of hasty military operations with the same "paranoia" hidden behind them.

We know that every psychic trauma has its own history, its own mythology and its irrational transformation. In any case it demand deeper investigation, and we can suggest such deeper approaches to these problems.

My main point is that we can't predict something irrational, so - we can't construct reconciliation process from outside. Reconciliation, which doesn't deprive hatred of its historical roots, would always be transient. Peace as a result of forceful intervention and authorities of superpowers will not last for long.

We do find spiritual leaders of terrorists in case of emergency. We do find suitable words and convincing arguments in these cases. Why can't we speak to them more often and try to understand them, their purposes, their and our delusions? Maybe we just have two different paranoias (from our side - as well)?

We speak about their atrocities but we don't discuss their ideas. We insist that they threaten our life style, but don't we threaten their?

I believe that in our complicated world we are doomed not to the escalation of violence but rather to dialogue and understanding.

I was a TERRORIST

Michel Mockers
Agent P2 (British Mission)
War Cross 1939-1945

Terrorism is a broad word, some kind of a bag into which a legitimate or illegitimate power can indiscriminately throw any insurgent or freedom fighter. It is very convenient to put the label "terrorists" on freedom fighters because it justifies killing them all. It is how, some sixty years ago, I found myself labeled "terrorist" during the German occupation of France. I was not considered to be a freedom fighter, which I really was, but as a "terrorist" because I opposed by force what was imposed upon me by force. Of course, we were not (I now write "we" because I was not alone) despicable terrorists killing women and children and blowing up public buildings, "using violence and threats to intimidate or coerce." We were outlaws living in the woods and attacking all that we could possibly attack, specially German military convoys. We became experienced in appearing, striking, and disappearing. We fought on roads and streets. We blew up all that we could blow up, such as bridges, railroad tracks, and electrical railroads lines. If there had been oil pipelines, we would have blown up oil pipelines too.

German armies occupied France for three years and for three years fought endlessly and uselessly - even when helped by French collaborators - against a growing resistance by "terrorists" who had to be "ground like meat for pat?," according to a German leaflet we found on the road. Germans were everywhere. They knew every street of every town and village. They arrested and tortured individuals, took hostages, tried to locate and eventually found radios by triangulation, set traps to catch terrorists off guard, and attacked groups of resistance, such as mine, with armed forces. They had all the weapons that armies have. Coming from the sky, ours weapons were of the lightest kind. Despite that handicap, Germans never succeeded in crushing the "resistance" because when they arrested or killed one "terrorist," more came in to replace him or her. In my own sector, we had three or four dozen people at the beginning, and more than two thousand at the end. Our other factor of success was that we were supported by a good part of the population. When people fight for a cause in which they believe, they are ready to give their life for it, as we were, as I was.

This is something nobody has the power to stop, and a terrorist is someone very difficult to catch. All the German armies were in charge all over Europe, in France as well as in Russia, to destroy any form of resistance. I was part of a British mission working in France to help the French resistance, all together 200 men and women working by groups of three or four with a radio. Our main job was to provide the Allieds with intelligence about German activities and to provide resistants with armaments delivered at night by parachutes, sometimes receiving from the sky special agents sent from England to accomplish special missions.

When we asked for the bombing of Vierzon switching station, we had knowledge of the accumulation of war supplies and guns of retreating German armies that were being transferred to Germany by secondary railroad lines. During the bombing of the station, the German flack shot down three

American bombers. "Pauline" (Pearl Corniolley), the commanding officer of our mission, was advised that three airmen were in hiding on a farm and had to be rescued. She had received no news about crewmembers of the other planes. She told me that she wanted those airmen brought directly to her.

I let her talk and asked no questions. Relations had to be by the book. Each member of the mission carried a small capsule of cyanide in case of capture. Moral scruples about suicide were out when eventually the lives of many people depended on the silence of one person.

"I do not have to suppose that it is an operation across field," I said. "No car, certainly much too dangerous in unknown territory, especially in that direction. Only legs. How far are they?"

The farm was south of the bombed railroad station. South was an advantage because it spared me what could have been the most perilous part of the rescue: crossing the railroad or the town itself to go north of it in heavily German controlled territory. For the rest, it was a 40-kilometer walk. Pauline gave me directions to reach the farm where the Americans were hiding. She had managed to have what she called a "relay" ready to provide me and then us on both ways with food and bed. The relay was about half the distance to the farm. The physical condition of the Americans was unknown except, supposedly, for the fact that none was wounded. Even so, 40 kilometers in insecure territory were a long stretch.

When I left the relay the next morning, the sun was already high in the sky. The sky was empty of any trace of clouds, a perfectly blue and empty sky. There was also no trace of wind, not even of a light breeze. The day was going to be warm. I walked across fields, crossing a road once in a while. At one of those crossings stood a small house with a sign on the door reading "Caf?." Good to know that there was a watering hole on the way. It would still be there on the way back if the Americans were thirsty. I wondered who the customers could be in such place. There was not even a house in sight.

It was near noon when I arrived at the farm. Mary's directions were good, and I did not get lost. I did not walk directly to the farmhouse but, hidden behind a tree, I observed the surroundings for half an hour. Nothing unusual aroused my suspicion. I got up and went straight to the open door of the farmhouse. A young woman with a baby on her arms welcomed me.

"Let me call my husband," she said. "He is in the back."

The husband came. "Here you are!" he exclaimed. "Happy to see you! We were not sure you were coming. Except for three men, I mean those who bought them here, nobody knows that they are here, but I cannot keep them forever. Some day, someone would find out. They are in the barn behind the house."

The three Americans came out when the farmer opened the door of the barn and called them. Their eyes crinkled in the bright sun for a few seconds and they looked at me. "They really look like three men out of an American movie," I thought looking at them. They were tall, two blond, one brunet, with smiling faces (as in the movies). Not a single word of French, of course. I thought that we could encounter a problem if we were to meet some Germans, or even some suspicious French. Never can tell. "We burnt their uniforms," declared the farmer, "and we found enough pants and shirts to clothe them. We also buried all they had on them. Their pockets are empty but they have no identification."

"It would make no difference," I answered, "except that now they can be accused of being spies instead of prisoners of war. They have however more chance with me in the clothing you gave them than walking on roads in American uniforms. But before we leave, let me give them a lesson in French language."

It had come to my mind that an elementary precaution would be to make believe that they were French by teaching them to pronounce "*Oui*" and "*Non*" correctly. I could act as if I was giving them an explanation that they would punctuate with some sonorous *ouis* and *nons*, of agreement and disagreement. After a few minutes of exercise, they were transformed into three French country boys.

The heat was of the type that makes you perspire heavily. After more than two hours of walking at a brisk pace, the small " Cafe" came into sight. I made a sign to stop and hide behind a short hedgerow. No movement was visible; no German vehicle was parked around the house. We could have a glass of wine or whatever with no problem. It would help us to go farther.

"Allons-y!" I said in French. There was no need for translation. We crossed the road and the space between the road and the house. Followed by my three Americans, I pushed and opened the door. My heart almost burst out of my chest when I entered the unique room that was the "cafe." It was full of German soldiers seating at small round tables, drinking and talking. There was only one thing to do: go in. Backing out would have been catastrophic. We would not have escaped immediate suspicion and there were some forty Germans to rush after us.

The three American understood the situation and their reflex was the same as mine. They remained cool and smiling. It was not our day to die. Providence helped us in that the only free table was the one closest to the door. We were not obliged to cross the room under the eyes of the Germans who certainly would have had a good look at us. I sat at the table and the three boys imitated me. The Germans had not really paid attention to us. I got up and crossing the room went straight to the bar. The man behind the bar, certainly the owner, looked at a French face that was not familiar with some curiosity in his eyes. He also looked, in purpose, at the table close to the door, but his curiosity did not go any further, and he gave me a bottle of wine with a smile. I never figured out the meaning of the smile.

Wine seemed to be the national French and German drink at that café. I paid and went back to the table with the bottle of wine and four glasses. And our conversation went on, cut once in a while by short silences, as would be a real conversation, and interrupted by sonorous "*ouis*" and "*nons*." I remembered, afterwards, kicking the foot of my right-hand partner to make him say *oui* or *non*. After the necessary time for a bottle of wine among countrymen talking about their cows, we got up to exit slowly. I turned my head to have a last glance at the bar. The bartender was looking at us, smiling. I turned back to the door and resumed the conversation, talking loudly while going through the door about that cow that I had bought a few months earlier.

We did not run, on the contrary, but exhaled lengthily as soon as we were out of sight and laughed. Good bye, cow, as nice as you were! We needed to release the enormous nervous pressure that had been ours during interminable minutes. We were very conscious that everything could have turned ugly.

Even with different circumstances, my short story is only an illustration of "the terrorists among us," there the French "terrorists" among their German hunters. It is, forever, a view of the mind, a mindless political statement, and a misleading of leaders to declare that they will go and get the terrorists wherever they are to bring them to justice or kill them.

This means, in unequivocal and logical terms, invading any country that gives asylum or support to terrorists. This means an enormous expenditure of soldiers' lives and money with no guaranty that "terrorists" would be caught. This also means that even smart bombs, as smart as they are, kill civilians and destroy civilian properties in the process, immediately transforming most of the population into enemies that will support "terrorism." The invading (delivering) force is soon considered to be an occupying force that has to be gotten rid of, creating a cause worth fighting for. The scenario will repeat itself in every country thus invaded.

One reason for the fall of the Roman Empire is that Rome was involved in too many military expeditions and ended up lacking both soldiers and money, money being - as everybody knows - the "nerve of war."

Of course, when violence strikes, force seems to be the only acceptable retaliation and answering violence with violence seems the most natural response, the gut reaction of a humiliated people. But peace never comes out of confrontation, and even less out of the barrels of guns. When problems are addressed with bullets, they are always answered with bullets in situations that transform philosophers into warriors.

The war against terror deserves its name. It is a war, and like any other war, it cannot be declared won until it is really over. The war against terrorism is not won so far because, despite all the modern means of detection, one single man or a small group can manufacture a primitive bomb carried in a primitive car to deliver a lethal primitive blow that the most powerful army in the world is unable to prevent. Because one single suicide bomber believes he is marching to his immortal reward like the first Christians marching to the lions, he can succeed at reaching his goal. But is there such a significant difference with the man or woman carrying a cyanide capsule, and ready to die instantly if necessary?

When we put two facts together, first that it is impossible to go and get all the terrorists wherever they are, and second, that it is impossible to continuously protect everybody and everything, we know that the war against terrorism will linger forever at the price of a great feeling of insecurity for the people at large, a great loss of human lives, and the spending of a tremendous amount of money.

Given that perspective, the future is as bleak as the past centuries were, full of wars and massacres. Neither Christ nor Mohammed brought relief in the history of humanity. Men still behave not like civilized creatures gifted supposedly with intelligence, not even like wild beasts. In nature, not all animals are killed. There is a balance. Humans are the only species that ignores that balance.

There are two possible solutions: the violent one, which we are proudly practicing, and the non-violent one, which we do not even think about, because it would require controlling our animalistic instincts and our superiority complex, so far it would be from the actual political rhetoric and one way thinking.

The non-violent solution is to "pull the rug from under the feet of the terrorists" by two actions:

Psychological: Stop repeating and demonstrating that we are the richest and the strongest. It only provokes reactions of irritation and hate, and can only generate intellectual, religious, political, and eventually military conflicts.

Practical: Give a job to the young unemployed man or woman who has nothing to expect from life but despair, and build a stadium where they may expend their energy instead of becoming terrorists. Bombing rogue countries with water pipes, bags of flour and butter, and medication for contagious diseases has never been tried. In more realistic words: buying the countries that are susceptible to being on the side of terrorism, addicting their inhabitants to the taste of bread and butter with a permanent threat of cutting the supply if. It would be much less expensive in blood and money than trying to overtake them by force. Terrorists could not survive the strong condemnations of populations unwilling to lose a better life in favor of more bombings aimed at biting the hands that would provide them with bread and butter.

Such a non-violent solution to the war on terrorism cannot be successful overnight. It will require time and effort by all the rich nations to gather together to provide bread, butter, and stadiums to poor nations. It will require that greed and selfishness, which are at the base of all human conflicts and practiced with no limit by the rich and powerful, be put aside. This will be the price to pay for more peace on earth.

Programme of the 12th Lomonosov Conference on Elementary Particle Physics and the 5th International Meeting on Problems of Intelligentsia

25 August, THU

08.00 - 09.30 **Registration** (Auditorium 1-31, Faculty of Physics, Moscow State University)

09.30 - 09.45 **Opening** (Conference Hall)
A.Studenikin, MSU
V.Belokurov, Vice Rector of Moscow Sate University
P.Kashkarov, Vice Dean of Faculty of Physics of MSU
A.Slavnov, Head of Department of Theoretical Physics of MSU
M.Panasyuk, Director of Institute for Nuclear Physics of MSU
V.Matveev, Director of Institute for Nuclear Research

09.45 - 13.40 **MORNING SESSION** (Conference Hall)
Chairman: V.Belokurov
09.45 A.Slavnov (Steklov Math. Inst & MSU) *Noncommutative theories and vacuum condensate in gauge models* (30 min)
10.15 V.Ilyin (SINP MSU) *GRID for exploration of high energy physics* (30 min)
10.45 M.Heldmann (Univ.of Freiburg) *Prospects for physics beyond the SM at LHC* (30 min)
11.15 T.Inagaki (KEK) *Status of E391a experiment for the rare decay* $KL \to \pi^0 \nu\nu$ (30 min)

11.45-12.10 Tea break
Chairman: S.Belayev
12.10 L.Okun (ITEP) *Einstein legacy in the Year of Physics* (30 min)
12.40 A.Kaidalov (ITEP) *Diffractive particle production: from hadrons to Higgses* (30 min)
13.10 E.Boos (SINP MSU) *Top quark physics* (30 min)

13.40 - 15.00 Lunch

15.00–18.20 **AFTERNOON SESSION** (Conference Hall)
Chairman: D.Rapin
15.00 J.Bernabeu (Univ.of Valencia) *Intrinsic neutrino properties* (30 min)
15.30 M.Panasyuk (SINP MSU) *Physics of cosmic rays: yesterday and tomorrow* (30 min)
16.00 V.Berezinsky (LNGS) *Ultra high energy cosmic rays (astrophysical and New Physics solutions)* (30 min)

16.30 – 16.50 Tea break
Chairman: V.Berezinsky
16.50 P.Picozza (Univ. of Rome-II) *The science of PAMELA space mission* (30 min)
17.20 A.Cherepashchuk (Steinberg Astr.Inst.MSU) *Black hole demography* (30 min)
17.50 J.W.F.Valle (IFIC, Valencia) *Neutrino oscillations and new physics* (30 min)

18.20 – 23.00 **SPECIAL SESSION** (40^0)
Reception banquet will be held on board of a ship that will stream along the river across the central part of Moscow

26 August, FRI

9.00 – 13.40 **MORNING SESSION** (Conference Hall)

Chairman: J.Bernabeu
9.00 T.Kobayashi (Inst. of Part. and Nucl. Studies & KEK, Japan) *Neutrino accelerator experiments* (35 min)
9.35 M.Lindner (Tech.Univ. of Munich) *Physics at future long baseline neutrino experiments* (30 min)
10.05 A.Olshevski (JINR), N.Savvinov (Univ. of Bern) *Search for neutrino oscillations with OPERA experiment* (30 min)
10.35 G.Domogatsky (INR) *Baikal neutrino experiment* (30 min)

11.05 – 11.25 Tea break

Chairman: J.W.F.Valle
11.25 C.Cardall (Oak Ridge National Lab.) *Supernova neutrinos* (30 min)
11.55 J.Shirai (Tohoku Univ.) *Reactor neutrinos and KAMLAND* (30 min)
12.25 R.Petti (CERN) *Cross-section measurements from the NOMAD experiment* (20 min)
12.45 M.Serin (Middle East Technical Univ.) *Structure functions results at CHORUS experiment* (15 min)
13.00 O.Shevchenko, A.Sissakian (JINR) *Next to the leading order in semi-inclusive deep inelastic scattering processes* (20 min)
13.20 M.Greco (Univ. of Rome-III) *Structure functions at small x: combining the leading logarithmic resummation with DGLAP* (20 min)

13.40 - 15.00 Lunch

15.00 – 18.40 **AFTERNOON SESSION** (Conference Hall)

Chairman: A.Dolgov
15.00 V.Lobashev (INR) *Direct methods of neutrino mass measurements* (30 min)
15.30 A.Barabash (ITEP) *Double beta-decay expriments* (30 min)
16.00 A.Ianni (LNGS) *Status of the Borexino experiment* (20 min)
16.20 O.Smirnov (JINR) *Physics results with the Counting Test Facility (CTF) of Borexino* (20 min)
16.40 A.Starostin (ITEP) *Measurements of neutrino magnetic moment* (30 min)

17.10 - 17.30 Tea break

Chairman: M.Vysotsky
17.30 J.Pulido (IST, Lisbon) *Solar neutrinos: spin flavour precession and LMA* (20 min)
17.50 V.Semikoz (Univ. of Valencia & IZMIRAN) *Generation of magnetic helicity by collective neutrino-plasma interaction in hot plasma of the Early Universe* (20 min)
18.10 A.Studenikin (MSU) *Neutrino quantum states in matter* (15 min)
18.25 A.Grigoriev (MSU) *Quantum theory of neutrino spin light in matter* (15 min)

18.40 - 22.00 Sight-seeing bus excursion in Moscow

27 August, SUT

9.00 – 14.00 **MORNING SESSION** (Conference Hall)
Chairman: A.Zakharov
9.00 A.Starobinsky (Landau Inst.) *Present status of the standard cosmological model* (30 min)
9.30 R.Bernabei (Univ. of Rome-II) *Dark Matter investigations* (30 min)
10.00 R.Volkas (Melbourne Univ.) *Mirror dark matter* (30 min)
10.30 V.Mitrofanov (MSU) *Search for gravitational waves by LIGO Scientific Collaboration* (30 min)

11.00 – 11.20 Tea break
Chairman: P.Picozza
11.20 D.Rapin (Univ. of Geneva) *The current status and future prospects of AMS experiment* (30 min)
11.50 C.Sbarra (INFN-Bologna) *Astroparticle Physics with AMS-02* (30 min)
12.20 V.Zhukov (Karlsruhe Univ.) *Indirect dark matter search* (25 min)
12.45 V.Dokuchaev (INR) *Dark energy and black holes* (15 min)
13.00 I.Khriplovich (BINP) *Quantized black holes, their spectrum and radiation* (20 min)
13.20 A.Zakharov (ITEP) *Observational properties of supermassive blackholes* (20 min)
13.40 Yu.Gnedin (Pulkovo Observatory, St.Petesburg) *Magnetic fields of supermassive black holes: current status of the observations of quasars and active galactic nuclei* (20 min)

14.00 – 15.00 Lunch
15.00 – 17.50 **AFTERNOON SESSION** (Conference Hall)
Chairman: A.Shabad
15.00 L.Gambicorti, P.Spillantini (INFN-Florence) *The optical system for the observation of the UHE neutrino showers from the orbit: possibilities, limits and technical problems* (20 min)
15.20 Yu.Gaponov (Kurchatov Inst.), V.Khruschov (Kurchatov Inst.), S.Semenov (Kurchatov Inst.), F.Simkovic (Comenius Univ.) *Spontaneous and induced 2β-processes* (20 min)
15.40 S.Banerjee (Univ. of Victoria) *Lepton flavour violation* (20 min)
16.15 A.Ali, A.Borisov, D.Zhuridov (MSU & DESY) *Neutrinoless double beta decay in theories beyond the Standard Model: electron angular distributions* (15 min)
16.30 K.Kouzakov (SINP MSU) *Beta decay of a neutron in a magnetic field: the role of bound-state mode* (15 min)
16.45 – 17.05 Tea break
17.05 M.Dvornikov (IZMIRAN & Univ. of Rostock) *Classical description of neutrino flavour oscillations in vacuum* (15 min)
17.20 A.Burinskii (Nuclear Safety Inst.) *The Kerr Solution, Dirac equation and structure of electron* (15 min)
17.35 E.Arbuzova (Int.Univ. of Dubna), V.Rodionov (Moscow State Geological Prospecting Acad.) *On influence of intensive stationary electromagnetic field on the behavior of fermionic systems* (15 min)

28 August, SUN

9.00–19.00 Bus excursion to Sergiev Posad

29 August, MON

9.00 – 13.45 **MORNING SESSION** (Conference Hall)
Chaiman: G.Diambrini-Palazzi
9.00 E.Rosenberg (Iowa State Univ.) *Review of BaBar results on CKM parameters* (30 min)
9.30 M.Pappagallo (INFN-Bari) *Charm spectroscopy at BaBar* (20 min)
9.50 V.Druzhinin (BINP) *ISR physics at BABAR* (20 min)
10.10 V.Azzolini (Univ. of Valencia) *Semileptonic decays at Babar* (20 min)
10.30 N.Nikitin (SINP MSU) *Rare radiative leptonic B-meson decays* (20 min)
10.50 A.Ali, A.Borisov, M.Sidorova (MSU & DESY) *Rare meson decays in theories beyond the Standard Model* (15 min)
11.05 D.Tlisov (SINP MSU) *Rare leptonic B-meson decays in external electromagnetic fields* (15 min)

11.20 - 11.40 Tea break
Chairman: A.Kataev
11.40 E.Shabalin (ITEP) *On CP effects generated by electroweak penguin diagrams in non-leptonic K decays* (20 min)
12.00 G.Fäldt (Uppsala Univ) *CP violation in* $K^{+/-}\to\pi^0\,\pi^0\,\pi^{+/-}$ *decays* (20 min)
12.20 V.Bolotov (INR) *Recent results of the kaon decays study* (20 min)
12.40 M.Krivoruchenko (ITEP) *Medium modifications of kaons in pion matter* (20 min)
13.00 F.Bezrukov (INR) *CP-violation in K-meson leptonic decays* (20 min)
13.20 M.Vysotsky (ITEP) $B\to\rho\pi,\ \rho\rho,\ \pi\pi$: *hunting for* α (25 min)

13.45 - 15.00 Lunch

15.00 - 18.15 **AFTERNOON SESSION** (Conference Hall)
Chairman: A.Starobinsky
15.00 D.Galtsov (MSU) *Superstring cosmology* (20 min)
15.20 M.Capaccioli, O.Khovanskaya, G.Longo, M.Paolillo, M.Sazhin (Sternberg Astron.Inst.MSU) *Possible observation of a cosmic string* (20 min)
15.40 J.Wudka (Univ. of California) *Light excitations in 5 dimensional gauge theories* (20 min)
16.00 M. Fil'chenkov (Peoples' Friendship Univ. of Russia) *Quantum cosmology and the global rotation problem* (15 min)
16.15 V.Burdyuzha (LPI) *Cosmology of gravitational vacuum* (15 min)
16.30 K.Belotsky (Moscow State Engineering Phys. Inst.), M.Khlopov (Univ. of Rome-I), K.Shibaev (Moscow Engineering Phys.Inst.) *Stable matter of 4th generation: Hidden in the Universe and close to detection?* (20 min)

16.50 – 17.10 **Tea break**
17.10 A.Nikishov (Lebedev Phys.Inst.) *Problems of energy-momentum tensor in gravity* (20 min)
17.30 V.Dmitriev, Yu.Grats (MSU) *Self-interaction of charged particles outside brane topological defects* (15 min)
17.45 O.Mozgunov (Adygya State Univ.) *Cosmological model with rotation for vacuum* (15 min)
18.00 E.Boos, Yu. Mikhailov, M. Smolyakov, I. Volobuev (SINP MSU) *Physical degrees of freedom in a stabilized Randall-Sundrum model* (15 min)

30 August, TUE

9.00 – 13.35 **MORNING SESSION** (Conference Hall)
Chairman: E.Rosenberg
9.00 Yu.Simonov (ITEP) *Dynamics of confinement and deconfinement* (20 min)
9.20 C.Maxwell (Durham Univ.) *Resummation of large logarithms within the method of effective charges* (20 min)
9.40 A.Kovalenko, S.Morozov, M.Polikarpov, S.Syritsyn (ITEP), V.Zakharov (MPI, Munchen) *Localization, low dimensional objects, confinement and chiral symmetry breaking on the lattice* (30 min)
10.10 A.Pivovarov (INR) $B_0 \leftrightarrow \overline{B_0}$ *mixing at three loops in QCD* (15 min)
10.25 A.Sidorov (JINR) *The QCD analysis of the polarized structure functions data and high twist terms* (15 min)
10.40 I.Narodetskii (ITEP) *Theoretical issues of pentaquarks* (20 min)
11.00 - 11.20 Tea break
Chairman: M.Polikarpov
11.20 M.Tokarev (JINR) *Z-scaling and Strange High-pT Particle Production in pp Collisions at RHIC* (15 min)
12.05 V.Khruschov (Kurchatov Inst.), V.Savrin (SINP MSU), S.Semenov (Kurchatov Inst.) *Mass spectra of radially and orbitally excited states of mesons* (15 min)
12.20 M.Nekrasov (IHEP) *Precision determination of the top quark mass from* M_{bl} *distribution in semileptonic decays* (15 min)
12.35 S.Baranov (Lebedev Phys.Inst.) *Tests of the kt-factorization approach at the LHC* (20 min)
12.55 I.Dremin (Lebedev Phys.Inst.) *Soft and hard processes in QCD* (20 min)
13.15 I.Aref'eva (Steklov Math. Inst) *Stringy models of dark energy* (20 min)

13.35 – 15.00 Lunch

15.00 – 17.00 **AFTERNOON SESSION** (Conference Hall)
Chairman: R.Volkas
15.00 K.Stepanyantz (MSU) *Summation of Feynman diagrams in N=1 supersymmetric electrodynamics* (15 min)
15.15 R.Nevzorov (Univ. of Southampton) E_6 *inspired supersymmetric standard model with extra U(1) factor* (15 min)
15.30 C.Das (IMSC, Chennai), L.Laperashvili (ITEP), H.B.Nielsen (NBI, Copenhagen) *Generalized dual symmetry of nonabelian theories, monopoles and dyons* (15 min)
15.45 A.Lobanov, E.Murchikova, V.Zhukovsky (MSU) *Radiative effects in theories with Lorentz symmetry violation* (15 min)
16.00 V.Zamiralov (SINP MSU) *New relations between the QCD sum rules for meson-baryon couplings* (15 min)
16.15 A.Safronov (SINP MSU) *The analytical approach to constructing effective hadron-hadron interaction operators and its application to nucleon-nucleon scattering at low and intermediate energies* (15 min)
16.30 V.Slad (SINP MSU) *Production of Heavy Baryons* (15 min)
16.45 K.Toms (SINP MSU) *New results for rare B-meson semileptonic decays with the ATLAS experiment* (15 min)

17.00 – 17.20 Tea break

17.20 – 19.05 **AFTERNOON SESSION** (Conference Hall)
Chairman: V.Ch. Zhukovsky
17.20 M.Chichikina (MSU) *Quantization of nonlinear fields on classical background* (15 min)
17.35 V.Mikhailin, K.Zhukovsky (MSU) *Spectral properties of planar bi-harmonic undulators and harmonic generation of an ultrarelativistic electron* (15 min)
17.50 Yu.Rylov (Instit. for Problems in Mechanics, RAS) *Dynamical methods of investigation in application to quantum systems* (15 min)
18.05 V.Bordovitsyn (Tomsk State Univ.), V.Bulenok (Tomsk State Pedagogical Univ.), T.Pozdeeva (Tomsk State Univ.) *On the wave zone of radiation of a point-charge in hyperbolic motion* (15 min)
18.20 V.Bordovitsyn, T.Pozdeeva (Tomsk State Univ.) *From radiation theory to dynamics of relativistic charged particles* (15min)
18.35 A.Shishanin (Moscow State Industr.Univ.) *Some aspects of multitrace matrix models* (15 min)

31 August, WED

9.00 – 14.00 **MORNING SESSION** (South Physics Auditorium)
Chairman: V.Semikoz
9.00 V.Markov (Petersburg Nucl.Phys.Inst.) *Casimir effect at micro and nanoscale* (15 min)
9.15 A.Shabad (Lebedev Phys.Inst.), V.Usov *Collapse of positronium and vacuum instability* (20 min)
9.35 A.Kataev (INR) *High order QED corrections to g-2 of muon* (20 min)
9.55 M.Passera (Univ.of Padua) *The (g-2) of charged leptons in the Standard Model* (30 min)
10.25 G.Rubtsov (INR) *Small-scale fluctuations of extensive air showers as the origin of energy estimation systematics* (15 min)

10.40 – 11.00 Tea break
Chairman: J.Pulido
11.00 I.Tkachev (CERN) *The quest for sources of ultra-high energy cosmic rays* (30 min)
11.30 A.Dolgov (INFN-Ferrara & ITEP) *Neutrino physics, BBN, LSS, and CMBR* (30 min)

Round Table Discussion on
"NEUTRINO AND ASTROPARTICLE PHYSICS"

12.00 – 14.00
Spokespersons: J.Bernabeu (Univ.of Valencia), C.Cardall (Oak Ridge National Lab.), A.Dolgov (INFN, Ferrara & ITEP), V.Semikoz (IZMIRAN), A.Starobinsky (Landau Institute), R.Volkas (Univ. of Melbourne), V.Zhukov (Karlsruhe Univ.)
14.00 – 15.00 Lunch

SIXTH INTERNATIONAL MEETING ON
PROBLEMS OF INTELLIGENTSIA:
"The Intelligentsia and Violence: Responses to Repression and Terrorism"

15.00 – 15.10 **Opening** (Conference Hall)
15.10 J.Bleimaier (Princeton) *Resisters & pacifists, the intelligentsia fights back* (25 min)
15.35 M.Reshetnikov (East Europ. Psychoanalytic Inst., St.Petersburg) *Psychopathology of terrorism* (25 min)

Closing of the 12th Lomonosov Conference on Elementary Particle Physics and the 6th International Meeting on Problems of Intelligentsia
SPECIAL SESSION (40^0)

List of participants of the 12th Lomonosov Conference on Elementary Particle Physics and the 6th International Meeting on Problems of Intelligentsia

Andrievsky Ivan	Adygehya State Univ.	s_tvb@zmail.ru
Arbuzova Elena	Int.Univ of Dubna	arbuzova@rambler.ru
Aref'eva Irina	Steklov Math. Inst.	arefeva@mi.ras.ru
Azzolini Virginia	Univ. of Valencia	azzolini@slac.stanford.edu
Bagrov Vladislav	Univ. of Tomsk	pecher@ido.tsu.ru
Banerjee Swagato	Univ. of Victoria	swaban@slac.stanford.edu
Barabash Alexander	ITEP	alexander.barabash@itep.ru
Baranov Sergey	Lebedev Phys.Inst	baranov@sci.lebedev.ru
Belokurov Vladimir	MSU	belokur@rector.msu.ru
Belotsky Konstantin	MEPhI	k-belotsky@yandex.ru
Belayev Spartak	Kurchatov Inst.	
Berezinsky Veniamin	LNGS	venya.berezinsky@lngs.infn.it
Bernabei Rita	DAMA	rita.bernabei@roma2.infn.it
Bernabeu Jose	Univ. of Valencia	jose.bernabeu@uv.es
Bezrukov Fedor	INR	fedor@ms2.inr.ac.ru
Bilokon Halina	LNF	halina.bilokon@lnf.infn.it
Bleimaier John	Princeton	bleimaier@aol.com
Bolotov Vladimir	INR	bolotovvn@mail.ru
Boos Edward	MSU	boos@theory.sinp.msu.ru
Bordovitsyn Vladimir	Tomsk State Univ.	bord@mail.tomsknet.ru
Borisov Anatoly	MSU	borisov@ave.phys.msu.su
Burdyuzha Vladimir	LPI	burdyuzh@asc.rssi.ru
Burinskii Alexander	Nuclear Safety Inst.	bur@ibrae.ac.ru
Cardall Christian	Oak Ridge National Lab	cardallcy@ornl.gov
Cherepashchuk Anatoly	SAI	cher@sai.msu.ru
Chichikina Margarita	MSU	chich@goa.bog.msu.ru
Diambrini Palazzi Giordano	Univ. of Rome-I	giordano.diambrini@roma1.infn.it
Dmitriev Vadim	MSU	cinq@rambler.ru
Dokuchaev Vladislav	INR	dokuchaev@inr.npd.ac.ru
Dolgov Alexander	INFN, Ferrara & ITEP	dolgov@fe.infn.it
Domogatsky Grigory	INR	domogats@pcbai10.inr.ruhep.ru
Dremin Igor	LPI	dremin@lpi.ru
Druzhinin Vladimir	BINP	druzhinin@inp.nsk.su
Dvornikov Maxim	Univ. of Rostock	maxim.dvornikov@uni-rostock.de
Fäldt Goeran	Uppsala Univ	goran.faldt@tsl.uu.se
Fil'chenkov Michael	Peoples' Friendship Univ. of Russia	fmichael@mail.ru
Galtsov Dmitry	MSU	galtsov@grg.phys.msu.ru
Gaponov Yuri	Kurchatov Inst.	gaponov2@imp.kiae.ru
Gnedin Yuri	Pulkovo Observatory	gnedin@gao.spb.ru
Grats Yuri	MSU	grats@string.phys.msu.ru
Greco Mario	Univ. of Rome-III	greco@fis.uniroma3.it
Grigoriev Alexander	MSU	ax.grigoriev@mail.ru
Heldmann Michael	Univ.of Freiburg	heldmann@cern.ch

Ianni Aldo	LNGS	aldo.ianni@lngs.infn.it
Ilyin Viatcheslav	SINP	ilyin@sinp.msu.ru
Inagaki Takao	KEK	inagaki@post.kek.jp
Ishhanov Boris	NPI-MSU	bsi@depni.npi.msu.su
Kadyshevsky Vladimir	JINR	kadyshev@jinr.ru
Kaidalov Alexei	ITEP	kaidalov@itep.ru
Kashkarov Pavel	MSU	pavel@vega.phys.msu.su
Kataev Andrei	INR	kataev@ms2.inr.ac.ru
Khriplovich Iosif	BINP	khriplovich@inp.nsk.su
Khruschov Vyacheslav	SINP, Kurchatov Inst.	khru@imp.kiae.ru
Kobayashi Takashi	Neutrino group, IPNS, KEK	takashi.kobayashi@kek.jp
Kouzakov Konstantin	SINP & MSU	kouzakov@srd.sinp.msu.ru
Krivoruchenko Mikhail	ITEP	mikhail.krivoruchenko@itep.ru
Laperashvili Larisa	ITEP	laper@heron.itep.ru
Lindner Manfred	TUM	manfred_lindner@ph.tum.de
Lobanov Andrei	MSU	lobanovmsu@yandex.ru
Lobashev Vladimir	INR	lobashev@al20.inr.troitsk.ru
Markov Vladimir	NPI, Petersburg	markov@thd.pnpi.spb.ru
Matveev Victor	INR	matveev@ms2.inr.ac.ru
Maxwel Christopher	University of Durham	c.j.maxwell@durham.ac.uk
Mikhailin Vitaly	MSU	vchzh@mail.ru
Mikhailov Yuri	SINP	mikhailov@theory.sinp.msu.ru
Mitrofanov Valery	MSU	mitr@hbar.phys.msu.ru
Mockers Michel	Phoenix	mockers@aol.com
Mozgunov Oleg	Adygya State Univ.	momphys@radnet.ru
Murchikova Elena	MSU	murchikova@yahoo.com
Narodetskii Ilya	IHEP	naro@heron.itep.ru
Nekrasov Maksim	IHEP	nekrasov@th1.ihep.su
Nevzorov Roman	Univ. of Southampton	nevzorov@phys.soton.ac.uk
Nikishov Anatoly	LPI	nikishov@lpi.ru
Nikitin Nikolai	SINP	nnikit@mail.cern.ch
Okun Lev	ITEP	okun@itep.ru
Olshevski Alexander	JINR	a.olchevski@cern.ch
Panasyuk Mikhail	SINP & MSU	panasyuk@sinp.msu.ru
Pappagallo Marco	Univ. of BARI, INFN	marco.pappagallo@ba.infn.it
Passera Massimo	Padova Univ.	passera@pd.infn.it
Petti Roberto	NOMAD	petti@cern.ch
Picozza Piergiorgio	PAMELA	picozza@roma2.infn.it
Pivovarov Alexei	JINR	aapiv@ms2.inr.ac.ru
Polikarpov Mikhail	ITEP	polykarp@itep.ru
Pozdeeva Tatyana	Tomsk State Univ.	pozdeeva@inbox.ru
Pulido Joao	Lisboa Technical Univ.	pulido@cftp.ist.utl.pt
Radchenko Vladimir	SINP	vrad@srd.sinp.msu.ru
Rapin Divic	Univ.of Geneva (AMS)	divic.rapin@physics.unige.ch
Rehar Giuseppe	Rubano	grehar@aliceposta.it
Reshetnikov Mikhail	EEPI	veip@yandex.ru
Rodionov Vasily	Moscow State Geological	physics@msgpa.msgpa.ru

	Prospecting Acad.	
Rosenberg Eli	Iowa State Univ.	redmount@iastate.edu
Rubtsov Grigory	MSU & INR	grisha@ms2.inr.ac.ru
Rylov Yuri	Inst. for Problems in Mechanics	rylov@ipmnet.ru
Safronov Arkady	SINP	safron@srd.sinp.msu.ru
Savochkin Alexander	MSU	alexandr.savochkin@gmail.com
Savrin Vladimir	SINP, Kurchatov Inst.	savrin@theory.sinp.msu.ru
Savvinov Nikolay	JINR	nsavvinv@lhep.unibe.ch
Sazhin Mikhail	SAI	sazhin@sai.msu.ru
Sbarra Cristina	AMS	sbarra@bo.infn.it
Semenov Sergei	Kurchatov Inst.	semenov@imp.kiae.ru
Semikoz Victor	Univ. of Valencia & IZMIRAN	semikoz@yandex.ru
Serin Meltem	East Technical Univ.	serinm@metu.edu.tr
Shabad Anatoly	Lebedev Phys. Inst.	shabad@lpi.ru
Shabalin Evgeny	ITEP	shabalin@heron.itep.ru
Shevchenko Oleg	JINR	shev@mail.cern.ch
Shibaev Konstantin	MEPhI	shibaev01@yandex.ru
Shinkevich Sergey	MSU	shinkevich@gmail.com
Shirai Jumpei	Tohoku Univ. (KamLAND)	shirai@awa.tohoku.ac.jp
Shishanin Andrei	Moscow State Industrial Univ.	shishandr@rambler.ru
Sidorov Alexander	JINR	sidorov@thsun1.jinr.ru
Sidorova Maria	MSU	mvsidorova81@mail.ru
Simonov Yuri	ITEP	simonov@itep.ru
Sissakian Alexey	JINR	sisakian@jinr.ru
Slad Vitali	SINP MSU	vslad@sinp.msu.ru
Slavnov Andrey	Steklov Math. Inst & MSU	slavnov@mi.ras.ru
Smirnov Oleg	JINR	osmirnov@jinr.ru
Spillantini Piero	INFN-Florence	spillantini@fi.infn.it
Starobinsky Alexei	LITP	alstar@landau.ac.ru
Starostin Alexander	ITEP	starostin@itep.ru
Stepanyantz Konstantin	MSU	stepan@m9com.ru
Studenikin Alexander	MSU	studenik@srd.sinp.msu.ru
Tkachev Igor	CERN	igor.tkachev@cern.ch
Tlisov Danila	SINP	tlisov-danila@yandex.ru
Tlyachev Vyacheslav	Adygehya State Univ.	tim1988@maykop.ru
Tokarev Mikhail	JINR	tokarev@sunhe.jinr.ru
Toms Konstantin	SINP	ktoms@mail.cern.ch
Trofimov Ilya	MSU	ilyatrofimov@mail.ru
Valle Jose	IFIC	valle@ific.uv.es
Volkas Raymond	Univ. of Melbourne	r.volkas@physics.unimelb.edu.au
Vysotsky Mikhail	ITEP	vysotsky@itep.ru
Wudka Jose	Univ. of California	jose.wudka@ucr.edu

Zakharov Alexander	ITEP	zakharov@vitep1.itep.ru
Zamiralov Valeri	SINP	zamir@depni.sinp.msu.ru
Zhukov Valery	Karlsruhe Univ, AMS	zhukov@physik.uni-karlsruhe.de
Zhukovsky Konstantin	MSU	vchzh@mail.ru
Zhukovsky Vladimir	MSU	zhukovsk@phys.msu.ru
Zhuridov Dmitri	MSU	jouridov@mail.ru